Signals, Systems and Sound Synthesis

Signals, Systems and Sound Synthesis

Martin Neukom

Translation from the German by
Gerald Bennett

Z ─ hdk
─
Zürcher Hochschule der Künste
Zurich University of the Arts

PETER LANG

Bern · Berlin · Bruxelles · Frankfurt am Main · New York · Oxford · Wien

Bibliographic information published by die Deutsche Nationalbibliothek
Die Deutsche Nationalbibliothek lists this publication in the Deutsche
Nationalbibliografie; detailed bibliographic data is available on the Internet
at ‹http://dnb.d-nb.de›.

British Library Cataloguing-in-Publication Data: A catalogue record for this book
is available from The British Library, Great Britain

Library of Congress Cataloging-in-Publication Data

Neukom, Martin, author.
 [Signale, Systeme und Klangsynthese. English]
 Signals, systems and sound synthesis / Martin Neukom.
    pages cm
 Accompanied by CD.
 Translation of: Neukom, Martin. Signale, Systeme und Klangsynthese. Bern : P. Lang, 2003.
 Includes bibliographical references and index.
 ISBN 978-3-0343-1428-2
 1. Computer music–Instruction and study. 2. Computer sound processing.
 3. Computer composition. I. Title.
 MT723.N4813 2013
 786.7–dc23
                         2013024500

Published with the support of the Swiss National Science Foundation.

Additional texts, emendations and programs for this book can be found at
the following address: www.icst.net/Signals_Systems_and_Sound_Synthesis.

ISBN 978-3-0343-1428-2 pb.   ISBN 978-3-0351-0609-1 eBook

© Peter Lang AG, International Academic Publishers, Bern 2013
Hochfeldstrasse 32, CH-3012 Bern, Switzerland
info@peterlang.com, www.peterlang.com

All rights reserved.
All parts of this publication are protected by copyright.
Any utilisation outside the strict limits of the copyright law, without the permission
of the publisher, is forbidden and liable to prosecution.
This applies in particular to reproductions, translations, microfilming, and storage
and processing in electronic retrieval systems.

Printed in Hungary

Short Contents

Preface	xxi
Acknowledgments	xxiii

1 How to Use This Book . . . 1
  1.1 Getting Started . . . 1
  1.2 Overview . . . 2
  1.3 Instructions for Using Specific Programs . . . 2

2 Fundamentals of Acoustics . . . 5
  2.1 Basic Physical Principles and Units . . . 5
  2.2 Vibration and Waves . . . 8
  2.3 Sound and Hearing . . . 30

3 Signals and Systems . . . 45
  3.1 Analog Signals and the Fourier Transform . . . 45
  3.2 Digital Signals, DFT, FFT . . . 63
  3.3 Systems and Filters . . . 78
  3.4 Dynamic Systems and Feedback Control . . . 109

4 Computer Programs and Programming Languages . . . 163
  4.1 Csound . . . 163
  4.2 Max . . . 176
  4.3 Mathematica . . . 179
  4.4 C / C++ . . . 190
  4.5 Processing . . . 203

5 Fundamentals of Sound Synthesis . . . 207
  5.1 Fundamental Techniques of Sound Synthesis . . . 207
  5.2 Additive Synthesis . . . 244
  5.3 Subtractive Synthesis . . . 259

6 Nonlinear Techniques . . . 267
  6.1 Modulation Techniques and Distortion . . . 267
  6.2 Nonlinear Systems . . . 297

7 Other Techniques for Sound Analysis and Synthesis . . . 317
  7.1 Granular Synthesis . . . 317
  7.2 Special Analysis Methods . . . 328

8 Physical Modeling . . . 337
  8.1 Mass-Spring Models . . . 338
  8.2 Wave Guides . . . 390

9　Sound and Space 417
　　9.1　Spatial Hearing. 417
　　9.2　Reflection and Reverberation 426
　　9.3　Sound Reproduction 435

10　Computers and Composition 455
　　10.1　Chance and Probability 455
　　10.2　Stochastic Processes 487
　　10.3　Other Techniques Used for Composition 508

Appendix A　　Fundamentals of Mathematics 545

Appendix B　　Tables 571

Bibliography 577

Index 587

Contents

Preface to the Original German Edition xxi
Preface to the English Edition xxii
Acknowledgements xxiii

1 How to Use This Book   1

1.1 Getting Started 1
1.2 Overview 2
1.3 Instructions for Using Specific Programs 2
    1.3.1 Using the Mathematica Notebooks 3
    1.3.2 Using the Csound Programs 3
    1.3.3 Using the C/C++ Programs 3
    1.3.4 Using the Max Patches 4
    1.3.5 Using the Processing Programs 4

2 Fundamentals of Acoustics   5

2.1 Basic Physical Principles and Units 5
    2.1.1 Path, Velocity, Acceleration 5
    2.1.2 Mass and Force 6
    2.1.3 Momentum, Work, Power, Energy 7
2.2 Vibration and Waves 8
    2.2.1 Harmonic Oscillation 8
        2.2.1.1 Definition and Mathematical Representation 8
        2.2.1.2 Damped Oscillation 10
        2.2.1.3 The Addition of Harmonic Oscillations 11
        2.2.1.4 Beats 13
        2.2.1.5 Natural Vibrations 13
        2.2.1.6 Driven Oscillation and Resonance 15
    2.2.2 Periodic Vibrations and their Spectrum 16
        2.2.2.1 The Definition of Periodic Vibrations 16
        2.2.2.2 Standard Examples 16
        2.2.2.3 Other Examples 16
        2.2.2.4 Constructing Periodic Oscillation from Harmonic Waveforms . 17
        2.2.2.5 The Spectrum of Periodic Oscillations 18
    2.2.3 Aperiodic Oscillation 20
        2.2.3.1 Non-harmonic Partials 20
        2.2.3.2 Noise. 20
        2.2.3.3 Pulses 21
        2.2.3.4 Quasi-periodic Oscillation 22
        2.2.3.5 Variable Frequencies 22
    2.2.4 Waves 23
        2.2.4.1 Definition and Examples 23
        2.2.4.2 Mathematical Description 24

		2.2.4.3	Superposition of Waves	25
		2.2.4.4	Propagation of Waves	27
		2.2.4.5	The Doppler Effect	29
2.3	Sound and Hearing			30
	2.3.1	Pitch		30
		2.3.1.1	Frequency Range and Octaves	30
		2.3.1.2	The Harmonic Series and Pure Intervals	31
		2.3.1.3	Intervals	32
	2.3.2	Timbre		33
		2.3.2.1	Periodic Vibration	33
		2.3.2.2	Formants	33
		2.3.2.3	Spectra of Natural Sounds	34
		2.3.2.4	Non-Harmonic Spectra	35
		2.3.2.5	Fusion	36
		2.3.2.6	Missing Fundamentals and Residue Pitch	36
	2.3.3	Loudness		37
		2.3.3.1	Sound Power and Sound Intensity	37
		2.3.3.2	Decibels	37
		2.3.3.3	Phons	38
	2.3.4	Interaction of the Parameters		38
		2.3.4.1	Frequency Groups	39
		2.3.4.2	Auditory Masking	39
		2.3.4.3	Combination Tones	40
		2.3.4.4	Particular Aspects of Pitch Perception	41
	2.3.5	Resonant Spaces		41
		2.3.5.1	Reflections and Reverberation	42
		2.3.5.2	Sound Localization	43
		2.3.5.3	Distance Perception	44

3 Signals and Systems   45

3.1	Analog Signals and the Fourier Transform			45
	3.1.1	Periodic Signals and Fourier Series		46
		3.1.1.1	Fourier Series	46
		3.1.1.2	Calculating the Coefficients	46
		3.1.1.3	Examples	48
	3.1.2	Complex Representation		50
		3.1.2.1	Complex Numbers	50
		3.1.2.2	Trigonometric Representation	51
		3.1.2.3	The Exponential Form	52
		3.1.2.4	Algebra With Complex Numbers	53
		3.1.2.5	Rotating Complex Vectors	55
		3.1.2.6	Fourier Analysis in Complex Representation	58
	3.1.3	Aperiodic Signals and Fourier Integrals		60
		3.1.3.1	Definition	60
		3.1.3.2	Examples	61

	3.1.4	Analog Systems	62
		3.1.4.1 Definition and Examples	62
		3.1.4.2 Linear Differential Equations	62
		3.1.4.3 Laplace Transform (→ CD)	62
3.2	Digital Signals, DFT, FFT		63
	3.2.1	Digital Representation of Signals	63
		3.2.1.1 Sampling	63
		3.2.1.2 Representations and Simple Examples	64
		3.2.1.3 Characteristics and Basic Transformations	64
		3.2.1.4 Quantization	66
		3.2.1.5 Aliasing	66
		3.2.1.6 Rotating Vectors	67
	3.2.2	The Discrete Fourier Transform DFT	68
		3.2.2.1 Calculating the Coefficients of Discrete Fourier Series	68
		3.2.2.2 Examples	68
		3.2.2.3 Signals With Unknown or Non-Integer Periods	69
		3.2.2.4 The Complex Form and the Characteristics of the DFT	70
		3.2.2.5 The Fast Fourier Transform FFT	73
	3.2.3	The Z-Transform	76
		3.2.3.1 Definition and Transformation of Simple Signals	76
		3.2.3.2 Characteristics and Examples	77
3.3	Systems and Filters		78
	3.3.1	Systems	78
		3.3.1.1 Definition and Examples	78
		3.3.1.2 General Properties of Systems	79
		3.3.1.3 Impulse Response and Convolution	79
		3.3.1.4 Properties of Systems in the Frequency Domain	80
		3.3.1.5 The Complex Representation	82
		3.3.1.6 Filters	84
		3.3.1.7 The Transfer Function and the Z-Plane	85
		3.3.1.8 Linear Phase Filters	89
		3.3.1.9 Filter Design	90
	3.3.2	Non-Recursive Filters / FIR Filters	91
		3.3.2.1 Definitions and Properties	91
		3.3.2.2 Implementation	91
		3.3.2.3 Designing Filters by Setting Zeros	93
		3.3.2.4 Fourier Approximation	94
		3.3.2.5 Windowing	95
	3.3.3	Recursive Filters / IIR Filters	96
		3.3.3.1 Definition and Properties	96
		3.3.3.2 Implementation	96
		3.3.3.3 Designing Filters by Setting Poles and Zeros	99
		3.3.3.4 Stability	100
		3.3.3.5 Special Filters	100
		3.3.3.6 Filter Design by Transforming Analog into Digital Systems	108

3.4　Dynamic Systems and Feedback Control. 109
　　3.4.1　Differential Equations 110
　　　　3.4.1.1　Introductory Example, Phase Space 110
　　　　3.4.1.2　The Types of Differential Equations and Further Definitions . 111
　　　　3.4.1.3　Stationary System Analysis, Characteristic Curves . . . 114
　　　　3.4.1.4　Numerical Methods, Difference Equations 116
　　　　3.4.1.5　The Pendulum 119
　　3.4.2　Fixed Points and Attractors 122
　　　　3.4.2.1　Fixed Points 122
　　　　3.4.2.2　Catastrophes 124
　　　　3.4.2.3　Attractors. 127
　　3.4.3　Chaos 128
　　　　3.4.3.1　Introductory Example 128
　　　　3.4.3.2　Conditions for Chaos 129
　　　　3.4.3.3　Chaos in Time-Discrete Systems 130
　　　　3.4.3.4　Chaos in Differential Equations 133
　　　　3.4.3.5　Multiple Solutions and Discontinuities 135
　　3.4.4　Techniques of Feedback Control 137
　　　　3.4.4.1　Introductory Concepts and Examples 137
　　　　3.4.4.2　Elements of Control Circuits 139
　　　　3.4.4.3　Feedback Circuits 144
　　　　3.4.4.4　Nonlinear Feedback Circuits 145
　　　　3.4.4.5　The Computation of Feedback Control 148
　　　　3.4.4.6　Control By Filters 149
　　　　3.4.4.7　Examples 150
　　3.4.5　Synchronization 153
　　　　3.4.5.1　Self-Sustained Oscillators 154
　　　　3.4.5.2　The Van der Pol Oscillator 154
　　　　3.4.5.3　Synchronization Using Periodic Excitation . . . 156
　　　　3.4.5.4　Mutual Synchronization of Weakly Coupled Oscillators . 158
　　　　3.4.5.5　Synchronization of Chaotic Oscillators 160

4　Computer Programs and Programming Languages　　163

4.1　Csound 163
　　4.1.1　The Syntax of the Csound Orchestra 163
　　　　4.1.1.1　The Structure of the Orchestra 163
　　　　4.1.1.2　Constants and Variables 164
　　　　4.1.1.3　Functions 165
　　　　4.1.1.4　Assignment, Operators, Expressions and Conditional Expressions　165
　　　　4.1.1.5　Control Flow 166
　　　　4.1.1.6　Simple Signal Generators (→ CD) 167
　　　　4.1.1.7　Signal Modifiers (→ CD) 167
　　　　4.1.1.8　Delay Lines 167
　　　　4.1.1.9　Sound Input and Output 167
　　　　4.1.1.10　Opcodes 168

	4.1.2	The Score		168
	4.1.3	Function Table Generators		169
	4.1.4	Generating Scores and Events		171
		4.1.4.1 Generating Events With Csound Instruments		171
		4.1.4.2 Generating Scores With Programs in *C*		172
		4.1.4.3 Generating Scores and Events Using *Processing*		173
		4.1.4.4 Generating Scores and Events With *Max*		175
		4.1.4.5 Generating Scores With *Python*		175
4.2	Max			176
	4.2.1	Fundamentals		176
	4.2.2	Feedback		176
	4.2.3	mxj-Objects		177
4.3	Mathematica			179
	4.3.1	Fundamentals		179
	4.3.2	Sounds		183
	4.3.3	Graphics		185
4.4	C / C++			190
	4.4.1	Fundamentals		190
		4.4.1.1 The Structure of a C Program		190
		4.4.1.2 Types		191
		4.4.1.3 Derived and Composite Types		191
		4.4.1.4 Operators, Expressions, Mathematical Functions		193
		4.4.1.5 Control Flow		194
		4.4.1.6 Functions		195
		4.4.1.7 Input and Output		197
		4.4.1.8 Classes and Objects (→ CD)		199
		4.4.1.9 Reading and Writing Binary Data		199
	4.4.2	Generating and Storing Sounds		200
		4.4.2.1 Raw Data		200
		4.4.2.2 Sound Files		201
4.5	Processing			203
	4.5.1	Fundamentals		203
	4.5.2	Simulations		205
	4.5.3	Libraries		206

5 Fundamentals of Sound Synthesis **207**

5.1	Fundamental Techniques of Sound Synthesis			207
	5.1.1	Overview		207
		5.1.1.1 Instruments and Their Schematic Diagrams		207
		5.1.1.2 Techniques for Sound Synthesis		208
		5.1.1.3 Programs and Programming Languages		209
		5.1.1.4 Tables		209
		5.1.1.5 Audio Signals and Control Signals		210
		5.1.1.6 Interpolation		210
		5.1.1.7 Program Control and Conditional Statements		211

	5.1.2	Unit Generators	211
		5.1.2.1 The Oscillator	211
		5.1.2.2 The Pulse Generator	214
		5.1.2.3 The Noise Generator	215
	5.1.3	Control Signals	217
		5.1.3.1 Simple Control Functions	217
		5.1.3.2 Splines	219
		5.1.3.3 Interpolation Filters	223
		5.1.3.4 Variable Control Signals	224
		5.1.3.5 Tempo Functions	226
		5.1.3.6 Synchronization	235
	5.1.4	Delay Lines	237
		5.1.4.1 Definition and Direct Implementation	237
		5.1.4.2 The Circular Buffer	237
		5.1.4.3 Delay Lines With Variable Delay	238
		5.1.4.4 Delay Lines With Feedback	242
		5.1.4.5 Applications	244
5.2	Additive Synthesis		244
	5.2.1	The Synthesis of Periodic Waveforms	244
		5.2.1.1 Basic Techniques	244
		5.2.1.2 Variable Parameters	245
		5.2.1.3 Fusion	246
		5.2.1.4 Data Reduction	247
		5.2.1.5 Acoustic Illusions	247
	5.2.2	Analysis-Resynthesis	249
		5.2.2.1 Introduction	250
		5.2.2.2 The Discrete Fourier Transform DFT	251
		5.2.2.3 Long-Term Fourier Transform LTFT	254
		5.2.2.4 Short-Term Fourier Transform STFT	256
		5.2.2.5 The Phase Vocoder	258
5.3	Subtractive Synthesis		259
	5.3.1	Filters	259
		5.3.1.1 Characteristics of Filters in the Frequency Domain	259
		5.3.1.2 Types of Filters	260
		5.3.1.3 Special Filters	261
		5.3.1.4 Combining Filters	261
		5.3.1.5 Effects in the Time Domain	262
		5.3.1.6 Variable Filters	263
	5.3.2	Applications	263
		5.3.2.1 Sound Sources	263
		5.3.2.2 Resonators and Formants	264
		5.3.2.3 Linear Prediction	266

6 Nonlinear Techniques — 267

6.1 Modulation Techniques and Distortion — 267

6.1.1	Amplitude Modulation and Ring Modulation		267
	6.1.1.1 Introductory Example		267
	6.1.1.2 Basic Techniques		268
	6.1.1.3 The Spectrum of Amplitude Modulated Waveforms		269
	6.1.2.4 Ring Modulation		271
6.1.2	Frequency Modulation and Phase Modulation		272
	6.1.2.1 Introductory Example		272
	6.1.2.2 The Basic Method		272
	6.1.2.3 The Spectrum of a Frequency Modulated Waveform		274
	6.1.2.4 The Proportion $fc : fm$		278
	6.1.2.5 Variable Spectra		278
	6.1.2.6 Synthesis Models and Examples		279
	6.1.2.7 Extensions of the Basic Method		280
	6.1.2.8 The Influence of the Phase		284
6.1.3	Nonlinear Distortion – Waveshaping		285
	6.1.3.1 Introductory Examples		286
	6.1.3.2 Waveshaping		288
	6.1.3.3 The Modulation Index		289
	6.1.3.4 Polynomials as Transfer Functions		290
	6.1.3.5 Chebyshev Polynomials as Transfer Functions		292
	6.1.3.6 Limiters, Compressors and Expanders		294

6.2 Nonlinear Systems 297

6.2.1	Non-Recursive Systems With a Single Input		297
	6.2.1.1 Functions of a Single Input Value		297
	6.2.1.2 Functions of More Than One Input Value		297
6.2.2	Non-Recursive Systems With More Than One Input Signal		302
	6.2.2.1 Functions Using a Single Value of Each Input		302
	6.2.2.2 Special Functions of Two Input Signals		303
	6.2.2.3 Functions of Several Values of Several Input Signals		307
6.2.3	Recursive Systems		308
	6.2.3.1 Functions of One Value		308
	6.2.3.2 Functions of Two Values		310
6.2.4	Time-Variant Systems		311
	6.2.4.1 Delimiting Systems		311
	6.2.4.2 Non-Recursive Systems With Constant Delay		311
	6.2.4.3 Non-Recursive Systems With Variable Delay		312
	6.2.4.4 Recursive Systems With Constant Delay		313
	6.2.4.5 Recursive Systems With Variable Delay		315

7 Other Techniques for Sound Analysis and Synthesis — 317

7.1 Granular Synthesis 317

7.1.1	Fundamentals		317
	7.1.1.1 Grains		317
	7.1.1.2 Techniques for Making and Controlling Grains		319
	7.1.1.3 Synchronous Granular Synthesis		321

		7.1.1.4 Asynchronous Granular Synthesis	322
	7.1.2	Applications	323
		7.1.2.1 FOF	323
		7.1.2.2 VOSIM	324
		7.1.2.3 Granulating Sampled Sounds	326
7.2	Special Analysis Methods		328
	7.2.1	Walsh Synthesis	328
		7.2.1.1 Walsh Functions	329
		7.2.1.2 Examples	329
	7.2.2	The Logarithmic Frequency Range in Spectral Analysis	330
	7.2.3	Wavelets	331
		7.2.3.1 Wavelets	331
		7.2.3.2 The Continuous Wavelet Transform	332
		7.2.3.3 The Discrete Wavelet Transform	335

8 Physical Modeling — 337

8.1	Mass-Spring Models		338
	8.1.1	Systems With One Mass	338
		8.1.1.1 Harmonic Oscillation	338
		8.1.1.2 Exciting the Oscillation	342
		8.1.1.3 Damped Harmonic Oscillation	342
		8.1.1.4 Exciting the Damped Oscillation	345
		8.1.1.5 Oscillation With Nonlinear Acceleration	347
		8.1.1.6 Calculations	348
	8.1.2	Systems With Two Masses	349
		8.1.2.1 The Oscillation of Two Coupled Masses	349
		8.1.2.2 Excitation and Damping	351
		8.1.2.3 Nonlinear Acceleration	352
		8.1.2.4 Computing the Frequencies of the Natural Resonances	353
		8.1.2.5 Calculations ($\rightarrow$ CD)	355
	8.1.3	The Linear Arrangement of Coupled Masses	355
		8.1.3.1 A Model With Three Masses	355
		8.1.3.2 The String	356
		8.1.3.3 Correcting the Dispersion Relation	358
		8.1.3.4 Damping and Nonlinearity	360
		8.1.3.5 Picking Up the Sound	361
		8.1.3.6 Exciting the String	362
		8.1.3.7 Harmonics	363
	8.1.4	Two-Dimensional Arrangements of Coupled Masses	365
		8.1.4.1 An Example With Three Masses	365
		8.1.4.2 Representing the Plane by a Regular Grid	366
		8.1.4.3 Resonant Frequencies of the Grid	367
		8.1.4.4 Objects With Curved Edges	369
		8.1.4.5 Objects With Freely Oscillating Edges	369
		8.1.4.6 Rigid Body Motion	370

Contents   xv

	8.1.4.7	Retroflex Surfaces	371
	8.1.4.8	A Grid With Unequal Distances Between the Masses	372
	8.1.4.9	Irregular Grids	374
	8.1.4.10	Irregular Density or Elasticity	376
8.1.5	Three-Dimensional Arrangements of Coupled Masses		377
	8.1.5.1	Longitudinal and Transversal Oscillations of a Mass	377
	8.1.5.2	Subdividing Space With a Regular Grid	378
	8.1.5.3	Bodies in Free Motion	380
	8.1.5.4	A Model With Fixed Points	383
	8.1.5.5	Variable Fixed Points	385
	8.1.5.6	The Effect of Gravity	386
	8.1.5.7	Damping	386
8.1.6	Arbitrary Configurations and Variations		387
	8.1.6.1	Coupled Strings	387
	8.1.6.2	Geometrically Impossible Shapes	388
	8.1.6.3	Spaces of More Than Three Dimensions	389

8.2 Wave Guides 390
8.2.1 Simple Delay Lines 390
- 8.2.1.1 Delay Lines 390
- 8.2.1.2 Simple Damping 392
- 8.2.1.3 Frequency 393
- 8.2.1.4 Nonlinearity 396
- 8.2.1.5 The Excitation 397
- 8.2.1.6 The Karplus-Strong Algorithm 398

8.2.2 Waveguides 399
- 8.2.2.1 The Ideal Waveguide 400
- 8.2.2.2 Reflection 401
- 8.2.2.3 The Advancing Wave Front as Solution of the Wave Equation . 402
- 8.2.2.4 Other Variables for Representing Waves . . 404

8.2.3. Sound Pickup and Excitation 405
- 8.2.3.1 The Postions of Pickup and Excitation . . 405
- 8.2.3.2 The Duration of the Excitation . . . 407
- 8.2.3.3 Excitation Without Feedback 408
- 8.2.3.4 Excitation With Feedback 410
- 8.2.3.5 Selective Reflection and Sound Radiation . 413
- 8.2.3.6 Harmonics 414

9 Sound and Space   417

9.1 Spatial Hearing 417
9.1.1 Sound Localization 418
- 9.1.1.1 Interaural Time Difference ITD . . . 418
- 9.1.1.2 Interaural Intensity Difference IID . . . 419
- 9.1.1.3 Head-Related Transfer Function HRTF . . 421

9.1.2 Distance 422
- 9.1.2.1 The Decrease of Sound Intensity With Distance . 422

		9.1.2.2	Proportion of Indirect Sound	423
	9.1.3		Movement of Sound in Space	424
		9.1.3.1	The Doppler Effect	424
		9.1.3.2	Additional Information Through Position Change	425
9.2	Reflection and Reverberation			426
	9.2.1		Reflections	426
		9.2.1.1	Geometrical Considerations	426
		9.2.1.2	Scatter and Absorbtion	430
	9.2.2		Reverberation	430
		9.2.2.1	The Nature of Reverberation	431
		9.2.2.2	Simple Reverberators	431
		9.2.2.3	Frequency Dependency	434
		9.2.2.4	More Complex Filters	435
		9.2.2.5	Convolution With an Impulse Response	435
9.3	Sound Reproduction			435
	9.3.1		Ideal Solutions	435
		9.3.1.1	Simulation of the Sound Source	436
		9.3.1.2	Sound Field Reproduction	436
		9.3.1.3	The Sound Wave in the Auditory Canal	437
	9.3.2		Practical Solutions	437
		9.3.2.1	Stereo	438
		9.3.2.2	Ambisonics	438
		9.3.2.3	Decorrelation	451

10 Computers and Composition   455

10.1	Chance and Probability			455
	10.1.1		Fundamentals of Combinatorics	455
		10.1.1.1	Introductory Examples	455
		10.1.1.2	Permutations	456
		10.1.1.3	Combinations	456
		10.1.1.4	Arrangements	457
		10.1.1.5	Ordering Permutations	458
		10.1.1.6	Binomial and Polynomial Coefficients	458
	10.1.2		Fundamentals of Probability Calculus	458
		10.1.2.1	Standard Examples and Definitions	458
		10.1.2.2	Combining Events	459
		10.1.2.3	The Terminology and Axioms of Probability Theory	459
		10.1.2.4	Conditional Probability and Stochastic Independence	461
		10.1.2.5	Examples	461
	10.1.3		Probability, Density and Distribution Functions of Random Variables	461
		10.1.3.1	Random Variables and the Probability Function	462
		10.1.3.2	Continuous Random Variables and Their Density Function	462
		10.1.3.3	Distribution Functions	463
		10.1.3.4	Continuous Density Functions and Their Distributions	464
		10.1.3.5	Functions of Random Variables	465

	10.1.3.6 Measuring Probability Distribution	.	465
	10.1.3.7 Parametric Control of Functions	.	467
10.1.4	Generating Random Numbers With a Given Density or Distribution.	.	467
	10.1.4.1 Pseudorandom Numbers	.	467
	10.1.4.2 Direct Methods for Generating a Given Distribution.	.	469
	10.1.4.3 Inverting the Distribution Function	.	470
	10.1.4.4 Rejection Sampling	.	472
	10.1.4.5 Tables of Elements With Specified Frequency of Occurrence	.	473
10.1.5	Particular Distributions	.	476
	10.1.5.1 Continuous Uniform Distribution	.	476
	10.1.5.2 Trapezoid Distribution .	.	476
	10.1.5.3 Binomial Distribution .	.	477
	10.1.5.4 Poisson Distribution	.	478
	10.1.5.5 Normal Distribution	.	479
	10.1.5.6 Exponential Distribution	.	480
	10.1.5.7 Gamma Distribution	.	481
	10.1.5.8 Weibull Distribution	.	481
10.1.6	Applications	.	483
	10.1.6.1 Sound Examples From Random Number Sequences.	.	483
	10.1.6.2 The Application of Random Numbers to Musical Parameters	.	484
	10.1.6.3 Applications Using Variable Distributions	.	485
	10.1.6.4 Choosing Among Random Values .	.	487
10.2 Stochastic Processes		.	487
10.2.1	Introductory Examples and Concepts	.	487
	10.2.1.1 Games of Chance	.	487
	10.2.1.2 White Noise	.	488
	10.2.1.3 General Formulation of the Concepts	.	489
10.2.2	Markov Chains	.	490
	10.2.2.1 Introductory Examples.	.	490
	10.2.2.2 Definition .	.	492
	10.2.2.3 Transition Matrix and State Vectors	.	493
	10.2.2.4 Applications	.	494
	10.2.2.5 Markov Chains With Variable Transition Probabilities	.	497
	10.2.2.6 Markov Chains With Variable States	.	499
	10.2.2.7 Other Examples .	.	499
10.2.3	More Stochastic Processes	.	502
	10.2.3.1 Processes With Independent Increments .	.	502
	10.2.3.2 Random Walk	.	503
	10.2.3.3 Describing a Process By Its Spectrum	.	505
	10.2.3.4 Processes Involving Previous Events	.	506
	10.2.3.5 Processes With Sieved Random Variables	.	507
10.3 Other Techniques Used for Composition.		.	508
10.3.1	Cellular Automata	.	508
	10.3.1.1 One-Dimensional Automata With Two States .	.	508
	10.3.1.2 One-Dimensional Automata With Many States	.	510

	10.3.1.3 Two-Dimensional Automata With Two States		513
10.3.2	The Golden Ratio		514
	10.3.2.1 Definition and Classical Construction		515
	10.3.2.2 Fibonacci Numbers		516
	10.3.2.3 Continued Fractions, Surds and Golden Ratio		516
	10.3.2.4 Fractals		518
	10.3.2.5 A Process of Natural Growth		518
	10.3.2.6 Applications		520
10.3.3	Chaos Theory		525
	10.3.3.1 Concepts		525
	10.3.3.2 The Theory of Dynamic Systems		526
	10.3.3.3 Self-Similarity and Fractals		527
	10.3.3.4 Applications to Music		528
10.3.4	Simulating Swarm Behavior		531
	10.3.4.1 The Classical Boids Algorithm		531
	10.3.4.2 Extensions of the Swarm Algorithm		533
	10.3.4.3 Implementations		533
10.3.5	Toward a Topology of Sounds		535
	10.3.5.1 Mental Representation		535
	10.3.5.2 Complex Topologies to Represent Pitch Space		537
	10.3.5.3 Time		539
	10.3.5.4 Timbre		540
	10.3.5.5 Position Space		541
	10.3.5.6 Mapping		542
	10.3.5.7 Systems		543

Appendix A Fundamentals of Mathematics 545

A.1	Numbers and Arithmetic Operations		545
	A.1.1 Numbers		545
	A.1.2 Rules of Algebra		545
A.2	Statements, Sets and Operations on Sets		546
	A.2.1 Statements		546
	A.2.2 Sets		547
	A.2.3 Subsets and Power Set		547
	A.2.4 Operations on Sets		547
	A.2.5 The Cartesian Product		547
A.3	Equations		548
	A.3.1 Definitions and Concepts		548
	A.3.2 Equivalence Transformations		548
	A.3.3 Algebraic Equations		548
	A.3.4 Systems of Linear Equations and Matrices		549
	A.3.5 Transcendental Equations		550
A.4	Functions		551
	A.4.1 Definition		551

	A.4.2	Properties of Functions and Graphical Representation	552
	A.4.3	Basic Functions	553
	A.4.4	Composite Functions	562
	A.4.5	Parametric Representation	562
	A.4.6	Functions of Several Variables	563
	A.4.7	Even and Odd Functions	564
	A.4.8	Sequences and Series	564
A.5	Calculus		565
	A.5.1	The Derivative of a Function	565
	A.5.2	Rules for Differentiation	566
	A.5.3	The Indefinite Integral of a Function	567
	A.5.4	Integration Formulas	567
	A.5.5	The Definite Integral	568
	A.5.6	Partial Derivatives	568

Appendix B Tables   571

B.1	Pitches	571
B.2	Formants of English Vowels	571
B.3	Constants	572
B.4	Fibonacci Numbers	572
B.5	Prime Numbers	573
B.6	Bessel Functions	574
B.7	Chebyshev Polynomials	575

Bibliography   577

Books (by Author)	577
Books (by Number of Reference)	580
Articles from the Computer Music Journal (CMJ)	584
Articles from Other Sources	585
Articles Available on the Internet	586

Index   587

Preface to the Original German Edition

The first digital sound synthesis dates back to 1957, when an IBM-709 Computer in New York calculated a 17-second long piece of music. The program that had produced these sounds was written by Max Mathews, who 12 years later would write the first book giving detailed information about the technology of digital sound synthesis. This book was of extreme importance. Interest in electroacoustic music, and particularly in music made by computer, had greatly increased since 1957, and Mathews' book contributed immeasurably to knowledge about digital sound synthesis and thus paved the way for further development in the field.

After 1969, great advances were made in the technology of sound synthesis and the digital treatment of sound: Frequency Modulation, Linear Predictive Synthesis, Granular Synthesis, Formant Synthesis, Synthesis in the Frequency Domain (for example the Phase Vocoder), to name only a few techniques. These developments took place outside the commercial world and were generally well documented in specialized journals and at conferences, so that the growing community of composers for whom digital sound synthesis had become important was able to remain abreast of the newest technologies.

The first international conference for computer music took place in 1974; the *Computer Music Journal* first appeared in 1977. By the middle of the 1980's there was enough known about digital sound synthesis that in 1985 Curtis Roads and John Strawn could publish *Foundations of Computer Music* [43], a collection of 36 technical articles, most of which had first appeared in the *Computer Music Journal*. In the same year, the first comprehensive book about techniques of computer music, *Computer Music: Synthesis, Composition, and Performance* by Charles Dodge and Thomas Jerse [2] was published. F. Richard Moore's book *Elements of Computer Music* [8] followed in 1990, Curtis Roads' 1234-page *Computer Music Tutorial* [11] in 1996, and a second, revised edition of Doge and Jerse's book was published in 1997. Each of these books added substantially to the body of acquired knowledge about the technology and aesthetics of computer music. Nowadays, courses in digital sound synthesis and the treatment of sound are offered in universities all over the world. These books, together with the *Computer Music Journal*, furnish the foundations for this teaching. Without the generosity of very many colleagues in the field who took the time to document their research and their achievements, and without the vision and magnanimity of the few companies that were willing to publish these writings, computer music as we know it today would surely not exist.

Signals, Systems and Sound Synthesis by Martin Neukom belongs in this group of definitive texts on computer music. Not only are all the important issues of sound synthesis discussed in the light of the most recent developments, but topics relevant to computer music but usually disregarded or mentioned only marginally in earlier books, like systems theory, room acoustics or programming for musicians, are treated at length here. The book is by no means simply a compendium of established techniques. Each theme is introduced and developed from a new point of view. The approach is often mathematical, and the resulting level of abstraction and change of perspective are refreshing and inspiring, even for mathematically less proficient readers. A CD-ROM is included, containing the entire text, many graphic animations and a great many sound examples, making of the book both an up-to-date documentation of what has been achieved in the field of computer and a handbook of areas of work for the future.

The publication of *Signals, Systems and Sound Synthesis* would have been a significant event in any country and language. It is our great good fortune that the book was published in German and thus became the first German-language textbook on digital sound synthesis. When I asked Martin Neukom in 1998 to write a short guide to sound synthesis so that we would not always have to direct our students to the English-language professional literature, I had no idea that I would receive such a rich, original and comprehensive text. The Zurich University of the Arts gave the project both moral and financial support, for which we are very grateful. With Signals, Systems and Sound Synthesis, Martin Neukom has given the computer music community a marvelous textbook. I am proud that the Zurich University of the Arts contributed to its realization.

Gerald Bennett

Preface to the English Edition

There is little to add to what I wrote nearly 10 years ago. In its German version, *Signals, Systems and Sound Synthesis* has become a standard work, the standard work in the field of computer music for German-language readers. The vision of a documentation which is both a conventional, albeit rather heavy, book and a complex interactive tool for learning has proven extremely fertile for teaching, and the book has been used for several years with success in courses at the Zurich University of the Arts. *Signals, Systems and Sound Synthesis* remains one of the few books to show in detail how to realize the techniques of contemporary computer music.

The time seemed ripe to extend the book's readership by issuing an English-language version. Martin Neukom has rewritten extensive sections of the book, bringing up to date both content and references. The enclosed CD-ROM contains much new additional material. Particularly the number of interactive programming examples has been greatly increased.

If I wrote the Foreword to the German Edition as supporter and defender of Martin Neukom's project, I now write as translator. Of course, the book's sheer bulk was a challenge, but not surprisingly I encountered the greatest difficulties with topics I knew little about. Despite 40 years' experience as a composer and researcher in computer music, I often found myself on paths which were new to me. I have tried to find factually and grammatically correct, felicitous English equivalents for the German text. I hope the reader will forgive me if she occasionally misses the specific jargon of a domain in which I sojourned for the first time.

Gerald Bennett

Acknowledgments

This book was commissioned by the Musikhochschule of Zurich, today the Department of Music of the Zurich University of the Arts (ZHdK) and was accepted by the University of Zurich as a doctoral dissertation in Musicology. The English translation was commissioned by the Institute for Computer Music and Sound Technology (ICST) of the Zurich University of the Arts. I wish to thank these institutions, and particularly Germán Toro-Pérez, Director of the ICST, for their trust and belief in the project and for their generous support. I am grateful to my colleagues at the ICST as well as to the students of the ZHdK for many interesting discussions about (and not about) electroacoustic music. Special thanks are due to Daniel Muzzulini, the editor of the English translation. The book has greatly profited from his thorough and meticulous reading and from his suggestions for the book's contents.

My particular thanks are reserved for Gerald Bennett, composer, teacher, founder of the ICST, and much more. He suggested that I write this book, and he accompanied its growth during the ten years from its first beginnings to the present final version. It has been great good fortune that Gerald Bennett agreed to translate the book into English, for he combines to an unusual degree professional knowledge, linguistic competence and generosity towards the ideas of others.

Martin Neukom

1 How to Use This Book

1.1 Getting Started

This book is a translation from the German original, *Systeme, Signale und Klangsynthese* (2003, 2005). The text for the English edition has been considerably revised, and whole sections have been rewritten. It has some specific features of format and layout that distinguish it from most other books, largely due to the fact that it is written using the computational software *Mathematica*. While graphically not as flexible as digital publishing software, Mathematica allows the text to be published simultaneously as a book and as a CD containing the complete text in computer-readable form, many directly executable programs as well as links to additional relevant passages within the text, example programs, animations and sound examples.

The book is stored on the CD in so-called *Notebooks* (see 1.3 below). It was decided that the printed text and the screen version of the Mathematica notebooks should look as much alike as possible. Because notebooks have no distinct pages, and hence no page numbers, reference within the book is not made to pages but to chapters and sub-chapters. The illustrations always follow immediately their explanations in the text and therefore have no captions. The mathematical formulas are not numbered as usual but instead are linked or repeated where necessary.

In the German original, Chapter 1 (this chapter) and the introductions to the programming languages Csound and C were quite long. In this edition these chapters have been drastically shortened since both programming languages are well documented in English. In Chapter 4 one will find short descriptions of the programming languages one will encounter in the text as well as explanations of some special applications and procedures used in the book.

The bibliography of the original edition contained a commented list of German and English books suitable for interested non-professional readers. In this edition, the books listed have been updated to their most recent editions, and most German references have been replaced by their English versions where these exist or by comparable English language literature. Most of the methods, formulas and algorithms used here belong to the common practice of their respective disciplines, and their origins cannot be determined with certainty. For this reason, no texts are quoted in the book and in general no mention is made of authors of formulas or techniques. In addition, reference is usually made to secondary literature and not to primary literature.

The CD contains:

– The complete text of the book as Mathematica-notebooks (*name*.nb) in the folder *Text*;
– Additional chapters, explanations and examples as hidden cells in the notebooks;
– C/C++ programs in source code (*name*.cpp) in the folder *CPP*;
– Java programs in source code (*name*.java) and Java classes (*name*.class) for the mxj-externals in the folder *Java*;
– Sound examples (*name*.wav) in the folder *Sounds*;
– Csound examples in the folder *Csound*;
– Max patches in the folder *Max*;
– Programs in the Processing language in the folder *Processing*.

1.2 Overview

This book gives an introduction to the techniques of digital sound synthesis and sound transformation. The relevant topics are presented using illustrations, animations of complex physical and mathematical relationships, sound examples and sample programs. Basic technical and mathematical principles will be explained where they are necessary for reading the specialized literature.
This first chapter gives an overview of the book's contents, the enclosed CD and the computer programs used in the book. In Chapters 2 to 4 the physical, mathematical and programming essentials for the rest of the book are developed. The short overview of acoustics in Chapter 2 is followed in Chapter 3 by a thorough presentation of signal theory and system theory. These theories provide the tools for a precise derivation of many techniques of sound synthesis, sound transformation and control theory. The text is written to give a clear and intuitive view of the material rather than a more abstract, general presentation. The chapters that follow are written so they can be understood in their broad lines without the theoretical material developed in Chapter 3. Traditionally, many programming languages and specific programs have been used in the treatment and synthesis of sound. In Chapter 4, only those languages are discussed which are later used in the book. Mathematica was chosen for the body of the text, because an editor, a high-level programming language and routines to generate illustrations, animations and sounds are integrated into the language. C/C++ was chosen as a general programming language because Csound, Max and many other programs are written in C and can be extended using additional routines written in C. Csound was chosen as the domain-specific language (DSL) for sound synthesis because it has a long history in computer music and because it has the flexibility of a general programming language. Max was chosen for interactive programming because it is the most frequently used language in live electronics. Max is intuitive to use and ideal for demonstrations because of its graphic interface. Several techniques of sound synthesis and sound treatment are introduced in Chapters 5 to 8. Nonlinear techniques and techniques that simulate the physical procedures of natural sound production are particularly emphasized because they give interesting results with only simple programming and modest computation times. Chapter 8 gives a comprehensive introduction to sound synthesis by physical modeling. At the same time it offers a first taste of digital filter theory. Chapter 9 discusses some of the many problems that arise in connection with the synthesis of sound in resonant spaces and presents suggestions for solutions. Chapter 10, finally, discusses techniques and aids for the composition of computer music.

1.3 Instructions for Using Specific Programs

The rest of this chapter contains instructions for using the data and the programs on the enclosed CD. Chapter 4 contains more detailed descriptions of the programs and programming languages Mathematica, C/C++, Csound, Max and Processing.

The text of this book is stored in Mathematica notebooks, which can be read with the *Wolfram CDF Player* or read and edited with the program *Mathematica*. The program Wolfram CDF Player can be downloaded from Wolfram Research (http://www.wolfram.com). The notebooks include closed *cells* containing additional text, the programs used to generate the illustrations, animations (→ Animation) and sound examples (→ Sound example) as well as additional cells. Longer sound examples are stored on the CD as WAVE files.

1.3 Instructions for Using Specific Programs

The *Csound program* (Version 5 or later) together with *Csound scores* and *orchestras* generates sounds. All the sounds synthesized by the Csound program in the course of the book can be found on the enclosed CD.

All the *C/C++ programs* can be compiled on a C++ compiler, but only some of the examples can be compiled on a classical C compiler.

The *Max programs* can be run using the freely available program *Max Runtime*. They can be executed, edited and extended using the Max program itself (http://cycling74.com).

The examples written in the language *Processing* can be executed, edited and extended using the freely available program Processing (http://processing.org).

1.3.1 Using the Mathematica Notebooks

After downloading and installing the Wolfram CDF Player, start the program. Now a notebook can be opened using the Menu item *File*. If the notebook *Contents.nb* or *Index.nb* is opened first, the other notebooks can be opened using hyperlinks (see below). A Mathematica notebook consists of cells that contain text, illustrations, sounds or other cells. The cells are indicated by brackets on the right edge of the screen. Nested cells are shown by nested brackets. By double-clicking on an outer bracket, whole groups of cells can be closed so that only the uppermost cell is visible. By double-clicking on an inner cell bracket, a group of cells can be closed so that only the selected cell is visible. Closed cells can be opend by double-clicking on a bracket with a hook. The text appears as in the printed book. The hidden cells contain the programs used to generate illustrations, animations and sounds as well as calculations and additional texts.

To start an animation, open the the group of cells whose visible cell has the comment (→ Animation) by double-clicking on the corresponding bracket. Most animations are interactive. Their parameters can be changed by graphic elements in and sliders next to the illustrations. The sliders can be animated by clicking on the plus sign next to the slider. In this way continuous animations can be generated.

The CD version of the book contains hyperlinks which give access to other cells in the current notebook or to cells in other notebooks. These hyperlinks are colored blue. Clicking on a hyperlink retrieves the corresponding cell.

1.3.2 Using the Csound Programs

Csound is a command-line oriented programming language designed for synthesizing and manipulating sound. Its name comes from the fact that it is written in C. For the examples in this book *QuteCsound* was used as a development environment within which programs can be edited and executed. QuteCsound has a highlighting editor with autocompletion, interactive features and integrated help files. In the Csound folder the orchestra and score programs have been combined to form files in so-called *Csound unified file format (name.csd)*.

1.3.3 Using the C/C++ Programs

C++ is an intermediate-level general-purpose programming language. It contains nearly all the features of the language C and adds object-oriented features, in particular, classes. This book does not always distinguish explicitly between programs in C and programs in C++. All the programs can be compiled with a C++ compiler, but only some can be compiled with a C compiler. In order to execute the programs that write sound files, the header file my_WAVE.h must be included in the program text using the command: #include "my_WAVE.h".

1.3.4 Using the Max Patches

The following explanations can be tested interactively by double-clicking on the Max patch *Instructions.maxpat* in the folder Max. Max patches are made up of elements such as generators, faders, inputs and outputs, etc. They can also contain subpatches. The audio settings can be changed by clicking on the word "Audio", which opens the window "DSP Status". The settings used for the examples in the book are stored under "Presets". Parameter values can be changed in the corresponding number fields and graphic elements and can then be stored as new presets. Clicking on the subpatch *p name* opens the subpatch and shows it in a new window. The subpatch *p comments* tells what the respective Max patch does and how to use it. The subpatch *p presets* explains the preset settings. Max objects can be programmed in C or Java as so-called *externals*. In this book only Java externals are used. Double-clicking on the object *mxj quickie name* displays the source code of the external *name*.

1.3.5 Using the Processing Programs

The examples written in the language Processing are stored as source code on the enclosed CD and must be executed using the freely available program Processing . Some examples require libraries that are not automatically installed with Processing (controlP5, oscP5, netP5) (http://processing.org). Instructions for and comments on the Processing examples can be found in the headers of the respective source code files.

2 Fundamentals of Acoustics

Further Reading: Musical Acoustics by Donald E. Hall [20] gives a comprehensive introduction to the general topic. Music Cognition and Computerized Sound by Perry R. Cook [41] is especially useful because of the enclosed CD with sound examples. The fundamentals of acoustic are summarized in several books on computer music, among them Computer Music by Charles Dodge and Thomas A. Jerse [2]. Genealogie der Klangfarbe by Daniel Muzzulini [77] presents a detailed discussion of the phenomenon of timbre.

2.1 Basic Physical Principles and Units

2.1.1 Path, Velocity, Acceleration

The distance between the points P_1 and P_2 is $|x_2 - x_1| = |x_1 - x_2|$. In Cartesian coordinates (straight perpendicular axes with the same unit of measurement) in two or three dimensions, the Euclidean distance d is calculated from the difference of the coordinates by

$$d = \sqrt{d_x^2 + d_y^2} \text{ or } d = \sqrt{d_x^2 + d_y^2 + d_z^2}$$

respectively (left figure below). If the points are defined by position vectors r_1 and r_2 from the origin, then the *shortest path* $\Delta r = r_2 - r_1$. The formulas above give the corresponding distance (right figure below).

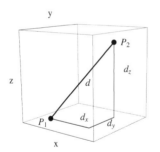
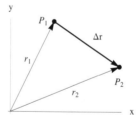

If it takes time Δt for a point to traverse the path Δr, then the point's mean *velocity* is $\bar{v} = \Delta r/\Delta t$. The direction of Δr and $\bar{v}$ are the same. The instantaneous velocity v is equal to the limit of $\Delta r/\Delta t$ when Δt approaches 0, that is, equal to the first derivative of the position with respect to time: $v = \lim_{\Delta t \to 0}(\Delta r/\Delta t) = dr/dt$. In contrast to the vector velocity v, we refer to the magnitude of the velocity as speed and write v or u. Acceleration is defined as the change in velocity over time: $a = \Delta v/\Delta t$. The instantaneous acceleration is defined as: $a = \lim_{\Delta t \to 0}(\Delta v/\Delta t) = dv/dt$. Correspondingly, velocity is calculated by integrating acceleration over time, position by integrating velocity over time (A5.3).

In the following example a mass is dropped from a height of 100 m. We know the position $x = 100$ m at time $t = 0$ s, the speed $v = 0$ m/s at time $t = 0$ s and the acceleration given by gravitation g. If the x-axis is pointing upward (figure to the right), the acceleration is negative: $g \approx -10$ m/s². Since the acceleration is constant, we calculate from $a = dv/dt$ the speed $v = tg$. For the speed after 3 seconds, we have

$$v(3) = -3 \text{ s} \cdot 10 \text{ m/s}^2 = -30 \text{ m/s}.$$

In general the speed is

$$v(t) = v_0 + \int_{\tau=0}^{t} g(\tau)\, d\tau.$$

At time $t = 3$ s the speed is

$$v(3) = v_0 + \int_{\tau=0}^{3} g(\tau)\, d\tau = 0 + \int_{\tau=0}^{3} -10 \text{ m/s}^2 d\tau = -30 \text{ m/s, as above}.$$

We calculate the distance covered at time t by

$$d = \int_{\tau=0}^{t} v(\tau)\, d\tau = \int_{\tau=0}^{t} \tau(-10 \text{m/s}^2) d\tau = -10 t^2/2$$

and the position x at time t by $x(t) = x_0 + d = 100 - 10t^2/2$. So after 3 seconds the mass is still $(100 - 10 \cdot 9/2)$ m $= 55$ m above the ground.

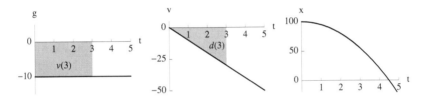

2.1.2 Mass and Force

Mass is a basic characteristic of matter. It is measured in kilograms (kg). Mass is different from weight, which arises from the action of a gravitational field on a body. The mass per volume unit of a body or material is called its density $\rho = m/V$ and is measured in kg/m³ or in kg/Liter = kg/dm³.

When a *force* acts upon a body, that body is either accelerated or deformed, or stress or pressure arises. Forces are defined by their magnitudes and directions; for this reason they are represented as vectors. If the forces F_1 and F_2 act simultaneously at the same point as in the illustration on the left below, the sum of the forces corresponds to the vector F.

The accelerating force acts in the direction in which a body can move. If a mass is on an inclined plane, as in the illustration to the right, the force of gravitation F pulls the mass straight down, but acceleration can only take place in the direction of the plane's inclination, shown here by the vector F_2.

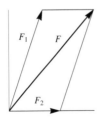
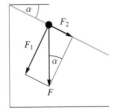

Force is measured in Newtons. One Newton is equal to the force necessary to accelerate a mass of one kilogram by one meter per second within a second:

$$1 \text{ N} = 1 \text{ m·kg/s}^2.$$

Newton's Second Law of motion states that force is proportional to a body's mass and to the acceleration acting upon that body.

$$F = ma$$

Example 1. The force of gravity at the earth's surface is equal to $F = mg$, where g is the acceleration due to gravity. Let the angle of inclination in the illustration to the right above be α, then the accelerating force $F_2 = F \cdot \sin(\alpha) = mg \cdot \sin(\alpha)$. For a mass of one kilogram, an angle of 30° and $g = 9.807$ m/s², that means:

$$F_2 = 1 \text{ kg} \cdot 9.807 \text{ m/s}^2 \cdot \sin(30°) = 4.9035 \text{ N}.$$

Example 2. The force with which a stretched ideal spring pulls back is proportional to the elongation of the spring x, that is, $F = -Kx$. The minus sign indicates that the force F acts in the direction opposite that of the elongation. The constant K depends on the properties of the spring, such as elasticity, mass, etc., and is called the spring or force constant.

2.1.3 Momentum, Work, Power, Energy

The *momentum p* of an object is defined as the product of its mass and its velocity.

$$p = mv$$

If no external force acts on an object, its momentum is conserved. If an external force F acts on an object, the object's momentum is changed according to:

$$dp/dt = F$$

The mechanical *work W* is the amount of energy transferred when a force F moves an object through the distance s and is given by:

$$W = \int_a^b F \, ds$$

Here F and s are vectors, whereas W is a scalar quantity, that is, a quantity without direction. If F and s are constant (with magnitude F and s) and include the angle β, then the work W is given by:

$$W = Fs \cdot \cos(\beta)$$

Work is measured in Joules: $1 \text{ J} = 1 \text{ m}^2 \cdot \text{kg/s}^2 = 1 \text{ N} \cdot \text{m}$.

Power P is defined as work performed over time.

$$P = W/t$$

Power is measured in Watts: $1 \text{ W} = 1 \text{ m}^2 \cdot \text{kg/s}^3 = 1 \text{ J/s}$. Specifications for electronic devices give some idea of how much a Watt is. Automobiles produce several tens of thousands of Watts of power (one horse-power is 735.3 W). The average overall performance of a human being in terms of power is about 100 Watts, whereas the sound power of a violin played fortissimo is only about 0.001 Watt.

Energy is a notion of great importance in physics. As work, energy is measured in Joules. In classical mechanics, the law of the conservation of energy states that the sum of all the various forms of energy in an isolated system remains constant. Forms of energy are mechanical energy, chemical energy and electrical energy as well as warmth.

Potential energy can be thought of as energy stored within a physical system. The potential energy E_P of an object results from the position of the object in space under the influence of certain forces. For objects in the earth's gravitational field the following relationship holds: $E_P = mgh$, where h is the object's height above a reference level. The potential energy E_P of a stretched spring is

$$E_P = \frac{K}{2} x^2$$

where x is the deviation of the spring from its initial position.

The *kinetic energy* E_k of an object of mass m and velocity v is:

$$E_k = \frac{1}{2} mv^2$$

2.2 Vibration and Waves

Further Reading: The book Waves *(Berkeley Physics Course, Vol. 3) by Frank S. Crawford [6] does not specifically treat musical acoustics, but it is nonetheless an excellent introduction to our subject. In addition, it describes numerous experiments that are easy to carry out.*

2.2.1 Harmonic Oscillation

2.2.1.1 Definition and Mathematical Representation

If the *excursion* of an oscillating point corresponds to a sine function in time, one speaks of *harmonic oscillation*. The figure below shows three snapshots taken from an animation and illustrates the relationship between the harmonic oscillation of a point, the rotation of a pointer on the left and a representation of the point's excursion as a function of time: $x = f(t) = \sin(t)$ on the right. (→ Animation)

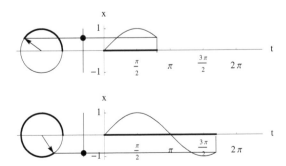

2.2 Vibration and Waves

The duration of the cycle of an oscillation, that is the interval between two corresponding states of the oscillation, is its *period T* (in seconds), the number of cycles per second is its *frequency f* (in Hertz). The following relationships hold:

$$f = \frac{1}{T} \quad \text{and} \quad T = \frac{1}{f}$$

In the illustration below, the period is 2/3 s, the frequency is therefore 3/2 Hz. The function $x = \sin(t)$ has the period 2π and the frequency $1/2\pi$. Hence, the so-called *angular frequency* of an oscillation is defined as $\omega = 2\pi f$, where f is the frequency. The maximum excursion from zero is the *amplitude*. The excursion is always a function of time; in the illustration below, the amplitude remains constant.

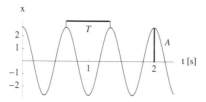

The instantaneous state of a vibration is described by the phase, or more precisely by the instantaneous phase. The *instantaneous phase* is defined either as the argument of a sine or cosine function (the *unwrapped phase*) or as this argument modulo 2π (the *wrapped phase*). Hence the phase is simply a number. In the representation of oscillation using a revolving vector, the phase corresponds to the angle α between the horizontal axis and the vector. So, for instance, the phase at a maximum point of a sine wave is 90° or $\pi/2$. Displacement along the time axis results from the zero phase angle or phase constant ϕ_0, that is, from the instantaneous phase at time $t = 0$. The oscillation above begins at a maximum point and thus has an initial phase $\phi_0 = \pi/2$. A harmonic oscillation is determined by its amplitude A, its frequency f or angular frequency ω, and its initial phase ϕ_0. The instantaneous phase of the oscillation is $\omega t + \phi_0$ or $(\omega t + \phi_0)_{\mathrm{mod}\,2\pi}$.

$$x = A \cdot \sin(\omega t + \phi_0)$$

In the oscillation above these parameters are $A = 2.7$, $f = 3/2$ and $\phi_0 = \pi/2$, which gives the equation: $x = A \cdot \sin(\omega t + \phi_0) = 2.7 \cdot \sin(3\pi t + \pi/2)$.

From the time when the sine function is zero, or more precisely from a time t_0, at which the phase is zero, we obtain the zero phase angle as follows. By substituting $t - t_0$ for t the graph of a function is displaced along the positive x-axis by t_0, giving the function equation

$$x = A \cdot \sin(\omega(t - t_0)) = A \cdot \sin(\omega t - \omega t_0))$$

At time $t = 0$, therefore, we have for the phase $\phi_0 = -\omega t_0$. In the figure below the frequency is 3/2, and we have for the phase constant

$$\phi_0 = -3/2 \cdot 2\pi(-1/6) = \pi/2 \quad \text{or} \quad \phi_0 = -3/2 \cdot 2\pi \cdot 1/2 = -3/2 \cdot \pi \;(\equiv \pi/2)$$

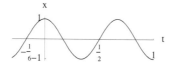

The frequency of a periodic signal is defined as the number of oscillations per second. In general, *instantaneous angular frequency* $\omega(t)$ is defined as the time derivative of the *instantaneous phase* $\phi(t)$ (2.2.3.5)

$$\omega(t) := \frac{d}{dt}\phi(t).$$

Conversely, we obtain the instantaneous phase $\phi(t)$ from the instantaneous frequency $f(t)$ by the relationship

$$\phi(t) = 2\pi \int_{\tau=0}^{t} f(\tau)\,d\tau.$$

Harmonic oscillation arises in systems in which reactive force is proportional to displacement and acts in opposition to it: $F = -Dx$. Newton's Second Law of movement (2.1.2.3) states that

$$F = ma = m\ddot{x}.$$

Hence, it follows that $\ddot{x} = -(D/m)x$. The solutions of these differential equations are harmonic oscillations. The total energy of an oscillating system consists of potential energy and kinetic energy. At the turning points of the oscillation, where the displacement is at its maximum and the velocity is zero, the potential energy

$$E_p = \frac{1}{2}Dx^2$$

is maximal (2.1.3.4) and the kinetic energy

$$E_k = \frac{1}{2}mv^2$$

is zero. At zero crossings of the oscillation the displacement is zero and the velocity is at its maximum, hence the potential energy is zero and the kinetic energy is maximal.

Systems in which a reactive force F acts on a mass m proportionally to the displacement (i.e. $F = -Dx$) oscillate harmonically: $x = A \cdot \sin(\omega t + \phi_0)$. The following relationships hold:

angular frequency $\omega = \sqrt{D/m}$
potential energy $E_p = \frac{1}{2}Dx^2$
kinetic energy $E_k = \frac{1}{2}mv^2$

2.2.1.2 Damped Oscillation

Free vibration in mechanical systems is always damped due to friction. In the simplest case damping causes an exponential reduction of amplitude, which means that the amplitude diminishes by the same amount at each period of oscillation. The ratio $A_1/A_2 = A_2/A_3 = ... = c$ is called the *damping coefficient* and its natural logarithm $\Lambda = \ln(c)$ the *logarithmic decrement*.

2.2 Vibration and Waves

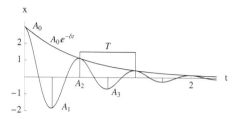

The damped oscillation can be described as the product of a sine or cosine function and a so-called *envelope*.

$$x = A_0\, e^{-\delta t} \sin(\omega t + \phi_0)$$

The function of damped oscillation solves the equation of motion of a system in which a reactive force is proportional to displacement and additionally friction is present, proportionally to the instantaneous velocity of the oscillating object: $R = -rv = -r\cdot dx/dt$, where the factor r is known as the *coefficient of friction*. The equation of motion of the system is: $F = -Dx + R = -Dx - rv$. The frequency of the sinusoid is constant but depends upon the friction. If the mass of the oscillating body is given by m and $\delta = r/(2m)$, then

> Frequency of the damped oscillation $f = \frac{1}{2\pi} \sqrt{(D/m - r^2/(4m^2))}$
> Logarithmic decrement $\Lambda = \delta T = 2\pi\delta/\omega$
> Energy $E = \frac{1}{2} D x^2 = \frac{1}{2} D A_0^2\, e^{-\delta t}$

The following figure shows two oscillations ($A_0 = 3$, $D = 2$, $m = 0.5$) with different damping coefficients ($r_1 = 0.1$, $r_2 = 0.7$). From the formulas above it follows that the frequencies are $f_1 = 0.3179$ and $f_2 = 0.2982$, and the corresponding periods are $T_1 = 3.146$ and $T_2 = 3.354$. After 10 time units the amplitudes are 1.104 and 0.00274.

2.2.1.3 The Addition of Harmonic Oscillations

When two or more oscillations are superposed, their displacements are added together:

$$x = x_1 + x_2 = A_1 \sin(\omega_1 t + \phi_{0,1}) + A_2 \sin(\omega_2 t + \phi_{0,2}).$$

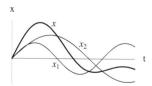

The addition can be simplified in straightforward cases. The sum of two oscillations of equal frequency and phase $x_1 = A_1\sin(\omega t)$ and $x_2 = A_2\sin(\omega t)$ is

$$x = x_1 + x_2 = A_1\sin(\omega t) + A_2\sin(\omega t) = (A_1 + A_2)\cdot\sin(\omega t).$$

The sum of two oscillations of equal frequency and amplitude but of different phase:

$$x = x_1 + x_2 = A\cdot\sin(\omega t + \phi_{0,1}) + A\cdot\sin(\omega t + \phi_{0,2})$$

can be written with the help of the formula $\sin(\alpha) + \sin(\beta) = 2\cdot\sin\left(\frac{\alpha+\beta}{2}\right)\cdot\cos\left(\frac{\alpha-\beta}{2}\right)$ as:

$$x = A\cdot(\sin(\omega t + \phi_{0,1}) + \sin(\omega t + \phi_{0,2})) = 2A\cdot\cos\left(\frac{\phi_{0,1}-\phi_{0,2}}{2}\right)\cdot\sin\left(\omega t + \frac{\phi_{0,1}+\phi_{0,2}}{2}\right).$$

The sum of the two oscillations is an harmonic oscillation of frequency ω whose initial phase is the arithmetic mean of the initial phases of the two partial oscillations

$$\phi_0 = \frac{\phi_{0,1} + \phi_{0,2}}{2}$$ and whose amplitude is

$$A_{\text{sum}} = 2A\cdot\cos\left(\frac{\phi_{0,1}-\phi_{0,2}}{2}\right).$$

The difference $\Delta\phi = \phi_{0,1} - \phi_{0,2}$ is called the *phase difference*. The following example uses the values $\phi_{0,1} = \pi/4$, $\phi_{0,2} = \pi/2$, $A = 1$ and $\omega = 1$.

As the phase difference of the two oscillations approaches π, the amplitude of their sum diminishes. When the phase difference is equal to π, the two oscillations cancel each other.

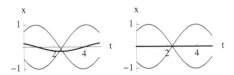

The sum of two oscillations of the same frequency but of different phase and amplitude is an harmonic oscillation whose phase and amplitude can be seen in a *pointer diagram*. If the oscillations are represented as circular motion (2.2.1.1), the pointer for their sum is the sum of the pointers of the two partial oscillations. In the illustration below, the bold curve represents the sum of two sine waves of different phase and amplitude.

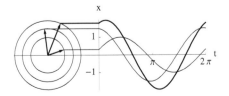

2.2 Vibration and Waves

2.2.1.4 Beats

When two oscillations of nearly equal frequency are superposed, so-called *beats* result. Because of the difference in frequency between the two oscillations, their phase difference gradually changes so that there are moments when they have the same phase and amplify each other and moments when the phase difference is equal to π and they cancel each other. The following illustration shows two oscillations with a frequency ratio of 6:7 and their sum (bold).

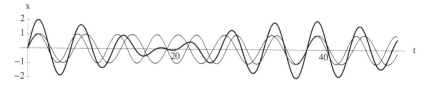

The sum of the oscillations

$$x = x_1 + x_2 = A \cdot \sin(\omega_1 t) + A \cdot \sin(\omega_2 t) = A(\sin(\omega_1 t) + \sin(\omega_2 t))$$

can be written as

$$x = 2A \cdot \cos\left(\frac{\omega_1 - \omega_2}{2} t\right) \cdot \sin\left(\frac{\omega_1 + \omega_2}{2} t\right)$$

or, since $\omega = 2\pi f$, as

$$x = 2A \cdot \cos\left(2\pi \frac{f_1 - f_2}{2} t\right) \cdot \sin\left(2\pi \frac{f_1 + f_2}{2} t\right).$$

The resulting function corresponds to a sinusoidal oscillation of frequency $(f_1 + f_2)/2$ and amplitude $2A \cdot \cos(2\pi t(f_1 - f_2)/2)$, which varies slowly because $(f_1 - f_2)/2$ is small. ($\rightarrow$ *Beats.maxpat*) ($\rightarrow$ *Beats_2.maxpat*) ($\rightarrow$ *Beats_Partials.maxpat*)

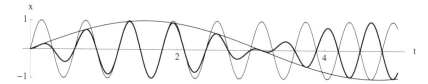

2.2.1.5 Natural Vibrations

Elastic bodies like strings, bars, gases, etc. oscillate if they are excited by an external force. If a body oscillates freely after the exciting force disappears, it does so only at certain frequencies. These oscillations are known as *natural oscillations* or *natural vibrations* of the body. When the body oscillates perpendicularly to the body's axis or surface, one speaks of *transversal* oscillation. Oscillation along the axis or surface is called *longitudinal*. Points whose displacement is always zero are called *nodes*, those whose displacement reaches a maximum are called *antinodes*. ($\rightarrow$ *natural_vibrations.pde*, *natural_vibrations_sum.pde*)

The following illustration shows the first three transverse natural oscillations of a string. Their frequencies are related as 1:2:3. ($\rightarrow$ Animation)

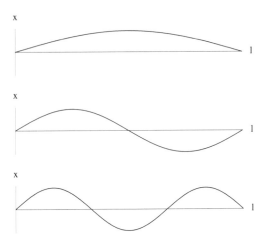

The attached ends of the string are always nodes. For the *n*th natural frequency there are $n - 1$ additional nodes which divide the string into n equal parts. The so-called *natural frequencies* or *eigenfrequencies* of the string can be calculated from the string's cross section S, from its normal stress σ and from its density ρ:

$$f_n = \frac{n}{2l}\sqrt{\frac{\sigma}{\rho S}}.$$

The following illustration shows the first three natural vibrations of a freely oscillating metal bar. The eigenfrequencies of metal bars are not harmonic (see [20]). (→ Animation)

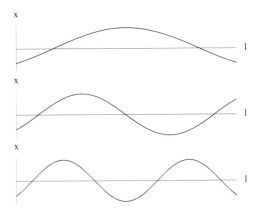

The oscillations which cause a column of air to sound are longitudinal. They can be described as variations of pressure or as displacement of the oscillating air molecules. In the animation below, an air column is modeled as a series of cells of equal volume. Air pressure is indicated by the intensity of the grey scale.

The next illustration shows the moment of maximum displacement for the air column's first three normal modes. If the distribution of pressure p is plotted as a function of position l, the same picture results as for the normal oscillations of the string shown above. (→ Animation)

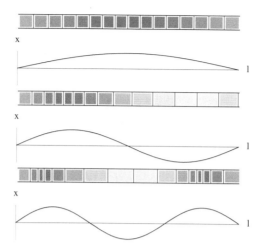

For tubes with one open and one closed end, there is always an antinode of the oscillation at the open end (that is, a pressure node) and an oscillation node (that is, pressure antinode) at the closed end. The following normal modes of oscillation are possible (→ Animation):

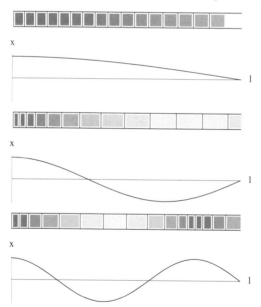

This explains the typical properties of stopped organ pipes (Stopped Diapason, Gedackt, Bourdon). The first eigenfrequency is an octave lower than for a pipe of the same length open at both ends, and only odd-numbered partial tones are produced. (The frequency ratios of the oscillations are 1 : 3 : 5 ...)
(→ *Natural_Modes_Open_Tube.maxpat*, *Natural_Modes_Closed_Tube.maxpat*)

2.2.1.6 Driven Oscillation and Resonance

When a periodically oscillating force acts upon an object capable of oscillation, that object, after a certain settling time, effects *driven oscillation*, the frequency of which is equal to that of the driving force. If the driving force is a sinusoid $F = F_0\sin(\omega t)$, then the driven oscillation is $x = x_0\sin(\omega t + \alpha)$. The amplitude of the driven oscillation x_0 depends on the ratio between the frequency ω of the exciting force and the eigenfrequency ω_e of the system. If the ratio ω/ω_e is close to 1, x_0 increases greatly and the system begins to resonate.

2.2.2 Periodic Vibrations and their Spectrum

2.2.2.1 The Definition of Periodic Vibrations

A repeated vibration or oscillation is called *periodic*. A particular vibration is determined by the duration of its period T, or by its frequency f, and by the form of the vibration. The pitch of the heard tone depends on the frequency $f = 1/T$. The vibration's form or shape influences only the tone's timbre. The curve corresponding to the vibration's form cannot have any discontinuities, because the vibrating point of a physical object cannot suddenly be in a different place. The curve cannot have any corners, because that would correspond to a sudden change of velocity. Thus the function describing the curve must be continuous and differentiable.

2.2.2.2 Standard Examples

The following waveforms, which are impossible in the real world because they have corners and discontinuities, are often approximated in electroacoustic music: triangle wave (a), square wave (b), sawtooth wave (c, d), impulses of different impulse length or duty cycle (e, f). The square wave corresponds to an impulse with a duty cycle of 50%.

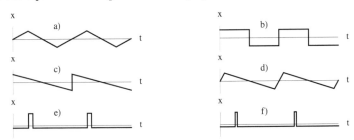

2.2.2.3 Other Examples

Periodic oscillation occurs whenever any waveform is repeated.

2.2 Vibration and Waves

The following example shows a sine wave of varying period. Because the variation is regular and repeats every 0.01 s, the result is a periodic oscillation with a fundamental frequency of 100 Hz (cf. Frequency Modulation 6.1.2).

```
ω=800*2*π;ω2=100*2*π;p1=Plot[Sin[ω*t+4*Sin[ω2*t+π]]...]
```

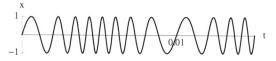

In the following example the amplitude of a sine wave varies regularly, causing again a periodic oscillation of frequency 100 Hz (cf. Amplitude Modulation 6.1.1).

```
ω=100*2*π;p1=Plot[Sin[ω*t]*Sin[8*ω*t]...]
```

2.2.2.4 Constructing Periodic Oscillation from Harmonic Waveforms

When one adds together harmonic oscillations whose frequencies are all multiples of a fixed *fundamental frequency* f_1, one obtains periodic oscillation, regardless of the amplitudes and initial phase angles of the components. Because the sine function is symmetrical and repeats after 2π, any sinusoidal oscillation can be written with positive amplitude and phase constant between 0 and 2π, or with positive or negative amplitude and a phase constant between 0 and π. In the first example, we add to a sine oscillation of 100 Hz (period = 0.01 s) and amplitude 2 a second sine oscillation of 900 Hz and amplitude –0.6.

```
ω=100*2*π;Plot[2*Sin[ω*t]-.6*Sin[9*ω*t]...]
```

If one varies the phase constant of a partial oscillation, the waveform of the periodic oscillation changes. In the second waveform the phase of the third harmonic is displaced by π.

```
Plot[2*Sin[ω*t]+2*Sin[3*ω*t]...]
Plot[2*Sin[ω*t]+2*Sin[3*ω*t+π]...]
```

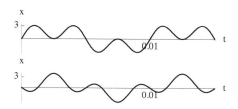

One can approximate a square wave by summing the odd partials with amplitudes inversely proportional to the partial number (1, 1/3, 1/5, ...). (→ Animation)

```
Plot[Sin[ω*t] + Sin[3*ω*t]/3 + Sin[5*ω*t]/5 +
    Sin[7*ω*t]/7 + Sin[9*ω*t]/9 + Sin[11*ω*t]/11 ...]
```

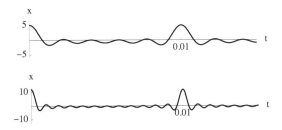

One can approximate a pulse train by adding together several harmonic cosine waves of the same amplitude. As the number of partials increases, the width of the pulse becomes narrower and the curve between the pulses becomes flatter. (→ Animation)

```
Plot[Cos[ω*t] + Cos[2*ω*t] + Cos[3*ω*t] ...]
```

(→ Timbre_and_Spectrum.maxpat)

2.2.2.5 The Spectrum of Periodic Oscillations

At the beginning of the 19th century, the French mathematician and physicist Jean Baptiste Joseph Fourier showed that any waveform with the period T could be expanded into a series of sine waves of frequency $f_1 = 1/T$, $f_2 = 2f_1$, $f_3 = 3f_1$, ... having suitable amplitudes A_1, A_2, A_3, ... and phase constants ϕ_1, ϕ_2, ϕ_3, ... The sinusoidal oscillations f_1, f_2, f_3, ... are known as the partials of the waveform; the overtones of a waveform are the partials without the fundamental frequency. The representation of the amplitudes A_1, A_2, A_3, ... as a function of frequency is called the amplitude spectrum of the periodic oscillation or sound. Although the phase constants are essential for determining the shape of the waveform of a sound, they are usually disregarded in the spectrum. The spectrum shows the partials present in a sound and hence informs one about the sound's timbre. The missing phase information is generally not a problem, because the phase constants do not as a rule influence the timbre. The following illustration shows the spectrum of a square wave.

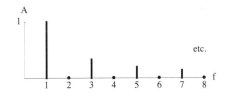

2.2 Vibration and Waves

Any harmonic sound of period T and corresponding angular frequency $\omega = 2\pi f = 2\pi/T$ can be produced by summing sine waves.

$$x(t) = \sum_{n=1}^{\infty} A_n \sin(n\omega t + \phi_n)$$

The identity $\sin(\alpha + \phi) = \sin(\alpha) \cdot \cos(\phi) + \cos(\alpha) \cdot \sin(\phi)$ implies that a sine wave of an arbitrary phase constant ϕ can be represented as a weighted sum of a sine wave and a cosine wave of the same frequency without phase constant:

$$x(t) = \sum_{n=1}^{\infty} (B_n \sin(n\omega t) + C_n \cos(n\omega t))$$
with $B_n = \cos(\phi_n)$ and $C_n = \sin(\phi_n)$

The calculation of the coefficients A_n (or B_n and C_n) in the formula above can be carried out exactly and without computer only in simple cases, like that of the square wave. There are, however, programs that can calculate the spectrum of a given sound using the so-called Fast Fourier Transform (cf. 3.2.2.5) and show the spectrum graphically. In this book, we will use the Fourier function in Mathematica to evaluate the spectrum of a list of values representing a waveform. In the following example, we calculate a square wave of 4 Hz and store 1000 points of this function in a list *list1*. Mathematica then evaluates the spectrum of the list.

```
list1=Table[Sign[Sin[ω*t/1000]]...]
ListPlot[list1]
```

```
ListPlot[Fourier[list1]]
```

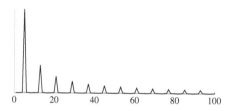

There are only odd partials (4, 12, 20, 28, 36, 44 Hz ...) whose amplitudes are inversely proportional to their frequencies.

In Chapter 2.2.2.4 a pulse train was approximated by adding several harmonic cosine functions of the same amplitude. To produce ideal pulses, infinitely many partials would have to be summed. In this case, the waveform of the pulse train and its spectrum would be identical.

2.2.3 Aperiodic Oscillation

2.2.3.1 Non-harmonic Partials

If we add two sine waves the ratio of whose frequencies is irrational, the resulting oscillation will not contain any periods. In the following example, two sine waves with the frequency ratio of the golden ratio 1 : 1.61803 ... (10.3.2) are combined. Although the resulting waveform looks simple and sounds smooth, in fact no section of it, taken at any order of magnitude, is ever repeated.

`Plot[Sin[ω*t]+Sin[1.61803*ω*t]`

The spectrum does not indicate that the oscillation is aperiodic, because one cannot tell whether the frequency ratio is rational or irrational.

If we change the second example of Chapter 2.2.2.3 so that the ratio of frequency variation to the sine wave's nominal frequency is irrational, an aperiodic oscillation results.

`ω=437.32171*2*π;ω2=100*2*π;`
`Plot[Sin[ω*t+3.5*Cos[ω2*t+π]],...]`

2.2.3.2 Noise

Random oscillation yields *noise*. The waveform in the illustration below was generated using a list (sequence) of random numbers between −1 and 1.

`l=RandomReal[{-1,1},100];`

The following figure shows two spectra, the first of a list of 200 random numbers, the second of a list of 2000 random numbers. Clearly, the first series will have a maximum of 100 oscillations, the second of 1000 oscillations. The spectra show partials in all frequency ranges. The amplitudes of the partials are randomly scattered around a certain mean value. In the case of light, this superposition of frequencies corresponds to the superposition of all the colors of the spectrum and gives rise to white light. Noise having these spectral characteristics is therefore known as *white noise*.

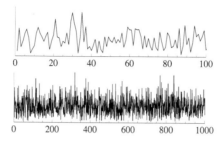

The spectrum of ideal white noise is a straight line.

If some frequencies are more strongly represented than others, one speaks of *colored noise* (5.3.2.1). White noise contains as much high-frequency as low-frequency energy, which is not the case for most natural sounds. Noise whose spectrum decreases exponentially with frequency is generally felt to be more natural. It is called pink noise and is often found in commercial synthesis devices.

2.2.3.3 Pulses

In Chapter 2.2.2.5 we saw how pulse trains can be made by summing cosine functions whose frequencies are multiples of a fundamental frequency. A single pulse can be considered to be the sum of cosine functions of all frequencies, since the functions $A \cdot \cos(\omega t)$ all take the value A at time $t = 0$, while at time $t \neq 0$ they have different values and cancel each other in the limit. The spectrum can be explained as follows. The spectrum of a pulse train is a line spectrum of infinitely many components of equal amplitude, separated by the frequency of the pulse train. As the pulse train's frequency decreases (and hence the length of its period increases), the spectral lines move closer to one another, blending together to form a single rectangle when the period T becomes infinitely long. Thus a single pulse has the same spectrum as does white noise.

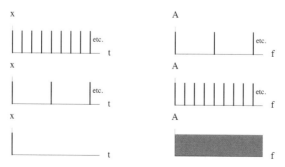

2.2.3.4 Quasi-periodic Oscillation

Various kinds of oscillation exist which are not strictly periodic but can be described as slowly varying or mildly perturbed oscillation. Beats, for instance, can be characterized as periodic oscillation with slowly varying amplitude, particularly when the ratio of the beat frequency to the oscillation's frequency is small (2.2.1.4). Whistle tones have a small noise component which makes them quasi-periodic, but to the ear they have a clearly defined pitch. The quasi-periodicity is easily recognizable in the waveform of a whistle tone below, and the tone's spectrum shows a frequency line which is only slightly broadened.

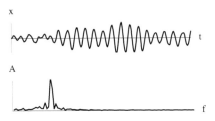

2.2.3.5 Variable Frequencies

The multiplier ω in the function $\sin(\omega t)$ should not be confused with the frequency if it is not constant, i.e., if $\omega = \omega(t)$. Using the Mathematica command below, we produce a function $\omega(t)$ which begins at $2\cdot 2\pi$, goes in one second to $4\cdot 2\pi$ and remains at $4\cdot 2\pi$ for another second.

```
Plot[Sin[If[t<1,(2+2*t)*2π*t,4*2π*t]]...]
```

At the end of the glissando, that is in the middle of the depicted waveform, the frequency is noticeably higher than 4 Hz. (→ Sound Example)

Using the formula above, one can only produce tones of constant frequency correctly. Because the frequency corresponds to the velocity of the changing argument of the sine function, that is, the phase, the instantaneous frequency is defined as the derivative of the phase with respect to time. When the argument c of $\sin(c)$ is constant, the frequency is $dc/dt = 0$; the function $\sin(\omega t)$ has the constant frequency $d(\omega t)/dt = \omega$. We calculate the phase by integrating the instantaneous time-dependent frequency $\omega(t)$. Thus in the first example, we have for the phase, instead of $(2 + 2t)\cdot 2\pi t = 4\pi t + 4\pi t^2$, the function

$$\int (2 + 2*t) *2\pi \, dt \quad = 4\pi t + 2\pi t^2$$

```
Plot[Sin[If[t < 1, 4πt+2πt², 4*2π*t]] ...]
```

Often the frequency $\omega(t)$ cannot be integrated analytically, or else the variation of frequency is not known. In such cases, integration has to be done numerically during the calculation of the waveform. (→ Sound Example) (→ *Variable Frequency.maxpat*)

2.2.4 Waves

2.2.4.1 Definition and Examples

The word "wave" originally referred to changes on the surface of a liquid. More generally, a wave is the propagation of a physical state through a medium. In an undisturbed medium, all parts of the medium are at rest. If one particle of the medium is moved out of its rest position by excitation, interaction among neighboring particles causes the excitation to be transmitted through the medium, forming a wave. One distinguishes between *transverse waves*, where the displacement of the particles is perpendicular to the direction of propagation, and *longitudinal waves*, where the displacement is in the same direction as the wave's propagation. The phase of a wave refers to its position in the vibration cycle. Points on a wave having the same phase form a *wave front*.

The first example shows a one-dimensional transverse wave caused by excitation at $x = 0$. One speaks of a *sine wave* or an *harmonic wave* if the waveform can be described by a sine function. Note that the horizontal axis in the graphic represents a spatial dimension and not time. Therefore, the following figures do not represent, as in case of oscillation, behavior over time. Rather, they are snapshots of instantaneous states. To illustrate the behavior of a wave over time, an animation can be made or several successive snapshots can be presented. The figure below shows three images from an animation. Considering two neighboring points, one can see that they go through the same motion at different times. Hence, a wave can be described as a set of coordinated oscillations of points in space. In the propagation of waves, no particle of the medium itself is transported. (→ Animation)

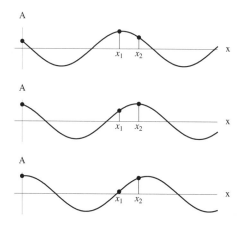

In the two-dimensional wave below, circular wavefronts move away from the point of excitation. Their amplitude diminishes with the distance from the excitation because the total energy is distributed over an ever-larger circle. (→ Animation)

If the excitation takes place along a straight line instead of at a single point, two-dimensional waves with straight wavefronts result. Waves excited at a single point also exhibit nearly straight wavefronts at considerable distance from the point of excitation. (→ Animation)

Longitudinal waves are more difficult to imagine and visualize. Experimentation with spiral springs can help the imagination here. In the following animation a longitudinal wave is created by a sinusoidal excitation in the direction of the wave's propagation. (→ Animation)

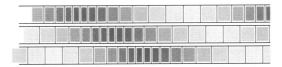

Three-dimensional waves cannot be illustrated even with animations. Sound waves are three-dimensional longitudinal waves of changing air pressure which propagate in all directions away from the sound source, giving rise to spherical wavefronts. All points on a wavefront have the same air pressure.

2.2.4.2 Mathematical Description

The displacement $y(t, x)$ of the oscillating particles of a wave depends on time t and position x of the particle. It is meaningful, for one-dimensional undamped waves, to speak of the wave's *amplitude* and *frequency*, because the oscillations of all the particles always have the same frequency and amplitude. The distance between the nearest particles of different fronts oscillating with the same phase is called the *wavelength* λ. The wavelength must not be confused with the period of an oscillation, which corresponds to the reciprocal of the frequency. The period is a duration in time, the wavelength a distance. The wave number k indicates the number of waves per unit of length, hence $k = 1/\lambda$ and $\lambda = 1/k$. The speed with which an excitation propagates in a medium is known as the *wave velocity* c (or more usually the *phase velocity*, referred to a specific phase of the wave). In acoustics, one speaks of the *speed of sound*. At the frequency f, a wavefront covers a distance of $f\lambda$ in one second. For the wave velocity we write: $c = f\lambda = \lambda/T$, where T is the wave's period.

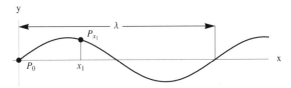

To derive the mathematical description of a wave, we begin by considering the point P_0 at $x = 0$. Its movement is described by the sinusoidal oscillation $y = A \cdot \sin(\omega t)$. Any point on the wave oscillates with the same amplitude and frequency but with a time delay dependent upon its distance from P_0 and upon the wave velocity c. This difference of phase is equal to $2\pi/\lambda \cdot x$.

2.2 Vibration and Waves

Hence, the equation for the harmonic wave giving the displacement of the particles as a function of their positions and of time is

$$y = A \cdot \sin\left(\omega t - \frac{2\pi}{\lambda} x\right) \quad \text{or}$$
$$y = A \cdot \sin\left(2\pi \frac{t}{T} - 2\pi \frac{x}{\lambda}\right) \quad \text{or}$$
$$y = A \cdot \sin\left(2\pi \left(\frac{t}{T} - \frac{x}{\lambda}\right)\right)$$

Here y = displacement, A = amplitude, T = period of the oscillation, λ = wavelength, t = time, x = distance from the origin $x = 0$.
If t is made constant, then $a = t/T$ is also constant, and the equation for the waveform becomes $y(x) = A \cdot \sin(2\pi(a - x/\lambda))$. If x is made constant, then $b = x/\lambda$ is also constant, and the equation for the oscillation of the point P_x becomes $y(t) = A \cdot \sin(2\pi(t/T - b))$.

2.2.4.3 The Superposition of Waves

The superposition of waves is given by adding their instantaneous displacements. If two waves of the same vibration direction and of equal frequency are superposed, one speaks of *interference*. The following figure shows two sine waves with the origins Z_1 and Z_2 and the wave resulting from the superposition of the two. The distance between the two origins Δx is called the phase difference.

The equations for the two waves are

$$y_1 = A \cdot \sin\left(2\pi\left(\frac{t}{T} - \frac{x}{\lambda}\right)\right) \text{ and } y_2 = A \cdot \sin\left(2\pi\left(\frac{t}{T} - \frac{x + \Delta x}{\lambda}\right)\right).$$

Their sum is

$$y = y_1 + y_2 = A \cdot \sin\left(2\pi\left(\frac{t}{T} - \frac{x}{\lambda}\right)\right) + A \cdot \sin\left(2\pi\left(\frac{t}{T} - \frac{x + \Delta x}{\lambda}\right)\right)$$

$$= A\left(\sin\left(2\pi\left(\frac{t}{T} - \frac{x}{\lambda}\right)\right) + \sin\left(2\pi\left(\frac{t}{T} - \frac{x + \Delta x}{\lambda}\right)\right)\right)$$

$$= 2A \cdot \cos\left(\pi \frac{\Delta x}{\lambda}\right) \cdot \sin\left(2\pi\left(\frac{t}{T} - \frac{x}{\lambda}\right) - \pi \frac{\Delta x}{\lambda}\right) \quad \text{(see 2.2.1.3)}$$

$$= 2A \cdot \cos\left(\pi \frac{\Delta x}{\lambda}\right) \cdot \sin\left(2\pi\left(\frac{t}{T} - \frac{x + \Delta x/2}{\lambda}\right)\right).$$

Only the term sin(...) depends on t and x, and so this equation describes a sinusoidal oscillation having the frequency of the two waves and the constant amplitude $2A \cdot \cos(\pi \cdot \Delta x/\lambda)$. The amplitude depends on the phase difference. If the phase difference is a multiple of the wavelength, the amplitude of the sum is maximal, if the ratio of phase difference to wavelength is 0.5, 1.5, 2.5, etc., the amplitude is minimal.

When circular waves are superposed, there are areas where the waves amplify each other and areas where they cancel each other. To calculate the points where the waves amplify each other maximally, the distance between the two origins Z_1 and Z_2 is called d, the distances of a point $P(x, y)$ to the two origins are called s_1 and s_2, and the difference between these distances Δs.

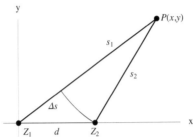

The greatest amplification occurs where Δs is a multiple of the wavelength λ. If we take the origin Z_1 as the origin of our coordinate system, we have

$$\sqrt{x^2 + y^2} - \sqrt{(x-d)^2 + y^2} = n + \lambda.$$

We solve the equation and show the solutions for $n = 1$, $n = 2$ and $n = 3$ graphically ($\lambda = 0.41$, $d = 1$). Here we solve the equation for x, because, as the figure below shows, the curves for $n = 1$ and $n = 2$ can only be written as unique functions in terms of y.

$$\text{Solve}\left[\sqrt{x^2 + y^2} - \sqrt{(x-d)^2 + y^2} == \lambda,\ x\right] \quad \rightarrow \quad \frac{d^3 - d\lambda^2 \pm \lambda \sqrt{d^2 - \lambda^2} \sqrt{d^2 + 4y^2 - \lambda^2}}{2d^2 - 2\lambda^2}$$

The following figure shows a snapshot of a corresponding animation.
(→ Animation) (→ Interference.maxpat)

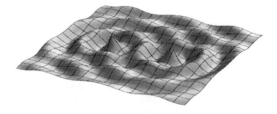

When two waves of the same frequency meet each other traveling in exactly opposite directions, a phenomenon occurs which is of particular importance for understanding the behavior of strings and air columns. The superposition creates points (in one-dimensional waves) or lines (in two-dimensional waves) where the waves' displacements are cancelled and points or lines where they are amplified. This phenomenon gives rise to *standing waves*, which correspond to the natural oscillations of strings and air columns. The following excerpts from an animation show two waves meeting each other head-on (dashed lines) and the resulting wave (bold line). (→ Animation)

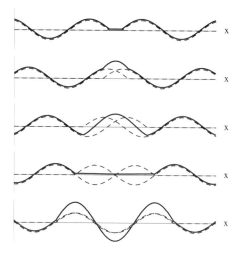

2.2.4.4 The Propagation of Waves

Many features of the behavior of waves can be explained by the *Principle of Huyghens*, named after the Dutch mathematician, astronomer and physicist Christiaan Huyghens (1629–1695). The Principle states that at any instant the wavefront of a propagating wave corresponds to the envelope of the spherical so-called wavelets or elementary waves emanating from every point on the wavefront at the prior instant. In the figures below, the inside (figure left) and the lower (figure right) lines show the prior wavefront and the outside and upper lines show the present wavefront as the envelope of the more finely drawn wavelets.

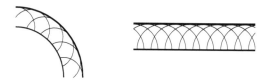

From the Principle of Huyghens we can derive the *law of reflection*, which says that the angle with which an incident wave strikes a reflective surface is the same as that of the reflected wave. The following figure shows a straight wavefront coming from the upper left (heavy dashed line) which is reflected on a horizontal surface. Four so-called normals, a, b, c, and d, are drawn perpendicular to the incident wavefront, meeting the surface at A, B, C and D respectively. By the time the wave reaches D, it has already reached A, B and C and provoked the wavelets K_a, K_b and K_c. The envelope of the wavelets is the straight line (heavy line) which includes the same angle to the surface as the incident wavefront. That the angles are the same can be seen from the normal b, which provokes the wavelet K_b at B, whose current radius is the same length as the normal from the incident wavefront to the point B.

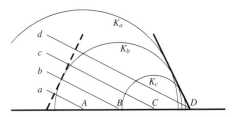

The displacement of reflected waves either keep their original direction or they are multiplied by −1. A wave in a wire which is fixed at one end turns negative when reflected at the fixed end and remains positive when reflected at the loose end (8.2.2.2).

In the discussion of reflection above, it may seem that the Principle of Huyghens offers a complicated description of a simple situation, but it provides a very elegant explanation for the phenomenon of *refraction*. Refraction is familiar from the change in direction of light when it passes from air to water. However, waves are always refracted when they pass from one medium to another having a different wave velocity. In the following figures the horizontal line represents the boundary between two media. At the bounding surface part of the wave is reflected, while another part continues in the new medium but in a different direction. In the example below, the phase velocity of the lower medium is lower than in the upper medium, hence the radius of the wavelet K_b is smaller than the length of the normal b between the wavefront (dashed line) and the point B. The simplified figure to the right will make the calculation of the relationship between the angle of the incident wave and that of the refracted wave easier. During time t the incident wave passes through the first medium with phase velocity c_1 from A_2 to B_2. During the same time, the refracted wave moves through the second medium with phase velocity c_2 from A_1 to B_1. From the illustration it follows that

$$\frac{\sin(\alpha)}{\sin(\beta)} = \frac{c_1 t/d}{c_2 t/d} = \frac{c_1}{c_2}.$$

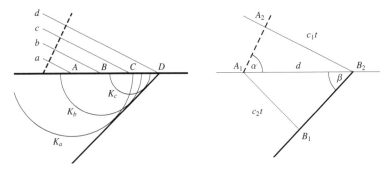

Refraction occurs not only at bounding surfaces but happens continuously in non-homogenous media. Sound waves, for instance, are skewed towards the ground when the air near the ground is colder than the air above, lowering the speed of sound close to the ground.

If a wavefront strikes a surface with a slit-like aperture, new circular waves, corresponding to the wavelets described above, propagate behind the aperture (figure left). By the same reason, there are no sharp shadows behind an obstacle struck by a wave: part of the wave is bent

2.2 Vibration and Waves

around the obstacle (figure right). This bending is called *diffraction*; its degree depends on the size of the obstacle and on the wavelength. The sound waves of low tones are more strongly diffracted than those of high tones.

The propagation speed of certain waves depends on their wavelength. The small ripples caused by the wind on the surface of water, for instance, move more slowly than the bow wave of a ship. The dependence of the propagation speed of a wave on its wavelength is known as *dispersion*. Non-sinusoidal waves suffer from dispersion, because the various spectral components have different wave velocities.

In the passage from one medium to another, but also in the propagation of waves in a homogeneous medium, energy is lost because of *absorption*. Absorption at bounding surfaces plays an important role in room acoustics, because the reverberation of a room depends on the absorptivity (9.2.1.2). Absorption by a medium not only contributes to a decrease of amplitude in circular and spherical waves but also causes a frequency-dependent change of spectrum (9.1.2.1). The decrease of energy caused by absorption is exponential: $W = W_0 \cdot e^{-kx}$, where W is the wave's energy after traversing the distance x, W_0 is the energy at the wave's origin ($x = 0$) and k is the *absorption coefficient*, which is dependent on the properties of the medium, the kind of wave and its frequency.

2.2.4.5 The Doppler Effect

If a sound source and a listener move relative to one another, the frequency of the sound generated and the frequency of the sound heard will not be the same. The heard frequency will be higher than the generated frequency when source and listener approach each other, lower when they move apart. The figure below illustrates the so-called *Doppler-Effect* (named after the Austrian physicist Christian Doppler, who described it in 1842) with the example of a moving sound source S and two stationary listeners H_1 and H_2. At time $t = 0$, the sound source was halfway between the two listeners and then began moving to the right. The circles represent the sound waves produced at different times, the dots indicate the position of the source at those times. The wavelength of the waves arriving at listener H_1 is greater than that produced by the source, and the perceived frequency is correspondingly lower. Conversely, the wavelength of the waves arriving at listener H_2 is smaller than that produced by the source, and the perceived frequency is correspondingly higher.

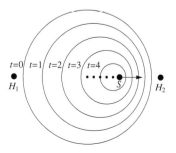

Let the speed of sound $c = 340$ m/s and the frequency f_0 be that of a sound source. Then there are f_0 cycles of sound distributed along a distance of 340 m $= c \cdot 1$ s. If the sound source moves with a velocity v_s towards the listener, the cycles of sound will be distributed along $(c - v) \cdot 1$ s meters. The wavelength of the compressed wave is therefore

$$\lambda = \frac{c - v_s}{f_0},$$

and since $c = \lambda \cdot f$, the resulting frequency is

$$f = \frac{c}{\lambda} = \frac{c}{\frac{c - v_s}{f_0}} = \frac{c f_0}{c - v_s} = \frac{f_0}{\frac{c - v_s}{c}} = \frac{f_0}{1 - v_s/c}.$$

If the sound source moves away from the listener, its velocity is negative. If we consider the case of a stationary source and a moving listener, we will see that $f = f_0(1 \pm v_e/c)$. These formulas can be summarized as

$$f = f_0 \frac{1 \pm v_e/c}{1 \mp v_s/c}$$

Here f_0 represents the original frequency of the source, v_s the velocity of the source and v_e the velocity of the listener.

Chapter 9.1.3.1 shows how the Doppler effect can easily be simulated, even when the sound source is not moving straight in the direction of the listener. (→ *Doppler-Effect.maxpat*, *doppler_effect.pde*)

2.3 Sound and Hearing

Further Reading: Musical Acoustics *by Donald E. Hall [20],* Music Cognition and Computerized Sound *by Perry R. Cook (including a CD),* Spatial Hearing – The Psychophysics of Human Sound Localization *by Jens Blauert [45].*

2.3.1 Pitch

2.3.1.1 Frequency Range and Octaves

Human beings perceive vibrations with frequencies between about 20 and 20,000 Hz as tones. The doubling of the frequency of a tone raises its perceived pitch by an octave. That is why an exponential increase of frequency means a linear increase in pitch. In Sound Example 2-3-1a one hears the frequencies 250, 500, 1000, 2000, 4000, 8000 and 16,000 Hz.

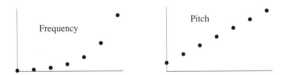

Conversely, halving a tone's frequency makes its pitch drop by an octave. In Sound Example 2-3-1b, we hear the frequencies 800, 400, 200, 100, 50, 25, 12.5 and 6.25 Hz. The last two

2.3 Sound and Hearing

frequencies are not audible, but they can be made visible by holding a flame in front of the loudspeaker. If we generate sequences of periodic impulses instead of sine waves (Sound Example 2-3-1c), the tone does not just become softer and finally disappear: even at less than 20 Hz, the impulses remain audible. The frequency at which single events are perceived as fusing into continuity is similar for audition and vision. The individual images of a film can be perceived singly only up to a frequency of about 20 images per second.

2.3.1.2 The Harmonic Series and Pure Intervals

In Sound Example 2-3-1d, one hears a sequence of many tones. The frequency of each tone is 100 Hz higher than that of the preceding one (100, 200, 300, 400, ... 4000 Hz). The distances between the tones in perceived pitch become smaller. A linear increase in frequency corresponds to a logarithmic increase in pitch.

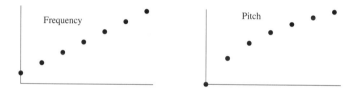

This succession of tones is called the *harmonic series*, or in a more general sense, an *overtone* or *partial series*. The first (lowest) tone is called the *fundamental*, the following tones the first, second, etc. overtone, or, rather confusingly, since the fundamental is considered the first partial tone, the second, third, etc. partial tone. Hence, the first overtone corresponds to the second partial tone, the second overtone to the third partial tone, etc. Starting from C^1 (or Pedal C, ca. 32.7 Hz), we have the following harmonic series (the seventh, 11th and 14 partial tones are noticeably lower than Bb3, F#4 and Bb4, and the 13th partial is noticeably higher than Ab4)

1	2	3	4	5	6	7	8	9	10	11	12	13	14	15	16
C1	C2	G2	C3	E3	G3	Bb3	C4	D4	E4	F#4	G4	Ab4	Bb4	B4	C5

The distance between two tones, known as an *interval* in music, is defined by the relationship of the frequencies of the tones. Those intervals appearing in the harmonic series are known as *pure intervals*, because they correspond to simple frequency relationships. From the table above we can read out the frequency relationships of the following pure intervals: octave 2:1, fifth 3:2, fourth 4:3, major third 5:4, minor third 6:5, major second or whole tone (major tone) 9:8, and the major second or whole tone (minor tone) 10:9. In Sound Example 2-3-1e, we hear tones having the frequencies 200, 400, 600, ... 3200 Hz. Notice the pure intervals, the two different major seconds (major and minor tone) and the tones we do not use in our musical system corresponding to the seventh and 11th partial tones.

Just Intervals	Octave	Fifth	Fourth	Major Third	Minor Third	Just Major Tone	Just Minor Tone
Frequency Ratio	1 : 2	2 : 3	3 : 4	4 : 5	5 : 6	8 : 9	9 : 10

The significance of the harmonic series lies in the fact that a tone's or a pitched sound's timbre is mainly determined by the relative strengths of fundamental and partials (2.3.2). The

harmonic series is also the basis for many elements of traditional music theory, like octave identity, the major triad, etc. The harmonics of string instruments and the intervals in which wind instruments can be overblown correspond to the intervals of the harmonic series.
(→ *Harmonic_Series.maxpat*)

2.3.1.3 Intervals

Because intervals are defined as frequency ratios, we add them by multiplying their ratios and subtract them by dividing their ratios. If we add a fifth and a fourth, we get the frequency ratio 2/1 from the product of 3/2·4/3. If we subtract a minor third from an octave, we get the major sixth 2/(6/5) = 5/3. If one tries to construct a circle of fifths using pure fifths, the circle does not quite close, because 12 pure fifths are a little more than seven octaves. This difference is called the *Pythagorean comma*. Twelve fifths, e.g. C1-G1-D2-...-B♯4 make an interval equal to $(3/2)^{12}$, seven octaves make an interval equal to 2^7. The difference in pitch between B♯4 and C5 is $(3/2)^{12}/2^7$ = 129.746 : 128 = 1.013643... Nor do three just major thirds give a just octave: $(5/4)^3$ = 1.95313. This discrepancy is called a *diesis*. Neither six major tones nor six minor tones give an octave. The difference between the just major and the just minor tone is known as the *syntonic comma* and corresponds to a frequency ratio of (9/8)/(10/9) = (81/80), or about 1/4 semitone. These discrepancies are the reason for the centuries-old search for a "good" tuning of the musical intervals. The discrepancies should be so distributed that they either are not noticeable or have a specific effect on the music played. Sound Example 2-3-1f shows that it is impossible to tune all the intervals even in a single key purely. After playing the degrees I–VI–II–V–I (5/3·2/3·4/3·2/3 = 81/80), we reach a note a syntonic comma below the beginning note. In so-called equal temperament, in which the octave is divided into 12 semitones of equal size, no interval except the octave is pure but rather all are defined by irrational ratios. If we call the frequency ratio of the equal tempered semitone x, then x^{12} must be equal to 2, which gives a frequency ratio of $2^{1/12}$ = 1.05946... Then we have for the major third $2^{1/3}$ = 1.25992..., for the whole tone $2^{1/6}$ = 1.12246... and for the fifth (seven semitones) $2^{7/12}$ = 1.49831. Small intervals are often indicated by cents. One *cent* is 1/100 of a tempered semitone and corresponds to the frequency ratio $2^{1/1200}$ = 1.00057779... The size of an interval in cents can be calculated from its frequency ratio using the following formula:

$$x \text{ (cents)} = 1200 \log_2 \frac{f_1}{f_2} = \frac{1200}{\log 2} \log \frac{f_1}{f_2}$$

Whether or not the pitches of two tones can be discriminated depends not only on the ratio of their frequencies, but also on the frequency range in which they occur and on the ability of the listener. This ability can be trained to some degree. Under ideal circumstances, most wind and string players can distinguish between tones whose frequencies vary by only a few cents. The literature indicates that the smallest Just Noticeable Difference JND for an average listener at frequencies around 100 Hz is 3% (50 cents) and is .8% around 2000 Hz (8.6 cents) (see [2] p. 33). Another representation (see [20] p. 113) shows a JND of ca. 1 Hz for frequencies below 1000 Hz (17 cents at 100 Hz, 1.7 cents at 1000 Hz) with a relatively rapid increase above 1000 Hz.

The frequencies of the chromatic scale can be calculated by using a pitch reference from which to start. Beginning with A = 440 Hz, we get the frequencies listed in Table B.1.

(→ Sound Examples) (→ *Intervals_1.maxpat*, *Intervals_2.maxpat*)

2.3.2 Timbre

Timbre is considerably more difficult to define than pitch or loudness. That is because timbre is determined by physical attributes which cannot be measured with a simple scale, as is the case for pitch and loudness. By the same token, there is no vocabulary in everyday language for timbre; depending on the sound in question, words for timbre can be borrowed from the visual world, like shrill or dull, or words with tactile associations are used, like rough or smooth.

Further Reading: Daniel Muzzulini, Genealogie der Klangfarbe *[77]*.

2.3.2.1 Periodic Vibration

We saw in Chapter 2.2.2 that every strictly periodic oscillation can be described by its spectrum. In Sound Example 2-3-2a, overtones are gradually added to a fundamental. As each overtone enters, one first hears it as a separate sound, but it soon blends with the fundamental, enriching the tone's timbre. In Sound Example 2-3-2b, only the odd overtones are added, so that the result gradually approaches a square wave (2.2.2.5). The following figure shows the waveform of this example in four snapshots (fundamental alone, fundamental and third partial, fundamental, third and fifth partial, and fundamental with third, fifth, ninth, 11th and 13th partials).

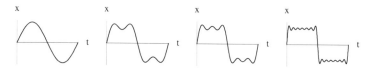

The next Sound Example shows that the phase constant hardly plays an important role for the timbre. The sound consists of the same odd partials at equal relative strength as before, but here every other partial has negative amplitude. Subtracting a partial is equivalent to adding that partial with a phase constant of π: $\sin(\alpha + \pi) = -\sin(\alpha)$ (Sound Example 2-3-2c). The four figures below show the same combination of partials as above.

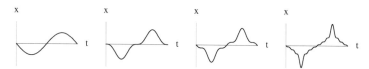

As the example shows, it is difficult to make predictions about how a given waveform will sound. Often it is not even possible to know whether the sounds corresponding to two different waveforms will have the same timbre. That is why sounds are usually described by their spectra and not by their waveforms. But even the spectrum permits one to draw only fairly general conclusions about a sound's timbre. Sounds with many partials are in principle brighter and brilliant, sounds with few partials are flat and dull, sounds with only odd partials are hollow and have a nasal character, etc. (→ *Timbre_and_Spectrum.maxpat*)

2.3.2.2 Formants

Various studies indicate, especially for the vowels of speech, that timbre is not determined by the relationship of the strengths of individual partials, but rather by so-called *formants* (B2), regions of frequency in which the partials are particularly strong. We notice especially that slowing down or speeding up the playback of a recording of the human voice not only changes the voice's pitch but also its timbre, although the spectral relationships are the same as for the original recording. In Sound Example 2-3-2d, a recording of the sung vowel "a" is played back first sped up and then slowed down, both times by a factor of 1.5, transposing it up and down respectively by a fifth. Conversely, the analysis of a vowel sung at various pitches shows the regions of strong partials essentially fixed at the same frequencies, regardless of the fundamental. This means that the relationships between the strengths of the partials differ at every pitch. If one draws a line connecting the amplitudes of the partials, one obtains an envelope that for a given timbre remains about the same for various fundamentals. The amplitudes of the partials in the first sound shown below have the relationships .85 : .65 : .19 : .16 : .38 ..., those in the second .87 : .32 : .16 : .41 : .17.

Sounds having the same spectral envelope arise from vibrations radiated by the same resonators. The shape of the spectral envelope reproduces the resonances of the sounding body. It is often possible to indicate which formant was produced by a particular part of the resonator. Our audition tries to draw conclusions about the object producing a sound from the sound's timbre. (→ *Formants.maxpat*)

2.3.2.3 Spectra of Natural Sounds

The properties of natural sounds are usually not constant, but rather are in constant flux. Even in sounds that seem to be held without change, loudness, frequency and spectrum change slightly. The changes are especially noticeable during the sound's attack (which can be between a few milliseconds and about 0.2 second long) and its release. Normally we can identify sounds within a fraction of a second, because the attack contains considerable information about how the sound was produced. Even the simplest models for the synthesis of instrumental sounds take into account the physical behavior of the sounding medium by distinguishing for amplitude and spectrum, at a minimum, the phases attack, sustain and decay.

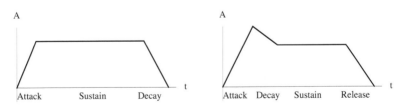

An evolving spectrum can be shown by displaying spectral snapshots made at regular intervals so that a three-dimensional image results, graphing the change in the amplitudes of the partials against time. The following figure shows typical behavior of the spectrum during the attack phase ($t = 0$ to $t = .12$).

2.3 Sound and Hearing

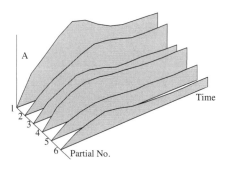

If one looks at this figure perpendicularly to the time-amplitude plane, one has a two-dimensional representation:

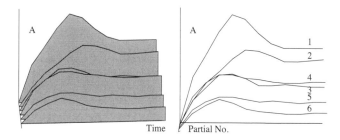

A temporal cross-section gives the spectrum at a specific moment (below $t = .02$ and $t = .05$).

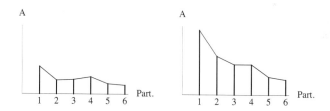

If one looks at the original figure from above, one has a spectrogram (or sonogram), in which the amplitudes of the of the partials can be estimated by the width or color of the lines.
(→ figure from above)

2.3.2.4 Non-Harmonic Spectra

Even the simplest bodies, like strings and air columns, produce sounds whose spectra are only approximately harmonic. For example, the frequencies of the upper partials of a loudly played string are somewhat high because of the string's stiffness. In pipes, the frequencies of the partials are not quite in integral relationships to each other because the length of the actually vibrating air column is not quite the same as the length of the pipe itself and is not quite the same for all partials. Objects in which vibrations can propagate in any direction, like plates and bells, have natural resonances whose frequencies can be in arbitrary proportions to each other.

2.3.2.5 Fusion

Whether or not sine waves fuse into one sound, depends on many factors. Although the audition essentially makes a Fourier analysis of the sounds it hears, as a rule it cannot perceive the individual partials of natural sounds. On the other hand, even the smallest differences in temporal patterns of sounds, or the smallest distance between sound sources, suffice to let us distinguish very similar sounds, for instance two violins playing in unison. Therefore, in the real world, we rarely have to ask which tones belong to which sounds. In electroacoustic music, especially when the sounds are relatively static, the situation is not always so clear.

The more isolated a partial is, the more likely it will be heard as a separate tone. In Sound Example 2-3-2e, we first hear the fundamental with amplitude 200 and then the eleventh partial with amplitude 12. Then other partials enter corresponding to the pattern in the figure below right. The eleventh partial only fuses with the fundamental when it becomes part of an amplitude envelope.

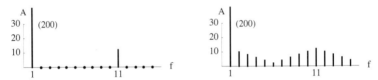

Normally the partials occur simultaneously, and changes of timbre are caused by changes in the envelope and not by changes in the strength of individual partials. That is why in Sound Example 2-3-2c in Chapter 2.3.2.1 the new tones are heard separately and only fuse when other new tones enter. If one interrupts playback for a moment and then resumes it, all the tones sound simultaneously, and it is considerably more difficult to hear the most recent partial. (See Chapter 5.2.1.3 for further examples.) (→ *Timbre_and_Spectrum.maxpat*)

2.3.2.6 Missing Fundamentals and Residue Pitch

When sine waves sound whose frequencies are multiples of a *missing fundamental* of frequency f, we perceive a tone with the frequency f, the so-called *residue pitch*, and we perceive the tones which actually sound as its overtones. In Sound Example 2-3-2f we first hear four equally strong tones with the frequencies 1600, 1800, 2000 and 2200 Hz, and then six tones with the frequencies 1200, 1400, 1600, 1800, 2000 and 2200 Hz and the amplitudes 1 : 2 : 3 : 3 : 2 : 1. The following figure shows the waveforms of the two sounds. It is easy to see a period in both waveforms which corresponds to the least common multiple of the individual period durations.

In the example above, it is difficult to decide whether the tone one hears is a phantom fundamental or a difference tone generated by the partials, because the frequency of the difference tone of two successive partials is also 200 Hz. Important features of combination tones are missing, however. In the first place, the fundamental can be heard even at low amplitudes, and in the second, there is no sensation of slight pressure in the ear (2.3.4.3). In Sound Example 2-3-2g we hear three tones with glissando (1600-1630 Hz, 1800-1830 Hz, and 2000-2030 Hz). The difference tones of the successive partials remain constant (200 Hz), while the missing fundamental rises slightly (from 200 to approximately 204 Hz), even though the frequencies are no longer integer multiples of the fundamental. If one listens to the example softly, one hears principally the slightly rising missing fundamental. If one increases the volume, the combination tones get louder and may even beat with the missing fundamental (see [20]). (→ *Timbre_and_Spectrum.maxpat*)

2.3.3 Loudness

2.3.3.1 Sound Power and Sound Intensity

Sound waves carry energy. The total energy radiated from a sound source per second is the source's *sound power* or *acoustic power*. It is measured in *watts* (W). The sound power of speaking is about .00001 W, that of a violin playing very loudly about .001 W and that of a grand piano playing loudly about 2 W. This relatively small amount of energy is dispersed in space so that very little of it actually reaches the listener's ear. The flow of sound power through a surface perpendicular to the flow is *sound intensity*. It is measured in watts per square meter, or more usually in watts per square centimeter. Even as small a sound intensity as 10^{-16} W/cm² can elicit in humans an auditory response. The threshold of pain is 10^{13} times higher at .001 W/cm².

Sound intensity is proportional to the square of the sound's amplitude. Doubling a sound's amplitude increases its sound intensity by a factor of four. The energy per surface (the sound intensity) carried by a spherical wave decreases inversely proportionally to the square of the distance traveled, because the sphere's surface increases with the square of its radius. Since the energy is proportional to the square of the amplitude, the amplitude decreases inversely proportionally to distance.

2.3.3.2 Decibels

When speaking about sound intensity, one is usually interested in comparing values. Because the comparisons can involve both very large and very small numbers, one uses a logarithmic unit of measure, the *decibel* (dB). The decibel measurement of the relationship between a given sound intensity J and a reference value J_0 is called the *sound intensity level L* and is derived from the following formula:

$$L = 10 \log_{10} \left(\frac{J}{J_0}\right) \text{dB}$$

For the proportion $J:J_0 = 1000$ we get L = $10 \cdot \log_{10}(1000)$ dB = $10 \cdot \log_{10}(10^3)$ dB = 30 dB. For the proportion $J:J_0 = 1/1000$ we get L = $10 \cdot \log_{10}(1/1000)$ dB = $10 \cdot \log_{10}(10^{-3})$ dB = −30 dB.

J/J_0	...	1/1000	1/1000	1/100	1/10	1	10	100	1000	1000	...
L in dB	...	−40	−30	−20	−10	0	10	20	30	40	...

Doubling the sound intensity raises the sound intensity level by $10 \cdot \log_{10}(2)$ dB = 3.0103... dB. Since the sound intensity is proportional to the square of the amplitude, a doubling of the amplitude raises the sound intensity level by $10 \cdot \log_{10}(2^2)$ dB = $10 \cdot 2 \cdot \log_{10}(2)$ dB = 6.02... dB.

$$L = 20 \log_{10} \left(\frac{A}{A_0}\right) \text{dB}$$

The reference value J_0 in psychoacoustics is the absolute *threshold of hearing* (ca. 10^{-16} W/cm²). This means that all audible sounds have sound intensity levels expressed in positive dB values. Since the threshold of pain is 10^{13} times higher, we can calculate the range of audibility as extending from $10 \cdot \log_{10}(J_0/J_0)$ dB = $10 \cdot \log_{10}(1)$ dB = 0 dB to $10 \cdot \log_{10}(10^{13})$ dB = 130 dB. The reference value can also be an ideal value, as is the case with electronic devices, giving rise to a scale with positive and negative decibel values. In computer music, the reference value is usually the largest amplitude the system can represent in binary form. As a result, all decibel value are negative. The difference between the lowest and the highest levels a system or a device can attain is called its *dynamic range*. The dynamic range of human language is about 50 dB, that of an orchestra about 70 dB. The range within which a device can operate reliably is also called its dynamic range.

2.3.3.3 Phons

The ear's sensitivity to amplitude is dependent on frequency. Let a pure tone be played at an amplitude so that it cannot be heard at 20 Hz. If the amplitude is kept constant and the frequency is raised, the tone will be perceived as becoming louder until about 4000 Hz, and then softer again, as the frequency continues to rise. Conversely, if a subject is given the task of matching the loudness of a tone of varying frequency to the loudness of a reference tone of 1000 Hz, the subject will raise the level of tones lower than the reference frequency and lower the level of tones higher than the reference frequency until about 4000 Hz, and then raise the level again. The trace of these levels plotted against frequency is called an equal-loudness contour, and the unit used for the measurement of perceived loudness is the *phon*. All tones on a given equal-loudness contour have the same phon value. The figure below show an equal-loudness contour for 50 phons. At 1000 Hz, the decibel and phon values are equal by definition (point A). At 100 Hz, a tone requires a level of 60 dB to be perceived by an average listener as equally loud, i.e. 50 phons (point B). (→ *Phon.maxpat*)

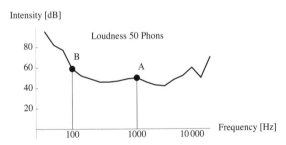

(→ Function for Approximating the Sound Intensity at the Threshold of Hearing)

2.3.4 Interaction of the Parameters

The rule that loudness is determined by intensity, pitch by frequency and timbre by waveform is a simplification. It will be shown in the following chapter that the perception of pitch, of loudness and of timbre interact with each other.

Further Reading: Donald E. Hall, Musical Acoustics *[20].*

2.3 Sound and Hearing

2.3.4.1 Frequency Groups

For various phenomena having to do with pitch, there is a critical frequency bandwidth within which our perception changes. For tones above 500 Hz this bandwidth is about 15 to 20% of a center frequency. If the frequencies of two tones f_1 and f_2 are close together, we do not hear the tones separately but rather as one tone beating at a frequency of abs$(f_1 - f_2)$. If the tones gradually move apart, we hear the beating accelerate until at a beat rate of about 20 Hz we hear a tone of uncertain pitch and characterized by a quality called *roughness*. Not until the two tones are no longer within the same critical band do we hear them as separate entities. In Sound Example 2-3-4a we hear two sine tones, both initially with a frequency of 1000 Hz. One tone rises until it reaches 1150 Hz, the other descends to 850 Hz. At a certain point, the loudness suddenly increases. One also hears the increase in loudness when the bandwidth of colored noise increases to greater than the size of critical band at that frequency. In Sound Example 2-3-4b we first hear noise with a constant center frequency of 1000 Hz and a bandwidth which begins at 10 Hz and increases to 200 Hz. To demonstrate that the increase in loudness is not related to the additional high frequencies of the first example, we hear a second example of noise whose center frequency descends from 1000 to 800 Hz while the bandwidth increases as before.

Frequency groups also influence timbre perception. If several partials of a sound are within one critical band, the timbre of the sound is not determined by the individual contribution of each partial but rather by their total energy. Sound Example 2-3-4c demonstrates this on five sounds. The first three have decidedly different timbres, because the partials 5, 6 and 7 differ in each sound and appear in different critical bands. The tones 3, 4 and 5 on the other hand can hardly be distinguished, because the partials 9 to 13 are in the same critical band and their total energy is the same in each sound. The table below shows the relative strength of the partials in each of the five sounds.

```
Sound 1   1 .1 .1 0  1  0       Sound 3   1 .4 .2 .1 .1 .1 .1 .2 .5 0 0 0 .5
Sound 2   1 .1 .1 1  0  1       Sound 4   1 .4 .2 .1 .1 .1 .1 .2 0 .5 0 .5 0
                                Sound 5   1 .4 .2 .1 .1 .1 .1 .2 0 0 1 0 0
```

2.3.4.2 Auditory Masking

Certain tones can affect the perception of other tones. If the tones' frequencies are decisive in the effect, one speaks of *spectral masking*. When two tones of different loudness sound simultaneously, the softer tone can only be heard when its frequency is sufficiently different from that of the louder tone. The principle of spectral masking is used in audio data reduction by recording only those tones which would not be masked. The following figure shows schematically the change in the threshold of hearing caused by a tone of 1000 Hz.

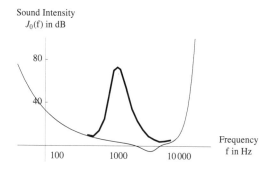

In Sound Example 2-3-4d there are two sine tones, one loud and of constant frequency of 1000 Hz, and a softer one whose frequency rises from 250 to 4000 Hz. Each tone is only on one stereo channel, so that the listener can experiment with the relative loudness of the two tones. When the rising tone comes within a critical band of the constant tone, it sounds even softer and even disappears for a while, although the two frequencies are the same only for an instant. In Sound Example 2-3-4e one again hears a soft sine tone rising, this time from 100 to 5000 Hz, and two bands of noise centered at 300 and 5000 Hz. (→ *Masking.maxpat*)

A tone can mask softer tones that sound just before or just after it. This is called *temporal masking*. The following figure shows schematically how a tone of 100 ms duration can briefly raise the threshold of hearing for other tones. (→ *Temporal_Masking.maxpat*)

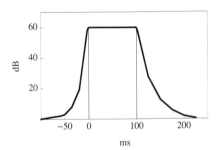

2.3.4.3 Combination Tones

At the beginning of the 18th century, the Italian violin virtuoso Giuseppe Tartini observed that when double stops were played loudly, a third tone could be heard. Several recorders playing together can also give rise to spurious tones. These so-called *combination tones* arise from non-linear superposition. The frequencies of possible combination tones arising from the frequencies f_1 and f_2 can be calculated as the sum and difference of multiples of the frequencies f_1 and f_2: $f = nf_1 \pm mf_2$. Particularly important are the *difference tone* $f_1 - f_2$ and the *sum tone* $f_1 + f_2$. The so-called *cubic difference tone* (2·lower − higher frequency) can sometimes sound as loud as the difference tone itself. Higher combination tones hardly ever occur. Combination tones are recognizable as such because they provoke a sensation of pressure within the ear.

Listen to the following examples with and without earphones, at various levels of loudness and, using the panoramic control, in various relationships of loudness between the stereo channels.
In Sound Example 2-3-4f, two tones sound with the frequencies 300 and 353.175 Hz. In intervals of three seconds, tones are added which could be possible combination tones: $f_1 - f_2$ (53.157 Hz), $f_1 + f_2$ (653.157), $2f_1 - f_2$ (256.843), $2f_2 - f_1$ (406.314), $2f_1 + f_2$ (953.157), $2f_2 + f_1$ (1006.314). Sound Example 2-3-4g is the same as the previous example except the two tones sound in separate channels.

f_1	f_2	$f_1 - f_2$	$f_1 + f_2$	$2f_1 - f_2$	$2f_2 - f_1$
353.157	300	53.157	653.157	406.314	256.843
971.758	700	271.758	1671.758	1243.516	428.242
2307.32	2000	307.32	4307.32	2614.64	1692.68

In Sound Example 2-3-4h one tone remains constant at 800 Hz while the frequency of the other decreases from 800 to 200 Hz. (→ *Combination_Tones.maxpat*)

2.3 Sound and Hearing

2.3.4.4 Particular Aspects of Pitch Perception

Frequency discrimination and recognition of intervals are reduced at both the high and the low end of the frequency range. Especially at the high end, intervals tend to be heard as too narrow (Sound Example 2-3-4i, cf. the Sound Example in 2.3.1.1). At high listening levels, high tones sound higher and low tones lower than they actually are. In Sound Example 2-3-4k all frequencies remain constant.

Only a few periods of a tone suffice to determine a tone's pitch. (Sound Example 2-3-4l)

Frequency	1000	1000	1000	400	400	400	3000	3000	3000	3000
Duration	.03	.01	.003	.01	.02	.04	.002	.004	.008	.016
Number of Periods	30	10	3	4	8	16	6	12	24	48

Frequency	200	200	200	800	800	800	800
Duration	.03	.06	.12	.002	.005	.01	.02
Number of Periods	6	12	24	1.6	4	8	16

The sound intensity influences the perception of timbre. Particularly, low notes with many partials of the same relative intensity sound very different at different levels of loudness (Sound Example 2-3-4m). That is because at high intensity all frequencies sound about equally loud, whereas at low intensity the perceived loudness of a tone varies considerably as a function of frequency. The effect can also be perceived using white noise.
(→ Timbre_and_Intensity.maxpat)

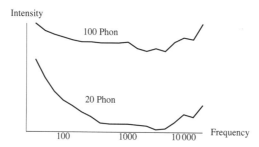

2.3.5 Resonant Spaces

The characteristics of the space in which sound is heard, as well as the position of both the sound source and the listener influence the sound in important ways. This influence must be taken into consideration when synthesizing and treating sounds. The spatial characteristics of sounds are often part of an electracoustic composition, and a good spatial representation of the sounds contributes greatly to their clarity and differentiation. It is not easy to realize spatial effects when playing back electroacoustic sounds, because the specific qualities of the sounds, the characteristics of the space in which they are played as well as the position of the loudspeakers and of the listeners all interact (see Chapter 9).

2.3.5.1 Reflections and Reverberation

Obstacles in space generally reflect sound waves. Especially in enclosed spaces, the reflected waves are an important part of the sound reaching the listener. Individually perceived reflections are heard as echoes, numerous reflections which cannot be perceived individually are heard as reverberation.

The law of reflection says that the angle between an incident ray and a flat surface is the same as the angle between the reflected ray and the surface. So to determine the path of a sound reflected by a wall from its source to the listener, one needs to find the point on the wall for which the two angles are the same (left figure below). In spaces with flat and straight walls, it is easy to indicate individual paths. The sound source must be reflected perpendicularly, as with a mirror, to the other side of the wall, which is where the listener localizes the sound. A line drawn from the listener to the reflected sound source subtends the wall at the point required.

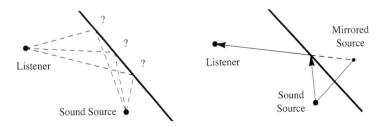

To determine the reflections from several walls, the sound source must be reflected as many times as there are walls (left figure below). For a curved wall, the reflection at point P corresponds to the reflection on the tangent plane at the point P (right figure below).

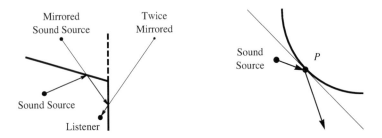

Reverberation consists of direct and indirect reflections. Especially in a large room, it takes a certain time for the first strong reflections, that is, the direct reflections from the wall, the ceiling and the floor, to reach the listener. Thereafter, it takes time for the reverberation to build to a maximum level. From then on, the level of the reverberation decreases nearly exponentially because the arriving sound waves have to travel farther and farther and are reflected more and more often. The reverberation time is the time it takes the sound level to decrease by 60 dB after the sound source is turned off. The reverberation times of large concert halls are usually between 1.8 and 3 s.

2.3 Sound and Hearing

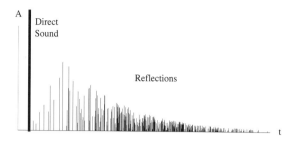

To measure the characteristics of the reverberation of a space, one generates an impulse (typically a shot or a clap) in the space. A recording of the impulse and the subsequent reflections shows a pattern similar to the figure above. Sound Example 2-3-5a illustrates the so-called impulse responses of a large space, a large church, a large space with echo and a simulated impulse response. It is also possible to apply the impulse response to sounds to reverberate them. Sound Example 2-3-5b illustrates a short melody reverberated by each of the impulse responses in the previous Sound Example (9.2.2.5).

2.3.5.2 Sound Localization

The audition determines the azimuth of a sound source (the angle between the sagittal plane and the direction of the source) from the difference in distance between the source and the left and right ears (x_l resp. x_r). This difference in distance produces a difference in both, the arrival times of the sound wave at the two ears, known as ITD (*interaural time difference*), and in the intensities at the two ears, known as IID (*interaural intensity difference*, or ILD *interaural level difference*). Ideally, the audition can determine the direction of a sound source to within about two degrees of azimuth.

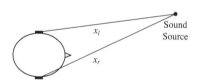

The difference in intensity comes partly from the difference in distance but more importantly from the occlusion of the laterally arriving sound, the so-called head-shadowing effect. The intensity difference at the two ears is the most important criterion for the localization of high tones. For low tones, where the diffraction around the head is more pronounced because of the longer wave lengths, the time difference is more important. Sound Example 2-3-5c presents in both channels tones of equal loudness but with different onset times. Sound Example 2-3-5d shows the same tones, both with the same onset time but with different loudness relationships ($A_l:A_r$). (Some effects can only be heard using a headphone.)

80 Hz	0	.001	−.001	.00025	.0003	.01	0	.0005	ITD in s
4000 Hz	0	.001	−.001	.00025	.0003	.01	0	.0005	ITD in s

50 Hz	1 : 1	100 : 1	1 : 10	3 : 2	2 : 3	$A_l : A_r$
2000 Hz	1 : 1	100 : 1	1 : 10	3 : 2	2 : 3	$A_l : A_r$

(→ *Sound_Localization.maxpat*)

The difference in a sound's arrival times at the two ears can be illustrated with a simple experiment. If one hold the ends of a thin tube about a meter long to both ears and taps or scratches at various places along the tube, the sound produced travels in both directions through the tube and reaches the ears with about the same intensity but with a delay corresponding to the distance traveled.

Differences in intensity and arrival times do not give information about whether a sound is behind or in front of the listener, nor about whether it is high or low in space. This information comes primarily from changes in the sound's spectrum due to the filtering effects of the head, ears and chest. The spatial effect in Sound Example 2-3-5e results from time and loudness differences as well as from changes in spectrum (9.1.1.3).

2.3.5.3 Distance Perception

The loudness of a sound diminishes rapidly with distance. One cannot, however, determine a sound's distance from the decrease in loudness alone, because there is no way to know whether the sound is soft and close or loud and far away. The high frequency energy of most natural sounds increases with loudness. Thus, by taking into account both amplitude and timbre, we can usually judge the distance of familiar sounds, like voices or musical instruments. But we cannot rely on our experience when listening to electronically reproduced sounds. The decrease in loudness of the spoken text in Sound Example 2-3-5f is not perceived as movement of the speaker away from the listener, but simply as the result of turning down a volume control. If two sounds are present, however, one of constant and the other of decreasing loudness, the decreasing loudness is heard as indicating greater distance (Sound Example 2-3-5g). Timbre, too, depends to some degree on distance, because air absorbs high frequency energy more strongly than it does low frequency energy (9.1.2.1). Like decreasing loudness however, the effect is not sufficient to allow one to determine distance reliably (Sound Example 2-3-5h). In enclosed spaces, the proportion of reflected sound (reverberation) to direct sound increases with distance (Sound Example 2-3-5i). Sound Example 2-3-5k shows the interaction of the three most important cues for distance perception: a) decrease in loudness, b) damping of high frequencies and c) the proportion of reflected sound to direct sound. (One hears at time 0" a) and b), at 11" b) and c), at 24" a) and c) and at 35" a), b) and c)).

3 Signals and Systems

Signals are optical, acoustic or electronic events that serve to transmit information. Systems are finite physical or virtual entities in which specific processes occur. Systems like living creatures, machines and electronic devices take in information, react to it and pass information on. Chapter 3.1 and 3.2 treat systems and signals in the narrower sense of communications technology, where a signal is a physical quantity whose variations in time contain information. Hence a signal can be represented by a one-dimensional time function. In what follows, we usually speak of audio signals, that is, of recorded oscillations as we described in the previous chapter. But what is said here is true for any kind of signal, for example for control signals like those frequently used in computer music.

It is the goal of this chapter to give the reader an introduction to the technical and mathematical fundamentals of digital signal processing so that he or she can follow the calculations that appear in the literature about specialized computer music applications.

Further Reading: Most books on digital signal processing are intended for electrical engineers and assume knowledge of technical mathematics and analog electronics. Understanding Signals and Systems *by Jack Golten [16] and* A Digital Signal Processing Primer *by Ken Steiglitz [32] are both books that can be understood and appreciated by non-specialists. There are also good introductions to digital signal processing in* Real Sound Synthesis for Interactive Applications *by Perry R. Cook [70] and in* The Theory and Techniques of Electronic Music *by Miller Puckette [87]. Especially recommended is* Struktur und Bedeutung *by Norbert Bischof [33], an introduction to systems theory as cybernetic theory, the science of communication and control in living creatures and machines. The book is intended for psychologists and accordingly treats above all psychological phenomena. It not only gives insight into signal processing and systems in a narrower sense but also into the processing of signals in more complex compound systems. Among the vast number of technical books available, we can mention:* Digital Signal Processing *by John G. Proakis and Dimitris K. Manolakis [65],* Digitale Signalverarbeitung *by Karl-Dirk Kammeyer and Kristian Kroschel [18],* Signale und Systeme *by Uwe Kiencke and Holger Jäkel [35],* Digital Signal Processing *by M. H. Hayes (Schaum's Outline Series) [36] and the online book* Digital Sound Generation *by Beat Frei [95] and [96]. Books on computer music usually have introductions to digital signal processing, for example:* Elements of Computer Music *by F. Richard Moore [8],* The Computer Music Tutorial *by Curtis Roads [11],* Real Sound Synthesis for Interactive Applications *by Perry R. Cook [70] and* The Theory and Techniques of Electronic Music *by Miller S. Puckette [87].* Fundamentals of Digital Filter Theory *by Julius O. Smith in* The Music Machine *edited by Curtis Roads [7] gives an excellent introduction to digital filters.*

3.1 Analog Signals and the Fourier Transform

Analog signals are continuous in time and amplitude. The functions to describe these signals are also continuous, that is, they show no discontinuities. The ear analyzes essentially the loudness and pitch of acoustic signals, where the pitch analysis is linked to the perception of the fundamental of a periodic signal but also to the perception of the timbre of that signal. The representation of a signal as a time function does not generally allow conclusions about the signal's "musical" quality. That is the reason why transformations of a signal from the so-called time domain to the frequency domain are so eminently important. A further advantage of a signal's frequency domain representation, its so-called spectrum, is that certain operations and manipulations are much easier to perform here than in the time domain.

Further Reading: In most of the books mentioned for Chapter 3, analog signals and the Fourier Transform are briefly treated. The practical applications of the Fourier Transform extend far beyond straightforward signal processing, however. One can find good presentations in mathematical handbooks, for instance: Schaum's Outline of Advanced Mathematics for Engineers and Scientists by Murray R. Spiegel (Schaum's Outline Series) [1] or Mathematics for Engineers and Scientists by Alan Jeffrey [89].

3.1.1 Periodic Signals and Fourier Series

3.1.1.1 Fourier Series

We saw in Chapter 2.2.2 that periodic oscillation arises from the addition of harmonic sine waves. Conversely, every periodic oscillation can be broken down into harmonic partial oscillations. The equation for an harmonic oscillation with the angular frequency ω, the initial phase angle ϕ and the amplitude Amp is $x(t) = Amp \cdot \sin(\omega t + \phi)$. Any sum of harmonic partials of the fundamental $\omega = 2\pi f = 2\pi/T$ can be written as

$$x(t) = \sum_{n=1}^{\infty} Amp_n \sin(n\omega t + \phi_n)$$

If follows from the identity $\sin(\alpha + \beta) = \sin(\alpha)\cdot\cos(\beta) + \cos(\alpha)\cdot\sin(\beta)$ that any sine oscillation with an arbitrary initial phase ϕ can be represented as the sum of a sine wave and a cosine wave of the same frequency.

$$Amp \cdot \sin(\omega t + \phi) = Amp(\sin(\omega t)\cdot\cos(\phi) + \cos(\omega t)\cdot\sin(\phi))$$
$$= Amp \cdot \cos(\phi)\cdot\sin(\omega t) + Amp \cdot \sin(\phi)\cdot\cos(\omega t)$$

Since ϕ is constant, we have

$$Amp \cdot \sin(\omega t + \phi) = A \cdot \sin(\omega t) + B \cdot \cos(\omega t)$$

with the amplitudes

$$A = Amp \cdot \cos(\phi) \text{ und } B = Amp \cdot \sin(\phi).$$

We can also write

$$x(t) = \sum_{n=1}^{\infty} [A_n \sin(n\omega t) + B_n \cos(n\omega t)]$$

If we include oscillations whose average displacement is not zero, we must add a constant term B_0. Since $\cos(0) = 1$, we have $B_0 = B_0\cos(0\cdot\omega t)$.

$$x(t) = B_0 + \sum_{n=1}^{\infty} [A_n \sin(n\omega t) + B_n \cos(n\omega t)]$$

3.1.1.2 Calculating the Coefficients

Fourier analysis provides the coefficients of the harmonic partial functions for a given periodic function $x = F(t)$. The term B_0 corresponds to the average deviation from a median value, the DC component, and results from dividing the signed area under the function by its period. In the example below, the area under the function $F(t)$ during the period $T = 4$ consists of $1.5 \cdot 2$ in the first half and $-0.5 \cdot 2$ in the second half, that is in total 2. Dividing this area by the period T gives $B_0 = 2/4 = 0.5$.

3.1 Analog Signals and the Fourier Transform

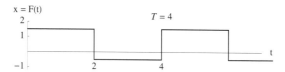

In general, B_0 is obtained by integrating the function $F(t)$ over a period and dividing by the duration of the period. From the final equation in Chapter 3.1.1.1, we can derive the result by integrating both sides of the equation over a period T, keeping in mind that the signed area under one period of a sine or cosine function is zero.

$$\int_0^T F(t)\,dt = \int_0^T B_0\,dt + \int_0^T \sum_{n=1}^\infty [A_n \sin(n\omega t) + B_n \cos(n\omega t)]\,dt$$
$$= \int_0^T B_0\,dt = B_0 T$$

This gives

$$B_0 = \frac{1}{T}\int_0^T F(t)\,dt.$$

We can calculate the coefficient A_m by multiplying both sides by $\sin(m\omega t)$ ($m \in \mathbb{Z}$) before integrating.

$$\int_0^T \sin(m\omega t) F(t)\,dt = \int_0^T \sin(m\omega t) B_0\,dt$$
$$+ \int_0^T \sum_{n=1}^\infty [A_n \sin(m\omega t) \cdot \sin(n\omega t)]\,dt$$
$$+ \int_0^T \sum_{n=1}^\infty [B_n \sin(m\omega t) \cdot \cos(n\omega t)]\,dt$$

The first term is zero because the integral of a multiple of the period of a sine function is zero. The term $\sin(m\omega t)\cdot\sin(n\omega t)$ can be transformed into

$$1/2\cdot\cos((n-m)\omega t) - 1/2\cdot\cos((n+m)\omega t)$$

and the term $\sin(m\omega t)\cdot\cos(n\omega t)$ can be transformed into

$$1/2\cdot\sin((n-m)\omega t) + 1/2\cdot\sin((n+m)\omega t).$$

The integrals of these terms disappear when $n \neq m$, since $n - m$ and $n + m$ are integers not equal to zero. Only terms with $n = m$ remain, and here the integral of

$$1/2\cdot\sin((m-m)\omega t) + 1/2\cdot\sin((m+m)\omega t)$$

also disappears because $\sin(0) = 0$. This leaves on the right side of the equation

$$\int_0^T \frac{1}{2}\cos((m-m)\,\omega t) - \frac{1}{2}\cos((m+m)\,\omega t)\,dt$$
$$= \int_0^T \frac{1}{2}\cos(0\cdot\omega t) - \frac{1}{2}\cos((2m)\,\omega t)\,dt = \frac{1}{2}T.$$

The equation above becomes

$$\int_0^T \sin(m\omega t) F(t)\,dt = \int_0^T A_m \frac{1}{2}\,dt = \frac{1}{2}A_m T.$$

Hence

$$A_m = \frac{2}{T}\int_0^T \sin(m\omega t)\, F(t)\, dt.$$

Similar calculations lead to

$$B_m = \frac{2}{T}\int_0^T \cos(m\omega t)\, F(t)\, dt.$$

The coefficients of the Fourier series for the periodic function $F(t)$ are calculated as follows:

$$\boxed{\begin{aligned} B_0 &= \frac{1}{T}\int_0^T F(t)\, dt \\ A_n &= \frac{2}{T}\int_0^T \sin(n\omega t)\, F(t)\, dt \\ B_n &= \frac{2}{T}\int_0^T \cos(n\omega t)\, F(t)\, dt \end{aligned}}$$

To obtain a spectrum which shows only the magnitudes of the partial functions and not their phases, we calculate the amplitude Amp_n from A_n and B_n.

When $A = Amp\cdot\cos(\phi)$ and $B = Amp\cdot\sin(\phi)$ (3.1.1.1), we obtain

$$\begin{aligned} A_n^2 + B_n^2 &= Amp_n^2\cos(\phi)^2 + Amp_n^2\sin(\phi)^2 \\ &= Amp_n^2(\cos(\phi)^2 + \sin(\phi)^2) = Amp_n^2. \end{aligned}$$

Hence

$$\boxed{Amp_n = \sqrt{A_n^2 + B_n^2}}$$

3.1.1.3 Examples

First we will calculate the Fourier coefficients of a square wave with amplitude 1, setting the duration of the period to 2π. The illustration below shows the square wave $F(t)$ and the function $\sin(\omega t)$ on the left and the product of the two functions and the areas I_1 and I_2 corresponding to the integral of the product on the right.

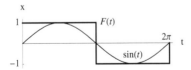 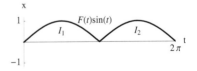

The areas I_1 and I_2 are each

$$\int_0^\pi \sin(t)\, dt = [-\cos(t)]_0^\pi = 2.$$

Consequently, $A_1 = \frac{2}{T}4 = \frac{4}{\pi} = 1.27324...$

The terms B_1 and A_2 disappear, since the areas under the functions $\cos(n\omega t)F(t)$ and $\sin(2\omega t)F(t)$ are equal to zero. The same is true for all B_n and for all A_n when n is even.

3.1 Analog Signals and the Fourier Transform

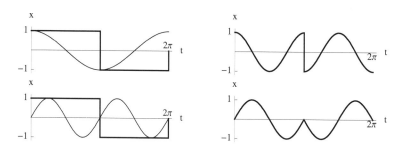

The area under the function $\sin(3\omega t)F(t)$ is $2I_1 = 2 \cdot 2/3 = 4/3$. It follows that $A_3 = 2/T \cdot 4/3 = 4/3\pi = 0.424413...$ The other coefficients, $A_5 = 4/5\pi$, $A_7 = 4/7\pi$ etc., can be derived similarly.

As a check, we make an approximation to the waveform by adding the odd partials up to $n = 15$.

```
Plot[∑⁷ᵢ₌₀ 4/((2*i+1) π) Sin[(2*i+1) t], {t, 0, 2 π}]
```

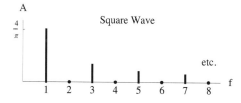

One can use these data to represent the function's amplitude spectrum graphically by calculating the amplitudes

$$Amp_n = \sqrt{A_n^2 + B_n^2} \quad (3.1.1.2)$$

and drawing the corresponding line spectrum.

(→ Second example)

3.1.2 Complex Representation

This chapter introduces complex numbers and explains them in detail. An understanding of complex numbers will be essential for the chapters ahead.

3.1.2.1 Complex Numbers

Because the square of any real number is always real, the square root of a negative number cannot be real. One defines

$$\sqrt{-1} \text{ as the unit } i$$

of a new set of numbers, the so-called *imaginary numbers*. Correspondingly $i^2 = -1$. Multiples of the imaginary unit, that is, products of real and imaginary numbers, are imaginary. If one adds an imaginary number and a real number, the result is a *complex number*. One uses $\mathbb{C}$ to represent the set of complex numbers. The real elements a and b of the complex number $z = a + bi$ are called the real and the imaginary parts of the number respectively. The real part is written as $\text{Re}(z) = a$, the imaginary part $\text{Im}(z) = b$. Two complex numbers are equal if and only if their real and imaginary part are equal. The number $\bar{z} = a - bi$ is called the *complex conjugate* of the number $z = a + bi$. Here $\text{Re}(z) = \text{Re}(\bar{z})$ and $\text{Im}(z) = -\text{Im}(\bar{z})$. The arithmetic rules for complex numbers are the same as for real numbers, whereby i^2 can always be replaced by -1.

Examples: $(2 + 3i) + (4 + 5i) = 6 + 8i$
$2(1 + i) = 2 + 2i$
$(1 + i)(2 + 3i) = 2 + 5i + i \cdot 3i = 2 + 5i + 3i^2 = 2 + 5i - 3 = -1 + 5i$
$(3 + 4i)^2 = 9 + 2 \cdot 3 \cdot 4i + (4i)^2 = 9 + 24i - 16 = -7 + 24i$

If one interprets the real and imaginary parts of a complex number as coordinates of a point on a plane, then the complex number $z = a + bi$ is associated with the point $P(a, b)$. The reference plane is called the *complex plane*, the *Gauss plane* or the *Argand plane* and is characterized by a real axis and an orthogonal imaginary axis. The complex number $z = a + bi$ is often represented not as a point but as position vector $P(a, b)$. This vector is often represented by an arrow (see below). The *modulus* or the *absolute value* of the complex number z is defined as the length of the arrow

$$|z| = \sqrt{a^2 + b^2}.$$

The vectors of the complex conjugates z and $\bar{z}$ are symmetrical to the real axis in the complex plane.

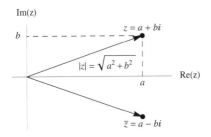

3.1.2.2 Trigonometric Representation

One can also represent the location of a number $z = a + bi$ on the complex plane by the so-called *polar coordinates* r and φ, where r is the modulus of the number z and the angle φ is the argument of z (written as $\varphi = \arg(z)$).

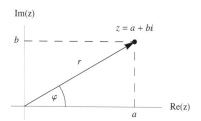

Since $a = r \cdot \cos(\varphi)$ and $b = r \cdot \sin(\varphi)$, we can write

$$z = a + bi = r \cdot \cos(\varphi) + r \cdot \sin(\varphi) i = r(\cos(\varphi) + \sin(\varphi) i)$$

This representation is ambiguous however, because the angles $\varphi, \varphi \pm 2\pi, \varphi \pm 4\pi$, etc. give the same results in the above formulas. Therefore, one often limits oneself to the so-called *principal value* of the argument, which lies either between 0 and 2π or between $-\pi$ and $+\pi$. By the same token, the polar coordinates r and φ can be calculated from the Cartesian coordinates a and b as follows.

$$r = |z| = \sqrt{a^2 + b^2}, \quad \tan(\varphi) = \frac{b}{a}$$

When calculating the angle φ from $\tan(\varphi)$, one must take account of the quadrant in which the complex number lies. If it is the first or the fourth quadrant, the rule is: $\varphi = \arctan(b/a)$, if it is in quadrants two or three, the rule is: $\varphi = \arctan(b/a) + \pi$. In Mathematica angles to all quadrants are obtained by using the function ArcTan[a, b]. In several programming languages however, the corresponding function is atan2(b, a).

ArcTan[1,-1] $-\frac{\pi}{4}$

ArcTan[-1,1] $\frac{3\pi}{4}$

The complex conjugate of $z = r(\cos(\varphi) + \sin(\varphi)i)$ is

$$\bar{z} = r(\cos(-\varphi) + \sin(-\varphi) i) = r(\cos(\varphi) - \sin(\varphi) i)$$

Example: The trigonometric representation of the numbers in the illustration below can be calculated as follows: From $z_1 = 4 - 1.8i$ we have

$$r_1 = \sqrt{4^2 + 1.8^2} = 4.386... \text{ and } \varphi_1 = \arctan(-1.8/4) = -0.4229...$$

The polar coordinates for z_3 and z_4 can be read from the illustration: $r_3 = 1.8$ and $\varphi_3 = \pi/2$, $r_4 = 2$ and $\varphi_4 = \pi$ or $-\pi$. The Cartesian coordinates a and b for the numbers z_2 and z_5 can be calculated from the polar coordinates:

$$a_2 = 2 \cdot \cos(\pi/4) = \sqrt{2} \text{ and } b_2 = 2 \cdot \sin(\pi/4) = \sqrt{2},$$
$$a_5 = 1.8 \cdot \cos(-.6\pi) = -.556... \text{ and } b_5 = 1.8 \cdot \sin(-.6\pi) = -1.7112...$$

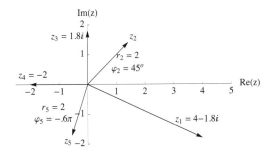

3.1.2.3 The Exponential Form

Using Euler's formula $e^{i\varphi} = \cos(\varphi) + i \cdot \sin(\varphi)$, where $e = 2.71828...$ (Appendix B3), the trigonometric representation can be transformed into exponential form.

$$\boxed{z = r\, e^{i\varphi}}$$

One can understand the relationship between the trigonometric and the exponential representation by considering their power series (Taylor expansions).

$$\sin(x) = x - \frac{x^3}{3!} + \frac{x^5}{5!} - \frac{x^7}{7!} + ...$$

$$\cos(x) = 1 - \frac{x^2}{2!} + \frac{x^4}{4!} - \frac{x^6}{6!} + ...$$

$$e^x = 1 + x + \frac{x^2}{2!} + \frac{x^3}{3!} + \frac{x^4}{4!} + ...$$

The complex conjugate of the number z, written $\bar{z}$, has the exponential form

$$\boxed{\bar{z} = r\, e^{-i\varphi}}$$

The numbers shown in the illustration above have the exponential form:

$$z_1 = 4.386 \cdot e^{5.860\, i}, \; z_2 = 2 \cdot e^{\frac{\pi}{4}i}, \; z_3 = 1.8 \cdot e^{\frac{\pi}{2}i} = 1.8i, \; z_4 = 2 \cdot e^{\pi i} = -2 \text{ and } z_5 = 1.8 \cdot e^{-.6\pi i}.$$

3.1 Analog Signals and the Fourier Transform

Complex exponentials can be decomposed into a product of a real exponential and an imaginary exponential.

$$\boxed{e^{a+bi} = e^a \, e^{bi}}$$

The imaginary exponential associates to every imaginary number $i\varphi$ the complex number

$$z = e^{i\varphi} = \cos(\varphi) + i \cdot \sin(\varphi).$$

The function is, in contrast to the exponential function with real exponent, periodic with the period $2\pi i$:

$$e^{(i\varphi + k \cdot 2\pi i)} = e^{i(\varphi + k \cdot 2\pi i)} = e^{i\varphi} \text{ for all } k \in \mathbb{Z}.$$

The modulus of any exponential function with a purely imaginary exponent is equal to 1.

$$\boxed{\left| e^{i\varphi} \right| = 1 \text{ for all } \varphi}$$

(→ Examples and Animations)

3.1.2.4 Algebra With Complex Numbers

Complex numbers are added and subtracted using Cartesian coordinates by adding or subtracting the real and imaginary parts separately: $a + bi \pm (c + di) = a \pm c + (b \pm d)i$. Geometrically this corresponds to the addition or subtraction of vectors.

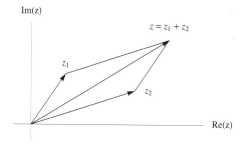

The product of two complex numbers can be calculated by expansion:

$$(a_1 + b_1 i)(a_2 + b_2 i) = a_1 a_2 + (a_1 b_2 + a_2 b_1)i + b_1 b_2 i^2 = a_1 a_2 - b_1 b_2 + (a_1 b_2 + a_2 b_1)i.$$

In trigonometric and exponential representation we have

$$\boxed{\begin{aligned} z = z_1 \, z_2 &= r_1(\cos(\varphi_1) + i \cdot \sin(\varphi_1)) \, r_2(\cos(\varphi_2) + i \cdot \sin(\varphi_2)) \\ &= r_1 \, r_2 (\cos(\varphi_1 + \varphi_2) + i \cdot \sin(\varphi_1 + \varphi_2)) \\ &= r_1 \, r_2 \, e^{i(\varphi_1 + \varphi_2)} \end{aligned}}$$

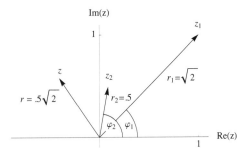

($\rightarrow$ Animation)

Correspondingly we have for the division of two complex numbers

$$z = \frac{z_1}{z_2} = \frac{r_1}{r_2}(\cos(\varphi_1 - \varphi_2) + i \cdot \sin(\varphi_1 - \varphi_2)) = \frac{r_1}{r_2} e^{i(\varphi_1 - \varphi_2)}$$

Complex numbers in polar coordinates are multiplied by multiplying the moduli and adding the arguments. They are divided by dividing the moduli and subtracting the arguments.
($\rightarrow$ Animation)

A complex number is raised to the nth power by raising its modulus to the nth power and multiplying its argument by n. ($\rightarrow$ Animation)

$$z^n = (r(\cos(\varphi) + i \cdot \sin(\varphi)))^n = r^n(\cos(n\varphi) + i \cdot \sin(n\varphi))$$
$$= \left(re^{i\varphi}\right)^n = r^n e^{in\varphi}$$

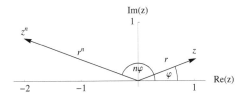

The fundamental theorem of algebra states that every single-variable polynomial has exactly as many roots as its degree. So the solutions of the equation $x^2 = 1$ are 1 and -1, and the solutions of the equation $x^4 = 1$ are $1, -1, i$ and $-i$. In order to find the roots of an arbitrary complex number $z = re^{i\varphi}$, we reverse the process and take the nth root of the modulus and divide the argument by n. This gives a first solution. The exponential function is periodic with period 2π, and so the other solutions can be found by replacing φ in the first solution by $\varphi + 2\pi, \varphi + 2 \cdot 2\pi, ..., \varphi + (n-1) \cdot 2\pi$.

$$\sqrt[n]{z} = \sqrt[n]{r(\cos(\varphi) + i \cdot \sin(\varphi))} = \sqrt[n]{r}\,(\cos(\varphi_k) + i \cdot \sin(\varphi_k))$$
$$\sqrt[n]{z} = \sqrt[n]{r e^{i\varphi}} = \sqrt[n]{r}\, e^{i\varphi_k}$$
$$\text{where } \varphi_k = \frac{\varphi + k \cdot 2\pi}{n} \quad (k = 0, 1, ..., n-1)$$

3.1 Analog Signals and the Fourier Transform

Example: Find the four roots of the equation $x^4 = 1 + i$.

The exponential form of $1 + i$ is $\sqrt{2}\, e^{i\cdot\pi/4}$. The modulus of the solutions is equal to

$$\sqrt[4]{\sqrt{2}} = \sqrt[8]{2} = 1.0905...$$

The arguments are

$\varphi_1 = \varphi/4 = \pi/16,\ \varphi_2 = (\varphi + 2\pi)/4 = \pi/16 + \pi/2,\ \varphi_3 = (\varphi + 4\pi)/4 = \pi/16 + \pi)$ and
$\varphi_4 = (\varphi + 6\pi)/4 = \pi/16 + 3\pi/2$.

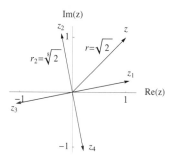

3.1.2.5 Rotating Complex Vectors

Continuous raising of a complex vector of modulus 1 to higher and higher powers causes the vector to rotate counterclockwise (figure at the left). The exponent can also be a fraction or negative (figure at the right).

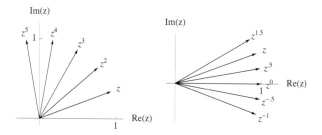

Using the exponential form of complex numbers, the periodic oscillation shown in Chapter 2.2.1.1 takes the following mathematical form. A vector of length r, rotating counterclockwise with an angular frequency ω can be described by

$$\boxed{r\, e^{i\omega t}}$$

Alternatively, using the terms of Chapter 2.2.1: A sinusoid wave $y = A\cdot\sin(\omega t + \varphi)$ can be represented as a complex vector $\mathbf{y}$ rotating around the origin with angular frequency ω

$$\boxed{\mathbf{y} = A\, e^{i(\omega t + \varphi)} = A\, e^{i\varphi}\, e^{i\omega t} = C\, e^{i\omega t}}$$

C is called the complex amplitude and corresponds to the vector at time $t = 0$. We obtain the oscillation from the vector representation as the imaginary part

$$y = \text{Im}(y)$$

In order to calculate the superposition of two sine waves of the same frequency $y_1(t) = A_1\sin(\omega t + \varphi_1)$ and $y_2(t) = A_2\sin(\omega t + \varphi_2)$, we first write the sine waves in complex form

$$y_1 = C_1 e^{i\omega t} \text{ and } y_2 = C_2 e^{i\omega t}.$$

Then we add the waves and get $y = y_1 + y_2 = (C_1 + C_2)e^{i\omega t} = C e^{i\omega t}$.

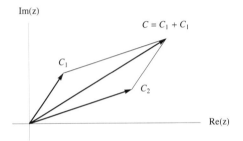

The *instantaneous phase* $\varphi(t)$ of a complex-valued function $x(t)$ is the real-valued function $\arg(x(t))$. The *instantaneous angular frequency* is the time derivative of the phase.

$$\phi(t) = \arg(x(t)) \qquad \omega(t) = \frac{d}{dt}(\phi(t))$$

The sum of two sine waves of nearly the same frequency produces beats (2.2.1.4). If the amplitudes of the two waves are the same, then its frequency is $(f_1 + f_2)/2$. If they are not the same, a variable frequency results.

In order to compute the frequency and the amplitude envelope of two superposed waveforms, we write the waves in complex form. Let the relation between the amplitudes be $1/c$; for the frequencies we shall write $f_1 = 1$ and $f_2 = 1 + \delta$. The sum of the two waves is given by:

$$z = z_1 + z_2 = e^{i \cdot 2\pi t} + c e^{i(1+\delta) 2\pi t}$$

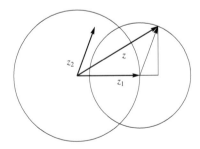

We first compute the amplitude envelope of the sum of the waveforms, that is, the absolute value of the vector z.

$$|z| = |z_1 + z_2| = |e^{i \cdot 2\pi t} + c e^{i(1+\delta) 2\pi t}| = |e^{i \cdot 2\pi t}(1 + c e^{i\delta \cdot 2\pi t})| = |1 + c \cdot e^{i\delta \cdot 2\pi t}|$$

3.1 Analog Signals and the Fourier Transform

or, as can be seen in the figure above,

$$|z| = \sqrt{(1 + c\cdot\cos(\delta\cdot 2\pi t))^2 + (c\cdot\sin(\delta\cdot 2\pi t))^2} = \sqrt{1 + c^2 + 2c\cdot\cos(\delta\cdot 2\pi t)}.$$

For the figure below, $\delta = 0.1$ and $c = 0.5$.

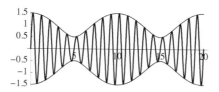

We can see that the frequency of z is not constant from the vector diagram. z_1 and z_2 revolve with constant velocities. In the time that z_2 makes a full revolution more than z_1, z revolves once around the smaller circle $1 + ce^{i\delta\cdot 2\pi t}$. z first moves faster, then slower, and then again faster than z_1. The following illustration shows the beating Im(z) for $c = .75$ (heavy line) and for comparison an oscillation with constant frequency 1 (light line). (→ Animation)

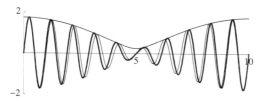

To compute the instantaneous frequency of the wave, we first determine the phase, that is, the argument of z. The vector z rotates with the angular velocity of $z_1 + z_2$.

$$\arg(z) = 2\pi t + \arctan\left(\frac{c\cdot\sin(2\pi t\delta)}{1 + c\cdot\cos(2\pi t\delta)}\right)$$

The time derivative is

$$\pi\left(2 + \delta + \frac{(c^2 - 1)\delta}{1 + c^2 + 2c\cos(2\pi t\delta)}\right)$$

The following figures show the instantaneous frequencies for $c = 0.75$ and $c = 0.999$ over the period of one beat. (→ Animation)

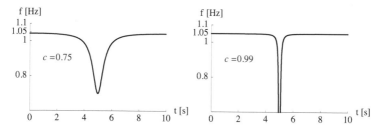

(→ Animation with adjustable frequency)

Now the question is, what frequency does one actually hear? In the low frequencies, one can hear the lowering of pitch although it occurs at minimum amplitude. The following illustration shows the progression of the instantaneous frequency over one period of beating for $f = 30$ Hz, $\delta = 0.05$ and $c = .8$ (Sound Example 3-1-2-5). In higher frequencies, either the lowering of the frequency is too small or the change in pitch too rapid to be heard. In any case, one does not hear the average of the instantaneous frequency f_{av}, for $f_{av} = f(1 + \delta/2)$ if $c = 1$, $f_{av} = f$ if $c < 1$ and $f_{av} = f(1 + \delta)$ if $c > 1$. In the Max/MSP patch *Beats_2*, the perceived frequency can be estimated using a reference pitch.

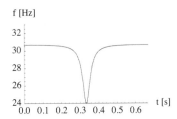

3.1.2.6 Fourier Analysis in Complex Representation

When writing Fourier series using complex exponentials, one usually assumes a periodic function $f(x)$ with a period of $2L$. The function then can be written as

$$f(x) = \sum_{n=-\infty}^{\infty} c_n\, e^{in\pi x/L}$$

with the coefficients

$$c_n = \frac{1}{2L} \int_{-L}^{L} f(x)\, e^{-in\pi x/L}\, dx$$

We can write the square wave function discussed in Chapter 3.1.1.3 as a sign function sgn(x) over the interval $[-L, L]$. We choose a period of 6, hence $L = 3$. In order to calculate the first seven partials, we choose $n = 7$. First we calculate the coefficients.

```
L = 3; c = Table[ 1/(2*L) ∫_{-L}^{L} Sign[x] * e^{i*n*π*x/L} dx, {n, -7, 7}]
```

$$\left\{-\frac{2i}{7\pi},\, 0,\, -\frac{2i}{5\pi},\, 0,\, -\frac{2i}{3\pi},\, 0,\, -\frac{2i}{\pi},\, 0,\, \frac{2i}{\pi},\, 0,\, \frac{2i}{3\pi},\, 0,\, \frac{2i}{5\pi},\, 0,\, \frac{2i}{7\pi}\right\}$$

Again the even partials disappear. For the fundamental frequency we obtain $-2i/\pi$ for $n = -1$ and $2i/\pi$ for $n = 1$. This means that the first partial consists of two complex vectors

$$z_{-1} = -\frac{2i}{\pi} e^{-i(-1)\pi x/L} \text{ and } z_1 = \frac{2i}{\pi} e^{-i\pi x/L}$$

rotating in opposite directions. To find the position of the vectors at time $x = 0$, we calculate the real and the imaginary parts of the vectors.

```
L = 3; n = 1; x = 0; z₁ = -(2*i)/π e^{i*n*π*x/L} ; z₂ = (2*i)/π e^{-i*n*π*x/L};
{Re[z₁], Im[z₁], Re[z₂], Im[z₂]}
```

$$\left\{0,\, -\frac{2}{\pi},\, 0,\, \frac{2}{\pi}\right\}$$

3.1 Analog Signals and the Fourier Transform

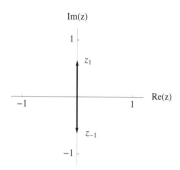

We add the first seven partials using this Mathematica command:

$$\text{Plot}\left[\sum_{n=-7}^{7} c[[n+8]] * e^{-i*n*\pi*x/L}, \{x, -5, 5\}\right]$$

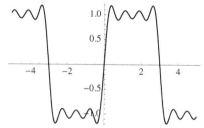

For clarity's sake, let us analyze another square wave, this one displaced along both axes. We choose the function $y = \text{sgn}(x - 1) + 0.5$ over the interval $[-L + 1, L + 1]$. The coefficients here are

```
L = 3; c = Table[
    1./(2*L) NIntegrate[(.5 + Sign[x - 1]) * e^(-i*n*π*x/L), {x, -L + 1, L + 1}], {n, -7, 7}]

{-0.079+0.045i,0.,0.11+0.064i,0.,0.-0.212i,0.,-0.551+0.318 i,
0.5,-0.551-0.318i,0.,0.+0.212i,0.,0.11-0.064i,0.,-0.079-0.045i}
```

All the even partials disappear, except the one for $n = 0$. This corresponds to the frequency 0 and yields the constant $0.5e^0 = 0.5$. The following illustration shows the position of the vectors at $x = 0$ for the components 0, 1, 3, and 5.

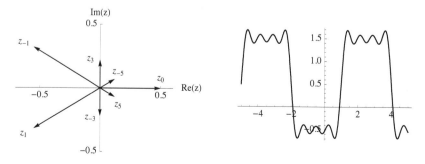

3.1.3 Aperiodic Signals and Fourier Integrals

3.1.3.1 Definition

An aperiodic function $f(x)$ cannot be developed into a Fourier series. One can however consider the entire domain of the function as a period P, thus obtaining a Fourier series with a low fundamental $1/P$ and a spectrum with lines correspondingly close together. In the limiting case, when the function is defined over $-\infty < x < \infty$, the fundamental tends toward zero and the spectral lines blend into a continuous spectrum (2.2.3.3). In this case, the sum in the formula for the Fourier series becomes an integral.

$$f(x) = \int_{-\infty}^{\infty}(A(\alpha)\cos(\alpha x) + B(\alpha)\sin(\alpha x))\,d\alpha$$

$$A(\alpha) = \frac{1}{\pi}\int_{-\infty}^{\infty} f(x)\cos(\alpha x)\,dx$$

$$B(\alpha) = \frac{1}{\pi}\int_{-\infty}^{\infty} f(x)\sin(\alpha x)\,dx$$

The right side of the first equation is called the Fourier integral of $f(x)$. The Fourier series decomposes a periodic function into a sum of harmonic oscillations; the Fourier integral, on the other hand, represents an aperiodic function as the sum of the oscillations of all frequencies between zero and infinity. The Fourier integral yields a continuous spectrum.

The Fourier integral of $f(x)$ in exponential notation is

$$f(x) = \int_{-\infty}^{\infty} F(\alpha)\,e^{-i\alpha x}\,d\alpha$$

Instead of the functions $A(\alpha)$ and $B(\alpha)$ we have one function $F(\alpha)$

$$F(\alpha) = \frac{1}{2\pi}\int_{-\infty}^{\infty} f(u)\,e^{i\alpha u}\,du$$

$F(\alpha)$ is the Fourier transform of $f(x)$, and $f(x)$ is the inverse Fourier transform of $F(\alpha)$. Usually the formulas for both operations are written with the factor $1/\sqrt{2\pi}$ to underline the symmetry of the transform. The definition given below is standard in all fields except engineering, where the signs of the exponents are transposed.

$$f(x) = \frac{1}{\sqrt{2\pi}}\int_{-\infty}^{\infty} F(\alpha)\,e^{-i\alpha x}\,d\alpha \qquad F(\alpha) = \frac{1}{\sqrt{2\pi}}\int_{-\infty}^{\infty} f(u)\,e^{i\alpha u}\,du$$

$$f(x) = \frac{1}{\sqrt{2\pi}}\int_{-\infty}^{\infty} F(\alpha)\,e^{i\alpha x}\,d\alpha \qquad F(\alpha) = \frac{1}{\sqrt{2\pi}}\int_{-\infty}^{\infty} f(u)\,e^{-i\alpha u}\,du$$

When $f(x)$ is even (A4.7), we have the so-called Fourier cosine transform

$$f(x) = \sqrt{2/\pi}\int_{0}^{\infty} F(\alpha)\cos(\alpha x)\,d\alpha \qquad F(\alpha) = \sqrt{2/\pi}\int_{0}^{\infty} f(u)\cos(\alpha u)\,du$$

When $f(x)$ is odd (A4.7), we have the so-called Fourier sine transform

$$f(x) = \sqrt{2/\pi}\int_{0}^{\infty} F(\alpha)\sin(\alpha x)\,d\alpha \qquad F(\alpha) = \sqrt{2/\pi}\int_{0}^{\infty} f(u)\sin(\alpha u)\,du$$

Because every function can be written as the sum of an odd and an even function, formula tables often only list Fourier sine and Fourier cosine transform.

3.1.3.2 Examples

The transition of the Fourier series to the Fourier integral mentioned above is illustrated in the example of a single pulse in Chapter 2.2.3.3. We calculate the Fourier transform of the function $f(x) = 1$ for $|x| < a$ and $f(x) = 0$ for $|x| > a$.

$$F(\alpha) = \frac{1}{\sqrt{2\pi}} \int_{-\infty}^{\infty} f(u)\, e^{i\alpha u}\, du = \frac{1}{\sqrt{2\pi}} \int_{-a}^{a} (1)\, e^{i\alpha u}\, du = [\frac{1}{2\pi} \frac{e^{i\alpha u}}{i\alpha}]_{-a}^{a}$$

$$= \frac{1}{\sqrt{2\pi}} \left(\frac{e^{i\alpha u} - e^{-i\alpha u}}{i\alpha} \right) = \sqrt{2/\pi} \cdot \frac{\sin(a\alpha)}{\alpha}$$

Because $f(x)$ is an even function, we can use the formula for the Fourier cosine transform.

$$F(\alpha) = \sqrt{2/\pi} \int_0^{\infty} f(u) \cos(\alpha u)\, du = \sqrt{2/\pi} \int_0^{\infty} (1) \cos(\alpha u)\, du$$

$$= \sqrt{2/\pi}\, [\frac{\sin(\alpha u)}{\alpha}]_{-a}^{a} = \sqrt{2/\pi}\, \frac{\sin(a\alpha)}{\alpha}$$

By decreasing a in the example above, we obtain the spectrum of an impulse. $F(\alpha)$ goes to zero, however. But if we let the function $f(x)$ increase for $-a < x < a$ by setting $g(x) = f(x)/a$, then we have

$$G(\alpha) = \sqrt{2/\pi}\, \frac{\sin(a\alpha)}{a\alpha}$$

Since $\sin(\gamma)/\gamma \to 1$ for $\gamma \to 0$, we obtain the constant function

$$G(\alpha) = \sqrt{2/\pi} = .797885...$$

The integral above can be calculated with Mathematica. The command ExpToTrig[] converts the exponential expression into trigonometric functions.

$$\frac{1}{\sqrt{2*\pi}}\; \texttt{ExpToTrig[Integrate[Exp[i*}\alpha\texttt{*u]/a, \{u, -a, a\}]]}$$

$$\frac{\sqrt{\frac{2}{\pi}}\; \texttt{Sin[a}\,\alpha\texttt{]}}{a\,\alpha}$$

It is not always possible to find a closed-form solution for more complicated functions. The commands Fourier[] and InverseFourier[] give numerical solutions for the Fourier transform and the inverse Fourier transform respectively of a list of values.

3.1.4 Analog Systems

This chapter gives a brief introduction to the realm of analog systems. Many digital techniques have an analog electronic model, and older, well-understood operations from analog electronics can be applied to digital signals.

3.1.4.1 Definition and Examples

In what follows, we shall only consider those systems whose behavior can be simply described mathematically and whose input and output can be represented as a function of time. These systems produce from an input signal $x(t)$ an output signal $y(t)$ according to certain rules. The input signal $x(t)$ is called the stimulus, the output signal $y(t)$ is called the response of the system to the stimulus. For this process we write $y(t) = G\{x(t)\}$.

Example 1: An attenuator reacts to an external force $F(t)$ either by increased strain or by changing position $y(t)$. In the simplest case, the deviation of position $y(t)$ is proportional to the force: $y(t) = cF(t)$.

Example 2: Let a mass be suspended from a spring. Spring and mass form a system which reacts to external excitation by movement of the mass. The equation of movement of the mass follows from Hooke's law $F = -Ky$ and from Newton's second law $F = ma$ (with F the force, K the spring constant, y the displacement, m the mass and a the acceleration). Since acceleration is equal to the second derivative of the displacement over time, we have

$$a = \frac{d^2 y}{dt^2} = \ddot{y} = \frac{K}{m} y.$$

Introducing an attenuation proportional to the velocity $b\dot{y}$ and an external force $x(t)$, we obtain

$$m\ddot{y} + b\dot{y} + Ky = x(t).$$

Example 3: An electromagnetic series LC circuit consists of a resistance R, a capacity C and an inductivity L connected in series. The circuit forms a system with an externally applied voltage $u(t)$ as input and the current $i(t)$ flowing through the system as output. We have

$$\frac{d^2 i}{dt^2} + 2\delta \frac{dy}{dt} + \omega_0^2 = \frac{1}{L} \frac{du}{dt},$$

where $\delta = \frac{R}{2L}$ is the attenuation factor and ω_0 is the circuit's resonance frequency.

3.1.4.2 Linear Differential Equations

Differential equations often crop up in discussions of mechanical and electrical systems. Linear differential equations are important. They are relatively easy to calculate and can often be used to approximately solve more complex equations. The general form of a linear differential equation of the nth order is

$$\boxed{a_0 y + a_1 \frac{dy}{dt} + a_2 \frac{d^2 y}{dt^2} + \ldots + a_n \frac{d^n y}{dt^n} = b_0 x + b_1 \frac{dx}{dt} + b_2 \frac{d^2 x}{dt^2} + \ldots + b_n \frac{d^n x}{dt^n}}$$

3.1.4.3 Laplace-Transform (→ CD)

3.2 Digital Signals, DFT, FFT

At the beginning of this chapter, concepts and problems in connection with the digitalization of signals are discussed. In the second part, we show how the Fourier transform is applied to digital signals. This leads to a discussion of the discreet Fourier transform DFT and of its optimized form, the fast Fourier transform FFT. Corresponding to the Laplace transform in the analog domain, the z-transform in the digital domain is a representation that does justice to the spectral periodicity of a sampled signal.

Further Reading: The Audio Programming Book *by Richard Boulanger and Victor Lazzarini [86], pp. 441-462.*

3.2.1 Digital Representation of Signals

In order to store and process signals digitally, the signals must be available as series of discrete values. Signals produced by computer are always discrete, whereas analog signals like sound waves or control voltages must be sampled and measured using an analog-to-digital converter (ADC).

3.2.1.1 Sampling

When sampling a signal, one usually measures and records the signal's value at regular intervals. The time interval is called the sampling interval T_s, and the number of sampling intervals per second is called the sampling rate or the sampling frequency f_s. The sampling rate determines the highest frequency that can be recorded. In the figure below, the sampling interval is 1 millisecond, the sampling rate therefore 1000 Hz. As the figure shows, when the signal's frequency increases, the samples represent the signal less and less well. At the far right of the figure, the signal's frequency is 500 Hz and the representation is particularly bad. The upper limit of the frequency range that can be correctly sampled is $f_s/2$. This limit is called the Nyquist frequency. To record all frequencies perceptible by the human ear, sampling rates over 40 kHz are usually used.

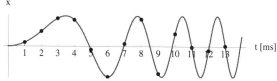

Not only the time axis is subdivided into discrete values, but also the function values can only be measured with a certain precision. The following figure shows the difference in precision according to the subdivision of the vertical axis. Measuring and storing with limited precision is called quantization. (→ *Digital_Signals.maxpat*)

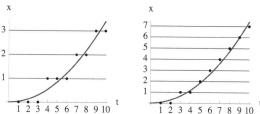

3.2.1.2 Representations and Simple Examples

A time-discrete signal is a series of numbers. The figure above (right) shows the sequence of numbers {0, 0, 1, 1, 2, 3, 4, 5, 6, 7}. There are several ways of notating such a sequence: $x_k(t)|_{t=kT}$ or $x_k(kT)$ or simply $x(k)$. We represent a sequence of number graphically as points, as points connected to form a continuous curve or, if there is no chance for misunderstanding, as a continuous curve.

The following figures illustrate three important simple signals:

The unit impulse (or unit sample) $\qquad \delta(k) = \begin{cases} 1 & \text{for } k = 0 \\ 0 & \text{for } k \neq 0 \end{cases}$

The unit step $\qquad u(k) = \begin{cases} 1 & \text{for } k \geq 0 \\ 0 & \text{for } k < 0 \end{cases}$

The real causal exponential sequence $\qquad e(a, k) = \begin{cases} a^k & \text{for } k \geq 0 \\ 0 & \text{for } k < 0 \end{cases} \qquad |a| < 1$

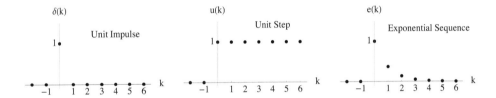

In what follows, the frequency f will often not be indicated in cycles per second or as the angular frequency $\omega = 2\pi f$, but rather as the normalized angular frequency $\Omega = \omega_s = 2\pi f / f_s$ (3.3.1.4). (→ Example)

3.2.1.3 Characteristics and Basic Transformations

Only those discrete signals are strictly periodic for which a sequence of numbers continuously repeats, that is, where $x(k) = x(k + N)$. The signal $x_1 = \cos(\pi k / 10)$ is periodic with the period $N = 20$, since $\cos(\pi(k + N)/10) = \cos(\pi k/10 + 2\pi) = \cos(\pi k/10)$. On the other hand, a periodic continuous signal whose frequency is an irrational fraction of the sampling rate cannot yield a strictly periodic digital signal. A periodic continuous signal whose frequency is a rational fraction of the sampling rate yields a periodic digital signal whose period does not correspond to that of the continuous signal. The period of the signal shown below is 20/3 sample intervals, hence the digital signal only repeats after three cycles of the analog signal.

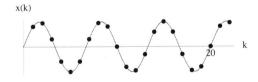

3.2 Digital Signals, DFT, FFT

Signals can be symmetrical in two ways. A signal is called even when $x(k) = x(-k)$ and odd when $x(k) = -x(-k)$. The signal $\cos(k\omega)$ is even, the signal $\sin(k\omega)$ is odd.

Every signal can be decomposed into an even and an odd component:

$x(k) = x_{even}(k) + x_{odd}(k)$, with

$x_{even}(k) = \frac{1}{2}(x(k) + x(-k))$ and $x_{odd}(k) = \frac{1}{2}(x(k) - x(-k))$.

The following figure shows a signal (bold) and its even (dotted line) and odd (fine line) components.

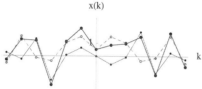

One way of processing a signal is to transform the independent variable: $y(k) = x(f(k))$. If we subtract a constant k_0 from k, the signal will be delayed by $k_0 T$, if we replace k by $-k$, the signal will be reversed in time. If we multiply k by a constant N, only every Nth sample will be kept and the signal will be shorted by a factor N. This process is known as downsampling. If we divide k by N, the signal will be stretched by a factor N. The gaps arising are filled with zeroes. This process is called upsampling with zero padding.

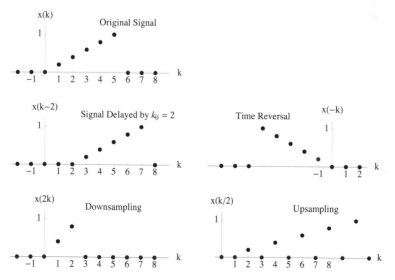

Signals can also be processed by transforming the dependent variable: $y(k) = f(x(k))$. The most important transformations are the addition and the multiplication of two signals and the multiplication of a signal by a constant (scaling).

3.2.1.4 Quantization

The precision of the numerical values of the samples depends on the number of bits used for storing them. A resolution of 16 bits means there are $2^{16} = 65536$ different values available. The values are indicated either as integers between -32678 and 32767 or scaled as rational numbers within the interval $[-1, 1[$.

The dynamic range of an electronic device or of a signal is defined as the ratio between the greatest and the smallest amplitude which can be represented. In a signal with 16 bit resolution, the greatest amplitude is 2^{15} and the smallest is 0.5 (e.g. $\{0, 1, 0, 1, ...\}$). This gives a dynamic range of $2^{16} = 20 \cdot \log(2^{16})$ dB $= 96.3296$ dB. (2.3.3.2)

3.2.1.5 Aliasing

When one tries to record an analog sinusoid whose frequency f^* is greater than the sampling rate f_s, one or more complete cycles of oscillation are missed. The result is a digital sinusoid with the frequency $f = f^* - f_s$. The first of the three figures below shows part of a sampled sinusoid with frequency $f = f_s/20$. The middle figure shows a sinusoid with frequency $f^* = f + f_s$, last figure shows a sinusoid with frequency $f^* = f + 2f_s$. We see that a time discrete signal represents not only an analog signal of frequency f but also an infinite number of other signals with frequencies $f + nf_s$. The occurrence of such additional frequencies is called aliasing.

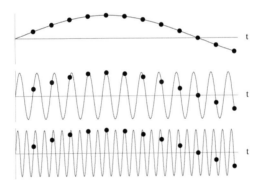

An analog sinusoid whose frequency f^* lies between the Nyquist frequency $f_s/2$ and the sampling frequency f_s results in the same time discrete signal as an oscillation with the frequency $f^* - f_s$. The higher frequency is said to fold back about the Nyquist frequency and so to appear in the range between 0 and the Nyquist frequency. The following illustration shows a sinusoid of frequency $f^* = 11/20\ f_s$ (above) and a sinusoid of frequency $f^* = -9/20\ f_s$ (below).

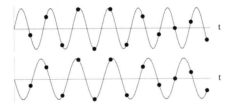

3.2 Digital Signals, DFT, FFT

All analog sinusoids with the frequencies $f^* = f \pm nf_s$ give the same sampled signal. To avoid disturbance through aliasing, all components above the Nyquist frequency $f_s/2$ are filtered out of a signal before analog-to-digital or digital-to-analog conversion. In the figures below, the bold line shows the increasing frequency of an analog signal, the dotted lines show the additional frequencies caused by aliasing ($f \pm nf_s$ to the left $|f \pm nf_s|$ to the right).

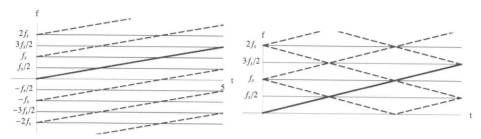

If the sampling rate is high in this example, one hears a tone rising to $f_s/2$, then descending to 0 and then rising again to $f_s/2$. If the sampling rate is low, one hears simultaneously rising and descending tones. (→ *Digital_Signals.maxpat*) (→ Effect on Sounds)

3.2.1.6 Rotating Vectors

In Chapter 3.1.2.5, the sinusoid was represented as rotating complex vector $z(t) = re^{i\omega t}$. If we substitute for t the discrete time values $t = nT$, we have the step-wise rotating vector $z(n) = re^{i\omega nT}$.

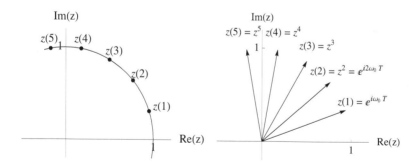

The identity of all digital sinusoids caused by aliasing with the frequencies $f \pm kf_s$ is given by the complex equation

$$e^{i\omega nT} = e^{i(\omega nT + 2\pi k)}.$$

The delay of a signal by $n_0 T$ corresponds to setting the vector back by n_0 steps. Mathematically this means multiplying the vector by $e^{-in_0 \omega T}$:

$$z(n - n_0) = re^{i\omega(n-n_0)T} = re^{i\omega nT} e^{-i\omega n_0 T} = z(n)e^{-i\omega n_0 T}.$$

3.2.2 The Discrete Fourier Transform DFT

3.2.2.1 Calculating the Coefficients of Discrete Fourier Series

Just as every periodic continuous waveform can be decomposed into a sum of continuous sinusoidal and cosinusoidal oscillations, every discrete periodic signal can be decomposed into a sum of simple discrete signals. We can derive a formula for discrete signals from the formula for the Fourier series for continuous waveforms (3.1.1.1) by substituting the terms $\sin(2\pi nk/N)$ and $\cos(2\pi nk/N)$ for the continuous partials $\sin(n\omega t)$ and $\cos(n\omega t)$.

$$x(k) = B_0 + \sum_{n=1}^{N} (A_n \sin(2\pi nk/N) + B_n \cos(2\pi nk/N))$$

We obtain the coefficients A_n and B_n for the spectrum of a discrete signal from the coefficients a_n and b_n for a continuous signal by substituting the number of points N for the period T and the sequence $x(k)$ for the continuous function $F(t)$. The integral over the period T becomes the sum of all N samples.

$$B_0 = \frac{1}{N} \sum_{k=0}^{N-1} x(k)$$

$$A_n = \frac{1}{N} \sum_{k=0}^{N-1} \sin(2\pi nk/N) \, x(k)$$

$$B_n = \frac{1}{N} \sum_{k=0}^{N-1} \cos(2\pi nk/N) \, x(k)$$

3.2.2.2 Examples

Let a four-value signal $x(k)$ be composed of a DC component $x_0 = -1$, a fundamental $x_1 = \sin(\omega t) + \cos(\omega t)$ and a second partial $x_2 = \cos(2\omega t)$. The signals of the three components are: $\{-1, -1, -1, -1\}$, $\{1, 1, -1, -1\}$ and $\{1, -1, 1, -1\}$, the complete signal $x = x_0 + x_1 + x_2 = \{1, -1, -1, -3\}$.

We can determine the Fourier coefficients A_n and B_n of the signal x by using the discrete Fourier transform. To calculate the value of the DC component B_0, we sum the x-values, divide by $N = 4$ and obtain $(1 - 1 - 1 - 3)/4 = -1$. For A_1 we have: $(\sin(1 \cdot 0 \cdot 2\pi/4) - \sin(1 \cdot 1 \cdot 2\pi/4) - \sin(1 \cdot 2 \cdot 2\pi/4) - 3 \cdot \sin(1 \cdot 3 \cdot 2\pi/4))/4 = (0 - \sin(\pi/2) - \sin(\pi) - 3 \cdot \sin(3\pi/2))/4 = (0 - 1 - 0 + 3)/4 = 1/2$. Similar calculations give for the remaining coefficients: $B_1 = 1/2$, $A_2 = 0$, $B_2 = 1$, $A_3 = -1/2$, $B_3 = 1/2$. Note that those components whose frequencies lie above the Nyquist frequency are reflected about the Nyquist frequency: $\sin(3 \cdot 2\pi k/4) = -\sin(2\pi k/4)$ and $\cos(3 \cdot 2\pi k/4) = \cos(2\pi k/4)$. Hence we have for the DC component the amplitude $B_0 = -1$, for the sine component of the fundamental $A_1 - A_3 = 1$, for the cosine component of the fundamental $B_1 + B_3 = 1$, and for the cosine component of the second partial $B_2 = 1$.
(→ Example 2) (→ Analysis of the Impulse)

3.2.2.3 Signals With Unknown or Non-Integer Periods

In previous examples, the length of the sequence analyzed corresponded to one period of the signal. When the period of the signal to be analyzed is not a multiple of the sampling interval, inaccuracy can occur. Let a sequence of length N contain 3.5 periods of the signal $x(k) = \sin(3.5 \cdot 2\pi k/N)$.

Although the signal to be analyzed is a sinusoid, the A-coefficients disappear because the truncation after half a period at the end of the sequence causes the signal to become even. The most important contributions to the spectrum come from the coefficients B_3 and B_4 and B_{N-3} and B_{N-4} respectively. This effect is called *spectral blurring* or *spectral leakage* (see [65] p. 489). (→ Calculation of the Coefficients)

If the analysis leads to the spectrum above, the signal's period is probably between $N/3$ and $N/4$. If one fits the length of the analyzed signal to the presumed period and sets $N = 27$, one obtains the following spectrum.

Another way to reduce spectral blurring is to fade in the beginning and fade out the end of the signal (3.3.2.5).

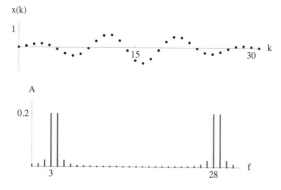

3.2.2.4 The Complex Form and the Characteristics of the DFT

A discrete signal $x(k)$ can be decomposed into a sum of exponential functions $e^{2\pi i n k/N}$.

$$x(k) = \sum_{n=0}^{N-1} X(n) \, e^{2\pi i n k/N} \qquad k = 0, 1, ..., N-1$$

For the complex amplitudes X_n the following holds:

$$X(n) = \frac{1}{N} \sum_{k=0}^{N-1} x(k) \, e^{-2\pi i n k/N} \qquad n = 0, 1, ..., N-1$$

In the discrete Fourier transform DFT, the second equation is used to transform a discrete signal $x(k)$ in the time domain to a discrete spectral function $X(n)$ in the frequency domain. The opposite operation, the transform of a spectral function $X(n)$ into the sequence $x(k)$, is called the *inverse discrete Fourier transform* IDFT.

Up to now, we have used the definition of Fourier transform that is standard in engineering and in the literature about digital signal processing. In other sciences, and hence in programs like Mathematica, the Fourier transform is defined with positive, the inverse Fourier transform with negative exponents. In addition, the scaling factor $1/\sqrt{N}$ is often used in the equations in both directions, which underscores the symmetry of the two transforms.

$$x(k) = 1/\sqrt{N} \sum_{n=0}^{N-1} X(n) \, e^{-2\pi i n k/N} \qquad k = 0, 1, ..., N-1$$

$$X(n) = 1/\sqrt{N} \sum_{k=0}^{N-1} x(k) \, e^{2\pi i n k/N} \qquad n = 0, 1, ..., N-1$$

Let us now calculate our first numerical example from Chapter 3.2.2.2 by hand. To calculate the value of the DC component $X(0)$, we set $k = 0$ so that the exponential factor becomes 1. We can then sum all the x-values and divide by $N = 4$, which gives $(1 - 1 - 1 - 3)/4 = -1$. The further components are

$$X(1) = (e^{2\pi i \cdot 1 \cdot 0/4} - e^{2\pi i \cdot 1 \cdot 1/4} - e^{2\pi i \cdot 1 \cdot 2/4} - 3 \cdot e^{2\pi i \cdot 1 \cdot 3/4})/4 = (1 - e^{\pi i/2} - e^{\pi i} - 3 \cdot e^{\pi i \cdot 3/2})/4$$
$$= (1 - i + 1 + 3i)/4 = 0.5 + .5i.$$
$$X(2) = (1 - e^{\pi i} - e^{2\pi i} - 3 \cdot e^{3\pi i})/4 = (1 + 1 - 1 + 3)/4 = 1.$$
$$X(3) = (1 - e^{3\pi i/2} - e^{3\pi i} - 3 \cdot e^{9\pi i/2})/4 = (1 + i + 1 - 3i)/4 = 0.5 - .5i.$$

We note that real values correspond to cosine functions and imaginary values to sine functions. In addition, the fourth component, which lies above the Nyquist frequency, is reflected, the cosine component is subtracted from the second component, and the sine component is added to it. To calculate the components with Mathematica, we use the command Fourier[] and compensate for the scaling factor.

```
Fourier[{1, -1, -1, -3}]/√4
```

```
{-1.,0.5+0.5 i,1.,0.5-0.5 i}
```

(→ 2. Example from Chapter 3.2.2.2)

3.2 Digital Signals, DFT, FFT

To illustrate some of the characteristics of the discrete Fourier transform, we introduce the following terms. For the two transforms we write $X(n) = \text{DFT}\{x(k)\}$ and $x(k) = \text{IDFT}\{X(n)\}$. We abbreviate the frequently occurring factor $e^{-2\pi i/N}$ as W_N.

The DFT is a linear transform, hence $\text{DFT}\{ax(k) + by(k)\} = aX(n) + bY(n)$.

If one shifts a periodic sequence $x(k)$ by i samples to the left, one obtains the sequence $x_1(k) = x(k + i)$, and the equality $X_1(n) = W_N^{-ni} X(n)$ holds. Thus $\{1, 2, 3, 4\}$ shifted by one sample gives $\{2, 3, 4, 1\}$.

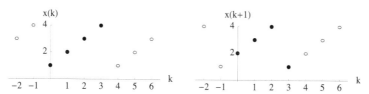

The Fourier analysis of the two sequences gives

    X = Fourier[{1,2,3,4}]

    {5.+0. I,-1.-1. I,-1.+0. I,-1.+1. I}

    X₁ = Fourier[{2, 3, 4, 1}]

    {5.+0. I,-1.+1. I,1.+0. I,-1.-1. I}

Using the equation above, we obtain the sequence X_1 from the sequence X by multiplying by the factor $e^{-in\pi/2}$ (n goes from 0 to 3, but the indices in the sequence X go from 1 to 4).

    Table[e^-i*1*n*π/2 X[[n + 1]], {n, 0, 3}]

    {5.+0. I,-1.+1. I,1.+0. I,-1.-1. I}

The discrete Fourier transform of the convolution (3.3.1.3) of two signals $x(k)$ and $y(k)$ corresonponds to the product of their discrete Fourier transforms (see [18] pp. 19-20 for the derivation):

$$\text{DFT}\{x(k)*y(k)\} = \text{DFT}\{\sum_{i=0}^{N-1} x(i)\, y(k - i)\} = X(n)Y(n).$$

The discrete Fourier transform of the product of two signals $x(k)$ and $y(k)$ corresponds to the convolution of their discrete Fourier transforms (see [18] p. 19 for the derivation):

$$\text{DFT}\{x(k)y(k)\} = X(n)*Y(n).$$

If one multiplies a sequence $x(k)$ by the exponential expression $W_N^{ki} = e^{-2\pi i k i/N}$, one gets for the DFT of the sequence

$$\text{DFT}\{W_N^{ki} x(k)\} = X(n + i).$$

In the following example, we consider the sequence $\{0, 1, 1, 0, -1, -1\}$, which describes a sinusoid in six samples. The DFT gives accordingly an imaginary part for the fundamental and zero for the remaining frequencies.

```
x = {0,1,1,0,-1,-1}; X = Fourier[x];

{0.,0.+1.41 i,0.,0.,0.,0.-1.41 i}
```

If we set $i = 1$, we obtain a complex sequence whose spectrum is shifted by 1.

```
x1 = Table[e^-i*1*k*2*π/6 x[[k + 1]], {k, 0, 5}]
```

$$\left\{0,\ e^{-\frac{i\pi}{3}},\ e^{-\frac{2i\pi}{3}},\ 0,\ -e^{-\frac{2i\pi}{3}},\ -e^{-\frac{i\pi}{3}}\right\}$$

```
X1 = Fourier[x1];       {0.-1.41 i,0.,0.+1.41 i,0.,0.,0.}
```

If we set $i = 3$, we obtain a real sequence, whose spectrum is shifted by 3.

```
x1 = Table[e^-i*3*k*2*π/6 x[[k + 1]], {k, 0, 5}]       {0,-1,1,0,-1,1}

X1 = Fourier[x1];       {0.,0.,0.-1.41 i,0.,0.+1.41 i,0.}
```

The following table summarizes the most important characteristics of the discrete Fourier transform ([65] p. 290).

Property/Operation	Time Domain	Frequency Domain
Linearity	$a\,x(k) + b\,y(k)$	$a\,X(\omega) + b\,Y(\omega)$
Time Shifting	$x(k - n)$	$e^{-i\omega n}\,X(\omega)$
Time Reversal	$x(-k)$	$X(-\omega)$
Convolution	$x(k) * y(k) =$ $\sum_{n=-\infty}^{\infty} x(n) * y(k-n)$	$X(\omega)\,Y(\omega)$
Frequency Shifting	$e^{i\omega_0 k}\,x(k)$	$X(\omega - \omega_0)$
Modulation	$x(k)\cos(\omega_0 k)$	$\frac{1}{2}X(\omega + \omega_0) + \frac{1}{2}X(\omega - \omega_0)$
Multiplication	$x(k)\,y(k)$	$\frac{1}{2\pi}\int_{-\pi}^{\pi} X(\lambda)\,Y(\omega - \lambda)\,d\lambda$
Differentiation in the Frequency Domain	$k \cdot x(k)$	$i\dfrac{d}{d\omega}X(\omega)$

3.2.2.5 The Fast Fourier Transform FFT

The computation described above is very time-consuming. There exist many algorithms for computing the discrete Fourier transform more quickly. These methods are called Fast Fourier Transform (FFT). Using FFT methods reduces the computation time for a sequence of length $N = 1024$ by about 99%. The best performance is achieved if N can be decomposed into many small equal factors, ideally when N is a power of 2.

We shall show first that a 2-point DFT is equivalent to a simple addition or subtraction of the two samples and that by re-grouping the addends in the Fourier transform a 4-point DFT can be decomposed into two 2-point DFTs, an 8-point DFT into two 4-point DFTs, etc. For this purpose, we write out the sum

$$X(k) = \sum_{n=0}^{N-1} x(n) e^{-2\pi i k n/N}$$

for the kth component:

2-point DFT: $\quad X(k) \;= x(0)e^{-0} + x(1)e^{-2\pi i k/2} \qquad\qquad k = 0, 1$
$\qquad\qquad\qquad\qquad = x(0) + x(1)e^{-\pi i k}$

Since $e^{-\pi i k} = 1$ for even k and $e^{-\pi i k} = -1$ for odd k, we get

$$X(0) = x(0) + x(1) \text{ and } X(1) = x(0) - x(1)$$

4-point DFT: $\quad X(k) \;= x(0) + x(1)e^{-\pi i k/2} + x(2)e^{-\pi i k} + x(3)e^{-3\pi i k/2}$
$\qquad\qquad\qquad\qquad = (x(0) + x(2)e^{-\pi i k}) + e^{-\pi i k/2}(x(1) + x(3)e^{-\pi i k/2})$

The re-writing shows that the 4-point DFT can be thought of as being made up of two 2-point DFTs, the first of which contains the even values $x(0)$ and $x(2)$ and the second of which contains the odd values $x(1)$ and $x(3)$. The coefficients of the second DFT are multiplied by $e^{-\pi i k/2}$ and are added to the values of the first DFT.

8-point DFT: $\quad X(k) \;= x(0) + x(1)e^{-\pi i k/4} + x(2)e^{-\pi i k/2} + ... + x(7)e^{-7\pi i k/4}$
$\qquad\qquad\qquad\qquad = x(0) + x(4)e^{\pi i k} + e^{-\pi i k/2}(x(2) + x(6)e^{\pi i k/2}) +$
$\qquad\qquad\qquad\qquad\;\; x(1) + x(5)e^{\pi i k} + e^{-\pi i k/2}(x(3) + x(7)e^{\pi i k/2})$

The re-writing shows that the 8-point DFT can be decomposed into two 4-point DFTs, each of which is made up of two 2-point DFTs.

We now formulate the decomposition for the general case and show the result graphically. We again make use of the abbreviated terms introduced above and write W_N for $e^{-2\pi i/N}$. Then $e^{-2\pi i k n/N}$ becomes $(W_N)^{kn} = W_N^{kn}$.

First we divide the sequence $x(k)$ into two sequences $g(k)$ and $h(k)$, each of length $N/2$. The sequence $g(k)$ contains even k, the sequence $h(k)$ odd k. The even indices can be written as $n = 2m$, the odd indices as $n = 2m + 1$. Following the definition of the DFT, the decomposition can be written as follows:

$$X(k) = \sum_{n=0}^{N-1} x(n)(W_N)^{kn} = \sum_{n=\text{even}} x(n)(W_N)^{kn} + \sum_{n=\text{odd}} x(n)(W_N)^{kn}$$

$$= \sum_{m=0}^{N/2-1} g(m)(W_N)^{k \cdot 2m} + \sum_{m=0}^{N/2-1} h(m)(W_N)^{k(2m+1)}$$

Since $(W_N)^{k \cdot 2m} = (W_{N/2})^{km}$ and $(W_N)^{k(2m+1)} = (W_N)^k (W_{N/2})^{km}$, it follows that

$$X(k) = \sum_{m=0}^{N/2-1} g(m)(W_{N/2})^{km} + (W_N)^k \sum_{m=0}^{N/2-1} h(m)(W_{N/2})^{km}$$

The first sum corresponds to the N/2-point DFT of the sequence $g(k)$, the second to the N/2-point DFT of the sequence $h(k)$.

$$X(k) = G(k) + (W_N)^k H(k)$$

The indices of $G(k)$ and $H(k)$ only run from 0 to N/2, but because of their periodicity, one can substitute $G(k - N/2)$ for $G(k)$ when $k > N/2$.

The following diagram shows the calculation schematically (W_N^k has been simplified to w^k).

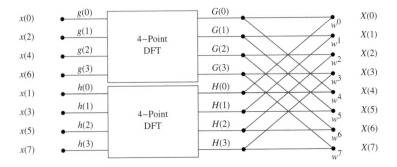

Using this technique, the number of computations can be reduced from N^2 complex additions and multiplications for the N-point DFT to $N^2/2 + N$, that is $2(N/2)^2$ for the N/2-point DFT and N for the multiplications by w^k and the addition of the two DFTs.

In the next step, the two 4-point DFTs are decomposed into two 2-point DFTs. The diagram show the operations for the sequence $g(k)$.

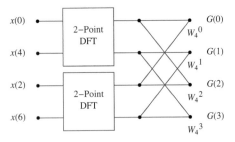

Finally, we obtain the two simple factors computed above: $W_2^0 = 1$ and $W_2^1 = -1$ by decomposing the 2-point DFTs.

$q(0)$ ——— $Q(0) = q(0) + q(1)$

$q(1)$ ——— $Q(1) = q(0) - q(1)$

3.2 Digital Signals, DFT, FFT

In the illustrations above we frequently find the diagram in the figure to the left below. By factoring out W_N^k before the branching and since $W_N^{r-N/2} = W_N^r / W_N^{N/2} = -W_N^k$, we obtain the simplified diagram to the right.

All the decompositions and simplifications described above are illustrated in the following diagram of an 8-point FFT.

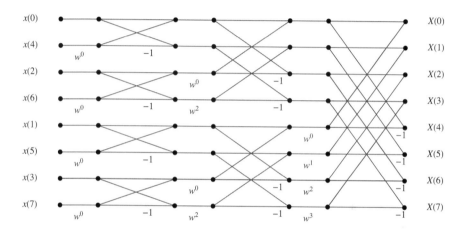

We now can determine the number of computations necessary for the FFT from the diagram above. We shall consider only the computationally expensive multiplications. For an 8-point FFT there are $\log_2(8) = 3$ branchings; in general there are $\log_2(N)$ branchings for an N-point FFT. Since all the factors before the first branching are $w^0 = 1$, the number of branchings with multiplication is reduced to $\log_2(N/2)$. At the branchings only half the lines show multiplications, hence there is a total of $N/2 \cdot \log_2(N/2)$ complex multiplications. In fact, the exact number is somewhat smaller because the factor w^0 also appears before other branchings. If we now compare the $N/2 \cdot \log_2(N/2)$ multiplications for the FFT with the N^2 multiplications for the DFT, we find for various values of N:

N	8	32	128	1024	32768
DFT : N^2	64	1024	16384	1048576	1073741824
FFT : $N/2 \cdot \log_2(N/2)$	8	64	384	4608	229376
FFT/DFT	.125	.0625	.0234	.00439	.000214

See [36] pp. 262-273, [17] pp. 271-312, [65] pp. 511-537.

3.2.3 The Z-Transform

The Fourier transform converts continuous signals from the time domain to the frequency domain. If problems of convergence arise, the Laplace transform can be used (3.1.4.3). In the same way, the z-transform can be used in place of the discrete Fourier transform DFT, as will be shown below.

See the detailed but not all too technical introductions in A Digital Signal Processing Primer *by Ken Stieglitz [32], pp. 173-195 and* Understanding Signals and Systems *by Jack Colten[16], pp. 230-240.* Struktur und Bedeutung *by Norbert Bischof [33], pp. 233-270 explains the z-transform in a way that is particularly easy to follow.)*

3.2.3.1 Definition and Transformation of Simple Signals

The bilateral z-transform maps discrete signals $x(k)$ to functions of the complex variable z.

$$X(z) = Z\{x(k)\} = \sum_{k=-\infty}^{\infty} x(k) z^{-k}$$

The z-transform of a causal signal, that is, a signal for which $x(k) = 0$ for $k < 0$, is unilateral.

$$X(z) = Z\{x(k)\} = \sum_{k=0}^{\infty} x(k) z^{-k}$$

If we write the complex variable z in polar coordinates: $z = re^{i\omega T} = re^{i\Omega}$, we obtain

$$X(re^{i\Omega}) = \sum_{k=-\infty}^{\infty} x(k) r^{-k} e^{-ik\Omega},$$

which is the discrete Fourier transform of the signal $x(k)$ multiplied by r^{-k}. When $r = 1$, that is, when z is on the unit circle, the Fourier transform and the z-transform are identical. The z-transform is a generalization of the Fourier transform and exists when this sum converges:

$$\sum_{k=-\infty}^{\infty} |x(k) z^{-k}| = \sum_{k=-\infty}^{\infty} |x(k) r^{-k}| < \infty.$$

This is generally true over a region $r_1 < r = |z| < r_2$. In the complex plane, the region of convergence corresponds to a ring around the origin.

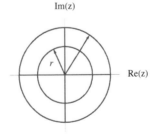

The inverse z-transform Z^{-1} converts a z-transform back into the original discrete sequence $Z^{-1}\{X(z)\} = x(k)$.

The z-transform of the impulse $\delta(k) = \{1, 0, 0, 0, ...\}$ is $Z\{\delta(k)\} = \sum_{k=0}^{\infty} \delta(k) z^{-k} = 1$ and is convergent for all z.

3.2 Digital Signals, DFT, FFT

By using the formula for geometric series (A4.8), we can show that the z-transform of the step function $u(k) = \{..., 0, u(0) = 1, 1, 1, ...\}$ is

$$Z\{u(k)\} = \sum_{k=-\infty}^{\infty} u(k) z^{-k} = \sum_{k=0}^{\infty} z^{-k} = \sum_{k=0}^{\infty} (z^{-1})^k = z/(z-1).$$

The sum is convergent for $1 < |z| \leq \infty$.

The z-transform of the exponential sequence $e_a(k) = \{..., 0, e_a(0) = 1, a, a^2, ...\}$ is

$$Z\{e_a(k)\} = \sum_{k=0}^{\infty} a^k z^{-k} = \sum_{k=0}^{\infty} (az^{-1})^k = z/(z-a)$$

and is convergent for $|a| < |z| \leq \infty$.

3.2.3.2 Characteristics and Examples

The z-transform is linear, that is $Z\{ax(k) + by(k)\} = aZ\{x(k)\} + bZ\{y(k)\} = aX(z) + bY(z)$.

Shifting a signal by l samples corresponds to multiplying its z-transform by z^l, since

$$\sum_{k=-\infty}^{\infty} x(k-l) z^{-k} = z^{-l} \sum_{k=-\infty}^{\infty} x(k-l) z^{-(k-l)} = z^{-l} X(z).$$

The convolution $x(k)*y(k) = \sum_{n=-\infty}^{\infty} x(n)y(k-n) = \sum_{n=-\infty}^{\infty} x(k-n)y(n)$ of two time series corresponds to the product of their z-transforms $Z\{x(k)*y(k)\} = X(z)Y(z)$.

The following table summarizes the most important characteristics of the z-transform.

Property/Operation	Time Domain	Z – Domain
Linearity	$a x(k) + b y(k)$	$a X(z) + b Y(z)$
Time Shifting	$x(k \pm l)$	$z^{\pm l} X(z)$
Convolution	$x(k) * y(k) = \sum_{n=-\infty}^{\infty} x(n) y(k-n)$	$X(z) Y(z)$
Exponentiation	$e^{akT} x(k)$	$X(e^{-aT} z)$
Multiplication	$x(k) y(k)$	$\frac{1}{2\pi i} \oint_C X(\omega) Y\left(\frac{z}{\omega}\right) \omega^{-1} d\omega$
Differentiation	$k x(k)$	$-z \frac{d}{dz} X(z)$

The table below lists several sequences and their z-transforms.

$x(n)$	$X(z) = Z\{x(n)\}$
1	$\frac{z}{z-1}$
$(-1)^n$	$\frac{z}{z+1}$
n	$\frac{z}{(z-1)^2}$
a^n	$\frac{z}{z-a}$
$\frac{1}{n}; n > 1$	$\ln\left(\frac{z}{z-1}\right)$

$x(n)$	$X(z) = Z\{x(n)\}$
$\frac{a^n}{n!}$	$e^{\frac{a}{z}}$
$e^{n\alpha}$	$\frac{z}{z - e^\alpha}$
$\sin(n\beta)$	$\frac{z \cdot \sin(\beta)}{z^2 - 2 z \cdot \cos(\beta) + 1}$
$\cos(n\beta)$	$\frac{z(z - \cos(\beta))}{z^2 - 2 z \cdot \cos(\beta) + 1}$

3.3 Systems and Filters

Examples of analog systems are living creatures, machines and mechanical devices, apparatuses and instruments. Digital systems can be digital devices, too, that is, the hardware, but the term usually refers to the structure of a system or to the mathematical function describing this structure. Hence digital systems are computer programs, algorithms or mathematical formulae. Any non-trivial system that produces an output different from its input can be called a filter. Hence the terms filter and system are often used as synonyms, as they frequently will be in this chapter. Nonetheless, in connection with non-specific situations we will generally use the term system, in connection with spectral variation we will speak of filters. First, we describe the characteristics of systems and filters and survey techniques for working with them. Then we present straightforward, mathematically clearly defined linear systems and their applications. nonlinear systems are difficult to describe mathematically and are usually avoided in digital technology when it is important to construct stable and predictable systems. Since nearly all natural and technical systems, but especially biological and cultural systems are nonlinear, nonlinear systems are considerably more important and interesting, at least for control functions in programs, synthesis instruments, algorithmic composition techniques, etc., than linear systems. Therefore we devote a special chapter to nonlinearity.

3.3.1 Systems

3.3.1.1 Definition and Examples

A *system* produces from an input signal $x(k)$ (or several input signals $x_i(k)$) an output signal $y(k)$ according to certain rules. The input signal $x(k)$ is also known as the excitation, the output signal $y(k)$ as the system's response to the excitation. We write for this process $y(k) = H\{x(k)\}$.

The basic transformations described in Chapter 3.2.1.3 can be interpreted as systems:

Example 1 Delaying a signal by d time units $y(k) = H_1\{x(k)\} = x(k-d)$
Example 2 Time-reversal of a signal $y(k) = H_2\{x(k)\} = x(-k)$
Example 3 Scaling a signal $y(k) = H_3\{x(k)\} = cx(k)$

Systems can process more than one input:

Example 4 Mixing several sounds $y(k) = H_4\{x_1(k), x_2(k), ...\} = c_1 x_1(k) + c_2 x_2(k) + ...$
Example 5 Ring-modulating two sounds (6.1.1) $y(k) = H_5\{x_1(k), x_2(k)\} = cx_1(k)x_2(k)$

Systems can use several values of an input to produce a single output value:

Example 6 By averaging two consecutive values of a signal high frequencies can be suppressed: $y(k) = H_6\{x(k)\} = 1/2 \cdot (x(k) + x(k-1))$.

Systems can use earlier output values to produce a new output (feedback):

Example 7 $y(k) = H_7\{x(k)\} = x(k) + cy(k-d)$

3.3.1.2 General Properties of Systems

A system is *linear* when the principle of superposition holds, that is when the response of a sum of arbitrary signals equals the sum of the individual responses. All of the examples above are linear except Example 5, which is nonlinear.

$$y(k) = H\{\sum_{i=1}^{n} c_i\, x_i(k)\} = \sum_{i=1}^{n} c_i\, H\{x_i(k)\} = \sum_{i=1}^{n} c_i\, y_i(k)$$

A system is *time-invariant* when the system response is independent of the moment of excitation. This means that the system always reacts in the same way.

$$y(k - d) = H\{x(k - d)\}$$

All of the examples above are time-invariant. This system is time-variant: $y(k) = H_8\{x(k)\} = x(k) + y(k - d)/k$.

A system is called *stable*, or more exactly *BIBO stable* (bounded-input bounded-output), when a limited excitation $x(k)$ produces a limited response $y(k)$. Example 7 is only stable for values $|c| < 1$.

$$|x(k)| < \infty \Rightarrow |y(k)| < \infty$$

A system is called *causal* if the output $y(k)$ is independent of future values $x(k + d)$. A system is called *recursive* if the output $y(k)$ depends on at least one earlier output value $y(k - d)$. A system is *non-recursive* if the output $y(k)$ does not depend on any earlier output value.

3.3.1.3 Impulse Response and Convolution

One often needs to know only the response of a system to a single impulse. The behavior of a linear time-invariant system (a so-called *LTI-system*) can be completely described by its *impulse response*. The impulse response is written as $h(k)$ and can be defined using the unit impulse $\delta(k)$ (3.2.1.2) as follows:

$$h(k) = H\{\delta(k)\}$$

The response $y(k)$ of a system to a signal $x(k)$ can be thought of as the sum of the responses to the individual values of $x(k)$. Below we compute the response of a system having the impulse response $h(k)$ shown at the right to the signal $x(k)$ shown at the left.

In the illustration below the values $x(k)$ are circled and their impulse responces are shown by the connected dots. The resulting signal $y(k)$, shown with larger dots, is given by the sum of the impulse responses.

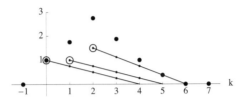

The impulse response to $x(0) = 1$ can be written as $h(k)$, the impulse response to $x(1) = 1$ as $h(k-1)$ and the impulse response to $x(2) = 1.5$ as $1.5h(k-2)$. The value of the resulting signal $y(k)$ at time $k = 3$ is $y(3) = h(3) + h(2) + 1.5h(1) = .25 + .5 + 1.125 = 1.875$.

If one excites an LTI-system with an arbitrary signal $x(k)$, the identity $y(k) = H\{x(k)\}$ holds. Hence, since every signal $x(k)$ can be written as $x(k) = \sum_{i=-\infty}^{\infty} x(i)\,\delta(k-i)$,

$$y(k) = H\{\sum_{i=-\infty}^{\infty} x(i)\,\delta(k-i)\}$$

and, because of linearity and time invariance,

$$y(k) = \sum_{i=-\infty}^{\infty} x(i)\, H\{\delta(k-i)\} = \sum_{i=-\infty}^{\infty} x(i)\, h(k-i) = \sum_{i=-\infty}^{\infty} h(i)\, x(k-i).$$

This sum is known as the *discrete convolution* $x(k)*h(k)$ of the signals $x(k)$ and $h(k)$.

$$\boxed{x(k)*h(k) = \sum_{i=-\infty}^{\infty} x(i)\, h(k-i) = \sum_{i=-\infty}^{\infty} h(i)\, x(k-i)}$$

Because the convolution of two long signals requires a great many computations, it is often useful to transform the signals into the frequency domain by the fast Fourier transform, do a simple multiplication of the spectra and transform the result back into the time domain (3.2.2.4). This procedure is called fast convolution and is used for high-order filters and for computing reverberation from the impulse response of a space to be simulated acoustically (9.2.2.5).

3.3.1.4 Properties of Systems in the Frequency Domain

To study the properties of a system in the frequency domain, we observe the system's response either to the sinusoid $x(k) = \sin(2\pi fTk) = \sin(\omega Tk) = \sin(\Omega k)$ or to the complex exponential sequence $x = e^{i\omega Tk} = e^{i\Omega k}$. In nonlinear systems new frequencies can appear. Upsampling, for example, multiplies the frequency of any sine wave. New frequencies also arise when two signals are multiplied together (6.1.1), and when there is nonlinear distortion, as in the system $y(k) = x^2(k)$, every sine wave gives rise to a sound whose spectrum depends on the amplitude of the signal $x(k)$ (6.1.3.1).

In the following discussion, we shall restrict ourselves to linear time-invariant systems. Here the frequencies remain the same (3.3.1.5). Depending on their frequencies however, the sine waves can change their amplitude and/or be shifted in time. The function describing these changes is called the *frequency response* or the *transfer function* of the system. The part of this function describing the change in amplitude as a function of the frequency ($A_{out}(f)/A_{in}(f)$) is called the *amplitude response*, that part describing the change in phase ($\phi_{out}(f) - \phi_{in}(f)$) the *phase response*. There are various ways to represent amplitude and phase response. For our calculations and illustrations we shall first use $\Omega = \omega T$ and then substitute either ωT or $2\pi fT$ to

3.3 Systems and Filters

obtain the more usual units of angular frequency and frequency in Hz respectively. The following table shows the usual representations for frequency together with the corresponding ranges for frequencies from 0 to the Nyquist frequency. Besides the frequency f and the angular frequency ω, the table also shows the *normalized angular frequency* Ω and the frequency as a fraction of the sampling frequency.

Symbol	Name		Range
f	Frequency in Hz		$[0, f_s/2]$
ω	Angular Frequency	$\omega = 2\pi \cdot f$	$[0, \pi \cdot f_s]$
Ω	Normalized Angular Frequency	$\Omega = \omega/f_s = \omega T$	$[0, \pi]$
f/f_s	Frequency as Fraction of the Sampling Frequency		$[0, 1/2]$

The frequency response for simple systems and specific frequencies, in particular $f = 0$ and $f = f_s/2$, can be computed by hand. A cosine wave of amplitude 1 and frequency 0 gives the signal $\{1, 1, 1, ...\}$, a cosine wave of amplitude 1 and frequency $f_s/2$ gives the signal $\{1, -1, 1, -1, ...\}$.

A system which delays a signal by one sampling period does not change the amplitude of an arbitrary sine wave. The following figure shows the signal $x(k) = \sin(\Omega k + \pi/2)$ and the delayed signal $y(k) = \sin(\Omega(k-1) + \pi/2)$, on the left for $\Omega = \pi$, on the right for $\Omega = .2\pi$.

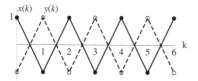
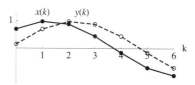

The amplitude response of the system is constant. The phase shift depends on the frequency. For the first signal, the phase shift is a semi-cycle $(-\pi)$; for the second signal, whose period is 10 sampling periods, the phase shift is $-\pi/5$. In general, the phase shift is $-\Omega$.

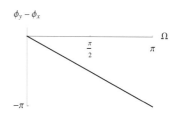

We obtain a system whose amplitude response is not constant by defining the output as the sum of the input and the input delayed by one sampling period: $y(k) = x(k) + x(k-1)$.

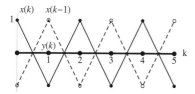
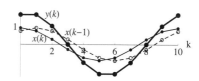

If the signal's frequency is $f_s/2$, the two addends cancel each other (left). If the frequency is 0, the amplitude is doubled. For any other arbitrary frequency the sum is between 0 and 2. Because the time delay is always one-half sampling period, the phase shift of the output to the input is always $\Omega/2$ (right). We can read off the amplitude of any frequency in this straightforward case at one-half sampling period after the maximum value of the input, that is at $k = 1/2$. The amplitude is $2\cdot\cos(-\Omega/2)$.

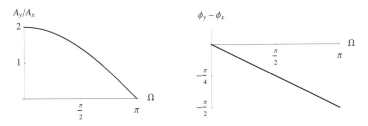

3.3.1.5 The Complex Representation

To study the behavior in the frequency domain of a linear time-invariant system whose impulse response is described by $h(k)$, we consider the system's response to excitation by the complex exponential sequence $x = e^{i\omega Tk} = e^{i\Omega k}$. By convolution we obtain

$$y(k) = h(k)*x(k) = \sum_{i=-\infty}^{\infty} h(i)\, e^{i\Omega(k-i)}.$$

Factoring out the time-dependent terms, we have

$$y(k) = e^{i\Omega k} \sum_{i=-\infty}^{\infty} h(i)\, e^{-i\Omega i}.$$

Since only the factor $e^{i\Omega k}$ is time-dependent, the system response $y(k)$ is a complex exponential sequence having the same frequency as $x(k)$. Only amplitude and phase are influenced by the system. The expression

$$H(e^{i\Omega}) = \sum_{k=-\infty}^{\infty} h(k)\, e^{-i\Omega k}$$

is called the transfer function or the frequency response of the system. The modulus of the vector $|H(e^{i\Omega})|$ is the amplitude response of the system, the argument $\arg(H(e^{i\Omega}))$ its phase response.

If we apply this to the system $H\{x(k)\} = y(k) = x(k) + x(k-1)$, we have for the impulse response $h(0) = h(1) = 1$ and $h(k) = 0$ otherwise. Hence the transfer function consists of only two components:

$$H(e^{i\Omega}) = h(0)\, e^{-0\cdot i\Omega} + h(1)\, e^{-1\cdot i\Omega} = 1 + e^{-i\Omega}.$$

For the amplitude response we have

$$|H(e^{i\Omega})| = |1 + e^{-i\Omega}| = \sqrt{\left(1 + \operatorname{Re}(e^{i\Omega})\right)^2 + \operatorname{Im}^2(e^{i\Omega})}$$

$$= \sqrt{(1 + \cos(\Omega))^2 + \sin^2(\Omega)} = \sqrt{2}\sqrt{1 + \cos(\Omega)}.$$

3.3 Systems and Filters

Because $\cos(\Omega) = \cos^2(\Omega/2) - \sin^2(\Omega/2)$, this result is equivalent to the result in Chapter 3.3.1.4. For the phase response we have

$$\text{Arg}(H(e^{i\Omega})) = \arctan(\text{Im}(e^{i\Omega}) / \text{Re}(e^{i\Omega})) = \arctan\left(\frac{\sin(\Omega)}{1 + \cos(\Omega)}\right),$$

or, using the formula above and $\sin(\Omega) = 2 \cdot \sin(\Omega/2) \cdot \cos(\Omega/2)$

$$\text{Arg}(H(e^{i\Omega})) = \arctan\left(\frac{2\sin(\Omega/2)\cos(\Omega/2)}{1 + \cos^2(\Omega/2) - \sin^2(\Omega/2)}\right) = \arctan(\tan(\Omega/2)) = \Omega/2.$$

As the example shows, even in the simplest cases considerable calculation is necessary to obtain a result in simple form. Therefore, in the examples which follow we will usually only show the results graphically.

```
Plot[Abs[1 + e^-i*Ω], {Ω, 0, π}]
Plot[Arg[1 + e^-i*Ω], {Ω, 0, π}]
```

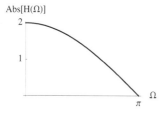
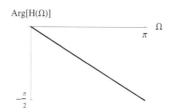

If in the formula above we replace the simple sum by a weighted sum $y(k) = x(k) + cx(k - 1)$, we can no longer compute amplitude and phase responses as we did in the preceding chapter. We obtain for the transfer function the complex expression $H(e^{i\Omega}) = 1 + ce^{-i\Omega}$. For the amplitude response we obtain

$$|H(e^{i\Omega})| = |1 + ce^{-i\Omega}| = \sqrt{|1 + c^2 + 2c \cdot \cos(\Omega)|}$$

For $c = 0.2$ we have the following frequency response.

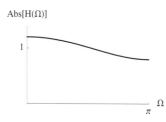
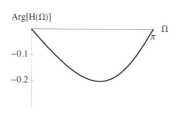

If we introduce an arbitrary delay nT, then the system equation is $y(k) = x(k) + cx(k - n)$, and the transfer function becomes

$$H(e^{i\Omega}) = 1 + ce^{-in\Omega}.$$

For the amplitude response we obtain

$$|H(e^{i\Omega})| = |1 + ce^{-in\Omega}| = \sqrt{|1 + c^2 + 2c \cdot \cos(n\Omega)|}.$$

The following figure shows the system's frequency response for $n = 4$ and $c = 1$ (see comb filter 3.3.3.5):

```
n = 4; p1 = Plot[Abs[1 + e^-i*Ω*n], {Ω, 0, π}]
n = 4; p2 = Plot[Arg[1 + e^-i*Ω*n], {Ω, 0, π}]
```

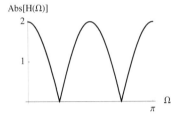
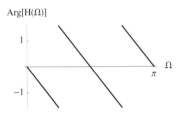

(→ Frequency response as a function of frequency f in Hz and as a function of f/f_s.)

3.3.1.6 Filters

Systems that change the spectrum of a signal are called filters. The systems described in the chapters above are typical examples of simple filters. *Linear causal filters* are systems whose output signal $y(k)$ is a weighted sum of the input signal and delayed in- and output signals.
A filter is called *recursive* when the output $y(k)$ depends on at least one earlier output value $y(k-d)$ *non-recursive* when it does not depend on an earlier output value $y(k-d)$.
A filter's structure can be shown in a so-called signal-flow graph. The illustrations below show the basic elements of linear filters, i.e., the weighted addition of an input to the delayed input $x \cdot z^{-1}$ (where $z^{-1} = e^{-i\Omega}$, 3.3.1.7).

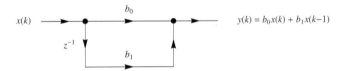

$$y(k) = b_0 x(k) + b_1 x(k-1)$$

and the weighted addition of the input to the delayed output.

$$y(k) = b_0 x(k) + a_1 y(k-1)$$

Generally, filters can be represented as weighted sums of delayed in- and output signals.

$$\boxed{y(k) + \sum_{i=1}^{m} a_i \, y(k-i) = b_0 \, x(k) + \sum_{i=1}^{l} b_i \, x(k-i)}$$ or

$$\boxed{\begin{aligned} y(k) &= b_0 \, x(k) + \sum_{i=1}^{l} b_i \, x(k-i) - \sum_{i=1}^{m} a_i \, y(k-i) \\ &= b_0 \, x(k) + \sum_{i=1}^{l} b_i \, x(k) \, z^{-i} - \sum_{i=1}^{m} a_i \, y(k) \, z^{-i} \end{aligned}}$$

3.3 Systems and Filters

The following figure shows the signal-flow graph of a filter with l delays of the input signal and m delays of the output signal.

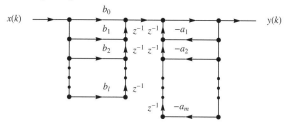

If we set $n = \text{Max}(l, m)$, we can write

$$y(k) = b_0\, x(k) + \sum_{i=0}^{n} (b_i\, x(k-i) - a_i\, y(k-i))\, z^{-i}$$

Any of the coefficients b_i or a_i can be zero.

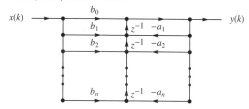

Filters can be represented clearly using signal-flow graphs. The graphs can help one find simplest realizations, i.e. simplest sequences of delays, multiplications and additions for a given task. Furthermore they show how much computation and computer memory a given filter requires.

3.3.1.7 The Transfer Function and the Z-Plane

Consider the representation below of a complex exponential sequence as a rotating vector in the complex plane (3.1.2.5). Here the frequency corresponds to the angle between the vector at time $t = 1$ and the real axis. Hence the normalized angular frequency Ω corresponds to the distance on the circle from point $(1, 0)$ to point z.

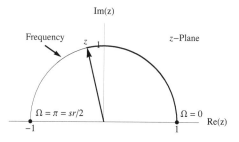

To determine the frequency response of a system, we calculate the system's effect on the exponential sequence

$$x(k) = e^{i\Omega k} = \cos(\Omega k) + i\cdot\sin(\Omega k)$$

and obtain for the filter $y(k) = x(k) - bx(k-1)$:

$$y(k) = e^{i\Omega k} - be^{i\Omega(k-1)} = e^{i\Omega k}(1 - be^{-i\Omega})$$

and substituting z for $e^{i\Omega}$:

$$y(k) = x(k)(1 - bz^{-1}).$$

Hence the transfer function is $H(z) = 1 - bz^{-1}$.

The amplitude response corresponds to the absolute value of the transfer function

$$|H(z)| = |1 - bz^{-1}| = \frac{|z-b|}{|z|} = |z - b|.$$

This value is the modulus of the vector $z - b$, that is, the distance between point $P(z) = \Omega$ and point $(b, 0)$. The phase response corresponds to the argument of the transfer function

$$\arg(H(z)) = \arg\left(\frac{z-b}{z}\right) = \arg(z - b) - \arg(z),$$

that is, the difference of the angles α and β.

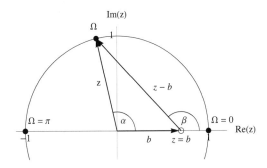

```
b = .8; Plot[Abs[1 - b*e^-i*Ω], {Ω, 0, π}]
b = .8; Plot[Arg[1 - b*e^-i*Ω], {Ω, 0, π}]
```

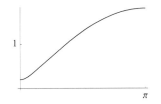
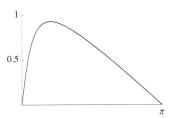

Similarly the filter $y(k) = x(k) - x(k-1) + x(k-2)$ gives the transfer function

$$H(z) = 1 - z^{-1} + z^{-2} = (z^2 - z + 1)/z^2$$

and the amplitude response

$$|(z^2 - z + 1)/z^2| = |z^2 - z + 1|.$$

3.3 Systems and Filters

The polynomial $z^2 - z + 1$ can be factorized to $(z - z_1)(z - z_2)$, where $z_1 = e^{i\pi/3}$ and $z_2 = e^{-i\pi/3}$ are the solutions of the equation $z^2 - z + 1 = 0$. Since the modulus of the product of two complex numbers equals the product of the moduli of the individual factors, we obtain

$$|H(z)| = |z^2 - z + 1| = |(z - z_1)(z - z_2)| = |(z - z_1)| \cdot |(z - z_2)|.$$

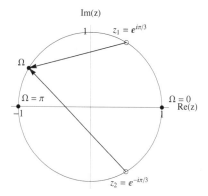

The absolute value of the transfer function $|H(z)|$ can be illustrated as a surface over the complex plane. The figure below shows the function $|H(z)|$ on a logarithmic scale, at the left for all z, at the right for the unit circle as the intersection between the surface $|H(z)|$ and a cylinder erected over the unit circle.

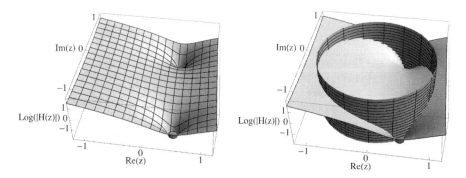

The following figure shows the function $|H(z)|$ on the unit circle for frequencies up to the sampling frequency ($0 < \Omega < 2\pi$).

```
Plot[20*Log[10, Abs[1 - e^-i*Ω + e^-2*i*Ω]], {Ω, 0, 2*π}]
Plot[Arg[1 - e^-i*Ω + e^-2*i*Ω], {Ω, 0, 2*π}]
```

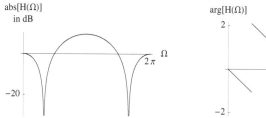

If we include only frequencies up to the Nyquist frequency, we have:

```
Plot[20 * Log[10, Abs[1 - e^-i*Ω + e^-2*i*Ω]], {Ω, 0, π}]
Plot[Arg[1 - e^-i*Ω + e^-2*i*Ω], {Ω, 0, π}]
```

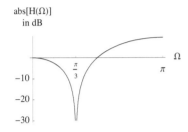
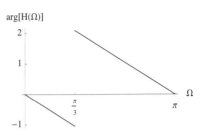

The recursive filter $y(k) + a_1 y(k-1) + a_2 y(k-2) = x(k) + bx(k-1)$ has the transfer function

$$H(z) = (1 + bz^{-1})/(1 + a_1 z^{-1} + a_2 z^{-2}).$$

If the denominator is equal to zero, $|H(z)|$ is infinitely large. Those values z_i for which the denominator is zero are called poles. We obtain them as solutions of the the equation

$$1 + a_1 z^{-1} + a_2 z^{-2} = 0,$$

as above. E.g., if $a_1 = -.5$, $a_2 = .7$ and $b = .85$, the poles $p_1 = .25 + .798i$ and $p_2 = .25 - .798i$ are obtained.

```
a1 = -.5; a2 = .7; b = .85; Solve[1 + a1 * z^-1 + a2 * z^-2 == 0, z]

{{z→0.25-0.798436 I},{z→0.25+0.798436 I}}
```

The filter has one zero, $z_1 = -.85$. Hence we can write

$$H(z) = \frac{1 + .85 z^{-1}}{1 - .5 z^{-1} + .7 z^{-2}} = \frac{z + .85}{(z - .25 - .798 i)(z - .25 + .798 i)}.$$

The figure below shows the poles (marked by points) and the zero (marked by a circle) in the Z-plane.

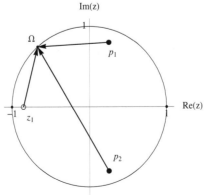

3.3 Systems and Filters

At the zero $|H(z)| = 0$ and $\log(|H(z)|) = -\infty$, at the poles $|H(z)| = \infty$. In the three-dimensional representation, there is a depression at the zero and two peaks at the poles.

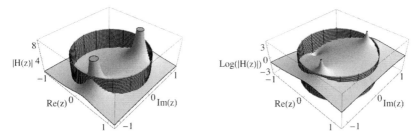

In general, from the equation

$$y(k) + \sum_{i=1}^{m} a_i\, y(k-i) = b_0\, x(k) + \sum_{i=1}^{l} b_i\, x(k-i)$$

we obtain using the z-transform (3.2.3) the transfer function

$$H(z) = \frac{Y(z)}{X(z)} = \frac{\sum_{i=0}^{m} b_i\, z^{-i}}{1 + \sum_{i=1}^{l} a_i\, z^{-i}}$$

$H(z)$ is the z-transform of the impulse response $h(k)$. In the same way as the impulse response informs us about a filter's behavior in the time domain, the transfer function informs us about its behavior in the frequency domain. (→ z-Plane.maxpat)

3.3.1.8 Linear Phase Filters

To say that a system has *linear phase* means that its phase response is a linear function of the frequency. The first example in Chapter 3.3.1.7 does not have linear phase, all the others do. Formally we can write

$$H(e^{i\omega}) = |H(e^{i\omega})| \cdot e^{-i\alpha\omega},$$

or more generally

$$H(e^{i\omega}) = |H(e^{i\omega})| \cdot e^{-i(\alpha\omega - \beta)},$$

where α and β are constant. A filter has linear phase when its impulse response is symmetric or antisymmetric, that is, when $h(k) = h(N - k)$ or when $h(k) = -h(N - k)$. If we distinguish further between systems with an even impulse response length and those with an odd impulse response length, we have the four basic types of linear phase filters shown below. Thanks to the symmetry of the impulse response, the computation time for linear phase filters is shorter than for comparable nonlinear phase filters.

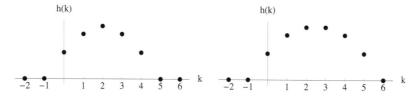

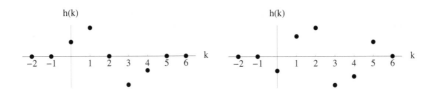

3.3.1.9 Filter Design

Thus far we have defined systems and examined their properties. Usually, however, one has certain ideas about a system's properties and tries to realize that system. There is considerable literature on filter design, particularly for linear time-invariant filters, because filters have many applications and have been thoroughly studied in analog electronics. Analog filters are in principle recursive; hence most books on digital signal processing treat recursive filters before non-recursive ones.

There are four basic types of filters. Those which pass low frequencies and attenuate high frequencies are *low-pass filters* (a in the figure below). Those which pass high frequencies and attenuate low frequencies are *high-pass filters* (b below). Those which pass or amplify frequencies within a certain frequency band and attenuate frequencies outside that band are *band-pass filters* (c). Band-pass filters can be thought of as a combination of low- and high-pass filters. *Band-stop* or *notch filters* attenuate a particular frequency band and pass frequencies outside that band (d). *All-pass filters* pass all frequencies and change only the phase of the signals' components. The figure below shows these four basic types schematically. The transition between the stop-band and the pass-band of any filter is continuous. Between the *edge* or *cutoff frequency* of a pass-band f_p and that of a stop-band f_s there is always a transition band. The amplitude response within a stop-band is not constant but oscillates. For this reason one indicates a tolerance for the values of pass- and stop-bands (the grey bars in the figure below show the areas in which the values for pass- and stop-bands are guaranteed; the rest is transition bands).

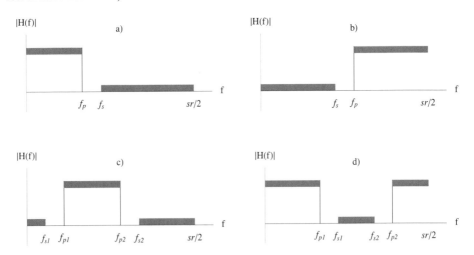

Filters in equalizers or in synthesis programs often use only the terms characterized by the following example. The figure below shows for a simple band-pass filter the *center frequency*

3.3 Systems and Filters

f_m of the pass-band and the cutoff frequencies f_{c1} and f_{c2}, defined as the frequencies at which the amplitude is attenuated by 3dB. The difference between the higher and lower cutoff frequencies is called the *bandwidth BW* of the filter. The filter's sharpness is given by the *Quality Factor Q*, defined as the relationship between the center frequency and the bandwidth: $Q = f_m/BW$.

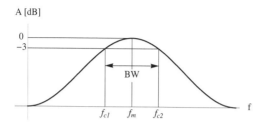

3.3.2 Non-Recursive Filters / FIR Filters

3.3.2.1 Definitions and Properties

Digital filters which use only the input signal and the delayed input to compute the output signal are called *non-recursive*. Hence we can write for time-invariant non-recursive filters

$$y(k) = b_0\, x(k) + b_1\, x(k-1) + b_2\, x(k-2) + \ldots = b_0\, x(k) + \sum_{i=1}^{l} b_i\, x(k-i)$$

$$y(k) = b_0\, x(k) + \sum_{i=1}^{l} b_i\, x(k)\, z^{-i}$$

Non-recursive systems have a finite impulse response, which is not necessarily the case for recursive systems. Non-recursive systems can be designed to have linear phase, which is not possible with recursive systems. Non-recursive systems are always *stable*, which means that their output cannot increase infinitely. For systems of nth order, the output becomes zero not later than n sample periods after the end of the input. For this reason, these systems are called *finite impulse response* (FIR) filters. Because of their guaranteed stability and their property of being bounded, they lend themselves well to so-called adaptive systems, in which coefficients can be changed during operation.

The impulse response of a non-recursive filter is

$$h(k) = b_0\delta(k) + b_1\, \delta(k-1) + b_2\, \delta(k-2) + \ldots = b_k, \text{ since } \delta(k-n) = 0 \text{ for all } k \neq n.$$

> The sequence of the coefficients b_k of a non–recursive filter corresponds to its impulse response $h(k)$.

(→ *FIR_Filter.maxpat*)

3.3.2.2 Implementation

The usual way to implement non-recursive filters is *direct structure*, where the signal is delayed by a number of sampling intervals and multiplied by the coefficient b_i corresponding to the value $h(i)$ of the impulse response. There can be any number of simultaneous delays and individual coefficients; the order of their delay and multiplication is of no importance.

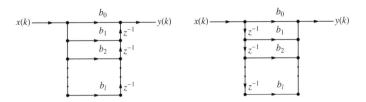

In what follows, we shall construct the filter $y(k) = x(k) - x(k-1)$ with Mathematica. In Mathematica a signal is a list. In order to show the effect of the filter, we choose as input a sine wave whose frequency increases over the duration of the signal.

```
x = Table[Sin[.005*k^2], {k, 0, 100}];
```

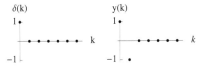

We define a list *y* for the output signal and a variable *xd1* for the input delayed by one sample period $x(k-1)$. We make the filter with the command $y(k) = x(k) - xd1$. By setting *xd1* equal to $x(k)$ at the end of each iteration, *xd1* corresponds to the delay $x(k-1)$ at the beginning of the next iteration.

```
y = Table[0,{k,0,100}]; xd1 = 0;
Do[y[[k]] = x[[k]]-xd1; xd1 = x[[k]];,{k,1,100}]
```

We can obtain the same result by subtracting $x(k-1)$ from $x(k)$.

```
x = Table[0,{k,0,6}]; x[[1]] = 1; xd1 = 0;
Do[y[[k]] = x[[k]]-xd1; xd1 = x[[k]];,{k,1,6}]
```

If we apply this filter to the pulse $\{1, 0, 0, \ldots\}$, we get an impulse response with the coefficients $b_0 = 1$ and $b_1 = -1$.

(→ Realization in Csound)

Because one often needs a high order to obtain good results with a non-recursive filter, one needs correspondingly long delays of the input signal. Instead of moving all the delayed values into a new memory location at each iteration, it is helpful to set up a so-called circular buffer, where only the final memory location, which is no longer needed for the next iteration, is overwritten by the new value of the input signal and a pointer to the memory address is shifted (5.1.4.2).

3.3 Systems and Filters

3.3.2.3 Designing Filters by Setting Zeros

Before discussing the systematic calculation of the coefficients of a non-recursive filter, let us see how simple filters can be realized intuitively. To implement a band-reject filter that attenuates frequencies between $\Omega = \pi/12$ and $\Omega = 4\pi/12$, we set zeros at $\Omega = 2\pi/12$ and $\Omega = 3\pi/12$ and, to get real-valued filter coefficients, at $\Omega = -2\pi/12$ and $\Omega = -3\pi/12$. We obtain the zeros shown below at the left and the amplitude response shown at the right.

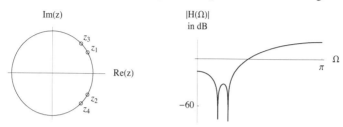

The amplitude response shows attenuation at the right place in the spectrum. One flaw of the filter is that it acts differently on the spectrum above and below the stop-band: low frequencies are attenuated, high frequencies amplified. We obtain the coefficients by expanding the product $(z - z_1)(z - z_2)(z - z_3)(z - z_4)$, where the z_i correspond to the zeros

$$z_1 = e^{i\pi \cdot 2/12}, \quad z_2 = e^{-i\pi \cdot 2/12}, \quad z_3 = e^{i\pi \cdot 3/12}, \quad z_4 = e^{-i\pi \cdot 3/12}.$$

Expand[(z - z₁) * (z - z₃) * (z - z₄) * (z - z₂)]

$$1 - \left(\sqrt{2} + \sqrt{3}\right) z + \left(2 + \sqrt{6}\right) z^2 - \left(\sqrt{2} + \sqrt{3}\right) z^3 + z^4$$

(→ Implementation)

In a second step, we try to correct the flaw described above and at the same time to flatten the amplitude response in both stop- and pass-band. To attenuate the amplification of the high frequencies and boost the low frequencies, we introduce in the high frequencies five additional zeroes. If then we put the original four zeroes not exactly on the unit circle, but rather a bit inside or outside of it, the deep notches in the spectrum vanish. The zeroes are now at $z_1 = .9e^{i\pi \cdot 2/12}$, $z_2 = .9e^{-i\pi \cdot 2/12}$, $z_3 = .9e^{i\pi \cdot 3/12}$, $z_4 = .9e^{-i\pi \cdot 3/12}$, $z_5 = 1.2e^{i\pi \cdot 6.7/12}$, $z_6 = 1.2e^{-i\pi \cdot 6.7/12}$, $z_7 = 1.3e^{i\pi \cdot 9.5/12}$, $z_8 = 1.3e^{-i\pi \cdot 9.5/12}$, $z_9 = 1.5e^{i\pi \cdot 12/12} = -1.5$.

which gives the amplitude response below right.

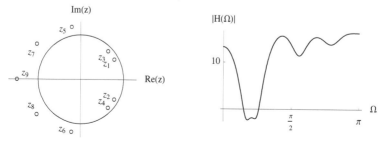

Again, we get the filter coefficients by expanding the product $(z - z_1)(z - z_2) \ldots (z - z_8)$. To compensate for the boost of about 15 dB in the pass-band, we divide the values of the computed signal by $10^{15/20} \approx 5.6$. (→ Implementation) (→ *FIR_Filter_Zeros.maxpat*)

3.3.2.4 Fourier Approximation

The frequency response of a system is periodic with the period $T_s = 1/f_s$. Like every periodic function, it can be decomposed into a Fourier series of complex exponential functions:

$$H(e^{i \cdot 2\pi fT}) = \sum_{n=-\infty}^{\infty} h(n) e^{-i \cdot 2\pi fnT} \qquad \text{or, with } 2\pi fT = \Omega,$$

$$\boxed{H(e^{i\Omega}) = \sum_{n=-\infty}^{\infty} h(n)\, e^{-in\Omega}}$$

The $h(n)$ are the coefficients of the Fourier series.

$$\boxed{h(n) = \frac{1}{2\pi} \int_{-\pi}^{\pi} H(e^{i\Omega})\, e^{in\Omega}\, d\Omega}$$

If we wish to obtain the frequency response $H_w(e^{i\Omega})$, the inverse Fourier transform gives us a sequence $h(0), \ldots h(m)$ which can be used as the impulse response of an FIR system. It can be shown that the system so produced is a least squares approximation (see [18] p. 177). However, there are two problems in this implementation. On the one hand, the Fourier series is infinite, on the other the filter is not causal, because $h(n)$ do not disappear for negative n. We solve the second problem by shifting the impulse response in time, which has no influence on the amplitude response of the system. (→ FIR_Fourier_Approximation.maxpat)

In what follows, we approximate a low-pass filter of order $m = 24$ with amplitude response

$$|H_w(e^{i\Omega})| = 1 \text{ for } |\Omega| < \pi/3 \text{ and } |H_w(e^{i\Omega})| = 0 \text{ otherwise.}$$

```
Hw[Ω_] = If[Abs[Ω]< π/3,1,0];
```

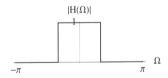

Using the following Mathematica command and the formula above, we compute the coefficients $h(-11)$ to $h(12)$ and store the values as elements 1 to 24 of a list h (left figure below). The amplitude response of the filter shows the typical ripple in both pass- and stop-band and the continuous, not very steep transition from one to the other (right figure below).

```
Do[h[[k]] = 1/(2π) ∫_{-π}^{π} Hw[Ω] * e^{i*Ω*(k-12)} dΩ, {k, 1, 24}]
Hr = Table[Re[h[[n]]] * e^{-i*Ω*(n-1)}, {n, 1, 24}]
Plot[Abs[Apply[Plus, Hr]], {Ω, 0, π}]
```

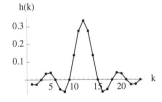
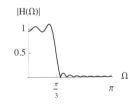

3.3 Systems and Filters

A higher-order filter ($m = 60$) gives us a noticeably steeper transition, but the ripple at the cut-off frequency is the same as before.

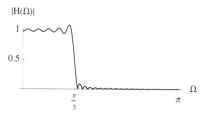

3.3.2.5 Windowing

The strong ripple can be avoided by fading the (in theory infinite) Fourier series in and out, so to speak, rather than truncating it. This is done with a so-called *window function* (also known as *apodization function* or *tapering function*). The figure below shows the triangular window, the Hann (incorrectly also called "Hanning") window, the Hamming window, the Blackman window and the rectangular window. The window function is defined for a window of length m as follows:

Rectangular window:	rect = 1
Triangular window:	tri = $1 - \text{Abs}[1 - 2i/m]$
Hann window:	hann = $.5(1 - \text{Cos}[2\pi i/m])$
Hamming window:	hamm = $.54 - .46\text{Cos}[2\pi i/m]$
Blackman window:	bl = $.42 - .5\text{Cos}[2\pi i/m] + .08\text{Cos}[4\pi i/m]$

If we apply these functions for the example from Chapter 3.3.2.4 by multiplying $h(n)$ by $f(n)$, we obtain the following amplitude responses:

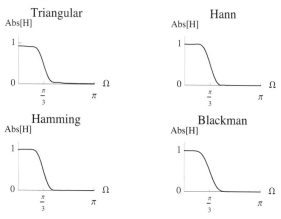

The logarithmic representations below show more clearly the differences in the windows' behaviors at high frequencies:

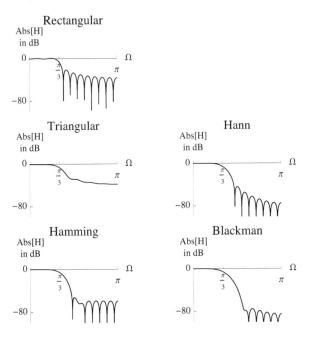

3.3.3 Recursive Filters / IIR Filters

3.3.3.1 Definition and Properties

Filters that compute their output using delayed samples of earlier output are called *recursive filters*. For linear time-invariant recursive filters we can write

$$y(k) + \sum_{i=1}^{m} a_i\, y(k-i) = b_0\, x(k) + \sum_{i=1}^{l} b_i\, x(k-i)$$

$$\begin{aligned} y(k) &= b_0\, x(k) + \sum_{i=1}^{l} b_i\, x(k-i) - \sum_{i=1}^{m} a_i\, y(k-i) \\ &= b_0\, x(k) + \sum_{i=1}^{l} b_i\, x(k)\, z^{-i} - \sum_{i=1}^{m} a_i\, y(k)\, z^{-i} \end{aligned}$$

A recursive system normally has an infinitely long impulse response. Such a filter is called an *Infinite Impulse Response* (IIR) filter. A non-recursive system cannot have an infinitely long impulse response (3.3.2.1). Systems with exactly linear phase cannot be realized with recursive filters. Recursive systems can be unstable, that is, the output can increase unlimitedly, or they can be bounded without converging to zero. (→ IIR_Filter.maxpat)

3.3.3.2 Implementation

There are various ways to implement filters. Those structures which require minimal storage are called *canonical structures*. The structure shown in Chapter 3.3.1.6 is the first canonical

3.3 Systems and Filters

structure (upper figure below). It can be shown that a graph transformation can generate a second canonical structure (lower figure below). The signal flow is reversed, and summing nodes become distribution nodes and vice versa.

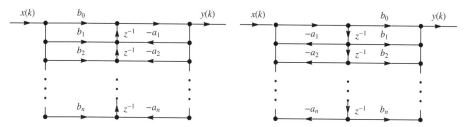

Numerator and denominator of the transfer function

$$H(z) = \frac{\sum_{k=0}^{q} b(k) z^{-k}}{1 + \sum_{k=1}^{p} a(k) z^{-k}}$$

can be written as products:

$$H(z) = \prod_{k=1}^{\max\{p,q\}} \frac{\beta_{0k} - \beta_{1k} z^{-1}}{1 - \alpha_{1k} z^{-1}}.$$

In general, the coefficients α_k and β_k are complex, but since, when not real, they occur as conjugate complex pairs, we can group two corresponding factors together and obtain a rational fraction of the second degree having real coefficients. If we denote the subsidiary systems by $H_i(z)$, we can decompose $H(z)$ into the product

$$H(z) = \prod_{i=1}^{p} H_i(z)$$

with subsidiary systems of first or second order (see biquad filters 3.3.3.5)

$$H_i(z) = \frac{1 - \beta_k z^{-1}}{1 - \alpha_k z^{-1}} \quad \text{or} \quad H_i(z) = \frac{\beta_{0k} + \beta_{1k} z^{-1} + \beta_{2k} z^{-2}}{1 - \alpha_{1k} z^{-1} + \alpha_{2k} z^{-2}}.$$

The third canonical structure generated in this way, the ladder network, has several advantages. Poles and zeros can be grouped in various ways, the sequence of the subsidiary systems is of no importance, and any imprecision in the parameters affects at most two poles or zeros.

$$x(k) \longrightarrow \boxed{H_1(z)} \longrightarrow \boxed{H_2(z)} \longrightarrow \cdots \longrightarrow \boxed{H_p(z)} \longrightarrow y(k)$$

By decomposing the transfer function $H(z)$ into partial fractions, one obtains the fourth canonical structure, which results in a parallel connection of the subsidiary systems $H_i(z)$ (see [18] p. 81).

Let us compare these various structures using the example of an IIR filter with three poles and three zeros. We set three arbitrary zeros $z_1 = -1 + i$, $z_2 = -1 - i$, $z_3 = -1$ and three poles $p_1 = .5 + .5i$, $p_2 = .5 - .5i$, $p_3 = .5$.

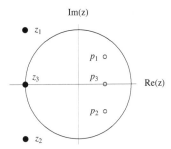

Since we start with poles and zeros, we can immediately write the transfer function as the product

$$H(z) = \frac{(z-z_1)(z-z_2)(z-z_3)}{(z-p_1)(z-p_2)(z-p_3)} = \frac{(z+1-i)(z+1+i)(z+1)}{(z-.5-.5i)(z-.5+.5i)(z-.5)}.$$

If we expand and reduce by z^3, we obtain

$$H(z) = \frac{2 + 4z + 3z^2 + z^3}{-.25 + z - 1.5z^2 + z^3} = \frac{2z^{-3} + 4z^{-2} + 3z^{-1} + 1}{-.25z^{-3} + z^{-2} - 1.5z^{-1} + 1}.$$

Since $H(z) = Y(z)/X(z)$, we have

$$Y(z)(-.25z^{-3} + z^{-2} - 1.5z^{-1} + 1) = X(z)(2z^{-3} + 4z^{-2} + 3z^{-1} + 1),$$

and from that

$$y(k) = x(k) + 3x(k-1) + 4x(k-2) + 2x(k-3) + 1.5y(k-1) - y(k-2) + .25y(k-3).$$

(→ Details)

We obtain the third canonical structure by decomposing the transfer function into two terms, grouping so that we only have real coefficients.

$$H(z) = \frac{(z-z_1)(z-z_2)(z-z_3)}{(z-p_1)(z-p_2)(z-p_3)} = H_1(z)H_2(z)$$

$$= \frac{(z-z_1)(z-z_2)}{(z-p_1)(z-p_2)} \cdot \frac{(z-z_3)}{(z-p_3)}$$

The transfer functions are

$$H_1(z) = \frac{2z^{-2} + 2z^{-1} + 1}{.5z^{-2} - z^{-1} + 1} \quad \text{and} \quad H_2(z) = \frac{z^{-1} + 1}{-.5z^{-1} + 1}.$$

For the first subsidiary system we have

$$y_1(k) = x_1(k) + 2x(k-1) + 2x(k-2) + y_1(k-1) - .5y_1(k-2)$$

and for the second subsidiary system, which uses $y_1(k)$ as input

$$y_2(k) = y_1(k) + y_1(k-1) + .5y_2(k-1).$$

(→ Details)

We obtain the fourth canonical structure by decomposing the transfer function in partial fractions, again regrouping so that only real coefficients occur:

3.3 Systems and Filters

$$H(z) = \frac{(z-z_1)(z-z_2)(z-z_3)}{(z-p_1)(z-p_2)(z-p_3)} = b_0 + H_1(z) + H_2(z)$$

$$= 1 + \frac{19.5}{-.5+z} + \frac{15-15z}{.5-z+z^2}.$$

The transfer functions are

$$H_1(z) = \frac{19.5}{.5z^{-1}+1} \text{ and } H_2(z) = \frac{15z^{-1}-15}{.5z^{-2}-z^{-1}+1}.$$

From them we obtain the equations for the subsidiary systems

$$y_1(k) = 19.5x(k) + .5y_1(k-1) \text{ and}$$

$$y_2(k) = -15x(k) + 15x(k-1) + y_2(k-1) - .5y_2(k-2). \qquad (\to \text{Details})$$

3.3.3.3 Designing Filters by Setting Poles and Zeros

We can improve the filter dicussed in Chapter 3.3.2.3 (to the left below) by not setting the zeros on the unit circle, and we can reduce the attenuation of the low frequencies by introducing two poles near the point $z = 1$. We choose $p_1 = .9e^{i\pi/12}$ and $p_2 = .9e^{-i\pi/12}$, which gives us the poles and zeros shown to the right in the figure below. ($\to$ IIR_z-plane.maxpat)

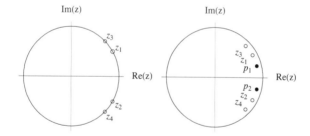

We obtain the filter coefficients by expanding the expression

$(z-z_1)(z-z_2)(z-z_3)(z-z_4)/((z-p_1)(z-p_2))$ with

$z_1 = .9e^{i\pi 2/12}, z_2 = .9e^{-i\pi 2/12}, z_3 = .9e^{i\pi 3/12}, z_4 = .9e^{-i\pi 3/12}, p_1 = .9e^{i\pi/12}, p_2 = .9e^{-i\pi/12}$:

`ExpandAll[ (z - z₁) * (z - z₂) * (z - z₃) * (z - z₄) / ((z - p₁) * (z - p₂)) ]`

$$\frac{0.656}{0.81-1.739\,z+z^2} - \frac{2.294\,z}{0.81-1.739\,z+z^2} + \frac{3.604\,z^2}{0.81-1.739\,z+z^2} - \frac{2.832\,z^3}{0.81-1.739\,z+z^2} + \frac{z^4}{0.81-1.739\,z+z^2}$$

Hence the transfer function is

$$H(z) = \frac{.66 - 2.29\,z + 3.60\,z^2 - 2.83\,z^3 + z^4}{.81 - 1.74\,z + z^2} = \frac{.66\,z^{-4} - 2.29\,z^{-3} + 3.6\,z^{-2} - 2.83\,z^{-1} + 1}{.81\,z^{-2} - 1.74\,z^{-1} + 1}.$$

This gives the filter

$$y(k) = x(k) - 2.83x(k-1) + 3.6x(k-2) - 2.29x(k-3) + .66x(k-4)$$
$$+ 1.74y(k-1) - .81y(k-2)$$

with the following amplitude response:

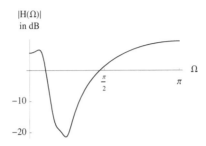

3.3.3.4 Stability

A systems is *stable* when its impulse response is limited, that is, when a limited excitation $x(k)$ induces a limited response $y(k)$ (3.3.1.2). In order to study the effect of the position of poles and zeros on a particular system, we shall consider the impulse response of that system. We obtain the impulse response $g(k)$ of the simple system $y(k) = ay(k-1) + x(k)$ by substituting for $x(k)$ the pulse train $\delta(k)$: $g(k) = 0$ for $k < 0$, $g(0) = 1$, $g(1) = a$, $g(2) = a^2$ etc. This sequence converges for $|a| < 1$. The transfer function of this system is $H(z) = 1/(z-a)$ and has a single pole at $z = a$. Since $|a| < 1$, the pole must be within the unit circle. The zeros can be anywhere, because they only influence the coefficients of the delays of the input signal.

> A system is stable when all its poles lie within the unit circle

3.3.3.5 Special Filters

The oscillatory behavior of simple physical systems (3.1.4.1 and 8.1.1) is described by the differential equation

$$\ddot{f} = -cf.$$

In a first approximation, we can transform this equation into a difference equation as follows. From three equidistant discrete values $f(t - 2T) = y(k-2)$, $f(t - T) = y(k-1)$ and $f(t) = y(k)$ we can approximate two sequential velocities

$$v(k - 1.5) = (y(k-1) - y(k-2))/T \text{ and } v(k - .5) = (y(k) - y(k-1))/T.$$

With T as the unit of time, we have

$$v(k - .5) = y(k) - y(k-1) \text{ and } v(k - 1.5) = y(k-1) - y(k-2),$$

and the acceleration

$$a(k-1) = v(k - .5) - v(k - 1.5) = y(k) - y(k-1) - (y(k-1) - y(k-2))$$
$$= y(k) - 2y(k-1) + y(k-2).$$

If we set the acceleration equal to $cy(k-1)$, add the input signal $x(k)$ and solve for $y(k)$ we obtain the equation of a so-called *resonator*:

$$y(k) = (2 - c)y(k-1) - y(k-2) + x(k).$$

3.3 Systems and Filters

Its transfer function is $H(z) = \dfrac{1}{1 - (2 - c)z^{-1} + z^{-2}}$.

The poles correspond to the zeros of the denominator.

`Solve[1 - (2 - c) z⁻¹ + z⁻² == 0, z]` $\left\{z \to \dfrac{1}{2}\left(2 \pm \sqrt{-4 + c}\ \sqrt{c} - c\right)\right\}$

The poles are on the unit circle (if $c < 4$), hence the amplitude response diverges at the system's eigenfrequency (resonant frequency Ω^*).

`Abs[(½ (2 - √(-4 + c) √c - c))]`   1.

The resonant frequency Ω^* of the system corresponds to the pole's argument. Now we calculate c from the resonant frequency Ω^*

`Solve[Arg[½ (2 + √(-4 + c) √c - c)] == ω, {c}]` $2 - 2\cos[\omega]$

and substitute the solution $2 - 2\cos(\Omega^*)$ for c in the equation of the resonator:

$$y(k) = 2\cos(\Omega^*)y(k - 1) - y(k - 2) + x(k).$$

The following figure shows the amplitude response of the resonator. ($\to$ Filter_fexpr.mapat)

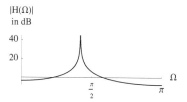

If we introduce a damping proportional to the velocity we get the differential equation

$$\ddot f = -cf - d\dot f.$$

If we assume for the velocity the average of the velocities $v(k - .5)$ and $v(k - 1.5)$, we have

$$v(k - 1) = .5(y(k) - y(k - 1) + y(k - 1) - y(k - 2)) = .5(y(k) - y(k - 2)).$$

From the differential equation above we now have the difference equation

$$y(k) - 2y(k - 1) + y(k - 2) = -cy(k - 1) - .5d(y(k) - y(k - 2)) + x(k)$$

and, when we solve for $y(k)$,

$$y(k) = ((2 - c)y(k - 1) - (1 - .5d)y(k - 2) + x(k))/(1 + .5d).$$

The transfer function is

$$H(z) = \dfrac{1}{(1 + .5d) - (2 - c)z^{-1} + (1 - .5d)z^{-2}}.$$

The poles are

Solve$\left[1 + d/2 - (2 - c)\, z^{-1} + (1 - d/2) * z^{-2} == 0,\, z\right]$ $\left\{z \to \frac{2-c\pm\sqrt{-4\,c+c^2+d^2}}{2+d}\right\}$

Hence the poles are within the unit circle and the amplitude response is finite even at the system's eigenfrequency.
($\to |z|$ for $0 < c < 4$) ($\to$ IIR_z-plane.maxpat) ($\to$ IIR_Filter_2.maxpat)

The eigenfrequency of the damped system is somewhat lower than that of the undamped system ($\to$ Figure). The bandwidth of the resonator depends on the damping d:

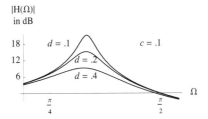

Now we calculate c from the resonant frequency Ω^*

Solve$\left[\text{Arg}\left[\frac{2-c+\sqrt{-4\,c+c^2+d^2}}{2+d}\right] == \omega,\, \{c\}\right]$ $2 - \sqrt{4 - d^2}\,\text{Cos}[\omega]$

and substitute the solution for c in the equation of the damped resonator:

$$y(k) = \frac{1}{1+.5\,d}\left((2-c)\,y(k-1) - (1 - .5\,d)\,y(k-2) + x(k)\right)$$

$$= \frac{1}{1+.5\,d}\left(\sqrt{4 - d^2}\,\cos(\Omega^*)\,y(k-1) - (1 - .5\,d)\,y(k-2) + x(k)\right).$$

As shown in Chapter 3.3.3.2, every filter can be decomposed into filters of first and second order. Filters with two poles and two zeros are called *biquad filters*. The transfer function for biquad filters can be written as follows, using the frequencies of the zeros f_z and the poles f_P (in Hz), the damping coefficients r_z and r_P and the scaling factor g.

$$H(z) = g\,\frac{1 - 2\,r_z \cos(2\,\pi \cdot f_z/f_s)\,z^{-1} + r_z^2\,z^{-2}}{1 - 2\,r_P \cos(2\,\pi \cdot f_P/f_s)\,z^{-1} + r_P^2\,z^{-2}}$$

From this we derive the system equation

$$y(k) = g(x(k) - 2r_z\cos(2\,\pi \cdot f_z/f_s)x(k-1) + r_z^2 x(k-2))$$
$$+ 2r_P\cos(2\,\pi \cdot f_P/f_s)y(k-1) - r_P^2 y(k-2)$$

The figure below shows the amplitude response for the values $f_z = 5000$ Hz ($\Omega = 0.71$), $f_P = 13000$ Hz ($\Omega = 1.85$), $f_s = 44100$ Hz, $r_z = 1.1$, $r_P = 1.03$. ($\to$ Manipulate)

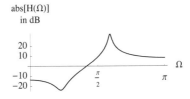

3.3 Systems and Filters

So-called *comb filters* are straightforward to realize and interesting for musical applications. By feeding the output signal delayed by n sample periods back into the system, one generates equally spaced resonances, that is, the filter produces a harmonic spectrum.

$$y(k) = b_0 x(k) + a_1 y(k-n)$$

The transfer function is $H(z) = \dfrac{1}{1 - a \cdot z^{-n}}$.

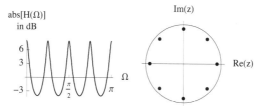

The poles correspond to the n roots of the equation $z^{-n} = 1/a$. When a is negative, the comb filter generates spectra having only odd partials.

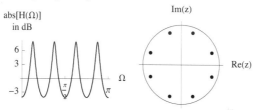

By using a so-called *notch filter*, it is possible to suppress individual frequencies completely while passing all other frequencies. This is done by placing a zero at the desired frequency on the unit circle and a pole on the line connecting this point to the circle's center. The zero suppresses the desired frequency and the position of the pole determines the size of the notch (the closer the pole is to the zero, the narrower the notch).
(→ Manipulate) (→ *IIR_z-plane.maxpat*) (→ *IIR_Filter_2.maxpat*)

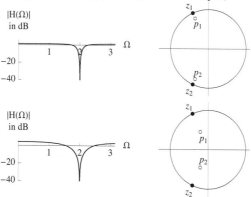

To compute the transfer function for the general case, we write the poles and zeros in polar coordinates: $z_1 = e^{i\Omega}$, $z_2 = e^{-i\Omega}$, $p_1 = be^{i\Omega}$ und $p_2 = be^{-i\Omega}$.

The transfer function is

$$H(z) = \frac{(z-e^{i\Omega})(z-e^{-i\Omega})}{(z-be^{i\Omega})(z-be^{-i\Omega})} = \frac{z^2-(e^{i\Omega}+e^{-i\Omega})z+e^{i\Omega}e^{-i\Omega}}{z^2-b(e^{i\Omega}+e^{-i\Omega})z+b^2 e^{i\Omega}e^{-i\Omega}}.$$

Since $e^{i\Omega}+e^{-i\Omega} = 2\,\mathrm{Re}(e^{i\Omega})$ and $e^{i\Omega}e^{-i\Omega} = e^{i\Omega-i\Omega} = e^0 = 1$, it follows that

$$H(z) = \frac{z^2-2\,\mathrm{Re}(e^{i\Omega})z+1}{z^2-2b\,\mathrm{Re}(e^{i\Omega})z+b^2} = \frac{1-2\,\mathrm{Re}(e^{i\Omega})z^{-1}+z^{-2}}{1-2b\,\mathrm{Re}(e^{i\Omega})z^{-1}+b^2 z^{-2}}.$$

The filter coefficients can be read out of the equation above:

$$y(k) = x(k) - 2\,\mathrm{Re}(e^{i\Omega})x(k-1) + x(k-2) + 2b\,\mathrm{Re}(e^{i\Omega})y(k-1) - b^2 y(k-2)$$
$$= x(k) - 2\cos(\Omega)x(k-1) + x(k-2) + 2b\cos(\Omega)y(k-1) - b^2 y(k-2)$$

With the real zero $z_1 = 1$ and a real pole p_1 near the zero, we obtain a so-called *DC blocker*.

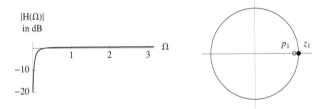

(→ Normalized DC blocker) (→ IIR_Filter_2.maxpat)

If the zeros are not placed on the unit circle but nearer to its center than the pole, we get a so-called *anti-notch* or *peak filter*.
(→ Manipulate) (→ IIR_Filter_2.maxpat) (→ IIR_z-plane.maxpat)

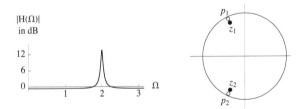

The filter's system equation is:

$$y(k) = x(k) - 2a\cdot\cos(\Omega)x(k-1) + a^2 x(k-2) + 2b\cdot\cos(\Omega)y(k-1) - b^2 y(k-2)$$

We obtain a combination of comb and notch or peak filters by producing zeros at the frequencies of the poles of the filter described above. The transfer function is then:

$$H(z) = \frac{1-az^{-n}}{1-bz^{-n}}.$$

3.3 Systems and Filters

The system equation is:

$$y(k) = (x(k) - ax(k-n) + by(k-n))$$

The illustration below shows the filter's amplitude response for $a = .85$ and $b = .98$ (left) and $a = .98$ and $b = .85$ (right). (→ Manipulate) (→ IIR_Filter_2.maxpat)

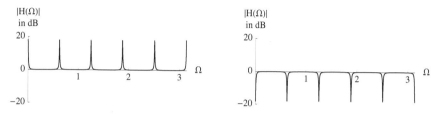

In certain situations filters are required that do not affect a signal's spectrum, that is, that have a constant amplitude response $|H(z)| = k$, but that affect a signal's phase. One way to construct a so-called *all-pass filter* is, for every pole of a filter within the unit circle, to place a zero outside the unit circle which exactly cancels the pole's effect on the filter's amplitude response. We can do this by choosing the same argument for the zero z_i as for the corresponding pole ($\arg(z_i) = \arg(p_i)$), that is, placing the zero on the extension of the straight line joining the center of the circle and the pole, and choosing the modulus of z_i to be the reciprocal of the modulus of p_i: ($|z_i| = 1/|p_i|$) (see [32] p. 114).
(→ Manipulate) (→ IIR_z-plane.maxpat)

In the following example, the pole is at $.7e^{i\pi/6}$, the corresponding zero is at $e^{i\pi/6}/.7$.

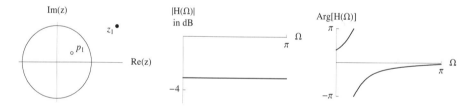

All-pass filters can have several poles and corresponding zeros:

$$p_1 = .7e^{i\pi/6}, z_1 = 1/.7 \cdot e^{i\pi/6}, p_2 = .2e^{i\pi/7}, z_2 = 1/.2 \cdot e^{i\pi/7}, p_3 = .37e^{i\pi/1.6}, z_3 = 1/.37 \cdot e^{i\pi/1.6}.$$

In both cases, the output has to be divided by the constant k to obtain an all-pass filter with the amplitude response $|H(z)| = 1$.

The following examples show that recursive filters do not necessarily give infinite impulse responses. (It can be shown that every FIR filter can also be implemented recursively, see

[65] p. 115). A system that computes the moving average of N successive values can be realized non-recursively as follows:

$$y(k) = \frac{1}{N}(x(k) + x(k-1) + x(k-2) + \ldots + x(k-N-1)) = \frac{1}{N}\sum_{i=0}^{N-1} x(k-i)$$

Instead of computing the average of the last N values at every point k, we can add to and subtract from the last average the $1/N$-th of the new value $x(k)/N$ and $1/N$-th of the $(k-N)$-th value $x(k-N)/N$ respectively.

$$y(k) = y(k-1) + \frac{1}{N}(x(k) - x(k-N))$$

(→ Amplitude Response) (→ *Filter_fexpr.mapat*)

We saw in Chapter 3.3.2.4 how we can compute the impulse response $h(n)$ of a non-recursive filter from a given frequency response (the frequency sampling method).

$$h(n) = \frac{1}{2\pi} \int_{-\pi}^{\pi} H(e^{i\Omega}) e^{in\Omega} \, d\Omega$$

If one defines the frequency response for M equally distributed frequencies $2\pi/M$, then the $h(n)$ are exactly the components of the IDFT of the frequency response.

$$h(n) = \frac{1}{M} \sum_{k=0}^{M-1} H(k) e^{i \cdot 2\pi k n/M}$$

If in the z-transform of the frequency response

$$H(z) = \sum_{n=0}^{M-1} h(n) z^{-n}$$

we substitute $h(n)$ by the expression above, we have

$$H(z) = \sum_{n=0}^{M-1} \left(\frac{1}{M} \sum_{k=0}^{M-1} H(k) e^{i \cdot 2\pi k n/M} \right) z^{-n}$$

and after rearranging we have

$$H(z) = \frac{1-z^{-M}}{M} \sum_{k=0}^{M-1} \frac{H(k)}{1 - e^{i \cdot 2\pi k/M} z^{-1}}.$$

This means that the filter can be realized as a comb filter with the frequency response

$$H_1(z) = \frac{1-z^{-M}}{M}$$

connected in series to M parallel recursive filters with the frequency response

$$H_2(z) = \sum_{k=0}^{M-1} \frac{H(k)}{1 - e^{i \cdot 2\pi k/M} z^{-1}}.$$

All those summands whose frequency response $H(k)$ at the given frequency is zero disappear.

In the following example, we realize a low-pass filter with a given frequency response $H(k)$ at $M = 20$ points. We first compute the impulse response $h(k)$.

```
M = 20; H = {1,1,1/2,0,0,0,0,0,0,0,0,0,0,0,0,0,0,0,1/2,1};
h =  1/√M  InverseFourier[H];
```

3.3 Systems and Filters

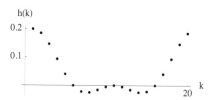

From this we can realize a non-recursive filter having the coefficients $h(k)$:

$$y(k) = h(0)x(k) + h(1)x(k-1) + \ldots + h(19)x(k-19)$$

(→ Amplitude response of the non-recursive filter)

Now we can realize the same FIR filter recursively as follows. The system equation of the (non-recursive) comb filter with the frequency response H_1 is:

$$y(k) = \frac{1}{20}(x(k) + x(k-20))$$

Because $H(k)$ is zero at most points, we have only the recursive partial systems (single pole filters) with the frequency responses:

$$H_{2,0} = \frac{1}{1-e^{-i\omega}}, H_{2,1} = \frac{1}{1-e^{-i\cdot 2\pi/M}e^{-i\omega}}, H_{2,2} = \frac{.5}{1-e^{-i\cdot 2\cdot 2\pi/M}e^{-i\omega}},$$

$$H_{2,18} = \frac{.5}{1-e^{-i\cdot 18\cdot 2\pi/M}e^{-i\omega}}, H_{2,19} = \frac{1}{1-e^{-i\cdot 19\cdot 2\pi/M}e^{-i\omega}}.$$

Because of the symmetry $H(k) = H(M-k)$, we can combine groups of two summands and realize them as two-pole filters:

$$\frac{H(k)}{1-e^{-ik\cdot 2\pi k/M}e^{-i\omega}} + \frac{H(M-k)}{1-e^{-i(M-k)\cdot 2\pi k/M}e^{-i\omega}} =$$

$$\frac{H(k) + H(M-k) - \left(H(k)\left(1-e^{ik\cdot 2\pi k/M}\right) + H(M-k)\left(1-e^{-ik\cdot 2\pi k/M}\right)\right)e^{-i\omega}}{1 - 2\cos(k\cdot 2\pi/M)e^{-i\omega} + e^{-2i\omega}}.$$

For our example we obtain

$$H_{2,1} + H_{2,19} = \frac{-1.902\, e^{i\omega} + 2\, e^{2i\omega}}{1 - 1.902\, e^{i\omega} + e^{2i\omega}}, H_{2,2} + H_{2,18} = \frac{-0.809\, e^{i\omega} + e^{2i\omega}}{1 - 1.618\, e^{i\omega} + e^{2i\omega}}.$$

(→ Calculation)

The following illustration shows the amplitude response of the filter.

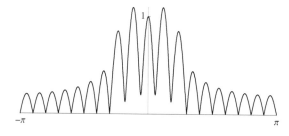

3.3.3.6 Filter Design by Transforming Analog into Digital Systems

In analog electronics, filters have been studied since the 1930's, and methods have been developed to design filters to given tolerances. The difference equations for digital filters can be transformed into algebraic equations with the z-transform, and the poles and zeroes of the transfer function $H(z)$ can be determined; in the same way the differential equations for analog filters can be transformed into algebraic equations with the Laplace transform (3.1.4.3), and the poles and zeroes of the transfer function $G(s)$ can be determined. By a further transform the poles and zeroes of $G(s)$ can be transformed into the corresponding poles and zeroes of $H(z)$. There are standard designs for Butterworth, Chebyshev and elliptic or Cauer filters. The following figures were generated by the MATLAB program "Lowpass Filter Design" using the following values: cutoff frequency of the pass band = 500 Hz; cutoff frequency of the stop band = 500 Hz; tolerance in the pass band 10 dB and attenuation in the stop band 70 dB. The program computes the minimum order of the filter necessary to fulfill the given requirements.

The Butterworth filter is very flat in the pass band, and the transition to the stop band is monotone. The filter requires a higher order than the others.

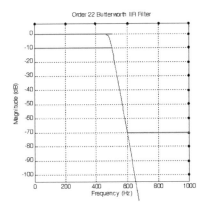

There are two types of Chebyshev filter. Both require here the same order, which is markedly lower than that required for the Butterworth filter. The first type is flat in the stop band, but there is ripple in the pass band. The second type has oscillation in the stop band but none in the pass band.

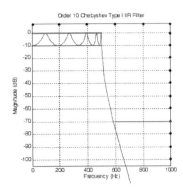
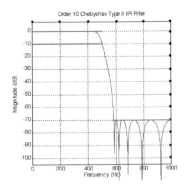

The elliptic filter needs the lowest order. There are oscillations in both the pass and the stop band (left figure below). For comparison the right figure below shows the design for an optimized non-recursive filter. It requires a somewhat higher order than any of the recursive filters.

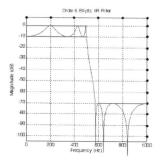
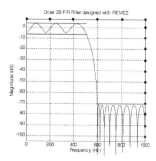

3.4 Dynamic Systems and Feedback Control

The behavior of systems that cannot be described with simple linear equations can often be simulated by computers. Complex physical processes whose principles have long been understood, but which are too complicated to compute, can be emulated and often visualized. An example is the three-body problem, where, given their initial masses and velocities, the positions of three bodies in space are to be calculated only taking account of gravitation and the laws of motion. The basic laws of gravitation and motion have been known since Newton, but with them only the simplest configurations can be exactly described mathematically. In all other cases, the evolution of the systems can only be predicted through simulation. However, nearly all natural and technical systems are considerably more complicated and are usually composed of many subsystems. In addition, the laws according to which such systems function are often only approximately or not at all understood.

Nonlinear systems occur in several areas of computer music. Early on, nonlinear techniques of sound synthesis like frequency and amplitude modulation (6.1) were used to generate complex oscillation with little computation. In the following chapters, we describe nonlinear systems that generate considerably more complicated waveforms and often show unpredictable and chaotic behavior. These waveforms can be used directly as sound but they can also be used as control signals to modify synthesis parameters. In complex composite systems, like living creatures or social systems, behavior is controlled by feedback. The system measures its own parameters and uses the results to control its functions. This feedback control is itself nonlinear (3.4.4.4).

Natural continuous systems are described using differential equations. For this reason, we begin the chapter with an introduction to differential equations. The second chapter contains explanations of concepts from chaos theory (10.3.3). Feedback control, an important application of nonlinear dynamics, is discussed in the fourth chapter.

Further Reading: Even simple nonlinear systems can show spectacular behavior. For this reason they have awakened considerable interest in the last years and there is a correspondingly large body of popular scientific literature, despite the complexity of the field. Especially recommended are the books by Heinz-Otto Peitge, Harmut Jürgens and Dietmar Saupe, for instance Fractals for the Classroom: Introduction to Fractals and Chaos *[22]. From Calculus to Chaos: An Introduction to Dynamics by David Acheson [46] is a good introduction to the topic. For applications of fractals to music, see* Fractals in Music *by Charles Madden [29].*

3.4.1 Differential Equations

In classical mechanics, the laws of physics can be used to calculate the state of a system at an arbitrary time. Many of these laws take the form of differential equations.

Further Reading: Textbooks on technical mathematics and collections of exercises may prove helpful, for example Mathematics for Engineers and Scientists *by Alan Jeffrey [89],* Handbook of Mathematics for Engineers and Scientist *by A. D. Polyanin and A. V. Manzhirov [90] and* Schaum's Outline of Advanced Mathematics for Engineers and Scientists *by Murray R. Spiegel [1]. For numerical methods see for example* Numerical Methods for Scientists and Engineers *by R. W. Hamming [91].*

3.4.1.1 Introductory Example, Phase Space

Using the standard example of an oscillating spring, we will first show how to derive an equation of movement from physical laws and how to represent the state of the system. Let the displacement of a mass suspended on a spring at time t be $x(t)$. Since the acceleration of the mass is proportional to its displacement but acts in the opposite direction, we have for the displacement $x(t)$ the equation $\ddot{x} = -kx$ (3.1.4.1). The solution of the equation is a linear combination of sine and cosine functions.

```
DSolve[x''[t]==-k*x[t],x[t],t]
```

$$\left\{\left\{x[t] \to C[2]\, \text{Cos}\left[\sqrt{k}\ t\right] + C[1]\, \text{Sin}\left[\sqrt{k}\ t\right]\right\}\right\}$$

This sum can also be written as a sine function $x(t) = c \cdot \sin(\omega t + \phi)$ with angular frequency $\omega = \sqrt{k}$, phase constant ϕ and amplitude c.

The instantaneous state of the system is described by the position and velocity of the mass. It is represented by the coordinates x and $v = \dot{x}$ in the so-called *state plane*. More generally, one speaks of *state space* or *phase space*. The representation in phase space allows a new interpretation of the sine function as the projection of circular motion. The following illustration shows in the middle a snapshot of the system, to the right a plot of the displacement against time and to the left the state plane with the so-called *trajectory* of the motion. The x-coordinate represents the displacement, which is described here by the equation $c \cdot \cos(t)$. The middle figure shows only the instantaneous displacement, the figure on the right shows the change of the displacement over time, and the figure on the left shows the phase space with displacement and velocity but not time.

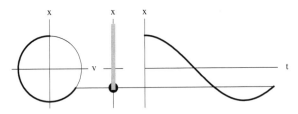

The trajectory's evolution over time can be shown for a one- or two-dimensional state space using so-called *extended phase space* with time as an additional coordinate. Other ways to show the trajectory's evolution are by animation or as a discontinuous plot, where the distance between dots gives an indication of the trajectory's velocity (3.4.1.5).

3.4 Dynamic Systems and Feedback Control

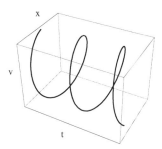

The following figure shows one moment of an animation representing a damped oscillation. The trajectory of the oscillation is a spiral. (→ Animation)

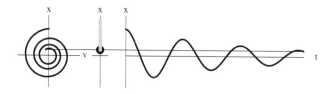

3.4.1.2 The Types of Differential Equations and Further Definitions

The example above shows that a dynamic system can be described using differential equations to define the relationship between the various state variables like displacement, velocity and acceleration, or temperature, volume and pressure. Because we are interested in the temporal evolution of dynamic systems, we shall only consider functions of time $x(t)$. In what follows, we will use Mathematica to solve the differential equations analytically or numerically. Therefore, we shall not discuss the extensive theory of the analytic solution of differential equations (see, for example, [1], [3] or [9]), but only indicate some of the important types of differential equations and their solution sets.

An equation containing derivatives up to the nth order of a function $x = x(t)$ is called an *ordinary differential equation of order n* or an ordinary nth-order differential equation. Equations containing partial derivatives of several variables are not ordinary and are called *partial differential equations*. The example above $\ddot{x} = -kx$ is an ordinary differential equation of order 2. In what follows, we shall only consider ordinary differential equations and leave off the term "ordinary". A first-order differential equation has the form $dx/dt = \dot{x} = f(x, t)$. The derivative $\dot{x}(t)$ is a function of the unknown function $x(t)$ and of time t. If the derivative depends only on x, the equation is called *autonomous*. It is easy to interpret the equation geometrically. Since dx/dt corresponds to the slope of the required function and is itself a function of x and t, we can draw the slope at every point in the x-t plane. The solutions of the equation are curves which follow this so-called *slope* or *direction field*.

A simple growth model can be described by a straightforward autonomous first-order differential equation. Let $x(t)$ be the size of a population. If the population increase dx per time $dx/dt = \dot{x}$ is proportional to the population x, we have the equation $\dot{x} = f(x) = \lambda x$, the solution of which is $x(t) = ce^{\lambda t}$, where c is an arbitrary constant.

```
DSolve[x'[t] == λ*x[t],x[t],t]
```

$\{\{x[t] \to E^{t\lambda} C[1]\}\}$

Three of the infinitely many solutions are shown in the figure below. If we substitute a number for c, we get a *particular solution*. If the initial value of the function $x(t)$, i.e. $x(0)$, is given, we speak of an *initial value problem*, if a value $x(t_1)$ at a particular time t_1 is given, we speak of a *boundary value problem*. In our example, we obtain the constant c which belongs to the initial value $x(0)$ by setting $t = 0$ in the equation above: $x(0) = ce^{\lambda \cdot 0} = c$.

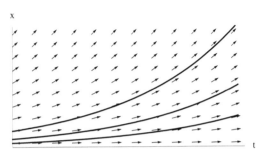

The solution of the first-order non-autonomous differential equation $\dot{x}(t) = f(x, t) = t/2 - x$ is $x(t) = -1/2 + t/2 + ce^{-t}$.

```
DSolve[x'[t]==t/2-x[t],x[t],t]
```

$$\left\{\left\{x[t] \to \frac{1}{2}\,(-1+t)\,+\,E^{-t}\,C[1]\right\}\right\}$$

The following figure shows the slope field and three special solutions of the differential equation. The constant c is computed from the initial value $x(0) = -1/2 + 0/2 + ce^{-0}$ to be $c = x(0) + 1/2$.

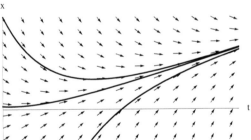

The nth derivative of $x(t)$ appears in differential equations of order n. Because n constants appear in the solution of an nth order differential equation, n initial or boundary values must be given to uniquely determine a special solution. In the introductory example the displacement was given an initial value and the velocity was set to zero, as the phase space drawing shows.

A nth order differential equation can be decomposed into a system of n first-order differential equations. For example the equation $\ddot{x} = -kx$ can be decomposed into the equations $\dot{x} = v$ and $\dot{v} = -kx$ (8.1.2.4).

Differential equations are called *linear* when they have the form

$$a_0 x + a_1 \frac{dx}{dt} + a_2 \frac{d^2 x}{dt^2} + a_3 \frac{d^3 x}{dt^3} + \ldots = R(t),$$

where the a_i are constants or functions of time $a_i = a_i(t)$. If $R(t) = 0$, one speaks of a homogeneous differential equation, otherwise the equation is inhomogeneous. Linear differential equations of arbitrary degree can be written as systems of first-order linear differential equations.

Depending on the coefficients a_1 and a_0, the trajectories of the equation $\ddot{x} + a_1\dot{x} + a_0x = 0$ trace different paths. The illustration below shows the various possible types of trajectory in phase space (so-called *phase portraits*), together with the names associated with the various behaviors. The phase portrait of closed trajectories like those of an undamped oscillation (3.4.1.1) shows a *stable center*. A damped oscillation gives a *stable spiral* (also inward spiral) with the trajectories moving in toward the center of the spiral, the *stable fixed point*. The trajectories of an unstable spiral move away from the fixed point. When the fixed point is unstable, the least movement disturbs the system's state of equilibrium. At a *node*, the trajectories approach a stable fixed point or move away from an unstable fixed point, without circling the fixed point in phase space. When the phase portrait shows a *saddle point*, the behavior of the trajectory near the fixed point depends on the trajectory's position in phase space. The fixed point acts as an *attractor* in some cases, in others it repels the trajectory (3.4.1.5).

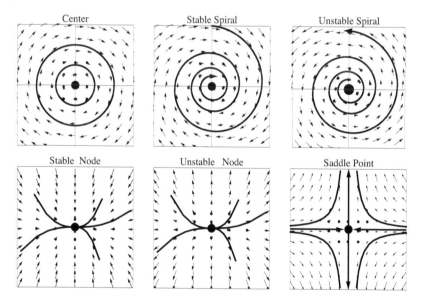

The following figure shows the behavior of the equation $\ddot{x} + a_1\dot{x} + a_0x = 0$ plotted against the factors a_1 and a_0. The equation of the undamped oscillation is

$$\ddot{x} = -kx \quad \text{or} \quad \ddot{x} + kx = 0,$$

which implies $a_1 = 0$ and $a_0 = k > 0$. The equation of the damped oscillation is

$$\ddot{x} = -kx - d\dot{x} \quad \text{or} \quad \ddot{x} + d\dot{x} + kx = 0,$$

which implies $a_1 = d > 0$ und $a_0 = k > 0$. If there is little damping, so that $a_0 > a_1^2/4$, the phase portrait shows a spiral. If there is greater damping, the system without oscillation approaches the rest position, and the phase portrait shows a stable node.

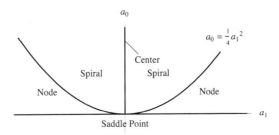

The function $R(t)$ in the linear differential equation $a_0 x + a_1 dx/dt ... = R(t)$ is known as a *forcing function*. The forcing function describes an external influence upon the system defined by the differential equation. The spring oscillation excited by an external sinusoidal force is described by the equation $\ddot{x} + kx = A \cdot \sin(\omega t)$.

3.4.1.3 Stationary System Analysis, Characteristic Curves

It is difficult to describe the dynamic behavior of complex systems like animals and social systems with formulas. Analysis often has to content itself with the description of the structure of the interaction of various elements of the system (see [33] pp. 106-130) and with the graphic representation of the stationary behavior of the subsystems. This representation is called the *characteristic curve* and shows the output values of the system as function of constant input values.

If one compresses a spring with a certain force, the spring is shortened by an amount proportional to the force applied. It is generally true that the reaction of a stable linear system to a constant excitation is both itself constant and proportional to the excitation. If the effect of the force occurs suddenly, the system needs a certain time to come to rest. The following figures show the response of the linear system $y(k) = x(k) - .8y(k-1)$ to the step functions $x(k) = 1$ and $x(k) = 2$ for $k > 0$.

The signals $y(k)$ converge toward $\bar{y} = c/2$ The function $\bar{y}(c)$ is called the *characteristic* of the system, its graphic representation is the *characteristic curve*. The characteristic curve of a linear system is a straight line.

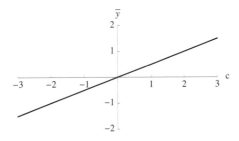

3.4 Dynamic Systems and Feedback Control

The following illustration shows characteristic curves of nonlinear systems.

In example a), negative input signals excite no response. Pressure sensors, for example, show such behavior. Example b) shows the characteristic curve of a system which only responds when a threshold is exceeded. Many sense organs, but also many sensors, demonstrate such a behavior. In systems with a logarithmic characteristic curve c), only positive input values are allowed. Again, there are examples from sensory perception for this behavior. In example d) the curve is linear in mid-range, while at the boundaries a kind of saturation occurs. An elastic spring becomes stiff under extreme tension or pressure. This kind of characteristic curve is used for audio limiters to avoid distortion. Characteristic curves can be step functions, as in example e). Example f) shows that even in stable systems there is not always a unique characteristic curve. Here there is so-called *hysteresis*, and the characteristic curve depends upon the direction from which an input value is reached. The current state depends on the system's history. In this example, the right fork of the curve is valid in mid-range when the input signal begins with negative values under a specific threshold, the left fork is valid when the input begins with positive values (3.4.2.2, 3.4.3.4).

The response of a linear system does not depend on the magnitude of the signals. That is why in the preceding chapters we could reduce the response of a system to its impulse response, that is to its reaction to a single pulse of magnitude 1. Nor does the frequency response of a system depend on the amplitude of the input signals. In nonlinear systems, on the other hand, even the response to a constant signal is dependent on the signal's amplitude. Other properties like stability, dynamic behavior and frequency response also depend on the input's amplitude.

Frequently the output of a system is a simple one-dimensional signal, that is, a continuous or discrete function $y(t)$ or a sequence $y(k)$, while the magnitude of the input is vectorial. The output of a microphone, for example, depends not only on the sound source recorded, but also on the distance between the microphone and the sound source and on the direction from which the sound reaches the microphone. The complicated nonlinear diagrams whose lines connect points of equal amplitude are typical examples of so-called *directional characteristics*.

If one knows how a system works, one can calculate the characteristic curve or determine it using simulations. When analyzing a system whose details one does not know, the task is just the opposite, that is, to describe mathematically an empirically found characteristic curve and from the mathematical description to infer details of the system's mode of functioning. The same process is used in the design of systems which should adhere to a given characteristic curve.

3.4.1.4 Numerical Methods, Difference Equations

Even simple nonlinear differential equations can often not be solved analytically, but approximately be computed point-wise.

We shall demonstrate the so-called *Euler method* on a first-order differential equation $\dot{x} = \lambda x$ (first example in 3.4.1.2). For the given initial value $x_0 = x(0)$, we can approximate the value x_{t_1} after time span h at time $t_1 = h$ by adding to x_0 the quantity $h \cdot f(x_0, 0)$, that is the slope of the curve at point $(0, x_0)$ multiplied by the duration h.

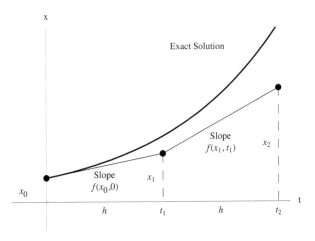

In this way, we obtain continuously from the last computed value and the slope given by the differential equation the next value:

$$x_1 = x_0 + h \cdot f(x_0, 0), \; x_2 = x_1 + h \cdot f(x_1, 1) \; \ldots \; x_{n+1} = x_n + h \cdot f(x_n, t_n).$$

Formally, this discretization of time derives a *difference equation* from the differential equation by substituting $(x_{n+1} - x_n)/h$ for $\dot{x} = dx/dt$. From $\dot{x} = f(x, t)$ we obtain

$$\frac{x_{n+1} - x_n}{h} = f(x_n, t_n) \text{ or } x_{n+1} = x_n + h \cdot f(x_n, t_n).$$

We can improve the method by computing the new value x_{n+1} not with the slope at time t_n but with the average of the slopes at times t_n and t_{n+1}, that is $.5(f(x_n, t_n) + f(x_{n+1}, t_{n+1}))$. We then obtain

$$x_{n+1} = x_n + \frac{1}{2}h(f(x_n, t_n) + f(x_{n+1}, t_{n+1})).$$

Since we do not yet know the value of x_{n+1}, we substitute the above approximation in the right side of the equation and obtain the *improved Euler method*:

$$x_{n+1} = x_n + \frac{1}{2}h(f(x_n, t_n) + f(x_n + h \cdot f(x_n, t_n), t_{n+1})).$$

If we write for the increase $h \cdot f(x_n, t_n)$ simply c_1 and x and x_{new} for x_n and x_{n+1} respectively, then the two methods and another improvement, the so-called *Runge-Kutta method*, can be represented as follows:

3.4 Dynamic Systems and Feedback Control

Euler's Method	Improved Euler's Method (or Heun's Method)	Runge–Kutta Method
$c_1 = h \cdot f(x, t)$	$c_1 = h \cdot f(x, t)$ $c_2 = h \cdot f(x + c_1, t + h)$	$c_1 = h \cdot f(x, t)$ $c_2 = h \cdot f\left(x + \frac{1}{2} c_1, t + \frac{1}{2} h\right)$ $c_3 = h \cdot f\left(x + \frac{1}{2} c_2, t + \frac{1}{2} h\right)$ $c_4 = h \cdot f(x + c_3, t + h)$
$x_{new} = x + c_1$	$x_{new} = x + \frac{1}{2}(c_1 + c_2)$	$x_{new} = x + \frac{1}{6}(c_1 + 2c_2 + 2c_3 + c_4)$

Using the equation $\dot{x} = f(x, t) = tx + t^2 + x$, let us examine the effect of various step sizes, comparing at the same time the different methods. We can use Euler's method to generate a list of successive function values with the initial value $x = -.31$ and using an increment $h = .3$ by giving the following Mathematica commands:

```
f[x_, t_] = t*x + t^2 + x; x = -.31; h = .3; t = 0;
l1 = Table[c1 = h*f[x, t]; t += h; x = x + c1; N[x, 3], {i, 0, 10}]
```

{-0.403,-0.533,-0.681,-0.826,-0.94,-0.969,-0.812,-0.244,1.24,4.79,13.2}

The figure below was generated with an initial value of $-.31$ and the increments $h = .1$, $h = .3$ and $h = .9$.

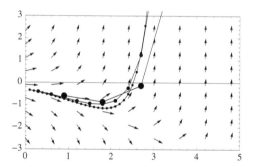

Even using small step sizes, it is possible to compute inacurate solutions, particularly at critical points in the function. The following figure again shows three numerical solutions of the differential equation above. The initial value is $x = -.325$, the increments are $h = .05$, $h = .1$ and $h = .15$.

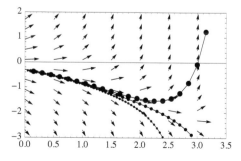

To compare the various methods, we generate lists of successive function values using the improved Euler's method and the Runge-Kutta method with the initial value $x = -.31$. The following figure shows that both methods perform better than Euler's method, even when their increments are twice as large.

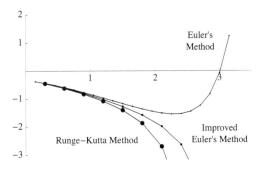

These methods can also be applied to nth-order differential equations (or equivalently, to systems of n first-order differential equations). Let us compare two ways to compute the solution of the differential equation of the undamped oscillation $\ddot{x} = -kx$ and of the equivalent system of differential equations $\dot{x} = v$; $\dot{v} = -kx$ from given initial values.

1) Because the second derivative corresponds to the velocity differences per time unit, we set

$$\ddot{x} = \frac{x_{n+1} - x_n}{h} - \frac{x_n - x_{n-1}}{h} = \frac{x_{n+1} - 2x_n + x_{n-1}}{h}.$$

Applied to the equation $\ddot{x} = -kx$ we obtain

$$x_{n+1} = (2 - hk)x_n - x_{n-1}.$$

2) Assuming the representation as simultaneous equations $\dot{x} = v$; $\dot{v} = -kx$, we obtain

$$v_{n+1} = v_n - hx_n \text{ and } x_{n+1} = x_n + hv_{n+1}.$$

The following figures each show one period of an oscillation. In the figure on the left the increment is one tenth of the period, in the figure on the right one hundredth. The dotted line was computed by the first method, the dashed line by the second.

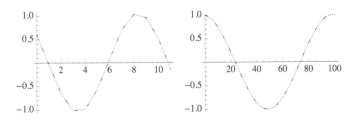

Difference equations describe discrete systems in the sense of Chapter 3.3. In the terms of Chapter 3.3.1.1, we can describe the system above as follows. The output of the system $y(k)$ corresponds to the displacement x_n. The initial displacement $x_0 = 1$ corresponds to the input $x(k)$. The input signal is 1 at time $t = 0$, 0 otherwise, and thus corresponds to the pulse $\delta(k)$ (3.2.1.2). Hence, the equation above, $x_{n+1} = (2 - hk)x_n - x_{n-1}$ with the initial value $x_0 = 1$, corresponds to the system $y(k) = (2 - c)y(k - 1) - y(k - 2) + \delta(k)$. In keeping with the usual representation, we have reduced the time indices by one. The oscillations produced above correspond to the impulse response of the IIR filter described in Chapter 3.3.3.5. As the example shows, differential equations can be transformed into difference equations describing digital systems. Because an equation's next value is always computed using one or more previous values, these systems are always IIR filters. The input signal $x(k)$ corresponds to the differential equation's forcing function $R(t)$. The impulse response of the filter corresponds to the solution of the homogenous differential equation $a_0 x + a_1 dx/dt \ldots = 0$ with the initial value $x(0) = 1$. The output produced by the filter from the digitized signal $x(k) \approx R(t)$ corresponds to a special solution of the non-homogeneous differential equation $a_0 x + a_1 dx/dt \ldots = R(t)$.

Mathematica offers various functions for numerical computation. Differential equations and systems of differential equations with given initial or boundary values can be solved with the command NDSolve[].

```
NDSolve[{x'[t] == t*x[t] + t^2 + x[t], x[0] == -.325}, x, {t, 0, 2.5}]

{{x→InterpolatingFunction[{{0.,2.5}},<>]}}

Plot[Evaluate[x[t] /. %], {t, 0, 2.5}]
```

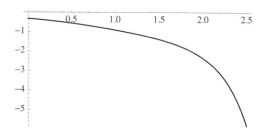

3.4.1.5 The Pendulum

Let us illustrate some properties of nonlinear differential equations using the example of the pendulum. Let a mass (or bob) m be fastened by a (massless) rod to a point Z so that it can turn freely in one plane about Z. There are two rest positions. The lower rest position is stable, the upper one is unstable. In the upper rest position, the slightest application of force causes the pendulum to oscillate. A pendulum's frequency depends on the amplitude of its oscillation. Hence, near the upper rest position, the pendulum almost stands still. When the pendulum oscillates fast enough, the bob goes over the top and its movement becomes circular. The faster the pendulum turns, the more regular the movement becomes.

To describe the pendulum's state, we determine the angle φ and the angular velocity $\dot\varphi$. The force F, which acts in the direction in which the bob can move, corresponds to the tangential component of the gravity force G, whose value is $-mg \cdot \sin(\varphi)$. Therefore, the acceleration is proportional to $-\sin(\varphi)$, and the related differential equation is

$$\ddot\varphi = -k \cdot \sin(\varphi).$$

The following figure shows on the left the pendulum and the acting forces and on the right the pendulum's potential energy V as a function of the displacement.

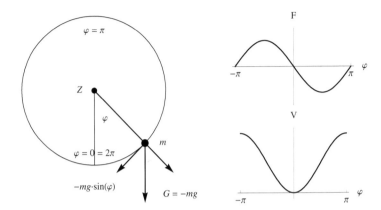

Using Euler's method we generate solutions of the nonlinear differential equation from above. In order to compare the various behaviors of the pendulum depending on its initial state, we make animations showing to the left the phase space, a representation of the system in the middle and to the right the horizontal displacement of the pendulum as a function of time. When the amplitude is small, the displacement nearly describes a sine wave. For small displacements, we can substitute φ for $\sin(\varphi)$ and we get the differential equation for harmonic oscillation $\ddot{\varphi} = -k\varphi$. (→ Animation)

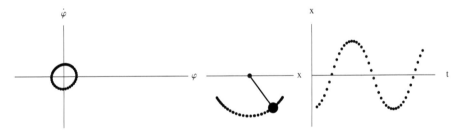

The second animation shows an oscillation having greater amplitude. The trajectory is not circular. The position of the dots in the illustration at the left shows how fast the trajectory moves in phase space. The period of the oscillation is greater than in the example above. (→ Animation)

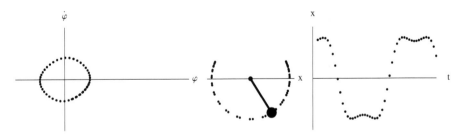

The third animation shows a pendulum whose bob goes over the top. There is no closed curve in phase space, because the angle φ is constantly increasing. The frequency of a pendulum whose bob goes over the top can be arbitrarily great. (→ Animation)

3.4 Dynamic Systems and Feedback Control

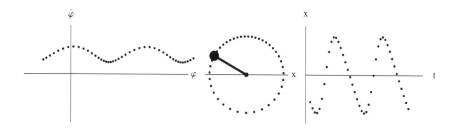

Because the system above describes a frictionless pendulum, its energy remains constant. The total energy of the system consists of the kinetic energy, which is proportional to the square of the velocity, and the potential energy, which is proportional to the height of the bob above the lowest point it can reach. This height is $-\cos(\varphi)$. If we make a three-dimensional representation of the energy as a function of $\dot{\varphi}$ and φ, we obtain a surface (to the left in the illustration below) whose isocontours (to the right) correspond to places of equal energy and thus to possible trajectories in phase space. In the middle figure, the surface has been truncated at two levels so that two isocontours form the edge of the surface.

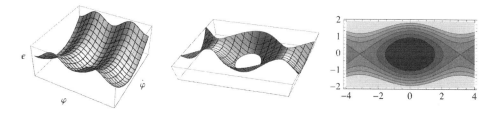

For a damped pendulum we obtain in the illustration above a continuously sinking curve. We introduce friction proportional to the velocity by adding the term $-r\dot{\varphi}$ to the differential equation. In the difference equation we introduce the corresponding term $-r(x_1 - x_0)$.

```
x2 = 2*x1-x0-dt*.2*Sin[x1]-r*(x1-x0)
```

For the animation below the initial values $x_0 = 9$ and $x_1 = -8.5$ were chosen. This gives the pendulum both an initial displacement and a relatively high initial velocity $x_1 - x_0 = .5$, which at first causes the pendulum's bob to go over the top. The movement is soon damped by friction, however. The frequency of the oscillation first gets lower until the bob no longer goes over the top and then increases again with diminishing amplitude. (→ Animation)

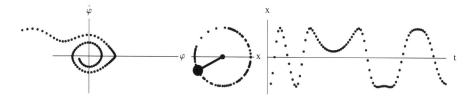

3.4.2 Fixed Points and Attractors

3.4.2.1 Fixed Points

It is often easy to determine the position and type of fixed points in simple systems. The pendulum has a stable lower rest position and an unstable upper rest point, a mass suspended on a spring has a stable middle position. In more complicated systems, it is not trivial to find the fixed points and to describe the system's behavior near the fixed points. Doing so, however, gives important insights into the system.

If one compresses an elastic rod longitudinally with increasing force, the rod will suddenly bend when a specific force is reached. The direction of the rod's bending depends on minimal deviations from the middle position. In the following example, we consider a mass that can only move vertically and is attached to springs at two anchor points. If the springs are longer than the distance of the mass to the points, the springs will be compressed. This makes the middle position unstable, and at the slightest deviation from the middle position, the mass will be pushed up or down. If the displacement is greater, on the other hand, the springs pull the mass to the middle position. If the velocity is great enough, the mass can swing freely through the middle position; if the velocity is too small, it oscillates about a point of equilibrium above or below the middle position. We shall assume that the reactive force of the spring system is proportional to $ax + bx^3$, where $a < 0$ and $b > 0$ are constant and x is the displacement. If we introduce friction proportional to the velocity, we can describe the system by the equation $\ddot{x} = -d\dot{x} - ax - bx^3$. (→ Animation)

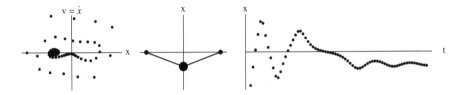

Fixed points in phase space correspond to rest positions in the system. At these points velocity and acceleration disappear. Let us call the unstable fixed point of the middle position x_0, the upper and lower rest positions x_1 and x_2 respectively. From the system equation

$$\ddot{x} = -d\dot{x} - ax - bx^3$$

we obtain for the fixed points the equation $0 = -ax_i - bx_i^3$, one solution of which is $x_0 = 0$. If $x_i \neq 0$, we can divide by x_i, obtaining $-a - bx_i^2 = 0$. Further,

$$x_i = \pm\sqrt{-a/b}, \text{ so } x_1 = \sqrt{-a/b} \text{ and } x_2 = -\sqrt{-a/b}.$$

The figure below shows to the left the function $f(x) = -ax_i - bx_i^3$ for the values used for the animation above, $a = -.5$ and $b = 1$. If the curve goes from upper left to lower right through a fixed point, the acceleration draws the mass to the rest position, and the fixed point is stable. If the curve goes from lower left to upper right, the acceleration draws the mass away from the rest position, and the fixed point is unstable. To determine the system's behavior, one only needs to take the derivative of $f(x)$ at the fixed points. If the derivative is negative, the fixed point is stable, if the derivative is positive, the fixed point is unstable (the linearization method, see e.g. [4] p. 43f.). The middle figure shows the function $f(x)$ for a normal spring with a nonlinear characteristic. The graph of the function f corresponds to the projection of the

3.4 Dynamic Systems and Feedback Control

characteristic curve onto the x-axis (3.4.1.3). If the first derivative $f'(x_i)$ at a fixed point is equal to zero, as is the case at the third fixed point in the figure at the right, then further derivatives have to be computed. So, for instance, if at x_i the second derivative $f''(x_i) \neq 0$, then x_i is unstable, if $f''(x_i) = 0$ and $f'''(x_i) < 0$, then x_i is stable.

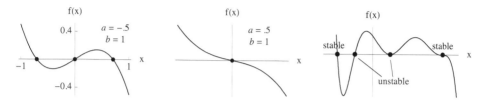

If we plot the fixed points x_i as a function of the parameter a ($b = 1$), we obtain a so-called *bifurcation diagram* for the spring system.

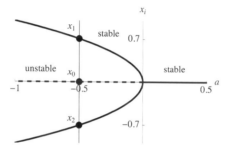

The system's behavior can be shown by plotting its energy as a function of displacement and velocity. In an undamped system, the energy remains constant and the trajectory moves along the isocontours. In a damped system, the energy diminishes constantly and the trajectory sinks continuously. The total energy of the system consists of the kinetic energy $E_k = mv^2$ plus the potential energy E_P. This corresponds to the work performed against the force of the spring $E_P = \int f(x)dx$. This rather informal description can be precisely stated and generalized by using a so-called *Lyapunov function* having the characteristics appropriate for the energy function above. If the fixed point being examined is at a minimum of this function, then the fixed point is stable (see e.g. [4] pp. 48-52).

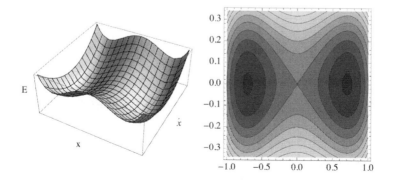

3.4.2.2 Catastrophes

Vibratory systems are called oscillators. The behavior of an oscillator acted upon by an external force c can be described by the equation for force

$$f(x) = ax_i + bx_i^3 + c$$

as well as by the potential function

$$V(x) = cx + \frac{a}{2}x^2 + \frac{b}{4}x^4.$$

To show the position of the fixed points, we plot the potential energy against the displacement. The resulting curve corresponds to the surface shown above left cut by the plane $\dot{x} = 0$. The following figure shows the cut lines for various values of c. The points mark stable positions of the system, the circles unstable ones.

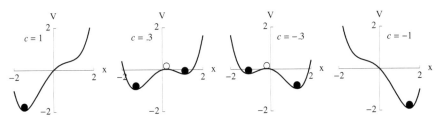

The following figure shows the position of the fixed points x_i against the constant external force c. The function is given implicitly by the equation

$$c = bx_i^3 - ax_i.$$

As the value of c gradually changes, the location of the rest position changes mostly continuously and remains on the same branch of the curve of solutions. But when c reaches a certain critical value, an abrupt transition to the other branch takes place. Because the system's current state depends on the history of the system, one speaks of a system with hysteresis (3.4.1.3).

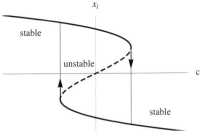

Such abrupt transitions are called *catastrophes*. Catastrophe theory is concerned with the systematic study of these transitions and with their mathematical description (see [34] or [4] pp. 171-182).

Let us consider the set of fixed points x_i for various system parameters. We shall begin with the simplest form of the potential function

$$V(x) = x^4 + ux^2 + vx.$$

3.4 Dynamic Systems and Feedback Control

This gives for the force $f(x) = 4x^3 + 2ux + v$ and for the fixed points $4x_i^3 + 2ux_i + v = 0$. We will use this representation for an overview of the elementary catastrophes, but for the moment we will continue to consider the example above, where we had for the force

$$f(x) = bx^3 + ax + c.$$

Because we are essentially interested in the relationships between the parameters a, b and c, we will set $b = 1$ and show the set of fixed points as a function of a and c. The surface implicitly defined by the equation

$$x^3 + ax + c = 0$$

is known as the surface of equilibrium and is shown below. In the following figure, one sees at the front edge of the surface the z-shaped curve illustrated above showing the dependency of the fixed points x_i on c when a remains constant. The projection of the surface of equilibrium onto the plane defined by a and c is called the *bifurcation set* and consists of those points at which the surface has a vertical tangent plane. At these points, the derivative of the surface with respect to x_i becomes zero. The point at which the two branches of the bifurcation set meet is called a *cusp*.

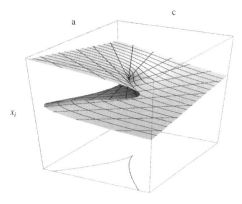

If we cut the surface with the plane $c = 0$ (the right boundary of the cube), we obtain the bifurcation diagram from Chapter 3.4.2.1, showing the dependency of the fixed points on the parameter a (below left). The figure on the right shows the bifurcation diagram when the cutting plane is $c > 0$.

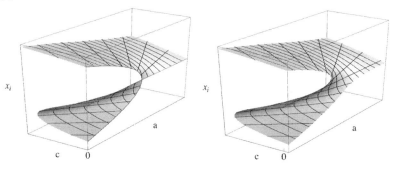

The following figures show several potential functions with constant $c = .1$ and various values of a. From left to right we see the sudden transition from the positive fixed point $x_1 > 0$ to the negative fixed point $x_2 < 0$. This corresponds, in the figure on the right above, to moving along the upper stable branch of the function and then falling down to the lower stable branch when $a = 0$.

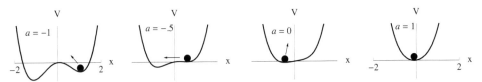

Catastrophe theory distinguishes seven fundamental types of catastrophe, defining each with the simplest form of the appropriate potential function. The first fundamental type is simpler than the cusp (which is the second fundamental type) and is called the fold catastrophe. Its potential function in its simplest form is

$$V(x) = x^3 + ux.$$

For the surface of equilibrium $V'(x) = 0$ the equation $3x^2 + u = 0$ holds. In this case, the "surface of equilibrium" is a curve in the x_i-u plane, and the bifurcation set is not a line but the point $(0,0)$.

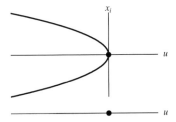

For catastrophes which are defined by higher-order potential functions, the sets of fixed points cannot be represented graphically, since they are surfaces in spaces of more than three dimensions. The third fundamental type of catastrophe is defined by the potential function $V(x) = x^5 + ux^3 + vx^2 + wx$. The implicit equation for the set of fixed points and the surface of equilibrium is $5x^4 + 3ux^2 + 2vx + w = 0$. The bifurcation set is a surface over the three-dimensional space u-v-w and is given by the equation $20x^3 + 6ux + 2v = 0$. Because of its shape it is called swallowtail catastrophe. The table below shows the names and potential functions of the seven fundamental catastrophes.

1	Fold Catastrophe	$x^3 + ux$
2	Cusp Catastrophe	$x^4 + ux^2 + vx$
3	Swallowtail Catastrophe	$x^5 + ux^3 + vx^2 + wx$
4	Butterfly Catastrophe	$x^6 + tx^4 + ux^3 + vx^2 + wx$
5	Hyperbolic Umbilic Catastrophe	$x^3 + y^3 + wxy + ux + vy$
6	Elliptic Umbilic Catastrophe	$x^3 - xy^2 + w(x^2 + y^2) + ux + vy$
7	Parabolic Umbilic Catastrophe	$y^4 + x^2y + wx^2 + ty^2 + ux + vy$

3.4.2.3 Attractors

Systems that lose energy through friction are called *dissipative systems*. A dissipative system constantly approaches a rest position without ever reaching it. Hence, the fixed points in the state space are called *attractors* for the trajectories. When energy is uniformly input to a dissipative system, the system approaches not a rest position, but rather the state of a bowed string, for example, an equilibrium between loss of energy and excitation. This state corresponds to a closed orbit in the state space, a so-called *limit cycle*. With the following Mathematica commands we can simulate the onset of an oscillation by moving one of the anchor points of the model from Chapter 3.4.2.1. The system is described by the differential equation

$$\ddot{x} = -d\dot{x} - a(x - A \cdot \sin(\omega t)/2).$$

We apply a damping factor $d = .2$ and set the constant for the reactive force to $a = .6$. The exciting oscillation is a sinusoid of amplitude $A = .35$ and frequency $\omega = .4$. ($\rightarrow$ Animation)

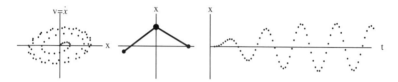

Numerical solutions of differential equations can be calculated with the Mathematica function NDSolve[]. We decompose the second-order differential equation into two first-order differential equations $\dot{x} = y$ and $\dot{y} = -dy - ax - A \cdot \sin(\omega t)$. Using the initial values $y(0) = x(0) = 0$, we pass the two differential equations to NDSolve[] and obtain the functions $x(t)$ and $y(t)$, with which we can represent the extended phase space (below left), the phase space (below middle) and the oscillation itself (below right).

```
s1 = NDSolve[{y'[t] == -.2*x[t] + Sin[.1*t] - .05*y[t],
    x'[t] == y[t], y[0] == 0, x[0] == 0}, {x, y}, {t, 0, 150}]

{{x→InterpolatingFunction[{{0.,150.}},<>],
  y→InterpolatingFunction[{{0.,150.}},<>]}}
```

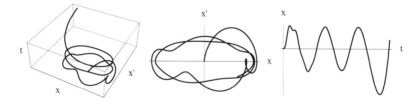

The early excitation of the higher eigenfrequency of the spring and the settling of the oscillation into the exciting frequency can be seen particularly well in the middle figure.

We can construct a system having as an attractor a unit circle as follows. We choose as state variables the angle $\varphi(t)$ and the radius $r(t)$ of a spiral. The angle should constantly increase

$$\dot{\varphi} = k_1.$$

The radius should converge to 1, so we set

$$\dot{r} = k_2 r(1 - r^2).$$

We pass the equations with various initial values to Mathematica and obtain approximations to the limit cycle, from the inside out for initial radii smaller than one, from the outside in for initial radii larger than one.

NDSolve$\left[\left\{\mathbf{r'[t]} == .2*\mathbf{r[t]}*\left(1 - \mathbf{r[t]}^2\right), \varphi\mathbf{'[t]} == 1, \mathbf{r[0]} == .3, \varphi[0] == 0\right\},\right.$
$\{\mathbf{r}, \varphi\}, \{\mathbf{t}, 0, 25\}]$

NDSolve$[\ldots \mathbf{r[0]} == 1.7, \ldots]$

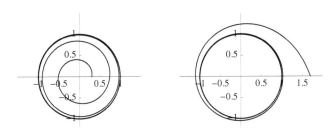

Under certain conditions, there can be several attractors in nonlinear systems. Which attractor a system's state approaches depends on the initial values given. A spring system with a cubic characteristic curve and without friction or excitation has only one fixed point. If we add friction and a sinusoidal excitation, more than one attractor can appear for certain parameters. The formula is

$$\ddot{x} = -d\dot{x} - ax^3 + A \cdot \sin(\omega t),$$

or written as a system of two first-order equations, $\dot{x} = y$ and $\dot{y} = -dy - ax^3 + A \cdot \sin(\omega t)$. Using the parameters $d = .08$, $\omega = 1$, $a = 1$ and $A = .2$, we have for the initial values 1, .2 and 0 the attractors shown in the figure below. The areas containing those points from which the various attractors are reached are separated by so-called *separatrices* (singular separatrix) (see [4]).

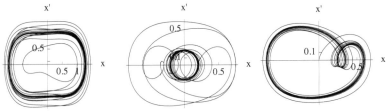

3.4.3 Chaos

3.4.3.1 Introductory Example

Nonlinear systems can behave in a chaotic manner. When they do so, they generate irregular, apparently random oscillations. The nonlinear spring system from Chapter 3.4.2.3, when excited by a periodic force, shows chaotic behavior at certain parameter values. The equation of the system in its simplest form is

$$\ddot{x} = -k\dot{x} - x^3 + A \cdot \sin(\omega t).$$

3.4 Dynamic Systems and Feedback Control

The following figure shows the oscillations of the system with the parameters $k = .05$ and $A = 7.4$. The initial values are $\dot{x} = x = 0$ (heavy line) and $\dot{x} = 0$ and $x = .01$ (light line). A comparison of the oscillations shows an essential characteristic of chaotic systems: their delicate dependence on their initial values. Even the smallest difference in the initial values can, sooner or later, lead to completely different evolutions of the same function.

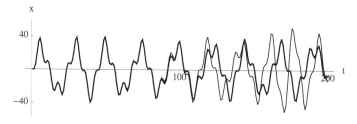

The illustrations below show the system's trajectory, to the left over a short time interval, to the right over a longer time. The trajectory does not approach a limit cycle but constantly inscribes new paths.

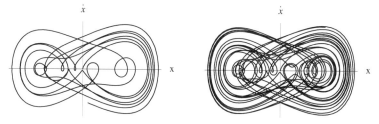

3.4.3.2 Conditions for Chaos

A given state of a system corresponds to a point in its phase space. A trajectory in a phase space cannot cross itself because at the intersection the system would follow two different developments. The trajectories of systems with two-dimensional phase spaces either approach a fixed point, form a closed curve or approximate a limit cycle (Poincaré-Bendixson theorem). Thus Chaos can only arise in a continuous system when its phase space is at least three-dimensional. For non-homogeneous differential equations, extended phase space must be used. Hence, the phase space of the systems in Chapter 3.4.1.1 and in Chapter 3.4.3.1 are three-dimensional (see e.g. [46], Chapter 11.3).

Trajectories can subtend each other in two-dimensional space, however, in discrete systems or if numerical computations are used to simulate continuous systems. Here individual points are computed, and a line is drawn between them which can subtend itself.

The simulation of a stable continuous system can behave chaotically if a too large increment is chosen for its computation. The non-homogeneous first-order differential equation

$$\frac{dx}{dt} = \dot{x} = t - x^2$$

has no simple solution. If we let Mathematica solve the equation numerically, we obtain a curve which as time t increases approximates the function $t^{1/2}$ (right), regardless of the initial values of x (left).

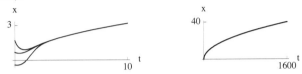

If we apply Euler's method with an increment of $dt = .06$, we obtain the same behaviour up to $t = 400$. But then the function begins to jump back and forth between two values. This process repeats faster and faster until for values $t > 500$ the function behaves chaotically (left figure below). The complete phase space $(x, \dot{x})$ is two-dimensional although the equation is non-homogeneous, because it is only of first order. The trajectory jumps back and forth with increasing amplitude ($t = 400$ to $t = 420$, middle figure below) until the lines connecting the computed points subtend each other ($t = 550$ to $t = 570$, right figure below).

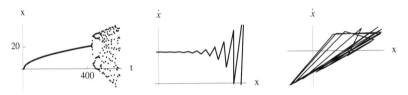

3.4.3.3 Chaos in Time-Discrete Systems

In previous chapters, we used difference equations to compute approximations for differential equations for which there are no analytical solutions. There exist, however, systems which are naturally discrete. The size of a population, for example, is always an integer. In games, system variables like advantage, points, etc. change discretely after each round. As the example in Chapter 3.4.3.2 shows, chaotic situations can suddenly appear in the simplest discrete systems. A well-studied and often described system is defined by the so-called logistic equation. The logistic equation (sometimes called the Verhulst model or logistic growth curve) is a model of population growth first published by Pierre Verhulst (1845, 1847). The original model is continuous in time, but we shall consider a discontinuous modification known as the logistic map. Using the recursive equation

$$x_n = r x_{n-1}(1 - x_{n-1}),$$

we calculate a sequence of numbers, starting from an initial value. If the initial value for x is between 0 and 1 and r is between 0 and 4, then the generated values will be between 0 and 1. We can generate the figures below using the values $r = 2.75$ and $x_1 = .1$ (left) or $x_1 = .9$ (right). Both initial values for x converge to the same limiting value, which is a stable fixed point in the map. (→ *Logistic_Map.maxpat*, *logistic_map.pde*)

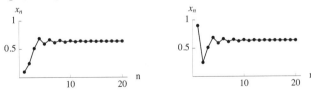

3.4 Dynamic Systems and Feedback Control

If one changes the value of the factor r, the limiting value changes as well. As r increases gradually from 0 to 3, the value of the fixed point increases slowly. The left figure below shows that from the value $r = 3$ on, the values for x begin to alternate between two values. There is not one fixed point but rather an oscillating path. As r increases, the computed values fork faster and faster. A cycle of four begins at about $r = 3.4495$ (middle figure below). The figure to the right shows the development from $r = 2.6$ to $r = 3.569$.

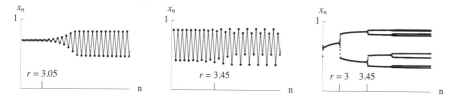

More so-called *bifurcations* lead to longer and longer cycles, until at $r = 3.6268$ chaotic behavior begins. Because the values approach the fixed points and the limit cycles regardless of the initial values, the set of points shown below is referred to as an *attractor*.

The figure below shows the attractor from $r = 3.4$ to $r = 4$. The generated values are not uniformly distributed between 0 and 1 until $r = 4$.

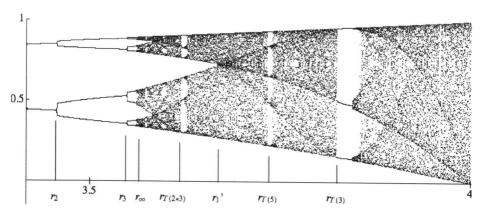

The forks into $2, 4, 8, \ldots, 2^n$ branches occur at the values $r_1 = 3$, $r_2 = 3.4495$, etc. At $r_\infty = 3.5699$ an "infinitely long period" begins. The proportion d_n/d_{n+1}, where $d_n = r_n - r_{n-1}$, converges to a universal constant. It is called the Feigenbaum constant after its discoverer, the American mathematical physicist Mitchell Feigenbaum. Its value is

$$\lim_{n \to \infty} d_n/d_{n+1} = 4.6692016\ldots$$

Between r_∞ and $r = 4$ cycles of arbitrary length generated by bifurcation occur until again chaotic behavior results. Periods of 3, 4 and 6 are clearly recognizable, and if one follows the diagram from right to left, points can be seen where the regions of x-values split. The first of these points is indicated by r_1' in the figure above.

r_1	r_2	r_3	r_∞	r_2'	$r_{T(2\cdot 3)}$	r_1'	r_{T5}	r_{T3}	r_0'
3	3.4495	3.5441	3.5699	3.5926	3.6268	3.6785	3.7379	3.8283	4

Some properties of this system can be shown geometrically. The graph of

$$f(x) = rx(1-x) = rx - rx^2$$

is a parabola opening down with zeros at $x = 0$ and $x = 1$. The function value of the vertex depends on r and is between 0 and 1 (for $r = 4$). Starting with x_0, we compute the value x_1. To use this value for the next step, we reflect it about the straight line $f(x) = x$. The entire procedure can be done geometrically by drawing a line from the starting value vertically to the parabola and then alternately drawing lines horizontally to the diagonal and vertically to the parabola. The middle figure shows how the trajectories approach each other for different initial values and small r ($r = 2.8$). The figure on the right shows how trajectories with initial values arbitrarily close to each other diverge for $r = 4$.

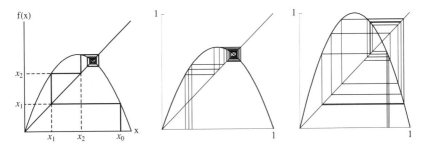

The geometrical approach shows that the intersection of the parabola with the diagonal is a fixed point. If the slope of the parabola at the intersection is greater than -1 (figures to the left and the middle above), the fixed point is stable. If the slope is less than -1 (figure to the right above), the fixed point is unstable.

To represent periodic solutions geometrically, we consider the repeatedly evaluated function $f(f...(f(x)))$. The function $f(f(x))$ gives for $r = 2.85$, as for $f(x)$ itself, only one intersection (figure at top left below). If we increase r, we obtain for $f(f(x))$ three intersections starting at $r > 3$ (middle figure bottom). The middle one of these corresponds to an unstable fixed point, the outer two to a cycle of two (middle figure top). At $r_{T3} = 3.8283$ we consider the function $f(f(f(x)))$. There are, besides an intersection which represents an unstable fixed point, three points which correspond to the cycle of three at r_{T3}.

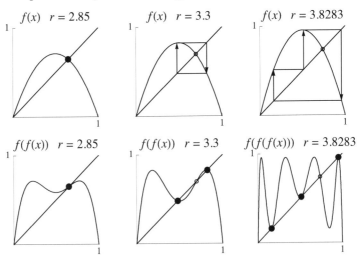

3.4 Dynamic Systems and Feedback Control

By using geometrical representation, one can often find a way to derive x or r from given conditions. For example, we can find the x-values for the cycle of two by setting $f(f(x)) = x$:

```
r = 3.3; f[x_] := r*x* (1 - x); NSolve[f[f[x]] == x, x]
```

$\{\{x \to 0.\}, \{x \to 0.479427\}, \{x \to 0.69697\}, \{x \to 0.823603\}\}$

When r is smaller than than r_{T3}, the logistic equation exhibits a behavior known as intermittence. Intermittence is characterized by long phases of nearly regular cycles, interrupted by short phases of chaotic behavior.

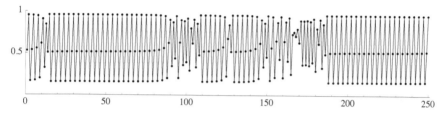

If we consider the repeated function $f(f(f(x)))$, there does not seem to be a trajectory approaching the three points of the cycle of three (figure to the left below). If we enlarge the areas marked by circles, however, we see that the function does not touch the diagonal and that therefore no fixed points occur. Trajectories approaching the function jump back and forth several times before they pass these narrow spots.

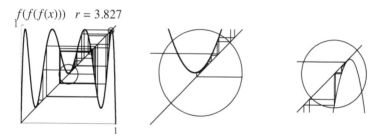

3.4.3.4 Chaos in Differential Equations

The chaotic behavior in the following examples is not due to the discrete treatment of the differential equations but rather is present in the underlying continuous system as well. Let us consider the Rössler system, a homogeneous system of three nonlinear differential equations named after its discoverer Otto E. Rössler:

$$\dot{x} = -y - z \qquad \dot{y} = x + ay \qquad \dot{z} = b + xz - cz.$$

The phase space is three-dimensional. When z is small, the trajectory is close to the xy-plane and the approximations $\dot{x} = -y$ and $\dot{y} = x$ hold. Hence it follows that $\ddot{x} = a\dot{x} - x$. This equation describes an oscillation with negative damping which in the phase space is a spiral moving outward from the center. When x becomes larger than c, the third equation causes z to increase exponentially, causing the trajectory to rise quickly out of the xy-plane. A large value for z makes $\dot{x}$ negative, so that x becomes smaller than c and the trajectory descends to the xy-plane again. The following figure shows a chaotic trajectory of the Rössler system.

```
a = .2; b = .2; c = 4.7; dt = .015; x = y = 1; z = 0;
Table[x0 = x; y0 = y; z0 = z; x += dt * (-y0 - z0);
  y += dt * (x0 + a * y0); z += dt * (b + (x0 - c) * z); {x, y, z}, {i, 0, 10 000}];
```

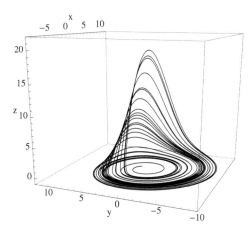

In what follows, we shall demonstrate three different ways to get a three-dimensional impression of the trajectory, all of which are valid analogously, if less obviously, for audition as well. First, we shall make use of the slightly different images the two eyes receive to create a stereoscopic representation of the trajectory. The pictures have to be looked at cross-eyed. Look at the book from a distance of about eight inches and hold a pencil between the eyes and the pictures. Two pencils will appear. Adjust the pencil so the two apparent pencils point to the two pictures. Now focus on the pencil, causing the two pictures to seem to move. Finally, focus slowly on the three-dimensional middle picture appearing behind the pencil.

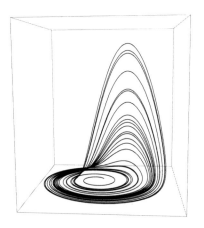
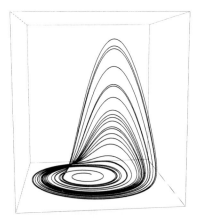

The second way to get a three-dimensional impression relates to the fact that a viewer receives more information about the form of an object when object and viewer move relative to each other than when there is no movement. The three-dimensional objects illustrated below can be moved on the computer screen with the mouse. Here one gets a spatial impression, even if one only watches with one eye. By the same token, spatial hearing can be improved by moving the head. The third possibility for visualization makes use of the fact

3.4 Dynamic Systems and Feedback Control

that changing figures can be more readily grasped than static ones. The following illustration shows three moments during the generation of the Rössler trajectory. (→ Animation)

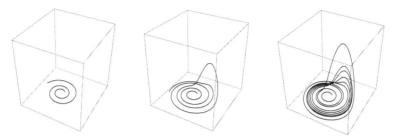

The following figures show the trajectories from above, that is, their projection in the xy-plane. The path to chaos again goes via period doubling. Now we keep a and b constant and vary c. If we chose $a = b = .2$, we get periodic solutions when c is small ($c < 3$). This means that the attractors are closed orbits (figure upper left). The figures above right and below left show a cycle of two and a cycle of four respectively. When c is greater than ~ 4.3, the Rössler attractor behaves chaotically.

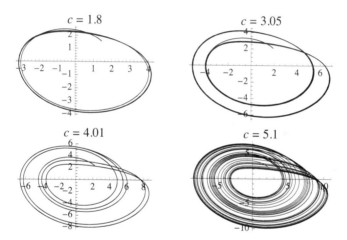

The first system in which chaotic behavior was discovered and described is the Lorenz model, which is named after climatologist Edward N. Lorenz, and simulates the coupling of convection and thermal conduction in gases or liquids. The Lorenz equations are:

$$\dot{x} = -sx + sy \qquad \dot{y} = rx - y - xz \qquad \dot{z} = xy - bz$$

(→ Stereoscopic Representation of the Phase Space)

3.4.3.5 Multiple Solutions and Discontinuities

A nonlinear system can oscillate in various ways depending on its initial conditions. We saw an example at the end of Chapter 3.4.2.3 in which the same parameters led to different limit cycles. In the following example we shall use the same system of two equations

$$\dot{y} + ky + x^3 = a \cdot \cos(\omega t) \quad \text{and} \quad \dot{x} = y.$$

The trajectories below were generated with the parameter values $a = 8$, $k = .2$ and $\omega = 1$. The figure to the left below shows a periodic solution with the initial values $y(0) = 0$ and $x(0) = 2$. The figure on the right shows a chaotic trajectory with the initial values $y(0) = 0$ and $x(0) = 1$.

```
NDSolve[
  {y'[t] + k y[t] + x[t]^3 == a Cos[ω*t], x'[t] == y[t], y[0] == 0, x[0] == 2}]
```

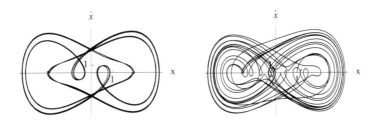

In the example that follows, identical parameters generate waveforms with different amplitudes, depending on the direction from which the parameter values were reached. We shall investigate the resonance of a nonlinearly damped oscillator which is excited by a periodic signal:

$$\ddot{x} + k\dot{x} + ax + bx^3 = a\cdot\cos(\omega t).$$

We begin with a low frequency and raise it two octaves. At first, the system resonates more and more strongly, only to collapse after about 1.2 seconds. But if we begin with a high frequency and lower it by two octaves, the resonance first gradually increases and then after about 1.5 seconds suddenly jumps to a maximum value. (→ Csound Example 3-4-3-5)

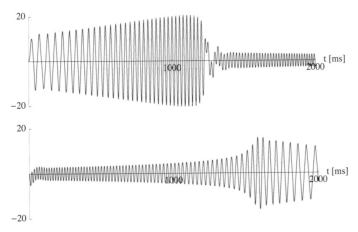

The following figure shows schematically the resonance curves of the nonlinearly damped oscillator for various amplitudes of the exciting signal. If the amplitude is very small, then the resulting amplitude is also small, and the nonlinearity plays almost no role (curve a). The resonance curve resembles closely that of a linear oscillator. As the amplitude of the excitation increases, so does that of the resulting oscillation (curve b). The eigenfrequency increases and the resonance curve takes on a shape typical for hysteresis (curve c). The amplitude of the resulting oscillation depends not only on the amplitude of the exciting frequency, but also on

3.4 Dynamic Systems and Feedback Control

whether that frequency was reached from larger or from smaller values. The diagram also shows that the decrease of amplitude in the first example and its increase in the second occurred discontinuously.

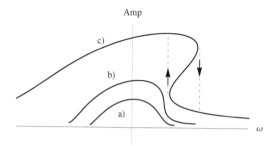

3.4.4 Techniques of Feedback Control

When movements or processes are modified according to predetermined values, one speaks of *control*. When the processes themselves influence the modifications, one speaks of *feedback control*. We will first discuss typical examples of feedback control like the control of water level and temperature and then demonstrate applications of feedback control in computer music. In non-realtime sound synthesis, changes in parameters are usually controlled, but controls often prove inadequate in unstable processes whose behavior cannot be foreseen in every detail. Feedback control is particularly necessary in works where musician and computer interact and sounds are recorded, transformed and played back. The most familiar example in electroacoustic music is the control of loudness. Feedback control is important in traditional music performance as well. Musicians like violinists and vocalists do not just play or sing the pitches and dynamics indicated by a score. Rather they first play approximately the indicated pitch of a note and then within a few milliseconds adjust the pitch by ear to match both their own inner pitch image and the other notes sounding at the same time. This rapid adjustment of parameters contributes much to the liveliness of instrumental and vocal music. In computer music, on the other hand, parameters obtained by control techniques often seem sterile and artificial, and the irregularities produced by random operations seem unnatural because they are not derived from processes involving feedback control.

Although this chapter discusses digital systems, the theory of Chapters 3.1 to 3.3 is not a prerequisite to what follows, because the literature of feedback control uses a quite different terminology. In the course of the chapter, however, we will occasionally point out similarities to the Chapter 3.1 to 3.3. Unlike most of the literature on feedback control which first discusses analog feedback and then mechanisms of sampling control, we shall begin with the latter and from time to time make reference to corresponding analog systems.

Further Reading: Schaum's Outline of Feedback and Control Systems *by Joseph DiStefano, Allan Stubberud and Ivan Williams [93] and* Nichtlineare Regelungen *by Otto Fölinger [38].*

3.4.4.1 Introductory Concepts and Examples

The components of a control circuit can be shown in a so-called fill level controller. Here the problem is to control the inflow of liquid into a tank so that a certain level is maintained despite loss of liquid through outflow. The following illustration shows a mechanical solution for the level control. A floater registers the difference of the current level $x(t)$ from the desired value w. A lever transmits this difference to a gate valve whose position $u(t)$ determines the inflow.

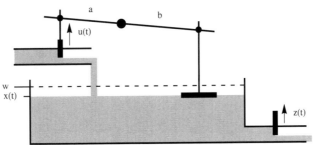

Many feedback controls have a similar structure, as does the classical example of feedback technique, the control of the temperature in a room. If the temperature of the radiator in a room is determined by information coming from a thermometer on the outside of the building, then we have a case of simple control, since the room temperature itself does not influence the temperature of the radiator. On the other hand, if we use an indoor thermometer, we have a closed circuit and control by feedback. For comparison, the amount of water in the illustration above corresponds to the current room temperature $x(t)$, the ideal level w corresponds to the desired room temperature. The floater, lever and gate valve are replaced in this case by a thermostat.

In the general case, the components of a feedback circuit are as follows. The state variable, that is, the current value of the system, is called the *controlled variable* $x(t)$, the ideal value is called the command variable or *set point* w. The difference between the ideal value and the current value of a system is called the *system deviation* or the *error variable* $e(t) = w - x(t)$, the value produced by the system controller is known as the *correcting variable* $u(t)$. A variable which disturbs the equilibrium of a system, like draining water in the first example or escaping heat in the second, is called a *disturbance variable* $z(t)$. The components of a feedback circuit usually belong to one of two sub-systems: the *controller* (or controlling computer) or the *control path*, which includes all other sub-systems (sometimes referred to simply as the system). This distinction, however, is fairly arbitrary. For example, a heater valve can be thought of as either part of the controller or of the control path, depending on whether it is a thermostatic valve or part of an electronically controlled heater. The unit of measure used for the controlled variable, the set point and error variable is the same. The correcting variable ordinarily uses a different unit of measurement or takes on values order of magnitude different from the other three. In addition, its unit of measurement depends on how the system is set up. For instance, in a heating system u can be the size of the opening of a valve, the voltage produced by the controller or the temperature of the water flowing into the system. The division into controller and control path is not only arbitrary, it is sometimes impossible to make. In particular, living organisms are too complex to be separated into individual parts. One can however examine the behavior of these systems in their entirety and draw conclusions about the controlling functions.

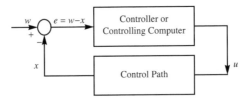

3.4.4.2 Elements of Control Circuits

Control circuits are usually composed of several elements, categorized, as we saw above, by the terms controller and control path. Subsystems are called *transfer elements* and are distinguished by their transfer characteristics. Analog transfer elements are usually nonlinear but can be approximated by linear equations. Digital transfer elements can be exactly linear. Another way to distinguish transfer elements is by their temporal behavior. Constant transfer elements are called *time-invariant*, changing elements are called *time-variant*. The most important group of transfer elements are the rational transfer elements, for their behavior can be calculated. They are identified by the letters P, T, I and D, where the letters indicate a proportional (P), integral (I) or differential (D) part, or show a time delay (T).

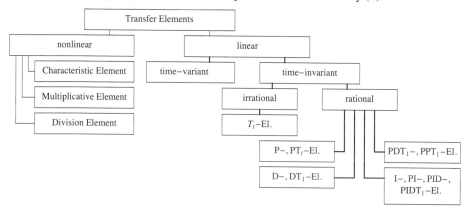

In this chapter we describe the most important transfer elements, indicate how to calculate the output values y from the input values x and discuss the properties of the transfer elements as controllers. The transfer elements will be characterized by their step response, that is, by their response to an abrupt change in the input value from 0 for $t < 0$ to 1 for $t \geq 0$. We will use the graph of the step response as the symbol for the respective element in connection diagrams.

A *proportional element* multiplies an input signal by a constant: $x_{out} = C_P x_{in}$. An analog example is the lever in the fill level controller in Chapter 3.4.4.1. The lever transforms a level change on one side into a change on the other side. If the lengths of the arms of the lever are a and b, we have $a x_{out} = b x_{in}$ or $x_{out} = C_P x_{in}$ with $C_P = b/a$. Ideal proportional elements do not exist in analog control paths because there is always a certain delay before the transformation is completed. To calculate the step response of this element, we set the input value u equal to 1, compute $x = C_P u$, giving the constant output value C_P. The resulting graph is the symbol of the proportional element in the following schematic diagrams.

```
c = .6; x = 0; u = 1; l = Table[x = c*u, {t, 0, 25}];
```

If this element is used as a controller, one speaks of a proportional or *P*-controller. We test the controller by making the set point equal to 1, using the error variable $e = w - x$ as the input and choosing the simplest control path $x = u$ (in other words, leaving out the control path altogether).

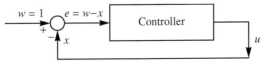

The result shows that the signal does settle into a constant value, but that this value does not satisfy the requirement $w = 1$. This constant error is typical for proportional controllers.

In an *integral element*, the input values are continuously summed. In continuous systems, the output signal is the integral of the input signal (possibly multiplied by a constant)

$$x(t) = C_I \int u(t)\, dt.$$

In digital systems, each new value x_k is computed by adding the input u_k multiplied by the sampling period and the factor C_I to the previous value x_{k-1}:

$$x_k = x_{k-1} + C_I T u_k.$$

The analog prototype is a water container like the one discussed in Chapter 3.4.4.1. We compute the step response of the element with the following command in Mathematica and get (figure left below) a constantly increasing function $x(t)$.

```
c = 0.05; x = 0; u = 1; l = Table[x += c*u, {t, 0, 25}];
```

Next we compute a sequence of x-values that starts at 0 and is adjusted to the set point w by an integral element as controller (integral or I-controller). To begin with, the sum s and the controlled variable x are both equal to 0. As set point we choose $w = 1$ and as time unit the sampling period T. We compute the new value

$$u_k = x_k = x_{k-1} + C_I T u_k$$

by the formula $x \mathrel{+}= k(w - x)$ (figure right below).

```
c = 0.2; x = 0; w = 1; l = Table[x += c*(w - x), {t, 0, 25}];
```

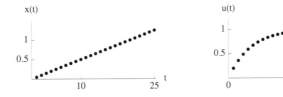

3.4 Dynamic Systems and Feedback Control

Now we introduce a disturbance variable $z(t)$ into the circuit. We divide the circuit into the control path $x(t) = u(t) + z(t)$, where the disturbance appears, and the controller $u \mathrel{+}= c(w - z)$.

```
s = x = 0; c = 0.2; w = 1; Table[u = s += c (w - x); x = u + z[[t + 1]], {t, 0, 25}];
```

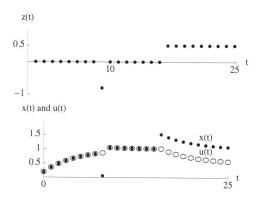

As the illustration shows, the first short disturbance $z(9) = -.8$ is slightly over-compensated so that the following x-values gradually approach the set point $w = 1$ from above. Later on, for $t > 16$, the disturbance variable is constant at .5 and the disturbance is gradually reduced as the correcting variable converges to $u(t) = .3$.

If we combine a proportional element and an integral element, we get the formula

$$x(t) = C_1 u(t) + C_2 \int u(t)\, dt$$

for the new element in analog form and the formula

$$x_k = C_1 u_k + u_{k-1} + C_2 T u_k$$

for the new element in digital form. The step response of this *PI*-element is a rising straight line that does not pass the origin of the coordinate system but is offset by C_1 (left). The illustration at the right shows how this element adjusts to the set point $w = 1$.

```
u = 1; s = x = 0; w = 1; c2 = 0.1; c1 = .3;
Table[s += c2 * u; x = s + c1 * u, {t, 0, 25}];
Table[s += c2 * (w - x); u = s + c1 * (w - x); x = u, {t, 0, 25}];
```

The next illustration shows the effect of a *PI*-controller on a control path having the same disturbance variable $z(t)$ as above.

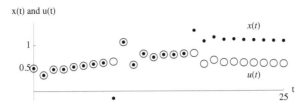

Elements that have recourse to earlier values for computing new values are called *proportional elements with delay*. If only the last value x_{k-1} is used ($x_k = C_1 x_{k-1} + u$), we speak of a proportional element with first-order delay (PT_1-element). If the last two values x_{k-1} and x_{k-2} are used, we speak of a proportional element with second-order delay (PT_2-element). The PT_1-element reacts with delay to the step function (left).

```
c = .4; x1 = x0 = 0; u = 1; Table[x0 = x1; x1 = u + c * x0, {t, 0, 25}];
```

The PT_2-element reacts with delay to the step-function and leads to oscillation (right).

```
c1 = 1.5; c2 = -.6; x2 = x1 = x0 = 0; u = 1;
Table[x0 = x1; x1 = x2; x2 = u + c1 * x1 + c2 * x0, {t, 0, 25}];
```

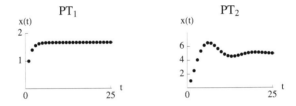

If we use these elements as controllers, we have, as we do with simple proportional elements, a constant control deviation.

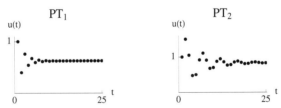

Physical systems are described using differential equations in the analog domain and difference equations in the discrete domain. The classical example of a mass suspended on a spring led us in Chapter 3.4.1 to the differential equation $a_2 \ddot{x} + a_1 \dot{x} + a_0 x = u(t)$ and to the difference equation $x(k) = c_1 x(k-1) - c_2 x(k-2) + u(k)$. Since the differential equation is of second order, the difference equation will have delays of two sampling intervals. Therefore PT_2-elements are systems that can oscillate. From the differential equation $a_1 \dot{x} + a_0 x = u(t)$ we get the difference equation of the PT_1-elements $x(k) = c_1 x(k-1) + u(k)$.

There are systems which react to excitation only after a certain delay. For example, if to control loudness we use a microphone that is $d = 5$ meters away from the loudspeaker to be controlled, there will be a delay of (5 m)/(340 m/s) = .0147 s. In technical systems like conveyor belts or water pipes, the delays can be much greater. These *reaction time systems* (PT_t-elements) are often responsible for oscillation and override, because the controller

3.4 Dynamic Systems and Feedback Control

continues to compensate during the reaction time. With the following command we produce a delay of five time units by defining five successive values and passing the value read in at *x5* one memory cell farther along at each computation cycle (cf. Delay Lines 5.1.4.1 and Ring Buffers 5.1.4.2). The illustration on the left shows the step response of a reaction time element.

```
x0 = x1 = x2 = x3 = x4 = x5 = 0; c = 1.27; dt = 1; u = 1;
Table[x0 = x1; x1 = x2; x2 = x3; x3 = x4; x4 = x5; x5 = u; x0, {t, 0, 15}];
```

If we control a path with the PT_1-element below, we get the succession of values illustrated on the right. The sequence oscillates with a period 6 (5 + 1 delays) and shows the deviation usual for proportional elements.

```
u = .4*u + .5*(w - x0)
```

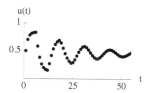

Elements that use the derivative of the input value to compute the output are called *differential element* (D-elements). In the simplest case, the output $x(t)$ is proportional to the derivative of the input: $x = C\dot{u}$ or $x_k = C(u_k - u_{k-1})$. The step response of the discrete D-element below consists of a pulse (left).

```
c = .9; x = 0; u0 = 0; u1 = 1; Table[x = c*(u1 - u0); u0 = u1; x, {t, 0, 25}];
```

A differential element with delay (DT_1-element) has the equation

$$K_1\dot{x} + x = K_2\dot{u} \quad \text{or} \quad x_k = K_1 x_{k-1} + K_2(u_k - u_{k-1}).$$

The illustration at the right shows the step response of the DT_1-element.

```
c1 = .9; c2 = .9; x1 = x0 = 0; u0 = 0; u1 = 1;
Table[x1 = c1*x0 + c2*(u1 - u0); u0 = u1; x0 = x1; x1, {t, 0, 25}];
```

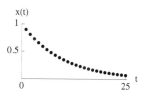

DT_1-elements are often used in controllers. They can increase the speed and stability of a feedback circuit. The first of the following illustrations shows the step response of a DT_2-element (step size .1) that oscillates without damping.

```
a1 = 1.9; a2 = 1; x2 = x1 = x0 = 0; u = .1;
Table[x2 = a1*x1 - a2*x0 + u; x0 = x1; x1 = x2; x2, {t, 0, 40}];
```

If we control this element with a proportional element, we get the undamped oscillation shown top right below.

```
a1 = 1.9; a2 = 1; x2 = x1 = x0 = 0; u = .1;
l = Table[x2 = a1 * x1 - a2 * x0 + u; x0 = x1; x1 = x2; x2, {t, 0, 40}];
```

This circuit does not dissipate disturbances. The entire circuit is excited to ever greater oscillation (lower left below).

```
x2 = a1 x1 - a2 x0 + u + RandomReal[{-0.3, 0.3}];
```

If we add a differential element, there is still a discrepancy, as there is with all proportional controllers, between the desired value and the value obtained, but the circuit is stable (lower right below).

```
a1 = 1.9; a2 = 1; x2 = x1 = x0 = 0; u = 0; k1 = .4; k2 = .3;
Table[x2 = a1 * x1 - a2 * x0 + u;
    u = k1 * (w - x2) - k2 * (w - x1); x0 = x1; x1 = x2; x2, {t, 0, 40}];
```

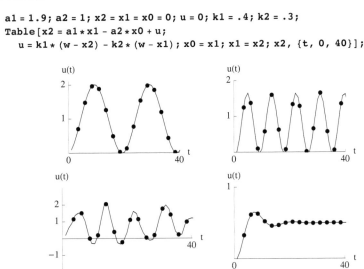

3.4.4.3 Feedback Circuits

In technical applications there are often complex networks of elements in the control circuits. These networks can be shown by flow diagrams, where the individual elements are shown by graphs of their step responses. A proportional element is described by its transfer factor, a linear element by its frequency response function (3.3.1.4), a nonlinear element by its characteristic curve. Elements can also be described by their transfer equations or their S- and Z-transforms (3.1.4.3 and 3.2.3 respectively). The circuit with a reaction time element and a PT_1-controller mentioned above can be described in one of the following ways:

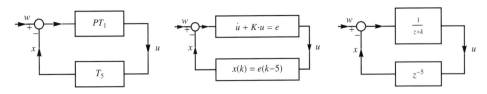

3.4.4.4 Nonlinear Feedback Circuits

The simplest practical control, the relay, is nonlinear. It can only be on or off, hence can only take one of two values. The illustration below left shows a circuit whose control path consists of an integral element, indicated by the sum sign Σ and a reaction time element T_3. The controller is a relay with a so-called *two-point characteristic* (right). The controller gives the value $u = b$ when the error signal is greater than 0 and $-b$ when the error signal is less than 0.

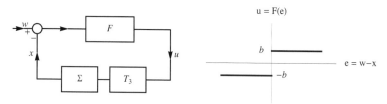

Although the control is very simple and the controlled variable $x(t)$ constantly oscillates if the system is without disturbance (left figure below), the circuit can be useful in certain situations. For instance, it can effectively compensate a strong disturbance random signal, which, when uncontrolled, leads to large changes in the system variable x. The right figure below shows the raw (upper figure) and the compensated signal (lower figure).

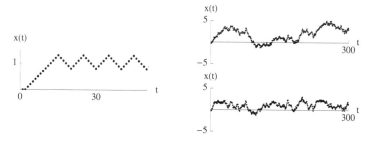

In Chapter 3.4.1.1 we studied the behavior of nonlinear systems by means of their representation in phase space. Phase-space representation can also help to understand nonlinear control circuits. In what follows we begin with a simple system and extend it gradually. Imagine a system with constant acceleration $\ddot{x} = -b$. We draw the vector field in the phase plane $(x, \dot{x})$ by choosing a set of states $(x_t, \dot{x}_t)$ and re-computing the states at a somewhat later time $t + T$. The new position of each point is computed as the sum $x_{t+T} = x_t + \dot{x}_t T$, the new velocity as the sum $\dot{x}_{t+T} = \dot{x}_t + T\ddot{x}_t = Tb$. The illustration shows the flow diagram of the system with two integral or sum elements and the relay described above as controller.

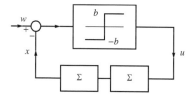

The left figure below shows the vector field of the differential equation for constant $b < 0$.

```
T = .15; b = .4; l = .3; Table[Arrow[{{x, y}, {x + T*y, y - T*b}}], ...];
```

Since the controller switches from b to $-b$ or from $-b$ to b when $w - x = 0$, in the phase-space representation we have the vector field shown on the right below with the line $x = w$ which separates the two states. This line is known as the *switching curve*.

```
w = .4; ....Arrow[{{x, y}, {x + T*y, y - T*If[x > w, -b, b]}}] ...
```

If we begin in an arbitrary state, that is at an arbitrary point in the phase plane, we get a closed trajectory as shown on the left below. The trajectories can be generated with the following commands:

```
T = .1; b = -.4; x = .4; y = .7; h = .2;
Table[{x += T*y, y += T*If[x > w, b, -b]}, {t, 0, 125}];
```

If we tip the switching curve (here a straight line) somewhat to the left, we can make x converge towards w. The equation for the switching curve is then $u = w - c\dot{x}$.

```
w = .2; c = .25; .. Arrow[{{x, y}, {x + T*y, y - T*If[x > w - c*y, -b, b]}}] ..
```

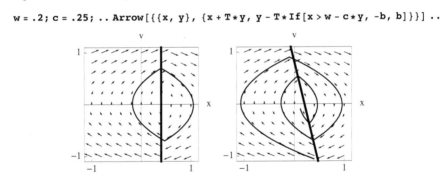

The addition of the term $-c\dot{x}$ corresponds to additional feedback in the circuit.

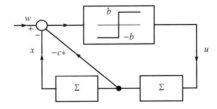

We can compute numerically examples containing equations that cannot be solved analytically. The illustration below shows trajectories of circuits having the same switching curve as above but with the nonlinear accelerations

3.4 Dynamic Systems and Feedback Control

$\ddot{x} = -b + ax^2$ (left), $\ddot{x} = -b - ax^3$ (middle) and $\ddot{x} = -b - ax\dot{x}$ (right).

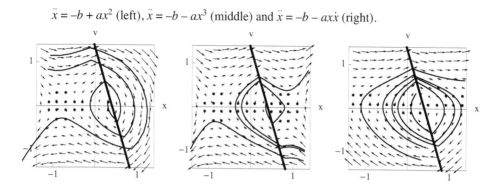

Let us analyze the properties of a controller exhibiting hysteresis in the control path $\ddot{x} = -b$. The illustration shows the controller's characteristic curve.

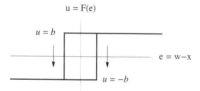

The switching curve in the phase plane is computed as follows: If $x > w$, then $u = b$; if $x < -w$, then $u = -b$. If $-w < x < w$, then we must determine from which direction we reach the value x. The direction is given by the sign of the velocity $y = \dot{x}$. The following command exhibits the desired behavior and, when used in the function for computing the arrows of the field and the trajectories, the illustration below on the left. The system turns out to be unstable, because all the trajectories move centrifugally. Systems with damping exhibit a limit cycle (3.4.2.3) to which all trajectories converge (below right).

```
y += T * If[x > w || (x > -w && y < 0), b, -b]
```

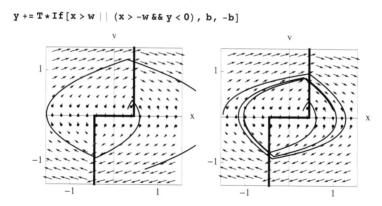

If we introduce a reaction time element $T_t = 5$ into the circuit with a relay and the simple control path $\ddot{x} = -b$, the trajectories continue on in the original direction for a while after crossing the switching curve before they begin to follow the new directional field.

```
z = 0; x = -.5; y = .1;
Table[{x += T*y, y += T*If[x < w, b, If[z < 5, z += 1; b, -b]]}, {t, 0, 25}];
```

The delay displaces the switching curve. It is no longer a single continuous straight line but consists of two separate and slightly offset lines. In general, the new switching curve is determined by the set of points at which the trajectories orient themselves to the new directional field. It is difficult to calculate in continuous time because the trajectory's path across the original switching curve has to be integrated over the reaction time. The second illustration was produced by computing for points on the straight line $x = w$ the corresponding points on the displaced switching curve and connecting them.

```
ii = 5; tt = 8;
Line[Table[x = w; y = .01 + i * .2;
    Do[y *= d; x += T*y; y += T*If[x > w, b, -b], {t, 0, tt}]; {x, y}, {i, 0, ii}]];
```

The following illustration shows a longer section of a trajectory.

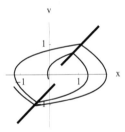

The illustrations above were computed using a bit of damping in the system. Without damping, the control circuit, like the hysteresis, is unstable (below left). The trajectory of the damped system approaches a limit cycle (below right).

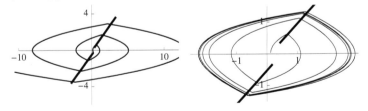

3.4.4.5 The Computation of Feedback Control

The literature on analog control circuits speaks mostly about linear circuits, since they can be calculated. The mathematical tools to do so are the same as for signal processing. The terminology and the choice of parameters however are adapted for the requirements of control theory. The transfer elements and the controllers in feedback control are systems. We can

3.4 Dynamic Systems and Feedback Control

consider control circuits as composite recursive systems in the sense of Chapter 3.1 to 3.3 and can calculate their behavior with the tools developed there. In linear transfer elements and feedback circuits one can define the frequency response and apply the s-transform for continuous time (3.1.4.3) or the z-transform for discrete time (3.2.3).

The proportional element $x_{out} = C_P x_{in}$ corresponds to the non-recursive system $y(k) = cx(k)$. A control circuit consisting solely of a P-controller corresponds to the recursive system $y(k) = c(w - y(k-1))$. A proportional element with second-order delay

$$x_k = C_1 x_{k-1} + C_2 x_{k-2} + u$$

corresponds to a recursive filter

$$y(k) = c_1 y(k-1) + c_2 y(k-2).$$

By choosing the parameters c_1 and c_2 appropriately, one gets a resonator with the resonance curve as amplitude response. In compound circuits, we can determine the frequency response of the individual components and, since the systems are linear, take their sum, or else combine the equations for the components and determine the frequency response for the entire system.

3.4.4.6 Control By Filters

Entire control circuits correspond to recursive systems, while single elements of these circuits can be either recursive or non-recursive. The computer musician usually is not acquainted with the terminology of control systems, but she does have experience with filters. Synthesis programs include various types of filters. We shall show in what follows how filters can be used for system control (3-4-4a.csd).

A recursive low-pass filter corresponds to the PT_1-element of control theory. Hence we can make a control circuit using the Csound unit generator tone. In the Csound orchestra below, we make a control path in which a disturbance signal az is simply added to the correcting variable. The PT_1-controller is realized in line 3: p4 is the correcting variable w and $p4 - ax$ is the error variable e. The parameter p5 is the filter's cutoff frequency (3.3.1.9). With it we can control the circuit's reaction time.

```
instr 1                              ;1
ax    init      0
au    tone      p4-ax,p5
az    oscil     1,1/p3
ax    =         au+az                ;5
      out       ax
endin
```

If we make a second instrument which is the copy of the instrument above, only substituting for line 3 the command below, we get a PT_2-controller corresponding to a resonator or bandpass filter. The center frequency p5 controls the reaction time and the bandwidth p6 controls the damping.

```
instr 2
....
au    reson     p4-ax,p5,p6,1
....
```

We create the disturbance signal *az* with the Csound function call below.

```
f1 0 1024 -7 1000 400 1000 4 -20000 200 -20000 20 15000 200 15000 200 0
i1 0 1 30000 5
i2 0 1 30000 2 20
```

The following illustration shows at the top the disturbing signal *az*. The lower two signals are the control variables *ax* of the two *PT*-controllers.

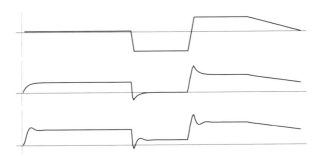

We can generate integral elements by adding to the correcting variable a value proportional to the error variable. The proportionality factor determines how rapidly the controller reacts.

```
au = au + p5 * (p4 - ax)
```

If we test the controller with the same disturbance signal as above, we do not get any remaining control deviation, but there are abrupt changes in the control variable when the disturbance signal changes rapidly (illustration on the left below). The controller is well-suited to controlling slowly changing disturbance signals (illustration on the right).

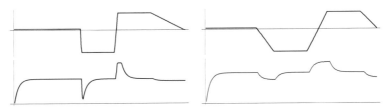

3.4.4.7 Examples

The loudness of a resonator excited by noise varies greatly. Hence it is difficult to fix a single value for the loudness. But we can control the loudness, for example by only turning on the exciting signal when the resonator's amplitude is smaller than a required value *w*. The control variable is the amplitude of the signal generated by the resonator and the correcting variable is the amplitude *u* of the noise exciting the resonator. The controller is a relay that sets *u* to *aexc* whenever the amplitude is less than *w* and turns *u* off whenever the amplitude is greater than *w*. We can determine the approximate amplitude by measuring the signal's greatest displacement, because the signal of a weakly damped and weakly excited oscillator only has a turning point at its maximal displacement.

3.4 Dynamic Systems and Feedback Control

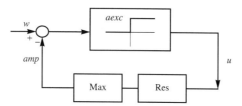

We can simulate the control process described above with the following program. At the outset, the set point is $w = 1$ and then diminishes geometrically during the process (lines 1-3). In the fourth line we make the resonator, which is excited by white noise with amplitude u that is determined in the fifth line. The terms $(x_i - x_{i-1})$ and $(x_{i+1} - x_i)$ have different signs only at the extrema of the signal, so that their product is negative.) The controller is represented by the conditional statement in line 6. Finally we store w, u and the displacement x (line 6).

```
i = 0; w = 1; c = .15; d = .0048; aexc = .036; amp = 0;            (*1*)
u = 0; x0 = x1 = x2 = 0; tt = 900;
l = Table[i += 1; If[i > 300, i = 0; w *= .6,];
    x2 = (2 - c) * x1 - x0 - d * (x1 - x0) + Random[Real, {-u, u}];
    amp = If[ (x1 - x0) * (x2 - x1) < 0, Abs[x1], amp];             (*5*)
    u = If[w - amp > 0, aexc, 0]; x0 = x1; x1 = x2;
    {w, u, x2}, {t, 0, tt}];
```

The following figures show (above) the diminishing set point s, the generated signal x and the correcting variable u (below). The set point takes on its second value before the amplitude reaches the original value $w = 1$. But because the amplitude is now greater than the new set point, the controller turns off the exciting noise. During the next two sections, the noise is turned on and off several times, and the amplitude oscillates fairly precisely about the set point.

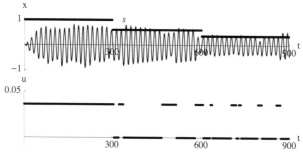

Let us test two other nonlinear characteristic curves with the same circuit. We choose the correcting variable u to be proportional to the error variable e when $e > 0$ and to be equal to zero when $e < 0$ (illustration below left). We change line 5 in the program above as follows:

```
u = If[w - amp > 0, aexc * (w - amp), 0];
```

In order to have a continuous amplitude for the exciting function, we use a controller with an exponential characteristic. This means that the exciting signal u will never be completely turned off (illustration below right).

```
u = aexc * Exp[w - amp];
```

Using the characteristic shown at the left above, the same parameter values as above and a proportionality factor *aexc* = .1, we get the results shown in the following figure. It is easy to recognize both where the exciting signal is turned off when the amplitude exceeds the required value w and where the excitation increases in proportion to the control deviation when the amplitude is less than *w*.

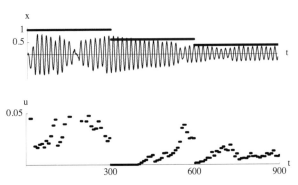

If we test the controller using the exponential characteristic and the same parameters, the results are no better than with the other controllers. With greater damping (*d* = .06) and strong excitation (*u* = .06·exp(*w* − *amp*)), the controller gives good results, even though the amplitude is no longer correctly measured because the signal now has relative extrema at other moments than those of maximal displacement.

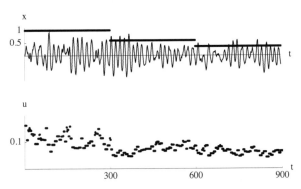

A parameter that is difficult to manage is the frequency of nonlinear oscillators because it depends on the oscillator's amplitude. The following illustration shows a damped vibration of a nonlinear oscillator.

```
x0 = 0; x1 = x2 = .3; d = .01; c = .23;
Table[x2 = 2*x1 - x0 - c*x1^3 - d*(x1 - x0); x0 = x1; x1 = x2; x2, {t, 0, 475}];
```

3.4 Dynamic Systems and Feedback Control

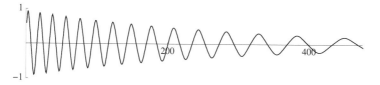

We can control the frequency by varying the factor c for the cubic restoring force according to the error signal. In this simple example, we can determine the oscillator's frequency from the last half-period td at each zero crossing. Only at these points the product of two successive displacements is negative. We determine the half-period td from the difference between the current time and the time $t1$ of the last zero crossing. At the outset, the period of the uncontrolled oscillation is a bit less than 20. So we set the half-period $td = w = 9$. On can see small irregularities in the amplitude and half-periods in the first part of the signal. But after only a few periods the amplitude decreases exponentially and the frequency remains constant. The error variable causes the actual frequency to be a bit lower ($td = 12$). The factor c has increased from .23 to 9.31 during the generation of the signal shown here.

```
... k = -.02; t1 = 0; td = w = 9; t1 = 0;
... If[x1*x0 < 0, td = t - t1; t1 = t]; ...
```

As the examples show, one does not only have to find suitable controllers for concrete situations, but also suitable methods for measuring control variables and control deviations. It is often considerably simpler to invent ad hoc control systems than to describe and solve a problem using the standard terminology of control theory. These chapters describe control systems in the broadest sense: Chapter 6.1.2.7 (stabilizing the frequency of frequency modulation by feedback), Chapter 8.1.3.6 (a physical model of the bowed string with a system to measure the speed of the string), Chapter 8.2.3.4 (excitation with feedback in wave guides), Chapter 5.1.4.4 (delay lines with feedback).

3.4.5 Synchronization

In this chapter, we consider oscillators that are synchronized either by an external force or by mutual influence. Examples of synchronization by an external force are the control of cardiac activity by a pace maker, the adjusting of a clock by radio signals and the synchronization of biological cycles through circadian rhythm. Examples for the mutual synchronization of oscillating systems are the synchronization of fireflies, the coordinated clapping of an audience and the synchronization of pendula suspended from the same point (cf. 5.1.3.6). These systems have in common that they are not linear and that they oscillate without external excitation. They are called *self-sustained* oscillators. The resonance of damped linear oscillators is not considered synchronization.

Further Reading: Synchronization: From Simple to Complex *by Alexander Balanov et al. [73]*, Synchronization in Oscillatory Networks *by Grigory V. Osipov et al. [74]*.

3.4.5.1 Self-Sustained Oscillators

Self-sustained oscillators have a natural frequency and compensate for energy loss by an inner energy source. The trajectory of the oscillation in phase space is a stable limit cycle (see periodic attractors 3.4.2.3). Oscillators that are characterized by successive phases of faster and slower movement are called *relaxation oscillators*. An example is the so-called *integrate-and-fire oscillator*, with which neuronal activities and oscillations in chemical reactions under the influence of catalysts can be simulated.

Let us first consider the behavior of a self-sustained oscillator from a purely qualitative point of view. The following illustrations show the simplest limit cycle: a circle. The state of the unperturbed oscillator in phase space $(x, \dot{x})$ (3.4.1.1) is described by a point rotating along the limit cycle (left figure). Let us now imagine a coordinate system rotating with the same angular velocity as the unperturbed oscillator. Then the system's state can be described as a stationary point. If the oscillator is perturbed, for example moved from state 1 to state 2 (right figure), the influence of the attractor (here the circle) gradually dissipates the amplitude change, but the phase shift remains (state 3). The fact that very weak external forces can perturb the phase is one of the main reasons why self-sustaining oscillators can synchronize themselves.

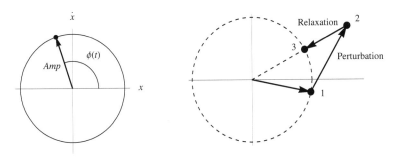

3.4.5.2 The Van der Pol Oscillator

While attempting to explain the nonlinear dynamics of vacuum tube circuits, the Dutch electrical engineer Balthasar van der Pol derived the equation

$$\ddot{x} = -\omega^2 x + \mu(1 - x^2)\dot{x}.$$

The equation describes a linear oscillator with an additional nonlinear term $\mu(1 - x^2)\dot{x}$. When $|x| > 1$, the nonlinear term results in damping, but when $|x| < 1$, negative damping results, which means that energy is introduced into the system. For the following calculation we write the differential equation above as a system of two equations (3.4.1.2).

$$\dot{x} = v$$
$$\dot{v} = -\omega^2 x + \mu(1 - x^2)v$$

The command NDSolve[] solves the equation for the initial values x_0, v_0. (→ Animation)

```
NDSolve[{v'[t] == -ω² * x[t] + μ * (1 - x[t]²) * v[t], x'[t] == v[t],
    v[0] == 0, x[0] == 1}, {x, v}, {t, 0, 50}, MaxSteps → 3000]
```

3.4 Dynamic Systems and Feedback Control

The limit cycle is quickly reached, regardless of the initial values.

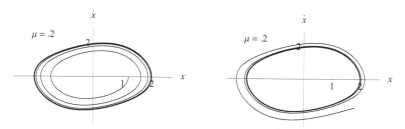

As μ increases, the limit cycle becomes more and more deformed. The maximum deflection x approaches 2, and the maximum velocity increases continuously.

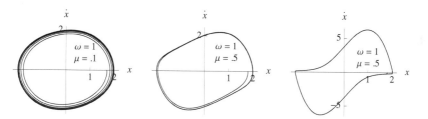

The velocity with which the system's state changes when the nonlinear term is large (that is, the velocity with which the point $(x, \dot{x})$ moves along the trajectory) varies. This is evident in the following numerical simulation. We transform the differential equation above into a difference equation and generate 100 discrete states $x(k)$. The illustrations below show the trajectory in phase space (x, v) and the corresponding time series $x(k)$.

```
v = 0; x = .4; ω = 0.1; μ = 0.25;
Table[v += (-ω² * x + μ * (1 - x²) * v); {x += v, v}, {100}]
```

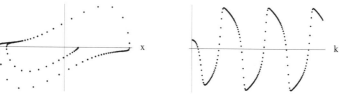

The following figure shows the spectrum of the oscillation for various values of μ.

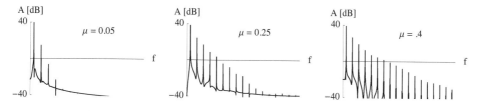

In the Max-patch Van_der_Pol_1, the Van der Pol oscillator is made using an mxj~ object. The input parameters are the frequency f in Hertz and the factor of nonlinearity μ. When μ is equal to zero, the frequency is $\omega = 2\pi f/sr$. As μ increases, the frequency goes down. The patch shows the waveform, the spectrum and the trajectory. The trajectory is generated using a scope object to show deflection and velocity (that is, the difference of two successive deflection values). In order to begin oscillating, the oscillator must be excited (e.g. by a click of amplitude 0.00001) and μ must be greater than zero.

3.4.5.3 Synchronization Using Periodic Excitation

First we describe a quasi-linear oscillator, that is, an oscillator with very little nonlinearity. Let this nonlinearity be an arbitrary function $n()$ of the state of the oscillator $(x, \dot{x})$:

$$\ddot{x} = -\omega_0^2 x + n(x, \dot{x})$$

The solution of the equation without the nonlinear term is (3.4.1.1):

$$x(t) = A \cdot \sin(\omega_0 t + \phi_0).$$

If we excite the oscillator with the periodic force $f(t) = \epsilon \cdot \cos(\omega t + \phi_0^e)$, we obtain the equation

$$\ddot{x} = -\omega^2 x + n(x, \dot{x}) + f(t).$$

The instantaneous phase of the exciting force is $\phi_e = \omega t + \phi_0^e$. The frequency ω generally differs from the frequency ω_0 of the autonomous oscillator. The difference of the two frequencies $\omega - \omega_0$ is called *detuning*. Because perturbances of amplitude decay rapidly, it suffices to consider the behavior of the phase. The limit cycle of the quasi-linear oscillator is a circle on which the point representing the phase rotates with the frequency ω_0. In a coordinate system revolving with the frequency ω of the exciting force, this point has the angular velocity $\phi - \phi_e$ (left figure below). The exciting force is represented as a vector of length ϵ (dotted arrows in the figure to the right below), acting at the angle $\phi^a = \phi_0^e + \pi/2$. The effect of the force depends on the phase difference $\phi - \phi_e$. At the points 1 and 2, the force acts perpendicularly to the trajectory and hence does not affect the phase. At all other points a force results that drives the points toward point 1. Hence point 1 is a stable fixed point, point 2 an unstable fixed point.

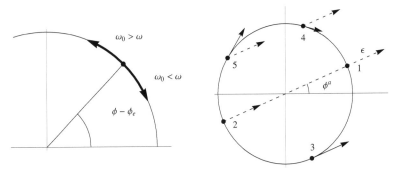

If the detuning is zero, any initial phase difference between the excitation and the quasi-linear oscillator is reduced until $\phi = \phi_e - \phi^a$ and phase locking between the two is obtained. If the detuning increases (e.g. when $\omega_0 > \omega$), then two tendencies are in competition with each other: rotation (thick arrows in the left figure below) and the force of the excitation. The phase difference between excitation and oscillator levels off at a certain value $\Delta\phi$ (function 1 in the right figure below). Their movements are synchronous but not identical. If the detuning

3.4 Dynamic Systems and Feedback Control

is large, the force offers insufficient resistance to the rotation. The result is a function $\Delta\phi(t)$ that remains constant for a certain time and then slips to make a quick full rotation (function 2 below). The phase point starts to rotate with the so-called beat frequency Ω_b.

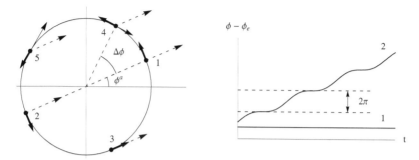

The illustration to the left below shows how the difference between the frequency Ω of the driven oscillator and the driving frequency ω depends on ω when ϵ is constant. The figure on the right shows the dependency when both ω and ϵ are variable. (For the calculation of the function see 3.4.5.4.)

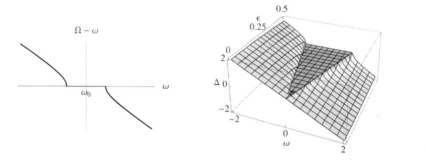

The Max patch Van_der_Pol_2 represents a Van der Pol oscillator with a natural frequency of ω_0 and a nonlinearity factor of μ (3.4.5.2). It can be excited by a sine wave of frequency ω and amplitude ϵ. The range of ω within which the oscillator is synchronized to the exciting frequency increases as μ and ϵ increase. The variation of the phase difference between excitation and oscillation, as well as the transitions between synchronous, beating and asynchronous behaviors, can be visualized by showing the sum of the excitation and the oscillation signals in a phase diagram. The following figures show to the upper left the waveform of the Van der Pol oscillator, to the lower left that of the excitation (normalized) and to the right the phase diagram of their sum. For these figures, the same values were always used for ω_0, μ and ϵ. Comparing the first two figures, one sees that the oscillator adopts the exciting frequency ω within a large frequency range. When the frequency is low (upper left), the phases of the two waves are nearly the same. Hence there is a large deflection along the x-axis in the phase diagram showing the sum of the waveforms. When the frequency is high, the phases are nearly inverted (upper right) and the phase diagram shows only a small deflection. The figure on the lower left shows the transition to asynchronous behavior (in the phase diagram of the oscillator alone the irregularity cannot be seen). If the proportion between the natural frequency of the oscillator ω_0 and the excitation frequency ω is approximately simple ($\omega/\omega_0 \cong m/n$), then within a certain range the frequency of the Van der Pol oscillator is synchronized so that $\omega/\omega_0 = m/n$. Here one speaks of higher order synchronization (figure to the right below).

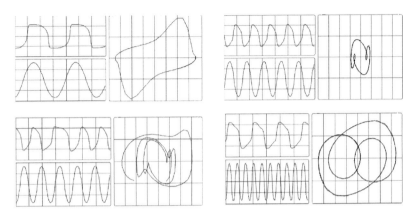

An even simpler nonlinear oscillator is used in the processing example *self_sust_osc_1*. We call the pointer of the driven nonlinear oscillator P and its phase ϕ. The pointer of the exciting sine wave is P_e and its phase is ϕ_e. The force component $\sin(\phi_e - \varphi)$ acts tangentially to the trajectory of the driven oscillator and is added to the phase ϕ at each step of the simulation. In order to have the unit circle as trajectory of the driven oscillator we set

$$r \mathrel{+}= ar(1 - r^2) + r_e\cos(\phi_e - \phi).$$

```
ϕ  += dϕ  + ampe*sin (ϕe - ϕ);
r  += a*r*(1 - r*r) + ampe*cos(ϕe-ϕ);
ϕe += dϕe;
```

(→ Illustration: Processing *self_sust_osc_1*)

3.4.5.4 Mutual Synchronization of Weakly Coupled Oscillators

Let us first consider two coupled limit-cycle oscillators. If we represent them as a system of first-order differential equations (3.4.1.2), we can write:

$$\dot{x}_1 = F_1(x_1) + \epsilon \cdot P_1(x_1, x_2) \text{ and } \dot{x}_2 = F_2(x_2) + \epsilon \cdot P_2(x_1, x_2).$$

Here the x_i are vectors, F_i and P_i are arbitrary functions and $|\epsilon| \ll 1$. The natural frequencies of the oscillators ω_i should be approximately equal. The behavior of the oscillators can be described by the following equations for their phases ϕ_i [see 72].

$$\dot{\phi}_1 = \omega_1 + \epsilon \cdot q_1(\phi_2 - \phi_1) \text{ und } \dot{\phi}_2 = \omega_2 + \epsilon \cdot q_2(\phi_1 - \phi_2).$$

3.4 Dynamic Systems and Feedback Control

The q_i are 2π-periodic functions. Then for the difference of the two phases $\theta = \phi_2 - \phi_1$ we have:

$$\dot{\theta} = \Delta - 2\epsilon \cdot q(\theta)$$

where $\Delta = \omega_2 - \omega_1$ and $q(\theta) = q_2(-\theta) - q_1(\theta)$. For there to be synchronization, the phase difference θ must be constant, that is $\dot{\theta} = 0$. Hence we have

$$q(\theta) = \frac{\Delta}{2\epsilon}.$$

For the simplest 2π-periodic function, the sine wave, we obtain the so-called *Adler equation* of the first degree $\dot{\theta} = \Delta - 2\epsilon \cdot \sin(\theta)$. $\dot{\theta}$ can only be made to disappear when $|\Delta| < |2\epsilon|$. If $|\Delta|$ becomes greater than $|2\epsilon|$, the coupled system begins to beat. The beat frequency Ω can be calculated from the equation

$$\dot{\theta} = d\theta/dt = \Delta - 2\epsilon \cdot q(\theta).$$

By solving for dt and integrating over one period of θ, we obtain the period's duration and from it the beat frequency:

$$\Omega = 2\pi \left(\int_0^{2\pi} \frac{1}{2\epsilon q(\theta) - \Delta} \, d\theta \right)^{-1}.$$

For $q(\theta) = \sin(\theta)$ we obtain $\Omega = \sqrt{\Delta^2 - 4\epsilon^2}$, which is the same function as for the synchronization of an oscillator using periodic excitation (illustration see 3.4.5.3). ($\to$ Computation)

Several coupled oscillators can be arranged simply as a one-dimensional chain in which each oscillator is connected to its two neighbors. As above, we can write N equations for the phases of N coupled oscillators. If the coupling constants and the coupling functions are the same for the whole chain, then the equations are:

$$\dot{\phi}_k = \omega_k + \epsilon \cdot q(\phi_{k-1} - \phi_k) + \epsilon \cdot q(\phi_{k+1} - \phi_k), \qquad k = 1, 2, ..., N.$$

Only in the simplest cases can we treat the behavior of several coupled oscillators analytically. Therefore, in what follows we will describe the qualitative behavior of arrays of coupled oscillators and provide Max patches for experimentation. In the Max patch V_d_Pol_lin_Array N oscillators are generated by the mxj~ object cm_v_d_pol_linarray and arranged in a circle. The frequencies ω_i and the nonlinearity factors μ_i of the individual uncoupled oscillators are randomly uniformly distributed within a range chosen by the user. An oscillator is coupled to its two neighbors by using a part of the sum of their velocities as its excitation:

$$\dot{x}_i = F(x_i) + \epsilon \cdot (\dot{x}_{i-1} + \dot{x}_{i+1})$$

The following code from the mxj~ object shows how the velocity v and amplitude x are calculated for the oscillators 1 to $n - 2$. The oscillators 0 and $n - 1$ at the beginning and end of the array are treated separately.

```
;mxj~ cm_v_d_pol_linarray
for(int k = 1; k < n-1; k++)
{   v[k] += (- c[k]*x[k] + mu[k]*(1.f - x[k]*x[k])*v[k]
            + fb*(v[k-1] + v[k+1]));
    x[k] += (v[k] + in[i]);}
```

In the Max patch *V_d_Pol_n-Array* N oscillators are generated, all coupled to one another. This is done by summing all the velocities to *vfb*, scaling this sum by the factor *fb* and feeding the result back into the oscillators as excitation.

```
;mxj~ cm_v_d_pol_n-array
for (int k = 0; k < n; k++) vfb += v[k];
for (int k = 0; k < n; k++)
{    v[k] += (-c[k]*x[k] + mu[k]*(1. f - x[k]*x[k])*v[k] + fb*vfb);
     x[k] += (v[k] + in[i]);}
```

The behavior of the coupled oscillators is easy to describe for extreme values of ϵ. When ϵ is zero, the oscillators are independent and oscillate at their natural frequencies. When ϵ exceeds a certain value, the oscillators are in synchrony. The transition from complete synchrony to asynchrony as ϵ diminishes takes place through bifurcation. The illustration below shows five Van der Pol oscillators as a function of their feedback factors. In this simulation over 25,000 samples the feedback is realized in the same way as in the Max patch. Over the duration *dur* of the simulation, the feedback factor *fb* sinks from 0.0067 to 0 ($fb(i) = 0.0067(1 - i/dur)^{0.69}$).

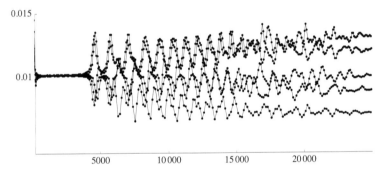

3.4.5.5 Synchronization of Chaotic Oscillators

First let us look at a very simple illustration showing chaotic behavior, the so-called *skew tent map*. On the unit interval it is defined as:

$$x(k+1) = f(x(k)) = \frac{x(k)}{a} \text{ if } 0 \leq x(k) \leq a, \frac{1-x(k)}{1-a} \text{ if } a \leq x(k) \leq 1.$$

The figure at the left below shows the function *f* for the parameter $a = 0.7$. The figure at the right shows the time series of 200 iterations of such a map with a random starting value.

```
f[x_] := If[x < 0.7, x/0.7, (1 - x)/0.3];
```

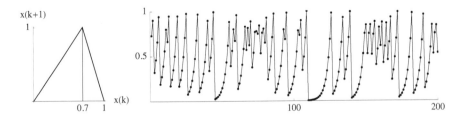

3.4 Dynamic Systems and Feedback Control

There are many ways to couple two functions $x(k + 1) = f(x(k))$ and $y(k + 1) = f(y(k))$. We choose one where the difference between the systems' states gets smaller and where the coupling disappears when the states are equal ($y = x$).

$$x(k + 1) = (1 - \epsilon) \cdot f(x(k)) + \epsilon \cdot f(y(k))$$
$$y(k + 1) = (1 - \epsilon) \cdot f(y(k)) + \epsilon \cdot f(x(k))$$

When ϵ is zero, the systems are independent of one another. When ϵ approaches 1/2, they become synchronous after just a few steps, although the (common) time series is chaotic. This state is known as complete synchronization.

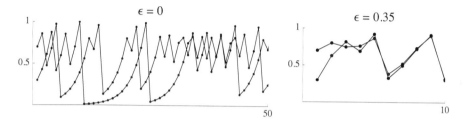

The critical value at which complete synchronization occurs is about $\epsilon_c = 0.228$. When ϵ is a bit smaller, nearly completely synchronized and unsynchronized phases alternate irregularly (see 3.4.3.3 on Intermittence). The next figure shows the difference $x(k) - y(k)$ of two time series for $\epsilon = 0.227$.

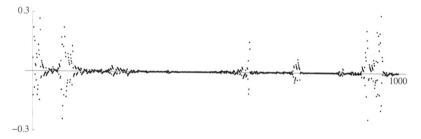

We can study the behavior of continuous oscillators using the Rössler oscillator (3.4.3.4). The following lines are from the mxj~ object cm_roessler, which is used in the following patches. The oscillation frequency depends essentially on the constant dt. In the examples which follow, $a = b = 0.2$ and the constant c is variable. When $c < 3$, periodic oscillation results, when $c > 3$, the periods are doubled, leading to chaotic behavior when $c \approx 4.3$.

```
;mxj~ cm_roessler
x += dt*(-y0 - z0);
y += dt*(x0 + a*y0);
z += dt*(b + (x0 - c)*z + in[i]);
```

The Max patch *Roessler_1* shows how the Rössler oscillator, within a certain range, takes on the frequency of an exciting oscillation. At the same time, it demonstrates how the influence of excitation can change originally chaotic behavior (figures on the left below) into periodic oscillation (figures on the right). The figures show at the top left the oscillation of the Rössler oscillator, at the lower left the excitation (considerably enlarged) and on the right the trajectory in the phase space $(x, \dot{x})$.

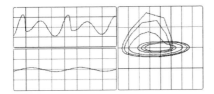 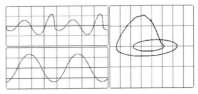

The Max patch *Roessler_2* demonstrates how coupling two Rössler oscillators having nearly identical frequencies leads to synchronization. If the oscillators' parameters are the same, the synchronization can be perfect.

Even the smallest differences in the initial conditions of chaotic systems lead rapidly to different trajectories (3.4.3.1). So it is astonishing that two identical uncoupled systems can be synchronized by being excited with noise. Chaotic systems can "forget", as it were, their initial conditions. Max patch *Roessler_3* shows the oscillations of two Rössler oscillators and their difference. The oscillators can be desynchronized by resetting one of them. Depending on the values of ω and μ, it can take a long time for the synchronization to become perfect again. We can see the same behavior in non-chaotic nonlinear systems. The following figure shows the time series of two uncoupled Van der Pol oscillators excited by noise ($\omega = .1$, $\mu = .25$). The asymmetrical waveform is due to the fact that the noise only has positive values.

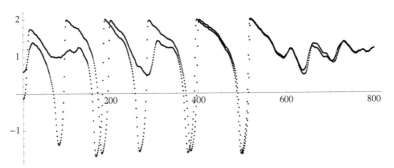

Within one period of oscillation, many self-sustained oscillators go through a phase of slow variation and a phase of rapid variation. For example, neurons slowly build up tension and then discharge it rapidly. Oscillators of this type are called integrate-and-fire or accumulate-and-fire oscillators. Examples are the Van der Pol oscillator with a large nonlinearity value μ or the skew tent map where a is almost equal to 1. The mxj~ object *cm_integrate_fire* realizes a simple oscillator by incrementing the variable x by a constant c and a random value until x is greater than 1. Then x is reset to zero, the feedback variable *sync* is set to 1 and the discharge is indicated by a "bang". The value of the variable *sync* is quickly reduced.

```
;mxj~ cm_integrate_fire
x += (c + drand*Math.random());
if(x + exc > 1)
{   sync = 1.f; x = 0 ; outletBang(2);}
outlet(1,sync); sync*=0.9; outlet(0,x);
```

Several such objects are coupled together in the Max patch *Integrate_and_Fire*. They produce more or less synchronous rhythms, depending on the parameter values used.

4 Computer Programs and Programming Languages

Commercial sound synthesis and sound treatment programs are generally well documented and will not be a topic here. But we will discuss a certain number of programming languages. C is a fundamental, flexible and common programming language. We present it here in order to demonstrate details of certain procedures that are usually combined into functions in synthesis programs. Most of the sound synthesis programs in this book are programmed in C and can be extended by C routines known as externals or plugins. Csound is a programming language specially developed for sound synthesis. In its division into scores and orchestras, it originally reflected a traditional vision of composition, although it has subsequently been greatly extended. The program Mathematica can perform numerical and symbolic computation, solve equations and can generate and visualize functions, lists and diagrams. Graphic sequences can be played as animations, and simple interactive *graphic user interfaces* (GUIs) can be made. In addition, lists of numbers can be played as sound and stored in various formats. Max is a program with a graphic interface that can generate, record, process and play sounds in real time. Processing is an object-oriented programming language with an integrated development environment for graphics, simulation and animation.

In this Chapter one will only find short descriptions of the programming languages used in this book. The explanations are restricted to special applications and procedures one will encounter in the text.

Further Reading: The Audio Programming Book *edited by Richard Boulanger and Victor Lazzarini [86]. A history of programming music and sound can be found in* The Cambridge Companion to Electronic Music *[83] by Julio d'Escrivan and Nick Collins, pp. 55-89.*

4.1 Csound

The sound-generating part of Csound, called the orchestra, plays the events, called notes, of a so-called score. The orchestra consists of instruments made up of elements such as oscillators, noise generators and filters and including instructions about the signal flow between the elements. Information like parameter values for the notes, tempo, etc. that the orchestra requires for the sound synthesis are taken from the score.

Further Reading: The Csound Book *[40] edited by Richard Boulanger and* The Audio Programming Book *edited by Richard Boulanger and Victor Lazzarini [86], pp. 581-626.*

4.1.1 The Syntax of the Csound Orchestra

4.1.1.1 The Structure of the Orchestra

A Csound orchestra consists of a header, where global parameters are defined, and instruments, where instructions are given that determine the sequence of program events and generate control and audio signals. In the header, sample rate *sr*, control rate *kr* and the number of channels *nchnls* are defined. The number of sample periods per control period *ksmps* (= *sr*/*kr*) can be given in place of the sample rate. The amplitude of the signals generated is between 0 and $32767 = (2^{15} - 1) = 0$ dB full scale. The statement 0dbfs = *n* sets the maximal amplitude to a different value (in particular 1.0).

```
sr     =    44100
kr     =    100
ksmps  =    441
nchnls =    2
0dbfs  =    1
```

An instrument definition begins with the statement *instr* followed by a number *n* (assigned by the user) and ends with the statement *endin*.

```
instr       n1
[ Commands ]
endin
instr       n2
[ Commands ]
endin
```

A command in Csound has this form:

```
[label:] result opcode argument1, argument2,...
[;comment][/*...comment...*/]
```

The label is optional and identifies the command in case of branching. Comments begin with a semicolon and extend to the end of the line. Arbitrarily long comments are enclosed between /* and */. Every command must be on a single line. Commands that are too long for one line must have a backslash at the end of the incomplete line. The opcode (operation code) is a pre-programmed subroutine that computes output values (*result*) from input values and parameters (*arguments*).

4.1.1.2 Constants and Variables

Numbers in Csound are either variables or constants and are always represented as floating point numbers. Variables can be changed at four different times: at setup time for the orchestra, at the initialization of an instrument (i-time), at the beginning of a control period (k-time) or at the beginning of every sample period (a-time, or audio-time). Variables and constants can be defined as local or global. The scope of local variables is one instrument; their names begin with i, k or a, depending on how often they are updated. The scope of global variables is the entire orchestra; their names begin with gi, gk or ga. Parameter values passed from the Csound score have names beginning with p followed by a number indicating the parameter's position in the score file. The header variables are reserved and can be used in expressions but cannot be changed after setup.

Type	When renewable	Local	Global
Reserved Symbols	–	–	r*symbol*
Score Parameters (pfields)	i – time	p*number*	–
Initialization Variables	i – time	i*name*	gi*name*
Control Signals	k – time	k*name*	gk*name*
Audio Signals	a – time	a*name*	ga*name*

Here number is a positive integer referring to a score parameter field and name is an arbitrary string of letters and/or digits.

Audio signals are stored as one-dimensional arrays of length *ksmps*. They are processed as a vector once per control period. This can sometimes lead to unexpected results. For instance, the assignment $a1 = -a1$ does not give a tone of frequency $sr/2$ but rather a square wave of frequency $kr/2$. By the same token, recursive filters can only be implemented by setting the control rate equal to the sampling rate (4.1.1.10).

A control signal can be changed into an audio signal using the function *upsamp*, and an audio signal can be changed into a control signal using the function *downsamp*. In downsamp, the (optional) variable *iwlen* gives the window length in samples over which the amplitude of the audio signal is averaged.

```
kres    downsamp    asig[, iwlen]
ares    upsamp      ksig
```

If the control signals change rapidly, the control rate should be high. Control parameters can also be defined as audio signals. In frequency modulation, for example, the modulating oscillator generates a control signal for the frequency of the carrier oscillation, but the signal must be defined as an audio signal.

4.1.1.3 Functions

Csound provides the most important mathematical functions as well as functions for converting amplitude and frequency into various units. Some of these functions are listed below, together with an indication of the argument types allowed with each function. For mathematical reasons, the range of values for arguments is often limited. Arguments can be not only numbers and variables but also expressions.

int(x)	returns the integer part of x (init- and control-rate)
frac(x)	returns the fractional part of x (init- and control-rate)
abs(x)	returns the absolute value of x (no rate restriction)
exp(x)	returns e^x (no rate restriction)
log(x)	returns the natural logarithm of x (no rate restriction), x > 0
sqrt(x)	returns the square root of x (no rate restriction), x >= 0
sin(x)	returns the sine of x (no rate restriction), x in radians
cos(x)	returns the cosine of x (no rate restriction), x in radians
tan(x)	returns the tangent of x (no rate restriction), x in radians
ftlen(x)	returns the length of stored function number x (only init-rate)
dbamp(x)	returns the decibel equivalent of the amplitude x (init- and control-rate)
ampdb(x)	returns the amplitude equivalent of the decibel value x (init- and control-rate)

(→ Additional Functions)(→ Amplitude in Decibels)

4.1.1.4 Assignment, Operators, Expressions and Conditional Expressions

In algebra, the expression $x = x + 1$ is an equation in x, to be sure, one without a solution. In computer programming languages it is an assignment which replaces the current value of x by the value of x augmented by one. The sign = is called an assignment operator. In Csound, variables can only be assigned values having the same or a lower rate.

```
ivar2   = ivar1
kvar2   = kvar1    or   kvar2   = ivar1
avar2   = avar1    or   avar2   = kvar1    or   avar2   = ivar1
```

The command *init* assigns a value to control- or audio-rate variable when an instrument is initialized.

```
kvar2      init     ivar1
avar2      init     ivar1
```

Csound provides the arithmetic operators +, −, *, / and ^ as well as the logical operators && (logical AND) and || (logical OR). Expressions are made up of functions, operators, variables and constants. (→ Rules of Precedence)

The computation of the individual terms of an expression is carried out in the rate of variables used. So in the expression $k1*p6 + abs(p5) - .01*i1 + 1/sqrt(2)$, the term $1/sqrt(2)$ is computed when the orchestra is initialized, the terms $abs(p5)$ and $.01*i1$ are computed when the instrument is initialized, and $k1*p6$ is computed at the beginning of each control period.

Conditional expression give different values depending on whether a logical condition is fulfilled or not. If the condition Q is fulfilled, then the expression (Q ? *var1* : *var2*) returns *var1*. Otherwise, it returns *var2*. These relations between the values a and b can be used to formulate the condition: >, <, >=, <=, == (equal), != (unequal). The values a and b can be expressions but they cannot contain audio-rate variables. (→ Examples)

```
( Q ? var1 : var2 )
```

4.1.1.5 Control Flow

The following is a list of instructions that can determine the order in which program sections are executed. *label* is a user-defined name whose scope is the current instrument. The instruction *tigoto* is used with slurred notes. The branch instruction *timout* is only executed during the first *idur* seconds after starting time *istrt*. (→ Example)

```
igoto         label
tigoto        label
kgoto         label
goto          label
if ia R ib    igoto label       ;ia R ib means: ia is in relation to ib
if ka R kb    kgoto label
if ia R ib    goto     label
              timout   istrt, idur, label
```

There are also instructions for reinitializing an instrument while it is playing. If the program encounters the instruction *reinit*, computation is interrupted and the program block between *label* and the instruction *rireturn* is re-initialized. Then computation continues from the place it was interrupted. The instruction *rigoto* is the same as *igoto* but must be used when re-initializing. (→ Examples)

```
reinit        label
rigoto        label
rireturn
```

An instrument can control its own duration with the instructions *turnon*, *ihold* and *turnoff*. *turnon* activates instrument number *insno* after a delay of *itime* seconds; it must be placed outside the orchestra's instrument definitions. The instruction *ihold* causes an instrument to be held indefinitely. The note can be turned off using *turnoff* or by playing another note with the same instrument. (→ Example)

```
turnon    insno[,itime]
ihold
turnoff
```

4.1.1.6 Simple Signal Generators (→ CD)

4.1.1.7 Signal Modifiers (→ CD)

4.1.1.8 Delay Lines

The unit generator *delay* delays the signal *asig* by *idlt* seconds, the unit generator *delay1* delays it by one sample period. The unit generator pair *delayr* and *delayw* creates a delay line of length *idlt* seconds. *delayr* establishes the necessary storage area, *delayw* writes the signal into it. The delayed signal can be read out by tapping the delay line with the unit generator *deltap* at time *kdlt* (or interpolating with *deltapi* at time *xdlt*). *iskip* indicates the initial state of the storage area (default 0). The unit generator *multitap* delays a signal by several invariable delay times *itime1*, *itime2*, etc. The delayed signals are scaled by the gain factors *igain1*, *igain2*, etc. before being added together and output. (Examples see 5.1.4.3)

```
ares      delay       asig, idlt[, istor]
ares      delay1      asig[, istor]
ares      delayr      idlt[, istor]
ares      deltap      kdlt
ares      deltapi     xdlt
          delayw      asig
ares      multitap    asig, itime1, igain1, itime2, igain2 . . .
```

4.1.1.9 Sound Input and Output

The unit generators *in*, *ins* and *inq* read audio data from the standard audio input buffer, the unit generator *soundin* reads data from a sound file.

```
a1                    in
a1, a2                ins
a1,..., a4            inq
a1                    soundin ifilcod[, iskptim][, iformat]
a1, a2                soundin ifilcod[, iskptim][, iformat]
a1,..., a4            soundin ifilcod[, iskptim][, iformat]
a1[,a2[,a3,a4]] diskin  ifilcod, kpitch
                      [, iskiptim][, iwraparound] [, iformat]
```

The unit generators of the family *outx* output any number of signals in any instrument. The opcode name has to agree with the declaration of the number of channels *nchnls* in the orchestra header. The unit generators *outs1*, *outs2*, *outq1*, etc. send signals to specific output channels.

```
out       asig
outs1     asig
outs2     asig
```

```
outs      asig1, asig2
outq1     asig
outq2     asig
outq3     asig
outq4     asig
outq      asig1, asig2, asig3, asig4
```

The unit generator *pan* distributes a signal to four channels.

```
a1,..., a4   pan   asig, kx, ky, ifn[, imode][, ioffset]
```

4.1.1.10 Opcodes

Csound also allows user-defined opcodes. There are many situations where a user-defined opcode is useful. One situation is implementing feedback, which requires one sample delay. One can either set the control rate of the entire orchestra equal to the sampling rate, or one can create an opcode with its own control rate and set *ksmps* = 1. The following example shows how to define an opcode for a simple recursive low-pass filter with the system equation $y[k] = c \cdot y[k-1] + x[k]$. The definition of the opcode *lp_filter* begins with the statement *opcode* and ends with the statement *endop*. After the opcode's name, the output and input signals are declared. Here, (... *a*, *ak*) means that the opcode generates an audio output and uses one audio signal and one control signal. (→ 4-1-1-10.csd)

```
...
ksmps = 100
...
opcode      lp_filter, a, ak
            setksmps    1
aout        init        0
ain,k1      xin
aout        =           k1*aout + ain
            xout        aout
endop

instr 1
a1          rand        .003
aout        lp_filter   a1, p4
            out         aout
endin
...
;           p4 = filter coeff. c
i1  0  4    .99
```

4.1.2 The Score

A Csound score contains the notes played by the instruments of an orchestra, a list of the tempi in which to play, functions to generate the tables used by instruments and delimitations of sections and of the end of the score. Each score statement begins with a new line and an opcode. The most common opcodes are: f, i, a, t, s and e. The opcode is followed by parameter values, either floating point numbers, alpha-numeric information within double quotes, expressions to be evaluated within square brackets [ ] or macros beginning with a dollar sign $. (Certain other characters can appear in the score. See the chapter "The Standard Numeric

Score" in the Csound manual.) One calls the first parameter value after the opcode *p1*, the second value *p2*, etc. The first parameter value can follow immediately after the opcode, all others must be separated by at least one space. A score instruction can continue on the next line by inserting a backslash at the end of a line.

```
opcode     p1  p2  p3  p4  ....      [;  comment]
```

Notes are indicated by the opcode i. The first three parameters have fixed meanings. The parameter *p1* indicates which instrument should play. A tied note from the same instrument can be indicated by an optional fractional notation. A negative values ends a tied note. *p2* shows the start time of the note in beats (or seconds if no tempo is given) and *p3* gives the note's duration in beats or seconds. If *p3* is negative, the note is held until the next call of the instrument turns it off (cf. *tigoto*) or the end of the score is reached. All other parameters are defined in the instruments.

One instrument can play arbitrarily many notes simultaneously. A note's start time is counted from the beginning of the current section and its duration can be set at initialization time or by the opcode *linenr*.

```
i     p1  p2  p3  p4  ....
```

Commands beginning with f call up a GEN-routine (4.1.3) that generates a table. An existing table can be erased either by using an f-command with a negative *p1* or by generating another table with the same number. An f-command with *p1* = 0 and *p2* > 0 does not create a table but an action time and can be used to lengthen a section.

```
f     p1  p2  p3  p4  ....
```

4.1.3 Function Table Generators

GEN-routines (function table generators) are procedures that generate tables whose values can be used by instruments. GEN-routines are called by f-commands. The first four parameter fields have fixed meanings. *p1* gives the function number, *p2* the time of generation, *p3* the table size (generally 2^n or $2^n + 1$) and *p4* gives the number of the GEN-routine to be used. The other parameters have a specific meaning for each GEN-routine. Most routines generate functions that are normalized by dividing all values by the largest absolute value generated, so that the values are between −1 and +1. If the number of the GEN-routine is negative, normalizing will be omitted.

```
f   #   time   size   Nr   p5   p6  ...
```

GEN01 reads the data from a sound file into a function table.

```
f   #   time   size   1   filcod   skiptime   format   channel
```

GEN02 reads the values of its parameter fields into a function table.

```
f   #   time   size   2   value1   value2   value3  ...
```

GEN03 generate a table by evaluating a polynomial (A4.3.3) with given coefficients over a specified interval $[x1, x2]$.

```
f  #  time  size  3  x1 x2  c0  c1  c2  ...  cn
```

GEN04 generates a normalizing table by examining an already existing table.

```
f  #  time  size  4  source#  sourcemode
```

GEN05 generates a function from exponential segments.

```
f  #  time  size  5  value1  length1  value2  length2  value3 ...
```

GEN06 generates a function from cubic polynomial segments.

```
f  #  time  size  6  value1  length1  value2  length2  value3 ...
```

GEN07 generates a function from linear segments.

```
f  #  time  size  7  value1  length1  value2  length2  value3 ...
```

GEN08 generates a piecewise cubic spline curve, which is the smoothest possible curve through the points given.

```
f  #  time  size  8  value1  length1  value2  length2  value3 ...
```

GEN09 and GEN10 generate functions from weighted sums of simple sinusoids.

```
f  #  time  size  9  partial1  amp1  phase1  partial2  amp2  phase2 ...
f  #  time  size  10 amp1  amp2  amp3 ...
```

GEN11 generates a function corresponding to a given set of cosine partials, similarly to the unit generators *buzz* and *gbuzz* (4.1.2.9, 5.3.2.1).

```
f  #  time  size  11  number_of_harmonics  [lowest_harmonic]
```

GEN12 generate the logarithm of a modified Bessel function of the second kind, order 0, for use in amplitude-modulated frequency modulation (6.1.2).

```
f  #  time  size  -12  xint
```

GEN13 and GEN14 generate polynomials from Chebyshev polynomials of the first and second kind respectively (6.1.3).

```
f  #  time  size  13  xint  xamp  h0  h1  h2  ...  hn
f  #  time  size  14  xint  xamp  h0  h1  h2  ...  hn
```

GEN15 generates two polynomial functions for use in phase quadrature.

```
f  #  time  size  15  xint  xamp  h0  phs0  h1  phs1  h2  phs2 ...
```

GEN16 generates a function of user-defined shape from a given starting value to a given end value.

```
f  #  time  size  16  beg  dur  type  end
```

4.1 Csound

GEN17 generates a step function.

```
f  #  time  size  17  x1  a  x2  b  x3  c  ...
```

GEN20 generates window functions (3.3.2.5) for use in spectral analysis (5.2.2.2) and in granular synthesis (7.1.1.1).

```
f  #  time  size  20  window  [  max  opt  ]
```

GEN21 generates functions with various random distributions (10.1.5).

```
f  #  time  size  21  type  lvl  arg1  arg2
```

GEN25 and GEN27 generate functions from exponential and linear segments respectively. The difference to GEN5 and GEN7 is solely in the format of the input parameters.

```
f  #  time  size  25  x1  y1  x2  y2  x3  y3
f  #  time  size  27  x1  y1  x2  y2  x3  y3
```

4.1.4 Generating Scores and Events

A Csound score can contain any number of notes. The scores of algorithmic and stochastic compositions, in particular, are generated by computer programs. Special programs and libraries have been developed for this purpose. The Csound package contains a program, Cscore, functions of which can be used in C programs to generate and alter Csound scores. In this chapter, we show some ways to generate scores and save them as text files, and we show how with the protocol OSC events can be generated in real time and sent to Csound.

4.1.4.1 Generating Events With Csound Instruments

Csound instruments can add notes to a score using the opcodes *event*, *schedkwhen*, *schedule* and *schedwhen*.

```
event "scorechar", kinsnum, kdelay, kdur, [, kp4] [, kp5] [, ...]
```

In the following example, every note of instrument 1 generates another note. To avoid generating a new note at every control period, a trigger is set to 1 at the beginning and after writing the note reset to 0. The only purpose of the function *f0*, generated at time 100, is to ensure that Csound runs that long. (→ *4-1-4-1.csd*)

```
instr 1
ktrig     init    1
          if      (ktrig == 0)    goto cont
kfreq     rand    600
          event   "i",1,.2,.1,.2,kfreq+1000
ktrig     =       0
cont:
kamp      expon   p4,p3,.001
aout      oscil   kamp,p5,1
          out     aout
endin
```

```
<CsScore>
f0 100 1024 0
f1 0 1024 10 1
i1 0 .1 .2 440
e
```

4.1.4.2 Generating Scors With Programs in C

This section describes the simplest way to generate a score using C. The first example shows the commands necessary for creating a file. The first line embeds the functions required for the input and output of data (stdio means "standard input/output"). The second line and the braces in lines 3 and 8 indicate the outer frame of the program. The lines 3, 4 and 5 define, open and close a file. The name of the new file in double quotes and the code "w" for "write" are passed to the command fopen(). The commands to generate and write values in the score take the place of the empty line 5.

```
#include <stdio.h>                    //1
int main()
{   FILE *fp;
    fp=fopen("4-1-4a.sco","w");
    .....                             //5
    fclose(fp);
    return 0;
}
```

The function fprintf(*file*, *string*) writes a character string to a file. If we put the following lines of code in place of the empty line 5 above,

```
fprintf(fp, ";4-1-4a.sco");
fprintf(fp, "\ni1 0 5");
```

we write into the file "4-1-4a.sco" these two lines, the first of which is a comment and the second of which is a note:

```
;4-1-4a.sco
i1 0 5
```

Notice that after the numbers 1 and 0 we have to insert a space. The control character "\n" is the newline character and moves the output following to the beginning of the next line.

Parameter values calculated by functions are defined as variables. The function fprintf(*file*, *format*, *Var*) writes the value *Var* to a file in a specified format. The following program uses a loop to make six successive notes of random duration and amplitude. Because we use the random function rand() and the exponential function exp(), we need to link the header files stdlib.h (for rand) and math.h (for exp) into our program (lines 2-3). In lines 6 and 11 we define the variables *p4* for amplitude, *n* for the number of notes and *i* as counter for the loop as integers, the variable *p3* for the duration and *p5* for the frequency as floating point numbers (line 7). After opening the file "4-1-4-2.sco", we write the file's name in the first line and a function assignment using the opcode *f* (line 10). The function rand() generates pseudo-random integers in the interval [0, RAND_MAX] (RAND_MAX = 32767 or 2147483647, depending on the system configuration). To get a fractional duration for *p3*, we divide the random numbers by RAND_MAX and multiply by 2.5 (line 12). We get random amplitudes between 0 and 20000 in the same way (line 12). The frequencies are exponentially distributed

4.1 Csound   173

between exp(5) = 148.41 and exp(5 + 2) = 1096.6 (line 13). Finally, the parameter values are written to the file in the proper order and with the proper spacing (lines 14). The placeholder for floating point numbers is indicated by "%f", the placeholder for integers by "%d".

```
// 4-1-4-2.cpp
#include <stdio.h>
#include <math.h>
#include <stdlib.h>
int main()                                                              //5
{    int p4,n=6;
     float p3,p5;
     FILE *fp;
     fp=fopen("4-1-4-2.sco","w");
     fprintf(fp,";4-1-4-2.sco");fprintf(fp,"\nf1 0 32768 10 1");       //10
     for(int i=0;i<n;i++)
     {    p3=2.5*rand()/(float)RAND_MAX; p4=20000.*rand()/(float)RAND_MAX;
          p5=exp(5.+2.*rand()/(float)RAND_MAX);
          fprintf(fp,"\ni1 +   %f    %d   %f",p3, p4, p5);
     }                                                                  //15
fclose(fp); return 0;
}

;4-1-4-2.sco
f1 0 32768 10 1
i1 + 0.000020    2630     672.640259
i1 + 1.146625    10655    229.962982
i1 + 0.117612    13577    577.433228
i1 + 2.336732    7670     419.403198
i1 + 2.077413    691      165.161423
i1 + 1.324250    13422    150.715866
```

4.1.4.3 Generating Scores and Events Using *Processing*

A Processing program consists of a function setup() that is executed at the beginning of the program, a function draw(), that is executed at a specified frameRate and functions that react to events (4.5). The program that follows creates a window 400 by 400 pixels large (line 4). When the right mouse button is pressed, a circle is printed in the window at the position of the cursor (line 13) and a note is written (line 14). When any key is pressed (line 7), the notes are written (line 8) to the output file (line 3), the file is closed and the program exits. Besides the commands particular to Processing, one can use any commands or classes from the Java programming language (4.5.1).

```
// Csound_score.pde
PrintWriter output;                                    //1
void setup() {
  output = createWriter("csore.txt");
  size(400,400);
}                                                      //5
void draw() {}
void keyPressed() {
  output.flush();
  output.close();
  exit();                                              //10
```

```
}
void mousePressed() {
  ellipse(mouseX,mouseY,10,10);
  output.println("i1"+"\t"+mouseX/10.f+"\t"+0.1f+"\t"+(600-mouseY));
}

i1  2.4    0.1  489
i1  8.5    0.1  519
i1  15.2   0.1  549
i1  27.7   0.1  447
i1  27.4   0.1  504
i1  27.4   0.1  571
```

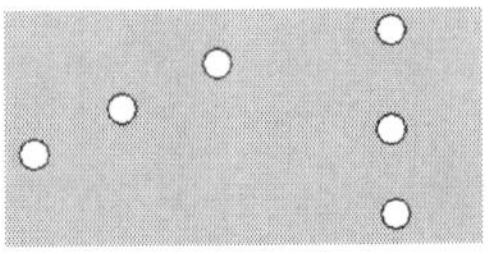

Using *Open Sound Control* (OSC) messages, data can be sent to Csound in real time. In the program *osc_send.pde* the logistic map (3.4.3.3) (line 20) generates new values at the frameRate that are then sent to the (arbitrarily chosen) port 3334 of the computer (lines 10 and 23).

```
//osc_send.pde
import oscP5.*;
import netP5.*;
OscP5 oscP5;
NetAddress myRemoteLocation;
float x = 0, y = 0.63454, r = 3.0;                              //5

void setup() {
  size(800,400);
  frameRate(15);
  oscP5 = new OscP5(this,3334);                                  //10
  myRemoteLocation = new NetAddress("127.0.0.1",3334);
}

void draw() {
  r = 2.9 + 1.1*mouseX/width; println(r);                        //15
  fill(231,10); rect(0,0,width,height);
  translate(5,0);
  fill(23); rect(x, height/2+250*(y-0.5),10,10);
  x+=10; if(x>width-10) x=0;
  y=r*y*(1-y);                                                   //20
  OscMessage myMessage = new OscMessage("/logistic");
  myMessage.add(y); /* add an int to the osc message */
  oscP5.send(myMessage, myRemoteLocation);
}
```

The first instrument of the Csound orchestra *4-1-4-3_osc_listen.csd* receives the output of the Processing program above and interprets it as the frequency of a note generated by the second instrument.

```
gihandle OSCinit 3334

instr 1
kf1       init       400
kf2       init       0
```

4.1 Csound

```
nxtmsg:
kk      OSClisten    gihandle, "/logistic", "f", kf1
        if (kk == 0) goto ex
        event        "i", 2, 0.5, .4, .1, 400+2000*kf1
        kgoto        nxtmsg
ex:
endin

instr 2
kamp    line         p4,p3,0
a1      oscil        kamp, p5, 1
        out          a1
endin
```

4.1.4.4 Generating Scores and Events With *Max*

The illustration below on the left shows a Max patch (*Csound_Score.maxpat*) that can accept and store as text four parameter values. The Max patch illustrated on the right (*OSC_Send.maxpat*), like the Processing program, generates numbers using the logistic map and sends them to the Csound program shown above (or to *OSC_receive.maxpat*).

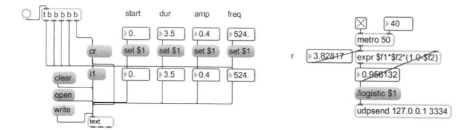

4.1.4.5 Generating Scores With *Python*

The following Python program creates the same score as the C program in chapter 4.1.4.2.

```
import random
import math

n = 6

outFile = open("4-1-5.sco","w")
outFile.write(";4-1-5.sco")
outFile.write("\nf1 0 32768 10 1")

for i in range(n):
    p3 = 2.5*random.random()
    p4 = random.randrange(20000)
    p5 = math.exp(5. + 2.*random.random())
    outFile.write("\ni1 + "+str(p3)+" "+str(p4)+" "+str(p5))

outFile.close()
```

4.2 Max

Max is a graphical development environment for real-time musical and multimedia applications. Objects like oscillators, filters, etc. are placed in a window called the *patcher* and are graphically connected to each other by *patch cords* indicating signal flow. Max produces graphic programs that resemble flow charts.

4.2.1 Fundamentals

The first version of Max was developed by Miller Puckette in the mid-1980's at IRCAM in Paris. First known as the Patcher editor, Puckette subsequently named the program for Max Mathews. It was used to create and process control signals. The program was extended to include MSP (Max Signal Processing) for sound synthesis and processing in 1997 and Jitter for video processing and three-dimensional graphics in 2002. Max/MSP quickly became the most used program in live electronics, thanks to intuitive design, graphic objects to show data and signal flow, easy ways to create new objects and a great variety of possible interfaces. Max/MSP is also used for measurement tasks, tests, audio-visual demonstrations and much more.

Max has graphic objects like number boxes, sliders, level meters, etc. as well as objects that are created simply by entering a name in a neutral object field. MSP ojects are indicated by a tilde (~) at the end of the name, Jitter objects have names that are prefaced by *jit*. Connections between objects for control signals are shown by simple lines, connections for audio signals by black and yellow lines. Complex data, like strings and lists, can also be transferred from object to object.

4.2.2 Feedback

Because the execution of Max objects is activated by placing a value in the object's left inlet, feedback like that shown at the left below leads to stack overflow and interrupts execution. The illustration on the right shows two functioning feedback loops: on the left a counter and on the right the logistic map. In both cases, the feedback goes to the right inlet, and the calculation of the next value is activated by the object metro.

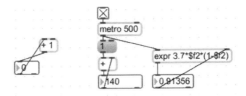

For the implementation of recursive systems, a delay of one sample is required. The object *delay~* produces a one-sample delay but does not allow feedback. The system shown at the left below can be made without getting an error message, but it has no output. A better feedback system can be made using *tapin~* and *tapout~* (illustration to the right below). The shortest delay for audio feedback depends on the current vector size, so the vector size needs to be set to 1 when implementing recursive systems. This can be done either in the menu "DSP Status" or by sending the message "sigvs 1" to the DSP. One sample is equivalent to a delay of .03 ms, but such a short vector size puts a heavy load on the CPU. The object *poly~*

can have its own vector size, which can be set in the object's inspector (or by sending the message "vs n", where n is the vector size). In the example to the right, the patch's vector size is 1024, which means that the delay shown of 0.12 ms (about 5 samples) is not possible. But we can implement the recursive system in a subpatch, giving it a vector size of 1 ms. Because of feedback, a comb filter is produced. (→ links) (→ *feedback.maxpat*)

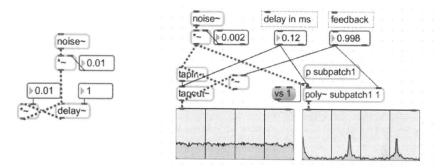

With the Max external *icst.fexpr~* feedback can be realized by writing a recursive sytem equation. The output signal delayed by n samples is $y[-n]$, the ith intup signal delayed by n samples is $xi[-n]$. With the following example we realize the system $y(k) = cy(k-5) + x(k)$. Disadvantages of this object are the restriction of the delay to constants and the computational cost. (→ icst/*icst.fexpr~.maxhelp*) (→ *Filter_fexpr.maxpat*)

The Max object *gen~* creates a signal processing routine from an embedded patcher that can contain arbitrary short delays with feedback. The following figure shows to the left a detail from a Max patcher and the embedded sub-patcher to the right. The maximum delay is given as argument in the delay object and the actual, variable delay is given as a paramter witch is labeled *del* in this example. With 1 sample delay and a feedback factor 0.998 this *gen~* object is a recursive low-pass filter. (→ *Feedback2.maxpat*)

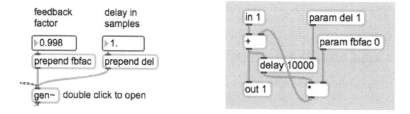

4.2.3 mxj-Objects

In Max, the user can program her or his own objects, so-called externals, in C or in Java. We will show here the quickest method to program externals, namely in Java using a simple programming environment included with Max. (This is how the externals in this book were programmed.) Externals written in C execute somewhat faster, but their code is longer and more complicated. Java externals can be written and tested while Max is running.

If we create a Max object *mxj quickie* named *cub_osc* and double-click on it, an editor opens with a default template for the new external's source code. The template can be set as the second argument of the object quickie (here mxj quickie cub_osc MSPOBJ_PROTO). The default template contains code for an object *gain*, however the name *cub_osc* has already been inserted in the name of the class (line 4) and the name of the constructor (line 9).

```
import com.cycling74.max.*;
import com.cycling74.msp.*;

public class cub_osc extends MSPPerformer
{                                                                           //5
    private float _gain = 0.f;
    private static final String[] INLET_ASSIST = new String[]{"input sig"};
    private static final String[] OUTLET_ASSIST = new String[]{"output sig"};
    public cub_osc(float gain)
    {   declareInlets(new int[]{SIGNAL}); declareOutlets(new int[]{SIGNAL});  //10
        setInletAssist(INLET_ASSIST); setOutletAssist(OUTLET_ASSIST);
        _gain = gain;
    }
    public void inlet(float f)  {   _gain = f;  }
    public void dspsetup(MSPSignal[] ins, MSPSignal[] outs) {  }            //15
    public void perform(MSPSignal[] ins, MSPSignal[] outs)
    {   int i;
        float[] in = ins[0].vec;
        float[] out = outs[0].vec;
        for(i = 0; i < in.length;i++)                                        //20
        {   out[i] = in[i] * _gain;
        }
    }
}
```

The example below shows the code for a cubic oscillator. We declare variables (line 6) for displacement (x), velocity (v), damping (d) and a factor for the restoring force (c). We give the text that will be shown when the cursor points to an inlet or an outlet of the object (lines 7-10), and we define a second inlet for a floating point control variable (line 12). If a number is passed to the mxj object at its instantiation, the number is assigned to the variable c (line 15). Because nonlinear oscillators can become instable, we introduce a reset method that is called if the object receives the message "reset" (line 17). We poll the control inlets 0 and 1 with an if-else test (lines 19-22). Finally, the *perform* method calculates the output signal out from the input signal in (lines 25-31). (→ *mxj_External.maxpat*)

```
import com.cycling74.max.*;
import com.cycling74.msp.*;

public class cub_osc extends MSPPerformer
{                                                                           //5
    private float x = 0.f, v = 0.f, d = 0.0001f, c = 0.f;
    private static final String[] INLET_ASSIST = new String[]{
        "input (sig) / float c","damping"
    };
    private static final String[] OUTLET_ASSIST = new String[]{"output (sig)"}; //10
    public cub_osc(float cin)
    {   declareInlets(new int[]{SIGNAL,DataTypes.FLOAT});
        declareOutlets(new int[]{SIGNAL});
        setInletAssist(INLET_ASSIST); setOutletAssist(OUTLET_ASSIST);
```

```
            c = cin;                                         //15
        }
        public void reset() {v = 0.f; x = 0.f; }
        public void inlet(float f) {
            if (getInlet() == 0) {
                c = f;                                       //20
            }
            else  d = f;
        }
        public void perform(MSPSignal[] ins, MSPSignal[] outs)
        {   float[] in = ins[0].vec;                         //25
            float[] out = outs[0].vec;
            for(int i = 0; i < in.length;i++)
            {   v += (-c*x*x*x + in[i]);
                v *= (1 - d);
                x += v;                                      //30
                out[i] = x;
            }
        }
    }
}
```

4.3 Mathematica

Mathematica is a fully integrated environment for technical computing. The program generates documents called notebooks. The notebooks consist of cells that can contain other cells, text, commands, numbers, pictures, etc. Computations and programming instructions can be carried out directly from a notebook. The results, which can be numbers, lists, graphics, sounds, etc., appear in cells of their own. Mathematica notebooks can be read by the freely available program Wolfram CDF Player, explained in Chapter 1. The introduction which follows gives a brief survey of those aspects of the program of particular interest for computer music and hence frequently used in this book.

4.3.1 Fundamentals

Mathematica's basic document is called a *notebook*. A notebook can contain texts, formulas, programs, pictures, sounds, etc., all of which are in individual *cells*. Each cell has a frame and is marked by a blue bracket at the right of the frame. Commands are executed by typing shift-return; their results appear in new cells. In addition to the cells containing executable calculations and those showing the output of these calculations, other cells can be defined with various text fonts and formats, background colors, etc. Double clicking on a cell alternately shows or hides subordinate cells. A cell can be given a cell tag, which can serve as a hyperlink address within or between notebooks.

When inputting, it is important to distinguish between various delimitation characters. Brackets [ ] are used in user-defined and Mathematica functions to group arguments. Braces { } group individual elements into lists. Parentheses ( ) are used to show priority in algebraic calculations. Additional character palettes extend the keyboard's input possibilities.

Mathematica can perform both numeric and symbolic calculations. Variables can be assigned either numerical values or terms containing other variables. A semicolon after a command means to execute the command without showing the result:

$a = 25; \sqrt{a}$ 5

If a variable has been assigned a value, this value is substituted for the variable in subsequent calculations. The command $a = .$ clears the variable a.

$$a =.; a \qquad\qquad a$$

Mathematica attempts to simplify expressions using transformation rules.

$$\sqrt{\frac{(\pi * e)^6}{\pi}} \qquad\qquad E^3 \, \pi^{5/2}$$

To get the numerical value of an expression one can write one number of the expression as a decimal value or use the command N[ ].

$$E^{3.0} \, \pi^{5/2} \qquad\qquad 351.365$$

$$N\left[E^3 \, \pi^{5/2}\right] \qquad\qquad 351.365$$

The following examples show some Mathematica commands for converting mathematical expressions.

$$\text{Expand}\left[(1 + x^2)\,(2 + x)^3\right] \qquad 8 + 12\,x + 14\,x^2 + 13\,x^3 + 6\,x^4 + x^5$$

$$\text{Factor}\left[8 + 12\,x + 14\,x^2 + 13\,x^3 + 6\,x^4 + x^5\right] \quad (2 + x)^3 \, (1 + x^2)$$

$$\text{TrigExpand}[\text{Sin}[2*x]] \qquad 2\,\text{Cos}[x]\,\text{Sin}[x]$$

A new function can be defined using a so-called *pattern*.

$$\text{dist}[x1\_, y1\_, x2\_, y2\_] := \sqrt{(x2 - x1)^2 + (y2 - y1)^2} \, ;$$

$$\text{dist}[1,1,2,2] \qquad\qquad \sqrt{2}$$

Sequences of numbers or expressions can be grouped into *lists*. One can get the *i*th element of the list l with the command $l[[i]]$. Indexing begins with 1.

$$l = \{10, 20, 30\}; \, l[[3]] \qquad\qquad 30$$

Functions can be applied to entire lists:

$$1/2 \qquad\qquad \{5, 10, 15\}$$

The following command stores in a table N the values of the function $f(x) = \sqrt{x}$ for $x = 1$ to $x = 5$:

$$\text{Table}\left[N\left[\sqrt{x}\right], \{x, 5\}\right] \qquad \{1., 1.41421, 1.73205, 2., 2.23607\}$$

The second argument determines the range and step size of a variable (here x).

$$\text{Table}\left[N\left[\sqrt{x}\right], \{x, 0, 1, .25\}\right] \qquad \{0., 0.5, 0.707107, 0.866025, 1.\}$$

4.3 Mathematica

One can define lists with arbitrarily many dimensions. The following command creates a list with 3·3 elements.

`l = Table[10*i + k, {i, 1, 3}, {k, 0, 2}]`

`{{10,11,12},{20,21,22},{30,31,32}}`

Mathematica provides many functions for list processing. These functions do not overwrite an existing list but rather create a new list.

`l = {}; Append[l, 0]; l`   `{}`

`l = {}; l = Append[l, 0]; l`   `{0}`

As in most programming languages, the equals sign is an *assignment operator*. Therefore, "$x = x + 1$" does not mean that x and $x + 1$ are equal, but rather assigns to the variable x the current value of x plus 1. To compare two values for equality, one uses the sign ==.

`2 + 2 == 4`   `True`

In addition to the equality sign ==, the following *conditional operators* can be used: != for inequality, < for less than, > for greater than, <= for less than or equal to, >= for greater than or equal to. Statements can be combined with these *logical operators*: ! for NOT, && for AND, || for OR and Xor[..,..,..] for Exclusive OR.

`2 < 5 && 7 < 5`   `False`

The command Solve[*eq*, *x*] creates a so-called rule list for the equation *eq*.

`Solve[a + 2 x^2 == 3, x]` $\left\{\left\{x \to -\frac{\sqrt{3-a}}{\sqrt{2}}\right\}, \left\{x \to \frac{\sqrt{3-a}}{\sqrt{2}}\right\}\right\}$

The command *x*/. gives a list of solutions from a rule list:

`x /. Solve[a + 2 x^2 == 3, x]` $\left\{-\frac{\sqrt{3-a}}{\sqrt{2}}, \frac{\sqrt{3-a}}{\sqrt{2}}\right\}$

The command Solve[{*eq1*, *eq2*..},{*x*, *y*, ...}] solves a system of equations.

`Solve[{2 x + 3 y == 0, x - y == 1}, {x, y}]` $\left\{\left\{x \to \frac{3}{5}, y \to -\frac{2}{5}\right\}\right\}$

The command Eliminate removes one variable from a system of equations and reduces the system by one equation.

`Eliminate[{2 x + 3 y == z, x - y == 1, a + x + y == z}, y]`

`2 x == 1 - a + z && 5 a == -1 + 3 z`

In many computations it is impossible to obtain exact symbolic solutions, or else the solutions are too complicated. In these case, one must fall back on numerical methods. The command N[*exp*, *n*] gives the numerical value of the expression *exp* to *n* decimal places.

$$N\left[\sqrt{2\pi}, 50\right]$$

2.5066282746310005024157652848110452530069867406099

The following apparently simple equation cannot be solved symbolically. FindRoot[ ] gives a numerical solution. One must indicate the value from which to begin searching for a solution.

```
x =.; Solve[Sin[x] == x - 1, x]
```

"The equations appear to involve the variables
to be solved for in an essentially non-algebraic way."

```
FindRoot[Sin[x] == x - 1, {x, -1}]
```
{x → 1.93456}

If one passes to the function Fit[ ] a list of function values and a list of functions, Mathematica will compute coefficients for the functions best yielding the required data.

```
data = {{0, 1}, {1, 0}, {3, 1}, {4, 2}};

parabola = Fit[data, {1, x, x^2}, x]
```
$0.9 - 1.03333 x + 0.333333 x^2$

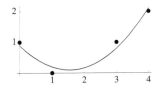

Rather than having each command execute separately, one can combine a sequence of commands (separated by semicolons) into programs or procedures. It is possible to program in many different ways in Mathematica. In this book, we shall limit ourselves to procedural programming.

Loops in Mathematica are made with the commands Table[ ], For[ ], While[ ] and Do[ ]. The command Do[ ] expects a sequence of commands and a list with an index and the index's range. The notation $\{i, 5\}$ means that the index i goes from 1 to 5, $\{k, 20, 30\}$ means that the index k goes from 20 to 30, $\{n, 0, 1, .1\}$ means that the index n goes from 0 to 1 in steps of 0.1. The following line of code computes the factorial of a number ($n! = 1 \cdot 2 \cdot 3 \ldots \cdot n$).

```
n = 10; fn = 1; Do[fn *= i, {i, 1, n}]; fn
```
3628800

The lists output by Mathematica cannot be directly used by other programs like Csound, because they are specially formatted. Lists in Mathematica have parentheses and commas and when copied into a text editor, the numbers are often shown with many decimal places.

```
t = RandomReal[{0, 1}, 2]
```
{0.603205, 0.125127}

{0.6032053649340716`, 0.1251269104807494`}

Numbers and text generated by loops have to be printed out with the command print and appear in separate lines. The values are not separated by commas or spaces, and mathematical expressions are not evaluated.

4.3 Mathematica

```
d = √3 ; Do[Print[i1 , 0, d], {2}]
```

i10√3
i10√3

By introducing quotes, spaces, etc. and by merging cells one can obtain whatever formatting one requires, but as soon as the results are copied into an editor, the program-intern formatting instructions appear.

```
Do[Print["i1 ", 0, " ", √3.0 ], {2}]
```

i1 0 1.73205
i1 0 1.73205

"i1" "0" "1.73205080756887692`

To avoid these difficulties, the data can be written as strings into an unformatted file using the command WriteString[*file_name*", data]. The file must be closed with the command "Close[*file_name*"]. The new file can be used directly by another program or displayed in Mathematica using the command *!!file_name* and then copied into an editor.

```
score = OpenWrite["score.sc"];
dur = 3; freq = N[440*¹²√2 , 6];
WriteString[score, "i1 ", "0", "\t", dur, "\t", freq, "\n"];
Close[score];

FilePrint[%]
```

i1 0   3   466.164

4.3.2 Sounds

In Mathematica, sounds are stored as lists of numbers. Accordingly, Mathematica can play any list of real numbers as a sound. In the following example, we make a list of 8000 random numbers between 0 and 1 and make a sound object of them with the command Listplay[ ].

```
noise = RandomReal[{-1, 1}, 8000];
ListPlay[noise, SampleRate → 8000]
```

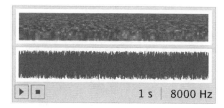

The sound object can be named and then played with the command EmitSound[ ].

```
snd = ListPlay[noise, SampleRate → 8000]; EmitSound[snd]
```

The sound's sampling rate can be changed, which changes its duration and its pitch.

```
sine = Table[Sin[t], {t, 1, 1000}]; ListPlay[sine, SampleRate → 6789]
```

To generate a tone of frequency f and sampling rate sr, we pass the argument $2\pi ft/sr$ to the sine function. If we synthesize N samples, the duration of the sound is N/sr seconds.

```
sr = 5000; f = 440; sine = Table[Sin[2 π * f * t / sr], {t, 1, 1000}];
snd = ListPlay[sine, SampleRate -> sr]
```

Function tables can be played as sounds by passing the function and the duration *tmax* to the function Play[ ]. To generate a tone of frequency 440Hz, we pass the argument $2\pi ft$ to the sine function, and Mathematica adjusts the values to fit the sampling rate.

```
Play[Sin[2 π * 440 * t], {t, 0, .5}]
```

Quantizing is done by the option SampleDepth.

```
Play[Sin[2 π * 440 * t], {t, 0, 1}, SampleDepth → 7]
```

Since the sounds are stored as lists, we can visualize them with the function ListPlot[ ] (left figure below). The range to visualize is indicated by the option PlotRange → {{*tmin*, *tmax*},{*xmin*, *xmax*}} (right figure).

```
snd = Table[Sin[.1 t] + RandomReal[{0, .01 t}], {t, 1, 100}];

lp1 = ListPlot[snd, Joined → True, Mesh → All, Ticks → {{50, 100}, {1}}]

lp2 = ListPlot[snd, PlotRange -> {{50, 70}, {-.8, 1.4}},
    Joined → True, Mesh → All, Ticks → {{60, 70}, {-.5, .5}}]
```

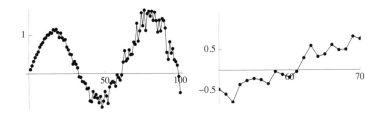

Play[ ] and ListPlay[ ] objects, as well as sound primitives can be passed to the command Sound[ ].

```
sound = Sound[
 {Play[Sin[2 π * 440 * t], {t, 0, .3}], ListPlay[RandomReal[{-1, 1}, 2400]]}]
```

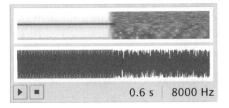

4.3 Mathematica

```
triad = Sound[SoundNote[{"C", "E", "G"}, .6, "Harpsichord"]]
```

Sound objects can be exported as sound files, and sound files can be imported as sound objects.

```
Export["exp_snd.wav", sound]
```

"exp_snd.wav"

```
snd = Import["exp_snd.wav"];
```

Sound objects consisting of SoundNote primitives can be exported as MIDI files.

```
Export["exp_snd.mid", triad]
```

exp_snd.mid

4.3.3 Graphics

Functions $f(x)$ can be visualized with the command Plot[f, {x, $xmin$, $xmax$}, $option \rightarrow value$]. There are many options available to fine-tune the image. The value of the option AspectRatio determines the ratio of the image's height to its width (default is the Golden Mean, 1.618). To make the units of both axes the same size, set AspectRatio $\rightarrow$ Automatic.

```
Plot[Sin[t], {t, 0, 2 π}, AspectRatio -> .2]
```

```
Plot[Sin[t], {t, 0, 2 π}, AspectRatio -> Automatic]
```

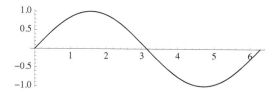

One can display a list of functions with one command. The example below shows several other options.

```
Plot[{Sin[t], Sin[2 t]/2, Sin[3 t]/3}, {t, 0, 2 π},
 AspectRatio -> Automatic, Frame -> True,
 AxesLabel -> {"t"}, FrameTicks -> {{π}, {-1, 1}},
 GridLines -> {{π/2, π, 2 π, 3 π/2}, {-1, -.5, .5, 1}},
 PlotLabel -> "Three Partials", PlotRange -> {{0, 7}, {-1.2, 1.2}}]
```

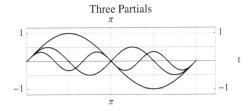

The command Show[*plot*] draws a plot. Show[{*plot1*, *plot2*, ...}] combines several plots. To draw an array of plots, one can use the command GraphicsGrid[{{*plot11*, *plot12*, ...},{*plot21*, *plot22* },...}].

One can display functions of two variables x and y as planes above the xy-plane with the command Plot3D[ ].

```
Plot3D[Sin[x*y], {x, -2, 2}, {y, -2, 2}, Ticks → None]
```

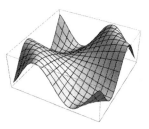

Plot3D[ ] only accepts a single function, but by using Show[ ], one can display several plots simultaneously.

```
p1 = Plot3D[Sin[x y], {x, -2, 2}, {y, -2, 2},
     PlotStyle → FaceForm[], Ticks → None];
p2 = Plot3D[Cos[2*x*y], {x, -2, 2}, {y, -2, 2},
     Mesh -> False, Ticks → None];
s1 = Show[p1, p2]; Show[GraphicsGrid[{{p1, p2, s1}}]]
```

Functions assign exactly one function value to every value of the independent variables. Therefore, curves like circles or spirals or surfaces like cylinders and spheres cannot be

4.3 Mathematica

represented as functions. But often we can show them using a parametric representation of the lines and surfaces, where the coordinates of the points (x, y) or (x, y, z) on the required curve or surface can be defined as functions of one or two parameters respectively.

```
p1 = ParametricPlot[{Sin[t], Cos[t]},
    {t, 0, 2 π - .5}, AspectRatio -> Automatic, Ticks → None];
p2 = ParametricPlot[{t*Sin[t], t*Cos[t]}, {t, 0, 20},
    AspectRatio -> Automatic, Ticks → None];
p3 = ParametricPlot3D[{t*Sin[20 t], t*Cos[20 t], t},
    {t, 0, 2}, AspectRatio -> Automatic, Ticks → None];
p4 = ParametricPlot3D[{u*Sin[10 u], u*Cos[10 u], v}, {u, 0, 2}, {v, 0, 3},
    AspectRatio -> Automatic, PlotPoints -> {40, 3}, Ticks → None];
```

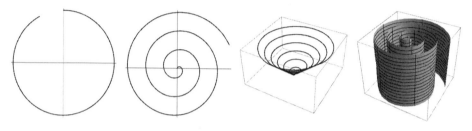

All images in Mathematica are composed of sets of *graphic primitives*, like Point[ ], Line[ ], Polygon[ ], etc., but also commands affecting the layout of the images like Thickness[ ], PointSize[ ] etc. as well as numerous graphic options. Lists of graphic primitives are passed to functions for generating graphic objects. In the example below, the function Plot first generates a graphic object *p*. Then we draw the object together with another object (Graphics[{...}]) using Show[ ] with a few options. If we do not set the option AspectRatio → Automatic, the figures are usually distorted. With functions like Plot[ ] and Plot3D[ ] we can insert graphic primitives using the options Prolog and Epilog (figure right).

```
p = Plot[-.3*Sin[π*t] - .3, {t, -1, 1}];
    Show[{p, Graphics[{PointSize[.03], Point[{-1, .5}], Point[{1, .5}],
        Thickness[.005],
    Line[{{-1.5, 0}, {-1.5, .5}, {1.5, .5}, {1.5, 0}}],
        Circle[{0, 0}, 1.5, {π, 2 π}]}]},
    PlotRange → All, AspectRatio → Automatic, Axes → False]

Plot[-Sin[t²], {t, 0, 5},
    AspectRatio -> .2, Ticks -> {{5}, {-1, 1}},
    Epilog -> {Thickness[.005],
    Line[{{.2, .3}, {1.5, .3}, {1.5, 1}, {.2, 1}, {.2, .3}}],
    Text["-Sin[t²]", {.8, .65}]}]
```

Points are given by three coordinates in three-dimensional graphic primitives. The following example combines a three-dimensional plot *p1*, a parametric three-dimensional plot *p2* and some three-dimensional primitives.

```
p1 = Plot3D[Sin[4*x*y] + y, {x, -1, 1},
    {y, 0, 2}, PlotPoints -> {30, 30}, ViewPoint -> {2, -2, .9}];
p2 = ParametricPlot3D[{Sin[10*t], t, t + u}, {t, 0, 2}, {u, -1, 1},
    PlotPoints -> {60, 5}, ViewPoint -> {2, -2, .9}];
Show[{p1, p2, Graphics3D[{AbsolutePointSize[12], Point[{-.9, .1, 2}],
    Thickness[.015], Line[{{-.9, .1, -1}, {-.9, .1, 2}}],
    Cuboid[{.6, .7, -1}, {1, 2, .5}]}]}, Ticks -> None]
```

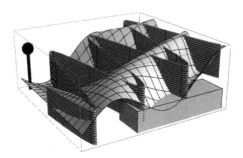

We can create new graphic primitives from those defined in Mathematica by using functions. The command below generates a function to draw the schematic diagram of a sine wave oscillator. The first two parameters define the midpoint of the object and the third parameter gives the absolute thickness of the drawing line.

```
Oscil[x_, y_, d_] := Graphics[{AbsoluteThickness[d],
    Line[Table[{t, -.3*Sin[π*t] - .3}, {t, -1, 1, .1}]],
    Circle[{x, y}, 2, {π, 2 π}],
    Line[{{x - 2, y}, {x - 2, y + 1}, {x + 2, y + 1}, {x + 2, y}}]}]

Oscil[0, 0, 1]
```

We can write interactive programs and graphic user interfaces using control objects and dynamic variables. The dynamic variable x in the example below can be changed using the control object Slider[ ].

```
Dynamic[x]                          0

Slider[Dynamic[x]]
```

Dynamic variable are global, that is, the slider above changes all dynamic variables named x in all currently open *Mathematica* notebooks. Local variables can be defined using the command DynamicModule[ ].

```
DynamicModule[{x = .7}, {Slider[Dynamic[x]], Dynamic[x]}]
```

4.3 Mathematica

We can write interactive programs in just a few lines using the command Manipulate[ ]. We pass to Manipulate[ ] an expression and the ranges of the variables in the expression. Manipulate automatically generates sliders for the variables and a window in which to display their values.

```
Manipulate[n/m, {n, 1, 20, 1}, {m, 1, 20, 1}]
```

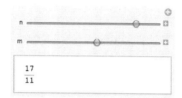

The expressions in Manipulate[ ] can be arbitrarily complex, and instead of sliders we can use check boxes, two-dimensional sliders, locators, etc. The following example shows left in the window the z-plane and right the amplitude response of a filter with two complex conjugate zeros z_1 and z_2. One zero can be moved using the locator (circle), the other is calculated and shown as a point. The amplitude response and the filter's coefficients are computed from the zeros. The amplitude response is shown in the image, and the coefficients are used to filter noise when one clicks on the Play button.

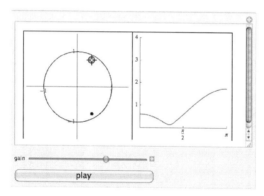

Manipulate[ ] can be used to generate animation, since the sliders are automated. Elaborate animations cannot be computed in real time, but one can combine several images in one cell and play them as an animation or export them as a film. We can make an animation out of individual images showing an event over time by using a Do[...] loop whose counter corresponds to the time. The frame rate of the animation can be set in the dialog field that appears when the animation is played. Since the animation's range usually changes when images or curves change, one should always indicate the required range with the option PlotRange[ ].
(→ Animation)

```
Do[Print[Plot3D[Sin[t x y], {x, -2, 2}, {y, -2, 2}, Ticks → None,
    PlotRange → {{-2, 2}, {-2, 2}, {-1, 1}}]], {t, 0.2`, 2, 0.2}]
```

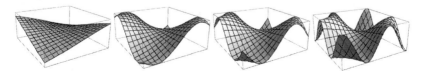

4.4 C / C++

C is a general-purpose, high-level programming language. Many operating systems have been programmed in C. Algorithms can be precisely and elegantly programmed in C. The language has a large standard library. Because C can be programmed at close to machine level, very efficient and fast applications can be realized.

There are many implementations of C, each with its own compiler and programming environment, so that the reader will have to use the corresponding manuals to learn the specific implementation at his or her disposal. In addition, this book will not discuss the programming of system-dependent functions, in particular those for in- and output.

In 1980, C was extended to become "C with Classes". In 1983 this extension was named C++. Almost all the features of C are available as a subset of C++. We will not always explicitly distinguish between classical C code and C++ code. All the examples can be compiled on a C++ compiler, but only some of the examples can be compiled on a classical C compiler.

Further Reading: The Audio Programming Book *edited by Richard Boulanger and Victor Lazzarini [86].*

4.4.1 Fundamentals

4.4.1.1 The Structure of a C Program

The source code of a simple C program consists of these parts: 1) The directive #include is used to embed external files into the source code or to establish links to libraries. For example, the control line #include math.h links the mathematical library of C into the source code. 2) The directive #define defines symbolic constants and macros. For example, after the control line #define pi 3.14159, the symbol pi can be used in the source text. 3) Variables and data types that should be available in all parts of the program are declared before the main program. 4) A C program contains exactly one main function and 5) any subordinate functions.

```
#include                          // Link to an external file
#define                           // Define constants and macros
Global variables and data types

int main()
{                                 // Main program
}
type function1(arguments)         // Function 1
{
}
```

The declarations of the library functions are grouped by domain and stored in so-called *header files* (*name*.h). The libraries required for a program are indicated by #include directives and are embedded into the program at compile time. The output of the compiler is called *object code* and is usually indicated by the extension .o. After compilation, the linker program combines all the object code and those library functions required by the source code into a program executable by the specific operating system being used.

The program below outputs "PI = 3.14159". To use the output function (std::cout) in a C++ program we have to include the library iostream. The main function of a C or a C++ program must return a value. In this example, main() is of type int and returns zero.

4.4 C / C++

```
#include <iostream>
#define PI 3.14159

int main ()
{   std::cout << "PI = " << PI;
    return 0;
}
```

4.4.1.2 Types

Every symbol used in a C program must be declared. The basic types are sub-divided into types for integers and types for floating point values. The length in bytes of the various types is dependent on the compiler. The actual length of a type for a given implementation can be found in the header file <limits.h>. INT_MAX, for example, gives the length of the type int in bytes. Integers in order of increasing length are: short int < long int < long long int. There are also unsigned data types whose names are preceded by "unsigned" and which require the same amount of memory as the corresponding signed types. Floating point types in the order of their length are: float, double, long double. The type char is used for alphanumeric data. One char is the smallest addressable unit in C.

The range of the integers that can be represented by a given type depends on the type's length in bytes. An integer of type short is between $-2^{15} = -32768$ and $2^{15} - 1 = 32767$. If the value of variable goes beyond these limits, the variable appears by convention at the other end of the range. (For example $(2^{15} - 1) + 1$ is represented as -32768.) For this reason, it is important not to go beyond the limits of the number range when working with sound (6.1.3.6). The qualifier "const" means that a variable can only be assigned a value at the time it is declared.

Floating point numbers are represented in up to six digits and a decimal point, for example 12.3413 and .001, or alternatively in so-called scientific notation by a number with one digit before the decimal point called the mantissa and an exponent base 10, for example 1.0e–4 for 0.0001 and 1.23413e2 for 123.413. The four bytes (32 bits) of a float variable are usually divided into 24 bits for the mantissa, 7 bits for the exponent and one bit for the sign. Although a greater range of numbers can be represented in floating point, there can be problems with precision. For example, in Csound and in Max the number .0000014 is read as .000001 and numbers smaller than .0000005 are read as 0. The problem can be avoided in some programs by entering the numbers in scientific notation or as products (for example, .0001*.0001 for .00000001). When the ratios between numbers in calculations are large, there can be considerable imprecision in the results. For example, because the floating value 1000.001 is rounded to 1000.0, the value of the differential quotient $(f(x + h) - f(x))/h$ for $x = 1000$ and $h = .001$ is 0 instead of 1. To test two floating point numbers for equality, one should not test whether $x == y$, but rather whether the numbers' difference is less than a given tolerance δ: $|x - y| < \delta$, for example $|x - y| < \delta = .00001$. Variables can be converted into another type using the cast operator (*type-name*), where the parentheses are required.

4.4.1.3 Derived and Composite Types

Sequences of variables of the same type are stored in arrays. An array's dimensionality is determined by the number of pairs of brackets following the array's name in the declaration and the number of elements in each dimension is given by numbers within the brackets. int *list*[10] declares a one-dimensional array or vector of 10 integers, the *n*th of which is referenced by *list*[*n*] (the first element in an array has the index 0). *table*[2][10] refers to a two-dimensional array of two rows and 10 columns. One can define new types using the statement *typedef.* The statement typedef float MATRIX[*n*][*n*] defines a type Matrix as a square two-dimensional array of floats. The name "MATRIX" can then be used to declare a specific instance of that type. The line: "Matrix m1;" does just that.

```
int main()
{    typedef float MATRIX[3][3];
     MATRIX m1;
     m1[1][1]=3.1415;
     return 0;
}
```

One of C's special features is the use and manipulation of pointers. In order to deal with a variable, the computer needs to know its address in memory. In the example that follows, programming statements are shown on the far left. Two variables, *x* and *y*, are declared and assigned values. These values are written to memory, here at the (arbitrary) locations 811 and 812. C can reference the address of a variable with the operator &. The address &x can be passed to a pointer *p*. The pointer must be declared of the type of the variable to which it points. Its name is prefixed by an asterisk.

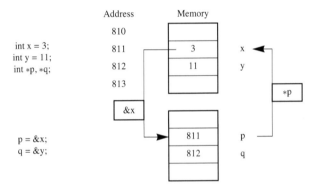

The name of an array is the address of its first element and hence a pointer to that element. The first member of the array *list*[10] can be referenced either as *list*[0] or *list.

Of the three combined types structure, union and bit-field, we will only speak of the structure. A structure combines variables of different types. The syntax for defining a structure is:

```
struct NAME { list of typed elements } [ obtional: object names ];
```

In the following example, a structure PLACE is defined with two floating point elements (*x*, *y*) and one alphanumeric element (*s*), after which a list of two instances of this type follows (*point1*, *point2*). Further instances of this type can also be declared later by giving the structure type and the variable name (PLACE midpoint). One accesses individual elements of a structure by giving the variable name and the element name, separated by a dot. the entire contents of a structure can be passed to another structure.

```
struct PLACE
{    float   x, y;
     char  s;
} point1,point2;
PLACE    midpoint;
point1.x = 0;    point1.y = 0;    midpoint = point1;
```

4.4.1.4 Operators, Expressions, Mathematical Functions

An expression consists of constants, variables and function calls that are linked by operators. The following (incomplete) list shows operators in C/C++ :

Unary Operators:

+, -	Sign (also negation)
*	De-referencing of a pointer (4.4.1.2)
&	Address of
!	Logical negation
++, --	increment, decrement
	(++ i increments the variable i by 1 before using it)
	(i++ increments the variable i by 1 after using it)
sizeof	Size of a variable in bytes
(*type*)	Cast operator (converts from one type to another; n = (int) x assigns to n the integer value of the float x)

Binary Operators:

+, -, *, /	Arithmetic operators
%	Modulo (remainder in integer division: 7 % 3 = 1)
^	Exclusive OR
&, |	Bitwise AND, bitwise OR
&&, ||	Logical AND, logical OR
=	Assignment operator
+=, -=, *=, /=	Assignment operators (i += 1 increments i by 1)
==, !=	Equality and inequality operators
<, <=, >, >=	Less than, less than or equal, greater than, greater than or equal

The table below shows the priorities of the various operators:

Operator	Rank	Associativity
() [] . ->	1	left to right
! ~ ++ -- * & (cast) sizeof ()	2	right to left
* / %	3	left to right
+ -	4	
<< >>	5	
< <= > >=	6	left to right
== !=	7	
&	8	left to right
^	9	
|	10	
&&	11	left to right
||	12	
? :	13	right to left
= *= += -= /= %= <<= >>= &= |= ^=	14	right to left
, (comma)	15	left to right

In order to use the mathematical functions of C/C++, the header file math.h must be included in the source code. The most important mathematical functions are:

abs(x)	Returns the absolute value of x (int)
acos(x)	Returns the arccosine of x (x in radians)
asin(x)	Returns the arcsine of x (x in radians)
atan(x)	Returns the arctangent of x (x in radians)
ceil(x)	Returns the next larger integer
cos(x)	Returns the cosine of x (x in radians)
cosh(x)	Returns the hyperbolic cosine of x (x in radians)
exp(x)	Returns e raised to power of x
fabs(x)	Returns the absolute value of x (float)
floor(x)	Returns the largest integer smaller or equal to x
fmod(x,y)	Returns the floating remainder of x divided by y
log(x)	Returns the natural logarithm of x
log10(x)	Returns the logarithm base 10 of x
pow(x,y)	Returns x to the power of y
sin(x)	Returns the sine of x (x in radians)
sinh(x)	Returns the hyperbolic sine of x (x in radians)
sqrt(x)	Returns the square root of x
tan(x)	Returns the tangent of x (x in radians)
tanh(x)	Returns the hyperbolic tangent of x (x in radians)

The function rand() generates a pseudo-random number between 0 and RAND_MAX. The same values are generated each time the program is run. One can avoid this by giving the random generator a starting value n with the command srand(n). To give a different starting value each time the program runs, one can use the function time() from the header file time.h. The command time(NULL) returns the system time as a long integer.

```
#include <time.h>
srand(time(NULL));
```

4.4.1.5 Control Flow

The if-statement produces the conditional execution of a command or a section of a program enclosed in braces { }. If the expression in parentheses is true, statements 1 will be executed, if not statements 2 will be executed (both else and statements 2 are optional).

```
if(expression)   {statements 1};
else             {statements 2};
```

The if-statement only allows two alternatives, but the switch statement passes control to any number of statements. If the value of the expression in parentheses (which is coerced into type int) is equal to a, then all subsequent statements are executed until a break statement is encountered, if it is equal to b, statement 2 is executed, etc. If the expression is equal to none of the values listed, then the statements following default is executed.

```
switch(expression)
{    case    a:   statements 1; break;
     case    b:   statements 2; break;
     ...
     default:     statements; }
```

Program loops can be realized in various ways. The for-loop uses a counter and repeats as long as expression 2 is true. Expression 3 gives the step size of the increment to the counter initialized in expression 1.

```
for( expression 1 ; expression 2 ; expression 3) {
    statements
}
```

Examples :

```
for ( z = 0;   z < 100;   z++ ) statement;
// The statement is executed 100 times
for ( z = 0;   z < N;   z += 2 ) statement;
// z is incremented by 2 at each repetition
```

The while-loop causes a part of a program to repeat as long as the expression in parentheses is true. To avoid the loop becoming endless, one has to make sure that the expression becomes false at some point. The while-loop is often used to control the exit from a program.

```
while( expression )
{   statements
};
```

If a loop should be executed at least once, even if the controlling expression is never true, one can use a do-loop.

```
do
{   statements
}   while( expression )
```

4.4.1.6 Functions

A function is an independent part of a program that performs specific operations, for instance the computation of a new value from argument values passed to it. The function's name and the types of the returned value and the arguments are specified. This declaration is known as the *function prototype*. The function itself is defined after the main program, now specifying the names of the passed arguments. If the definition of the function precedes the main function no prototype is needed. The program statements are written between braces, and the return-statement is followed by the value to be passed back to the controlling program.

```
typ function_name ( typ, ...)            // function prototype

void main()
{
...
}

typ function_name (typ argument1, ...)   // function definition
{   statements;
    return   var;
}
```

```
float maximum(float, float);          // function prototype

int main()
{   std::cout << max(-3.,4.);         // function call
    return 0;
}
float maximum(float a, float b)       // function definition
{   if(a > b) return a;
    else return b;
}
```

All variables defined within a function are *local variables*, that is, they are known only to that function. If variables need to be accessed by functions and by main(), they must be declared as *global*. In the example that follows, we define a function reset() with no arguments which does not return a value. It sets variables used in the main program to zero.

```
...
float x,y,z;                          // globally defined variables
void reset(void);                     // prototype of the function "reset"
int main()
{
//  ...
    x = 1;
    reset();                          // this is the function call
                                      // setting variables to zero
    std::cout << x << std::endl;
    return 0;
}
void reset(void)                      // the definition of the function
{   x = y = z = 0;
}
```

C and C++ compilers include many functions. The function declarations are in header files which must be linked into the source code with #include. The list below shows some header files found in all compilers.

assert.h	diagnosis
ctype.h	char functions
errno.h	error numbers to text
float.h	floating implementation
limits.h	implementation limits
locale.h	setlocal function
math.h	mathematical functions
setjmp.h	non–local jumps

complex.h	complex numbers
signal.h	signal functions
stdarg.h	variable numbers of arguments
stddef.h	standard definitions
stdio.h	standard in–and output
stdlib.h	general functions
string.h	string functions
time.h	time and date functions

A call to a function by the function itself is referred to as *recursion*. To avoid an infinite loop, one must be sure that the recursion will be interrupted at some point. The following example shows how to calculate the sum of the number from 1 to n using the recursive function sum(). The declaration of the function is the same as any non-recursive function, but the definition of the function includes a conditional call to the function itself. When the variable m has been decremented to 0, the function exits.

```
#include <iostream.h>
int main()
{   int n = 10, sum(int);
    std::cout << sum(n);
    return 0;
}
int sum(int m)
{   if (m-- > 0)            return sum(m)+m+1;
    else                    return 0;
}
```

4.4.1.7 Input and Output

The standard input and output of alphanumeric characters with keyboard and monitor is rather complicated in C. It has been simplified in C++ through the commands std::cin >> and std::cout <<. After std::cin >> x, the program waits for input and assigns the number input to the variable x. The sequence: std::cout << "x squared:" << std::endl << x*x outputs a string, moves to the next line (std::endl) and writes a number.

Characters and numbers can be written to a file as follows. The declaration FILE *fp results in a file pointer (the standard type FILE is declared in the header file stdio.h). The statement fp = fopen("*name*", "w") opens a file named *name* to which *fp* points. The statement fclose(fp) closes that file.

```
FILE *fp;
fp = fopen("example.sc","w");
fclose(fp);
```

The second argument of fopen() indicates how the file will be used.

```
"r"         // open an existing file to read
"w"         // open a new file to write to
"a"         // open an existing file to append to
"r+"        // open an existing file to read and write
"w+"        // open a new file to read and write
"a+"        // open an existing file to read and append
```

The following functions write to an open file:

```
int fputc(char,file)                    // write a character
int fputs(string,file)                  // write a string
int fprintf(file,format,var1,var2,...)  // write one or more variables
                                        // in specified formats
```

The format has the form "%conversion-character, %conversion-character, ..." The most common conversion characters are listed below:

```
%d (or i)   // signed integer
%f          // float
%e          // float in exponential form
%c          // single character
%s          // character string
%li         // long integer
%lf         // long float
```

```
%m.nf         //  (m, n integers) float with m decimal places
              //  of which n after the decimal point
\n            //  new line
\t            //  tabulator
```

The program excerpt below shows how to use the writing functions to begin a score for Csound. It is important not to output numbers with more decimal places than necessary, especially in large scores. This excerpt first writes the name of the score as a comment at the beginning of the new file using fputs(). Using fputs() again, we write the character 'i' on the next line (\n) and then use fprintf() to write variable values, a number and a comment. The example shows that a tab stop can be generated either with the control character '\t' or by using the tabulator within the format statement. The variable values are output with three places after the decimal point, the number is rounded to four places after the decimal point.

```
int p1=1;
float p2=1.1234,p3;
FILE *fp;
fp=fopen("4-4-1.sco","w");
fputs(";4-4-1.sco",fp);
fputs("\ni",fp);
fprintf(fp,"%d\t%4.3f    %.4f    %s",p1,p2,.98765432,";first note");

;4-4-1.sc
i1  1.123    0.9877              ;first note
```

The following program shows how to check whether a filename is already taken. If the name is taken, the program asks if we want to overwrite the existing file. We declare a character char c for the answer, initializing it with the letter 'w'. At the same time, we declare a character array "filename" of length 12. Then the program asks for the filename to check and stores it in the character array. To check whether the file already exists, we try to open a file of that name for reading with fopen(filename, "r"). If fopen() succeeds, it returns a non-zero number. The program reports that the file already exists and asks if we want to overwrite the existing file. The answer is stored in the variable c. If the user types 'y' or if c still has its initialized value 'w', the new file is created and a report to that effect is output.

```
#include <iostream>

int main(){
    char c = 'w', filename[12];
    std::cout << "input file name" << std::endl;
    std::cin >> filename;
    FILE *fp;
    if((fp = fopen(filename,"r"))!=NULL){
        std::cout << "file " << filename
                  << " already exists." << std::endl;
        std::cout << "overwrite ? y/n" << std::endl;
        std::cin >> c;
    }
    if(c == 'y' || c == 'w'){
        fp = fopen(filename,"w");
        fprintf(fp, .....);               // print something
        fclose(fp);
```

```
            switch(c){case 'j': {std::cout << "file overwritten"
                                            << std::endl; break;}
                      case 'w': std::cout << "file created" << std::endl;}
    }
    return 0;
}
```

4.4.1.8 Classes and Objects (→ CD)

4.4.1.9 Reading and Writing Binary Data

Binary data transfer is unformatted, that is, it is read and written as it is represented in computer memory and not converted into sequences of ASCII characters. Binary data cannot be read with a conventional text editor, but it requires less memory because formatting information is not necessary, and reading and writing are faster than with formatted data. For binary data transfer one uses the functions read() and write(). To write a binary file, we set a pointer to the beginning of the data to be written, first having cast the pointer to type char* (this is necessary because char corresponds to one byte). We also give the number of bytes to be transferred. In the example that follows, we generate 10 random numbers and write them in binary form to the file int.dat, read the numbers back in and display them on the screen. The header file fstream.h contains the prototypes and the function declarations necessary for writing binary data. The statement ofstream (output file stream) creates a file object with the arbitrarily chosen name stream1. The method open() creates a file whose name is the first argument passed to the method. The second argument says how the data will be written. In the for-loop, the 10 random numbers are generated and written to the file. First a number r is generated, and then the pointer to r is cast into type char with the cast operator (char *) and passed to the function along with the size of r in bytes (sizeof(r)). After closing the file int.dat, we define a new object stream2 using the statement ifstream (input file stream), read the numbers from the binary file and display them on the screen, where they appear as a column of 10 numbers. Because the formatted numbers are only on the screen and not in a file, text editors cannot interpret or use them.

```
#include <fstream>
#include <iostream>

using namespace std;

int main(){
    int r, s;
    ofstream stream1;
    stream1.open("int.dat", ios::binary | ios::out);
    for(int i = 0; i < 10; i++){
        r = rand();
        stream1.write((char*)&r,sizeof(r));
    }
    stream1.close();
    ifstream stream2;
    stream2.open("int.dat", ios::binary | ios::in);
    while(stream2.read((char*)&s,sizeof(s)))
        cout <<s << endl;
    stream2.close(); return 0;
}
```

4.4.2 Generating and Storing Sounds

4.4.2.1 Raw Data

Sound files are binary files (4.4.1.9). They consist of a header and the actual data (4.4.2.2). Since many synthesis programs, software sequencers and sound editing programs can read raw data, one can often simply write the raw data into a binary file. One must tell the reading program sampling rate, quantization and number of channels of the file. The program below writes a stereo file with a sine wave of frequency 440 Hz in the left channel and a sine wave of frequency 660 Hz in the right channel. 16-bit samples are (signed) short integers, that is numbers between -32768 and 32767. In stereo files, the samples for the left and the right channels are written alternately.

```
#include <iostream>
#include <math.h>
#include <fstream>

float PI2 = 2*3.141592;

using namespace std;

int main(){
    int sr = 44100;
    short int r;
    ofstream stream1;
    stream1.open("int.raw", ios::binary | ios::out);
    for(int i=0; i<44100;i++){
        r = 32000*sin(440*PI2*i/sr);
            stream1.write((char*)&r,sizeof(r));
        r = 32000*sin(660*PI2*i/sr);
            stream1.write((char*)&r,sizeof(r));
    }
    stream1.close();
```

In the program above, each sample was written into the file individually. It is considerably faster to compute large blocks of sound, or even the entire sound, and then to write it to the disk.

```
...
int sr = 44100;
int dur = 2*sr;
short int smp[2*dur];
ofstream Stream1;
Stream1.open("int.dat", ios::binary | ios::out);
for(int i = 0; i < 2*dur; i++){
    smp[i] = 32000*sin(440*PI2*i/sr);   i++;
    smp[i] = 32000*sin(660*PI2*i/sr);
}
Stream1.write((char*)&smp,sizeof(smp));
Stream1.close();
...
```

4.4.2.2 Sound Files

Let us show what a header looks like and how to write one into a binary file using the example of a minimal WAVE file format. The WAVE specification requires three segments of information. The first declares that the file uses the Resource Interchange File Format (RIFF). Then follow the file length in bytes and the indication that this is a "WAVE" file. This section is called the "RIFF" chunk descriptor. The next section is the "fmt" (format) sub-chunk and contains information about the format of the data. The third section is called the data sub-chunk. It begins with the word "data" and the size of the of the sound in bytes. Then follows the actual data of the sound.

Offset	Length in bytes	Contents	Commentary
0	4	'RIFF'	
4	4	Data length – 8	Length of the remaining file
8	4	'WAVE'	
12	4	'fmt '	Space after fmt !
16	4	16	Length of the fmt data–chunk
20	2	1	Format : 1 = PCM
22	2	Number of channels	1 or 2
24	4	Sampling rate	eg. 44 100
28	4	Bytes per second	Sampling rate * Block size
32	2	Block size	Number of channels * Bytes per sample
34	2	Bits per sample	8 or 16
36	4	'data'	
40	4	Length of the data sub–chunk	in bytes
44		data	(alternating left–right for stereo)

The example that follows shows how to write the header. We first define the sound's duration *dur*, the sampling rate *sr* and the number of channels *chnls*. These values determine the length *len* of the data-chunk. Next the variables for the information in the table above are declared, 4-byte numbers as long and 2-byte numbers as short. The ASCII code-words are defined as strings char*, that is, as pointers to the beginning of the strings. Note the space after "fmt" making it a 4-byte string like the others. We declare the array smp[] for the computed values. Then we open the binary file 4-4-2.wav, into which we write the numbers and the code-words in the order given in the table above. Since we defined the code-words as pointers to strings, we can pass them directly to the method write(). The second parameter in write() is the length of the string in bytes. To write the numbers, a pointer of type char* pointing to the variable name of the numbers to be written must be passed. The sound itself is computed in the for-loop and then written to the file.

```
#include <iostream>
#include <fstream>
#include <math.h>
float PI2 = 2*3.141592;

using namespace std;
```

```
int main(){
    float dur = 2.;
    long sr = 44100;
    short chnls = 2;
    const long len = (long) dur*sr;
    long len_8 = chnls*len*sizeof(short)+44-8;
    long len_fmt = 16;
    short pcm = 1;
    short bits_spl = 16;
    short block_al = chnls*sizeof(short);
    long bytes_sec = sr*block_al;
    char* data = "data";
    long len_data = chnls*len*sizeof(short);
    short smp[chnls*len];

    ofstream wavefile;
    wavefile.open("4-4-2.wav", ios::binary | ios::out);

    wavefile.write("RIFF",4);
    wavefile.write((char*)&len_8,4);
    wavefile.write("WAVE",4);
    wavefile.write("fmt ",4);
    wavefile.write((char*)&len_fmt,4);
    wavefile.write((char*)&pcm,2);
    wavefile.write((char*)&chnls,2);
    wavefile.write((char*)&sr,4);
    wavefile.write((char*)&bytes_sec,4);
    wavefile.write((char*)&block_al,2);
    wavefile.write((char*)&bits_spl,2);
    wavefile.write(data,4);
    wavefile.write((char*)&len_data,4);

    for(int i = 0; i < chnls*len; i++){
        smp[i] = 32000*sin(440.*PI2*i/sr);   i++;
        smp[i] = 32000*sin(660*PI2*i/sr);
    }
    wavefile.write((char*)(&smp),len*sizeof(short));
    wavefile.close();
    return 0;
}
```

Let us simplify things a bit by moving the writing of the WAVE file into a header file my_wav.h. We include the necessary files and define the function writewav(). The parameters passed to the function are the file name, sampling rate, number of channels, duration of the sound and a pointer to a buffer containing the samples. The values for the file's header are computed from these parameters. We write the buffer into the WAVE file using the statement wavefile.write().

```
// my_wav.h
#include <iostream>
#include <fstream>

using namespace std;
```

```
int writewav(char* name, int sr, short chnls, float dur, short* smp){
    const long len = (long) dur*sr;
    .....
    long len_data = chnls*len*sizeof(short);

    ofstream wavefile;
    wavefile.open(name, ios::binary | ios::out);

    wavefile.write("RIFF",4);
    ...
    wavefile.write((char*)&len_data,4);
    wavefile.write((char*)smp,chnls*len*sizeof(short));
    wavefile.close();
    return 0;
}
```

We include the file my_wav.h in the file used to generate the sound, compute the sound and pass it to the function writewav().

```
#include <math.h>
#include "my_wav.h"

void main(){
...
    for(int i=0; i<len;i++){
        snd=35000*sin(440*i*2*3.141592/44100);
        smp[i]=cntrl(snd);
    }
    writewav("4-4-2.wav",1,len,smp);
}
```

Various libraries are available in the Internet (e.g. libsndfile) that provide functions for reading and writing sound files to and from many different formats.

4.5 Processing

Processing is an object-oriented programming language with an integrated development environment for graphics, simulation and animation. The open source project had its beginning at the Massachusetts Institute of Technology in 2001. Processing is a considerably simplified version of the Java programming language with specialized libraries for video, graphics, graphic formats, sound, animation, typography, simulation, data access and transfer and network protocols.

4.5.1 Fundamentals

Programs written in the Processing language consist of a *setup* and a *draw* function and functions that react to events. In the setup function the size of the window generated is declared, as well as background color, frame rate for animations, etc. Even the empty structure setup() {} generates a window as output. Using the following commands we generate a window 400 pixels wide by 300 pixels high with a grey background (the function background expects a grey scale value between 0 and 255). The command rect(*x*, *y*, *width*, *height*) draws a rectangle (the origin of the coordinates is the upper left corner of the window). Processing provides many functions for two- and three-dimensional graphic primitives, coloring and defining graphic rendition. For example, the command ellipse(*x*, *y*, *width*, *height*) draws an ellipse. The command smooth( ) smoothes the edges of crooked lines or curves.

```
void setup() {
    size(400, 350);
    background(120);
    rect(100, 50, 200, 250);
    smooth();
    ellipse(56, 46, 55, 55);
}
```

The drawing function draw() {} draws a new image at the default frame rate or at the frame rate defined in the setup structure. The command triangle(*x1*, *y1*, *x2*, *y2*, *x3*, *y3*) draws a triangle. In the example below, we draw triangles on the baseline (50, 340), (300, 340) having a random third vertex. The command background(120) resets the display window for each frame.

```
void draw() {
    background (120);
    triangle (50, 340, 300, 340, random (400), random (350));
}
```

Besides the commands particular to Processing, one can use any commands or classes from the Java programming language. In the following example, we draw the *Sierpinkski Triangle* using a simple algorithm: given are the vertices of an equilateral triangle and an arbitrary point. A new point is defined to lie halfway between the old point and one of the (randomly chosen) vertices of the triangle. The example shows the declaration and initialization of variables (*sizex*, *sizey*, ...) and arrays (*px*[ ] und *py*[ ] for the three vertices of the triangle). Using a for-loop, we compute and draw points according to the algorithm above. Since the range *ni* of the counter constantly increases, more and more points are drawn in each frame.

```
int sizex = 450;
int sizey = 400;
int n;
float ni = 1;
float x = random(sizex), y = random(sizey);
float[] px = new float[3];
float[] py = new float[3];

void setup(){
    size(sizex, sizey);
    frameRate(10);
    px[0]=10; py[0]=sizey-10;
    px[1]=sizex-10;  py[1]=sizey-10;
    px[2]=sizex/2;   py[2]=sizey - 10 - sqrt(3.0)*(sizex-20)/2;
}
void draw(){
    ni += 0.2;
    for(int i = 0; i < ni; i++){
        n = (int)random(3);
        x = (x + px[n])/2;
        y = (y + py[n])/2;
```

4.5 Processing

```
        point(x, y);
    }
}
void mousePressed() {
save("sierpinski.tif");
}
```

Using functions reacting to keyboard input or mouse movements, parameters can be changed and commands can be triggered, like the writing to disk of the current frame as a TIFF-file in the example. The figures below show the image after 1000 and 50000 frames (on the left and right respectively). (→ *Sierpinski.pde*) (Variations on the Sierpinski Triangle → Processing/_fractals/)

 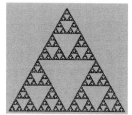

4.5.2 Simulations

In the following example we simulate the motion of a mass connected to two fix points with elastic bands (8.1.1.1). The acceleration a is proportional to the displacement y ($a = cy$), the velocity changes according to the acceleration and the displacement changes according to the velocity ($v \mathrel{+}= a, y \mathrel{+}= v$).

```
float c, y, v, a;
void setup() {
    c = -0.01;  y = 130;  v = 0.0;  a = 0.0;
    size(400, 400);
    smooth();
}
void draw() {
    background(200);
    ellipse(200,200 + y,20,20);
    line(100, 200, 200, 200 + y);
    line(200, 200 + y, 300, 200);
    a = c*y;  v += a; y += v;
}
```

In the example below, 36 masses are connected into a chain (8.1.3.2). Displacement, velocity and acceleration are defined as arrays. The computation of new values for the three parameters is done in a for-loop, but keeping the values for the end masses constant, since they do not move. The command v[i] *= 0.998 causes a slight damping proportional to velocity. To avoid discontinuities in the exciting vibration when the frequency changes, the excitation is not computed by the formula $exc = amp*\sin(fr*t)$, but rather by incrementing the phase. Frequency and excitation can be controlled by the mouse.

```
int ni = 36, dx = 27;
float c = -0.01;
float amp_exc, fr_exc, phase = 0;
```

```
    float[] y = new float[ni];
    float[] v = new float[ni];
    float[] a = new float[ni];

    void setup() {
        size(1000, 400);
        background(102);
        smooth();
    }
    void draw() {
        background(200);

        for(int i=1; i<ni-1; i++) {
            ellipse(20+i*dx,200 + y[i],15,15);
            line(20+(i-1)*dx, 200 + y[i-1], 20+i*dx,200 + y[i]);
            a[i] = c*(2*y[i]-y[i-1]-y[i+1]);
        }
        line(20+(ni-2)*dx, 200 + y[ni-2], 20+(ni-1)*dx,200);

        for(int i=1; i<ni-1; i++) {
            v[i] += a[i];
            v[i] *= 0.998;
            y[i] += v[i];
        }
        phase += 0.0001*fr_exc;
        y[1] = 0.1*amp_exc*sin(phase);
    }
    void mouseMoved() {
        fr_exc = mouseX;
        amp_exc = mouseY;
    }
```

A sinusoidal excitation of the first masses causes waves to propagate along the chain that are reflected at the end and then overlap with the waves coming from the left. Standing waves can arise at certain excitation frequencies. (→ Illustration)

4.5.3 Libraries

Processing's functionality can be extended by libraries that are included with the program or have been provided by third parties. Processing includes libraries to use serial interfaces, Quick Time, OpenGL, Arduino, etc. The library *Minim Audio* is a simple but flexible audio library that uses the JavaSound API. It lends itself particularly for reading in and playing sounds, although it can also be used for sound synthesis and treatment.
The example file *self_sust_osc_1.pde* (3.4.5.3) uses the library *controlP5*, which provides knobs, sliders, etc., and the example file *osc_send.pde* uses the libraries *oscP5* and *netP5* (4.1.4.3). The code example below shows how libraries are linked into a Processing program.

```
    import controlP5.*;
    ControlP5 controlP5;
    ....
    import oscP5.*;
    import netP5.*;
    OscP5 oscP5;
```

5 Fundamentals of Sound Synthesis

This chapter is primarily addressed to musicians who want to generate sound with the computer, whether for pedagogical purposes or for musical composition. The chapter does without complicated calculations and instead concentrates on fundamental concepts in order to give the reader an idea of various synthesis techniques.

Further Reading: Computer Music *by Charles Dodge and Thomas A. Jerse [2], a standard work in the field, discusses (pp. 72-108) the elements of digital sound synthesis thoroughly and clearly without going into technical details. F. Richard Moore's* Elements of Computer Music *[8] is less sparing of technical details (pp. 152-315). The comprehensive article* A Tutorial on Digital Sound Synthesis Techniques *by Giovanni De Poli can be found in the* Computer Music Journal *[1-2] and in* The Music Machine *[7], pp. 229-249. Another comprehensive overview with programming examples on CD is in Eduardo Reck Miranda's book* Computer Sound Synthesis *[15].*

5.1 Fundamental Techniques of Sound Synthesis

5.1.1 Overview

5.1.1.1 Instruments and Their Schematic Diagrams

In computer music, an *instrument* is an algorithm that generates individual *events*. Such an event can be a simple tone, as in traditional music, but it can also be arbitrarily complex. In the extreme case, an entire composition can consist of a single "note" of a complex instrument. Normally a program generates sounds by taking the required *parameters* from a list. In the synthesis language Csound, this list is called a *score* and the algorithms are called the instruments of an *orchestra*. Orchestra and score are passed to the actual synthesis program for processing as two text files. In real time applications, parameters are often read into the program through MIDI interfaces or OCS protocols.

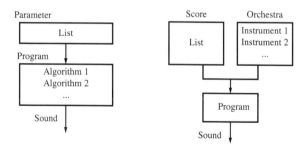

The instruments are constructed from simple elements called *unit generators*. Most unit generators have several *inputs* for parameter values or control signals and at least one *output* providing either a control signal for another unit generator or an audio signal. An instrument can be represented in a schematic diagram showing the connections between its unit generators. The following illustration shows the essential features of such diagrams. In general for each symbol, the inputs are shown above, the output below. The form of the symbol often indicates the unit generator's function. The signs for addition and multiplication are circles with the corresponding symbols inside. Subtraction is written as addition with a minus sign in front of the subtrahend. Conditional branching is shown by a lozenge, the output is a small circle.

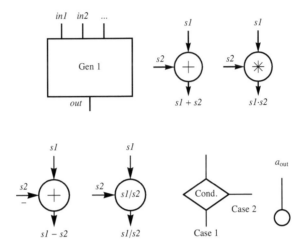

The direction of flow, that is, the order in which processing takes place, is usually from top to bottom, although feedback will obviously flow in the opposite direction. Arrows help avoid ambiguity of flow. At branching nodes, a signal is not split, as in schematics in analog electronics; instead, the entire signal is sent unchanged in both directions.

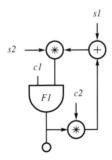

Programs like Max and pd offer graphic elements with which these schmatics can be directly realized.

5.1.1.2 Techniques for Sound Synthesis

In the following, we shall point out the most important techniques of sound synthesis and explain some of the concepts on which they are based. Every sound can be represented by its spectrum. Hence, in *additive synthesis* sounds are generated by adding sine waves of varying amplitude and frequency. In *subtractive synthesis*, on the contrary, one begins with spectrally rich material and filters out certain parts. Most synthesis techniques try to realize an acoustical model of the sound desired. Other techniques, on the contrary, model the physical production of the sound and are known as techniques of *physical modeling*. The vibration of a physical body can be modeled using points (*mass-spring systems*, see Chapter 8.1), lines (*wave guides*, see Chapter 8.2) or entire spatial elements (*finite element method* FEM). The expression *acoustic modeling*, although not in current use, could be used to subsume synthesis techniques which consider only the final sound, regardless of the physical process of its genesis, like additive synthesis and the modulation techniques described below. *Granular synthesis* is more difficult to characterize. It is based on the physical observation that many natural sounds are created by the pulse-like excitation of resonating bodies. In granular synthesis, however, the pulses and the resonances do not depend on physical models. Here, pulse-like waveforms (*grains*) are synthesized directly with the desired acoustic characteristics.

Techniques which use special methods to generate spectrally rich sounds and sound transformations with little computation form a class of their own. These techniques usually involve the nonlinear manipulation of simple waveforms. In *amplitude modulation* and *ring modulation*, the oscillation of a simple waveform is multiplied by another signal. In *frequency modulation*, the frequency of a signal is periodically modified so rapidly that the change is not perceived as pitch variation but as timbral variation. In the technique known as *waveshaping*, waveforms are distorted so that new sounds are generated. These techniques were particularly important in the early days of computer music, when processing power and computer memory were limited and expensive. Modulation techniques tend to use sine waves as their basic waveform, because only then can the acoustic characteristics of the resulting sounds be calculated. In principle, however, any waveform can be used.

Wave table synthesis uses non-sinusoidal waveforms to generate periodic signals with rich spectra. The example of *Walsh synthesis* shows that arbitrary sounds can be generated using other waveforms than sine waves.

5.1.1.3 Programs and Programming Languages

Sounds are stored as lists of numbers in sound files. The sound file's header contains information about the file's format. Such numerical lists can be created without difficulty using higher programming languages like C or Basic, but the transformation of such a list into a sound file requires detailed knowledge about the operating system used and the formats of sound files (4.4.2.2). The advantages of using a higher programming language for sound synthesis are that one is not limited to particular techniques, that one has a wide selection of functions at one's disposal and that the compiled programs can run very fast (see the C programs in Chapter 8). The disadvantage is that usually no specialized functions for sound synthesis are available, so that one has to program elements like oscillators and filters oneself. Commercial programs for working with sound make it possible to generate sound without having to worry about the details of the synthesis, for example, with a graphically controlled oscillator provided by many programs. In general, however, the possibilities for sound synthesis with commercial programs are limited. Programs that are addressed through schematic diagrams are adequate for simple applications and for teaching, but for more complex tasks they are often much less easy to program and understand than a well-written text file. Text-oriented synthesis programs like CLM (Common Lisp Music) or Csound are well-suited for many applications. They are based on all-purpose programming languages but handle many functions and system-specific tasks necessary for sound synthesis like storage, playing back sound, etc. Technical programs like Mathematica and Matlab are particularly suited for teaching. They too handle system-specific tasks and provide many mathematical and graphical functions.

5.1.1.4 Tables

When composing with the computer, it is often necessary to carry out the same calculations on several values. If these calculations can be expressed as a function of one variable, it is often expedient to make a table of the function $y = f(x)$ for all expected values of x. Then the value does not need to be recomputed for each new x but can simply be read from the table. Many functions cannot be described by simple mathematical formulas but are available in graphical form. These functions can be digitalized and stored in tables or lists.

The procedures for writing to and reading from a table are dependent on the program used, but the operations are generally as follows. First the size of the table, that is, the number N of values to be stored, is determined and the required space in computer memory is allocated. Then the function values for N x-values are either computed or measured in a graphical representation or control signal and written to the table. Synthesis programs like Csound provide numerous routines to construct tables (4.1.3).

5.1.1.5 Audio Signals and Control Signals

To ensure good sound quality, sounds are recorded digitally with a sampling rate of at least 40,000 Hz (3.2.1.1) and an amplitude resolution of at least 16 bits. Because parameters like pitch, loudness etc. vary slowly in time compared to the audio sampling rate, synthesis programs can generate control signals whose sampling rate is considerably lower than that of audio signals.

In traditional scores, the pitch and loudness of the individual notes are usually constant, but in the realization by a player, they vary all the time. For instance, pitch can vary due to a portamento at the beginning of a note, a vibrato later on and many small corrections during the note. In carefully written scores, one often finds several performance instructions for a single note, but even so, the performer's contribution to shaping the music is great. In computer music, instrument and performer are identical, and the composer must define all the performance details in the instrument. For this reason, there are often many more control signals than audio signals in an instrument. In the following descriptions, we will not distinguish between the notions of *control signal*, *envelope* and *control function*. In contrast to the audio signals, which in most cases have a sample rate of 44,100 Hz, the control signals have sample rates dependent on their use. In the early days of computer music, low control rates saved valuable computing time. In programs written in C or Mathematica, audio and control rates can be implemented by separate loops.

(→ Framework of a C program with separate loops for audio and control signals)

5.1.1.6 Interpolation

In digital sound synthesis, all signals and functions are discrete and hence discontinuous. Therefore it is frequently necessary to interpolate between two given values. We will only discuss linear interpolation here; for other techniques see Chapters 5.1.3.2 and 5.1.3.3. The figure to the left below shows two given values of a function $x(t)$ at times t_1 and t_2 and a value between them at time t_i. Because of the similarity of the two triangles we have

$$(x(t_2) - x(t_1)) / (t_2 - t_1) = (x(t_i) - x(t_1)) / (t_i - t_1).$$

Hence we derive

$$x(t_i) = x(t_1) + \frac{(t_i - t_1)(x(t_2) - x(t_1))}{(t_2 - t_1)}.$$

5.1 Fundamental Techniques of Sound Synthesis

If we divide the time $(t_2 - t_1)$ in n equal parts d (figure to the right below), we have for the kth intermediate value $x(t_i + kd)$

$$x(t_i + kd) = x(t_1) + k \frac{x(t_2) - x(t_1)}{n}.$$

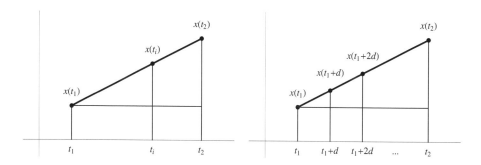

5.1.1.7 Program Control and Conditional Statements

Control flow in a program is determined by conditional statements. Statements like *if* and *switch* test certain conditions. The outcome of such tests controls the further flow of the program. For example, by using conditional statements, control function values can be reset and instruments can turn themselves on or off or be instructed to influence one another.

5.1.2 Unit Generators

Unit generators are the smallest building blocks in many computer music programs. Unit generators can generate audio or control signals or process existing signals. Hence they usually have one or more inputs for control or audio signals. In this chapter we will only mention three of the most important generators for sound generation.

Further Reading: There is an excellent technically detailed introduction to oscillators in the online book Digital Sound Generation: Part 1 *by Beat Frei [95].*

5.1.2.1 The Oscillator

The fundamental tone generator in electronic music is the *oscillator*, a device or a program which generates a periodic signal. Its parameters are the waveform to be generated, an amplitude A, a frequency f and optionally the initial phase. A simple sine wave generator computes for every moment in time t the value $A \cdot \sin(2\pi f t)$. However, most programs have oscillators that read the sine values from a stored table (wavetable oscillator) (5.1.1.4) rather than computing them on the fly. The values are read from the table so as to give the required frequency (see below) and are multiplied by the amplitude A. The number of values N in the table is usually a power of two. In the following illustration, 16 values of a sine wave are stored in a table. To read out a sine wave with frequency f, values of the table will usually have to be skipped or repeated, depending on the sampling rate of the system, the table length and the desired frequency. For example, to generate a sine wave of 800 Hz at a sampling rate of 44,100 Hz, the oscillator has to read the entire table once every 44100/800 = 55.125 sample periods. That means that at every sample the oscillator should advance 16/55.125 = .2902 values in the table. But since the oscillator can only read the stored values but not between, the resulting sine wave is very imprecise, for some values will be repeated three times, others four. The amount by which the reading point advances each sample is called the increment and can be calculated from the sample rate sr, the frequency f and the table length N as $inc = f \cdot N/sr$.

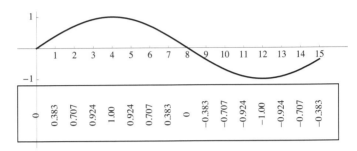

Let us work through the example above and see the difficulties which arise. First we fill a table with 16 values of a sine wave and name the table $f1$.

```
f1 = Table[N[Sin[2*π*t/16], 4], {t, 0, 15}]
```

In Mathematica, the pointer to step through a table has 1 as its first value ($t = 1$). For each sample, the pointer is incremented by .29025 ($t \mathrel{+}= .29025$). The integer part of the pointer t (IntegerPart[t]) can be used to read out the next value of the table ($f1[[...]]$). This value is multiplied by the amplitude A and the result is stored in a sound file, the list snd.

```
Amp = 10; t = 1; snd = Table[Amp*f1[[IntegerPart[t += 0.29025]]], {i, 0, 30}]
```

The figure to the left below shows the first 30 values of the sine wave. If an initial phase ϕ is required, the pointer is set to begin at $\phi N/2\pi$. In the example at the right, we set the initial phase $\phi = .9\pi$ to $.45 \cdot 16 = 7.2$.

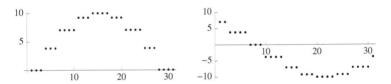

The waveform can be made more precise by using a larger table or by interpolating between the values (5.1.1.6). When the pointer reaches the end of the table, it has to be set back. We check at each sample whether the pointer has past $N + 1$, and if it has, we subtract N from the pointer's value. In the following example, we choose a frequency of 4000 Hz and compute a corresponding increment of 1.45.

```
Table[Amp*f1[[IntegerPart[If[(t += 1.45) < 16 + 1, t, t -= 16]]]], {30}]
```

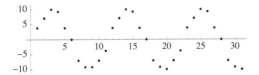

5.1 Fundamental Techniques of Sound Synthesis

Any waveform whatsoever can be stored in the wave table and the waveform can be generated in any convenient way. In general, the waveform remains unchanged for the duration of a note. The most important techniques for generating waveforms are the summation of sine waves to achieve a desired spectrum (here *f2* with fundamental and fifth partial) and the direct generation of the desired waveform (here *f3*, a sawtooth wave).

```
n = 128; f2 = Table[N[.7*Sin[2*π*t/n] + .3*Sin[5*2*π*t/n]], {t, 0, n}];

n = 128; f3 = Table[2*t/n - 1, {t, 0, n}];
```

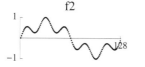
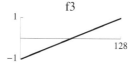

Using these waveforms, we generate two tones of one second duration each. The sample rate is 2000 Hz, the frequencies 3 Hz and 3.3 Hz respectively.

```
snd1 = Table[Amp*f2[[IntegerPart[If[(t += inc) < n + 1, t, t -= n]]]], {sr}];
snd2 = Table[Amp*f3[[IntegerPart[If[(t += inc) < n + 1, t, t -= n]]]], {sr}];
```

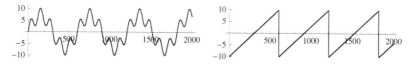

In what follows, we discuss alternative methods of sinusoidal synthesis (→ *Oscillators.maxpat*). A sine wave can be defined as the projection of a rotating pointer (2.2.1.1). In Chapter 3.1.2.5 we generated rotating pointers in the complex plane. We can achieve the same effect by multiplying a vector (x_i, y_i) in the *xy*-plane by the rotation matrix for the angle ϕ.

$$\begin{pmatrix} x_{i+1} \\ y_{i+1} \end{pmatrix} = \begin{pmatrix} \cos(\phi) & -\sin(\phi) \\ \sin(\phi) & \cos(\phi) \end{pmatrix} \begin{pmatrix} x_i \\ y_i \end{pmatrix}$$

Because the components of the rotating pointer are coupled sine and cosine waves, we speak of a *coupled-form oscillator*. The oscillator reacts continuously to changes in frequency. The amplitude may change minimally over time because of rounding errors in the computations.

```
x1 = 1; y1 = 0; φ = .1; cc = Cos[φ]; cs = Sin[φ];
snd = Table[x = x1; y = y1; x1 = cc*x - cs*y;
            y1 = cs*x + cc*y; sin[[i]] = x; cc[[i]] = y, {i, n}];
```

The so-called *Chamberlin oscillator* also generates two coupled sine waves. Its transfer matrix is

$$\begin{pmatrix} (1-c^2) & -c \\ c & 1 \end{pmatrix} \text{ with } c = 2 \cdot \sin(\phi/2).$$

The Chamberlin oscillator is robust against quantizing errors and does not change amplitude when the frequency changes. The two signals x_i and y_i are practically orthogonal at low frequencies ([95] 2.1.3).

```
x1 = 1; y1 = 0; ϕ = 0.1; c = 2*Sin[ϕ/2];
snd = Table[x = x1; y = y1; x1 = (1 - c²)*x - c*y;
    y1 = c*x + y; sin1[[i]] = x; sin2[[i]] = y;, {i, n}];
```

We obtain the so-called *direct form oscillator* by transforming the equation $x[k] = \sin(k\varphi)$. Using the identity

$$\sin(a+b) + \sin(a-b) = 2\cdot\sin(a)\cdot\cos(b)$$

we can write $x[k+1] - x[k-1] = 2x[k]\cdot\cos[\varphi]$, and shifting by one time unit we get

$$x[k] = 2x[k-1]\cdot\cos[\varphi] - x[k-2].$$

```
x1 = 1; x2 = 1; ϕ = .1; c = Cos[ϕ];
snd = Table[x = x1; x1 = x2; x2 = 2*cc*x1 - x; sin[[i]] = x;, {i, n}];
```

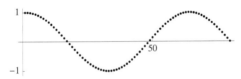

The direct form oscillator is the most efficient sine wave generator. It has the disadvantage however that if the frequency changes, the amplitude also changes. Correcting this feature is computationally costly ([95] 2.1.2).

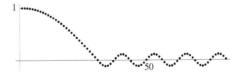

The digital *sinusoidal oscillator* ([65]) is a two-pole resonator. If the poles are on the unit circle, the damping is zero and the quality factor infinite. When the resonator is excited by a pulse, it resonates sinusoidally for an infinitely long time (3.3.3.5 and 8.1.1). (→ Example)

5.1.2.2 The Pulse Generator

Periodic pulse trains generate sounds with many partials (2.2.2.5) and are often used as the source material for subtractive synthesis. The spectrum of a periodic pulse train is not band-limited, which means that frequencies above the Nyquist frequency are also generated. For this reason, we cannot use a table for pulse synthesis having 1 as its first value and 0 for all other values. We can generate a band-limited pulse by summing harmonic cosine functions.

5.1 Fundamental Techniques of Sound Synthesis

The sum of the first N partials at equal amplitude is

$$\frac{A}{N}\cos(2\pi f_0 t) + \frac{A}{N}\cos(2\pi \cdot 2 f_0 t) + \frac{A}{N}\cos(2\pi \cdot 3 f_0 t) + \ldots$$

$$= \frac{A}{N}\sum_{k=1}^{N}\cos(2\pi k f_0 t)$$

which can be simplified to

$$= \frac{A}{2N}\left(\frac{\sin((2N+1)\pi f_0 t)}{\sin(\pi f_0 t)} - 1\right).$$

The following illustration shows pulses constructed by summing 5, 10 and 20 partials of equal amplitude respectively.

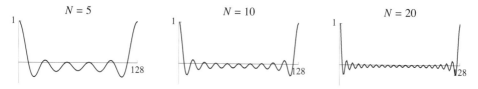

Similarly, pulses can be constructed with a spectrum whose components have exponentially decreasing amplitudes. If the fundamental has the amplitude 1, then the successive other partials have the amplitudes $a, a^2, a^3, \ldots$ ($a < 1$). The sum of the infinitely many partials is

$$\sin(2\pi f_0 t) + a \cdot \sin(2\pi \cdot 2 f_0 t) + a^2 \cdot \sin(2\pi \cdot 3 f_0 t) + \ldots$$

$$= \sum_{k=0}^{\infty} a^k \cdot \sin((k+1) 2\pi f_0 t)$$

which can be simplified to

$$\frac{\sin(2\pi f_0 t)}{1 + a^2 + 2 a \cdot \cos(2\pi f_0 t)}.$$

With the following command line we generate one period of a pulse train. The figures show that the pulses become narrower as a becomes larger, that is, as the partials become stronger.

```
f4 = Table[Sin[2*π*t/n] / (1 + a² - 2*a*Cos[2*π*t/n]), {t, 1, n}];
```

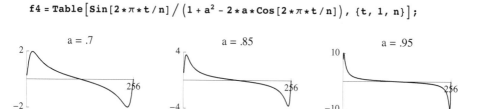

5.1.2.3 The Noise Generator

The waveform of *noise* is irregular and random. In *ideal white noise* all frequencies are present in equal quantity. Hence, the spectrum of white noise does not consist of individual lines, but is constant (2.2.3.2). White noise arises naturally from random movement like the thermal movement of molecules in the air. It is generated digitally using a random generator

(10.1.4.1 and 10.2). The computer does not produce really random numbers, but rather deterministic sequences of numbers called pseudo-random numbers which fulfill certain conditions. All programming languages provide random generators. The most important of these are so-called recurrence generators, which use previously computed numbers z_{i-k} through z_i to compute the next number $z_{i+1} = f(z_{i-k}, ..., z_i)$. The most commonly used algorithm uses only the last number computed, multiplies it by a factor a, adds a constant b and takes the remainder after division by another constant m as the next "random number"

$$z_{i+1} = az_i + b \pmod{m}.$$

The following command uses this algorithm to generate random number between 0 and 9.

```
z = 0; a = 7; b = 3; m = 10; Table[z = Mod[a*z + b, m], {i, 0, 10}]
```

{3, 4, 1, 0, 3, 4, 1, 0, 3, 4, 1}

Here we only get four of the possible ten numbers, which shows that the algorithm does not guarantee a good random generator. If we set $a = 13$, $b = 7$ and $m = 16$, we get all the numbers between 0 and 15. (see [26] pp. 252-253).

```
z = 0; a = 13; c = 7; m = 16; Table[z = Mod[a*z + c, m], {i, 0, 20}]
```

{7, 2, 1, 4, 11, 6, 5, 8, 15, 10, 9, 12, 3, 14, 13, 0, 7, 2, 1, 4, 11}

We can get real valued random numbers between 0 and 1 by dividing the natural random numbers from the algorithm by the maximum m.

```
z = 0; a = 13; c = 7; m = 16; Table[z = Mod[a*z + c, m]; N[z/16], {i, 0, 20}]
```

{0.4375, 0.125, 0.0625, 0.25, 0.6875, 0.375, 0.3125, 0.5, 0.9375, 0.625, 0.5625, 0.75, 0.1875, 0.875, 0.8125, 0, 0.4375, 0.125, 0.0625, 0.25, 0.6875}

A maximum period length (that is, the pseudo-random sequence does not repeat until the $(m + 1)$-th value) does not in itself guarantee random behavior. So, for instance, a pseudo-random generator can yield first predominantly small numbers and then predominantly large numbers. For this reason, algorithms have to pass a number of tests (see [26] pp. 253-254). In other words, pseudo-random sequences are anything but random: on the one hand, the sequence is determined, and in many programs the random generator always gives the same sequence, on the other the values show a most improbably uniform distribution. The probability that in a random sequence of N numbers each value appears once and only once is very small: the number of combination without repetition divided by the total number of combinations is $N!/N^N$. For five numbers we have a probability that each number appears once and only once of $24/625 = .0384$, for 1024 numbers a probability of $1.537 \cdot 10^{-443}$. In order to ensure that successive calls of a random generator will always provide different sequences of numbers, many random generators can be passed a starting value, known as a *seed*. Some programs automatically produce a starting value from the system clock of the computer (4.4.1.4).

The following illustration shows to the left the first one hundred values of white noise generated by a pseudo-random generator, to the left the signal's spectrum.

```
11 = Table[RandomReal[{-1, 1}], {i, 0, 2000}];
```

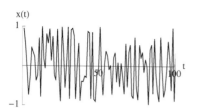
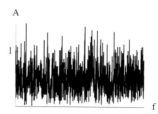

The formula $z_{i+1} = 69069 z_i + 1 \pmod{2^{32}}$ produces good white noise. (→ Sound example and spectrum)

Some applications (4.1.1.6) provide random generators that run at a slower rate f_r than the sample rate f_s, outputting the last random value until it is time to generate the next one. We can produce such a sequence with the command below. If we set $f_s = 2000$ Hz and $f_r = f_s/10$, we get a new random number every 10 sample periods.

```
z = 0; l3 = Table[If[Mod[i, 10] == 0,
                  z = RandomReal[{-1, 1}], Null]; z, {i, 0, 2000}];
```

The following illustration shows to the left the first 200 samples of the random signal, to the right the signal's spectrum.

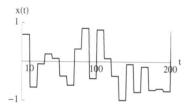
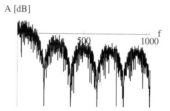

(→ Computing the spectrum)

See [2] p. 95-99, [8] pp. 408-413 and [26] pp. 252-254

5.1.3 Control Signals

Programs for sound synthesis and processing have functions for changing parameters like pitch, volume, panorama position, etc. A succession of control values for such a function is called a *control signal*, its course is called a *control function* or an *envelope*. Control signals can be defined as functions or given in tables, they can be read in real time by an interface or they can be calculated from previous values of the current processes.

5.1.3.1 Simple Control Functions

Besides functions that can be defined for an entire note, piece-wise defined functions are often used for control signals. The following figure shows an amplitude envelope made of linear segments. The segments are called *attack* (d_1), *decay* (d_2), *sustain* (d_3) and *release* (d_4). Synthesis programs have functions that take the values for amplitude a_i and duration d_i from stored lists. Most programs provide not only functions made up of linear segments, but also functions made of exponential or cubic segments (4.1.1.6).

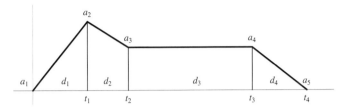

We perceive pitch, loudness and duration essentially as relationships between values, not as absolute values. Hence exponential functions are a particularly important kind of control function (A4.3.5). To create exponential segments, given times t_i and values a_i, we set for each segment

$$f(x) = e^{xk+d},$$

where $k = \ln(a_i/a_{i+1})/(t_i - t_{i+1})$ and $d = \ln(a_i) - t_i k$ (A4.3.5). Note that the boundary values can be arbitrarily small but not zero. The given value a_2 in the example below right gives the amplitude at time $t = 5$ and determines how quickly the function approaches zero.

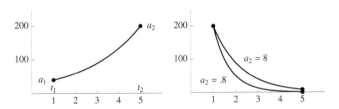

The function's decrease can be described by its half-value period, that is, the time it takes the function to decrease by half. Call this duration d_h. In time d the value of the function decreases by half d/d_h times. Hence,

$$a_2 = a_1 \cdot .5^{d/d_h} \text{ or } d_h = -\ln(2)/\ln(a_2/a_1) = -.6931 d/\ln(a_2/a_1) \text{ and } d = -1.44 d_h \cdot \ln(a_2/a_1).$$

So, for example, to make a tone of amplitude $a = 30,000$ die away in $d = .3$ seconds with a half-decay time of .03 seconds, we calculate the final value $a_2 = 30,000 \cdot .5^{.3/.03} = 29.3$. To make a tone with amplitude $a = 30,000$ die away with a final value of $a_2 = 1$ takes $d = -1.44 \cdot 0.03 \cdot \ln(1/30,000) = .445$ seconds.

To produce a glissando whose perceived pitch (not its frequency) increases exponentially, we use an exponential function of an exponential function (left figure below). One can obtain a similar result by adding a constant value to an exponential function (right figure below).

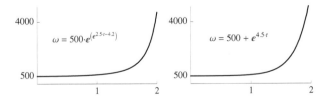

5.1.3.2 Splines

One often does not have precise mathematical expressions for functions, curves and surfaces but only knows individual points. For example, faders but also real-time drawing programs only transmit or record values at certain time intervals. To derive continuous functions from this data, or to compute other points within the function, one uses interpolation techniques and so-called *splines*. "Spline" originally meant slat. In the late 19th century the word referred to a piece of wood or hard rubber used by draftsmen to draw curved lines between fixed points. Today, "splines" refer to the lines themselves. If the splines pass through the control points, one speaks of *interpolation*, otherwise one speaks of *approximation* or *smoothing*, that is, finding the best fit of a curve to given data. Here we will only consider interpolating splines, which are of particular importance in computer graphics.

To define a function passing through n points, n parameters are necessary. The parameters a and b of the straight line $y = ax + b$ are determined by two points, the parameters a, b and c of the parabola $y = ax^2 + bx + c$ by three points. In addition to polynomials of the form

$$y = c_n x^n + c_{n-1} x^{n-1} + \ldots + c_0,$$

exponential functions and trigonometric functions (especially for periodic successions of points) are used for interpolation. From n points we can compute the coefficients of a unique polynomial of degree $n - 1$ or less. In polynomials of high order there is often strong oscillation between the points, as the following illustration shows. The heavy line shows the interpolation with a polynomial of the fifth order, the dashed line the interpolation using cubic splines (see below). (→ Animation)

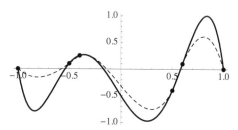

To avoid this oscillation, polynomials of a low order are usually used to compute piece-wise segments. Using polynomials of the first order, one obtains the linear interpolation between two points (5.1.1.6). As a rule, the function has kinks at the points, that is, it is not differentiable at the points. In order to have smooth transitions, the function must be differentiable everywhere, that is, it must have the same slope on both sides of a point. Hence the slope of the combined function changes continuously. If one further requires that the combined function have a second derivative everywhere, then its curvature (the second derivative) changes continuously. Third-order polynomials are determined by two points and the slopes at these points. The piece-wise combination of the functions generated by these polynomials is called a *cubic spline*, a *cubic Hermite spline* or *cspline*. We obtain the ith polynomial

$$a_i + b_i(x - x_i) + c_i(x - x_i)^2 + d_i(x - x_i)^3$$

connecting the points (x_i, y_i) and (x_{i+1}, y_{i+1}) (where $x_1 < x_2 < \ldots < x_n$) from the two equations for the polynomial and the two equations for the slopes m_i and m_{i+1} at the so-called *nodes* (x_i, y_i) and (x_{i+1}, y_{i+1}). From

$$y_i = a_i, \; y_{i+1} = a_i + b_i(x_{i+1} - x_i) + c_i(x_{i+1} - x_i)^2 + d_i(x_{i+1} - x_i)^3,$$

$$m_i = b_i, \quad m_{i+1} = b_i + 2c_i(x_{i+1} - x_i) + 3d_i(x_{i+1} - x_i)^2$$

we obtain

$$a_i = y_i, \quad b_i = m_i,$$

$$c_i = \frac{(2m_i + m_{i+1})(x_i - x_{i+1}) - 3y_i + 3y_{i+1}}{(x_i - x_{i+1})^2}, \quad d_i = \frac{(m_i + m_{i+1})(x_i - x_{i+1}) - 2y_i + 2y_{i+1}}{(x_i - x_{i+1})^3}$$

The following figure shows a cubic spline where the slope at the nodes is always zero. In this way, one can force the minima and maxima of curves whose y-values jump back and forth to be located at the nodes. Where the y-values increase or decrease monotonically, the spline is also monotonic, because a cubic polynomial has at most two turning points and these are located at the given nodes.

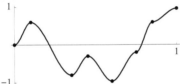

The following figure shows the same succession of points with a given slope $m = -2$ at all points. The piecewise functions are shown extending beyond their respective interpolation points. (→ Animation)

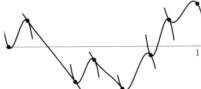

There are many ways to adapt the slope at the points to the course of a function. For example, we can define the slope at point i as:

$$m_i = \frac{(y_{i+1} - y_i)}{(x_{i+1} - x_i)} = \text{the slope of a line to the next point } i + 1,$$

$$m_i = \frac{1}{2}\left(\frac{(y_i - y_{i-1})}{(x_i - x_{i-1})} + \frac{(y_{i+1} - y_i)}{(x_{i+1} - x_i)}\right) = \text{the average of the slopes before and after point } i,$$

$$m_i = \frac{(y_{i+1} - y_{i-1})}{(x_{i+1} - x_{i-1})} = \text{the slope of the line between the neighboring points of } i.$$

The following figure illustrates the last of these three definitions. (→ Animation)

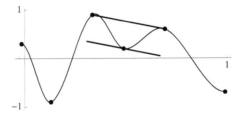

5.1 Fundamental Techniques of Sound Synthesis

The method described above of setting the slope at the interpolation points to zero to get monotonic splines results in a staircase-like shape for the function. The cubic spline increases monotonically when its derivative is positive. This is the case when we define the slope as follows ([81] p. 69):

$$m_i = \min(\tfrac{1}{2}(s_i + s_{i-1}), 2s_i, 2s_{i-1} - m_{i-1}), \text{ for } i > 1 \text{ and } m_1 = s_1,$$

where s_i is the slope from point i to point $i+1$. (→ Animation)

Curves in two or three dimensions can be defined parametrically. Then every coordinate of a point on the curve is a function of the parameter (A4.5). If we choose the time t as parameter, the curve $u(t)$ can be interpreted as the trajectory of a particle. If we call the coordinates in two dimensions u_1 and u_2 and in three dimensions u_1, u_2 and u_3, then the general parametric representation is: $u(t) = (u_1(t), u_2(t))$ for two dimensions and $u(t) = (u_1(t), u_2(t), u_3(t))$ for three dimensions. In what follows we will consider curves on a plane (i.e. in two dimensions) and we call the interpolation points p_i. The tangents are now defined as tangent vectors m_i. The parameters of the curve segments pass through areas of length t_i.

In the spline functions x was the independent, y the dependent variable, or in other words, y was parametrized by x. To derive the so-called *Hermite blending functions*, we consider the following polynomial from $x_i = 0$ to $x_{i+1} = t_i$:

$$y(x) = a_i + b_i(x - x_i) + c_i(x - x_i)^2 + d_i(x - x_i)^3$$

having the coefficients

$$a_i = y_i, \; b_i = m_i, \; c_i = \frac{(2m_i + m_{i+1})(x_i - x_{i+1}) - 3y_i + 3y_{i+1}}{(x_i - x_{i+1})^2}, \; d_i = \frac{(m_i + m_{i+1})(x_i - x_{i+1}) - 2y_i + 2y_{i+1}}{(x_i - x_{i+1})^3}.$$

Substituting the parameter t for x and p_i for y_i we obtain

$$a_i = p_i, \; b_i = m_i, \; c_i = \frac{3(p_{i+1} - p_i)}{t_i^2} - \frac{2m_i + m_{i+1}}{t_i}, \; d_i = \frac{2(p_i - p_{i+1})}{t_i^3} - \frac{m_i + m_{i+1}}{t_i^2}.$$

If we call the normalized parameter t/t_i, then the polynomial for the coordinates is

$$u_i(s) = a_i + b_i s + c_i s^2 + d_i s^3.$$

Regrouping, we obtain a representation of the curve using functions of s, the Hermite blending functions, multiplied by the values p_i, p_{i+1}, m_i and m_{i+1}.

$$u_i(s) = (1 - 3s^2 + 2s^3)p_i + (3s^2 - 2s^3)p_{i+1} + (s - 2s^2 + s^3)t_i m_i + (s^3 - s^2)t_i m_i.$$

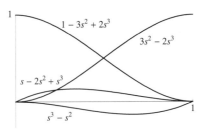

In the following example we calculate the tangent vector m_i as the difference of the vectors p_{i-1} and p_{i+1}. The vector's direction determines the direction of the tangents, and its length shows how far beyond an interpolation point the curve follows the vector. By introducing a parameter which multiplies the vectors m_i, we can control how tensely the spline is spanned between the points. In the left figure this parameter has the value 1, in the right figure it has the value 0.2. (→ Animation)

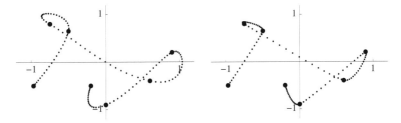

The illustration below shows the same type of spline in a three-dimensional implementation. (→ Animation)

If we always use the same number of intermediate points between two neighboring interpolation points for our calculations, the distances between the intermediate points will vary (below left). If we want the intermediate points to be more evenly distributed, we need to adjust the number of intermediate points to the distances between the interpolation points (below right). Calculating the exact distances on the curve is not trivial, but a first estimation can be made from the length of the straight line between two interpolation points. (→ Animation)

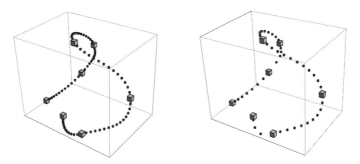

(→ Kochanek-Bartels spline)

Instead of indicating the slopes of the tangents numerically or calculating them from neighboring points, they can also be defined by the endpoints of the tangent vectors (P_2 and P_3 in the figure below) to the interpolation points (P_1 and P_4 below). The resulting curves are called *Bezier splines* and are equivalent to cubic Hermite splines. (→ Animation)

5.1 Fundamental Techniques of Sound Synthesis   223

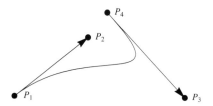

The following figures show a geometrical method for constructing Bezier splines of degree n. A Bezier spline of first degree is a straight line between two points, here traced by point P moving from point P_1 to point P_2 as a function of parameter t (figure left). In the construction of a second-degree Bezier spline, two auxiliary control points are introduced, here Q_1 and Q_2 (middle figure below). As a function of parameter t, Q_1 and Q_2 move from P_1 to P_2 and from P_2 to P_3 respectively. At the same time, the point P, which traces the Bezier spline, moves from Q_1 to Q_2. By introducing more control points (also called handles), Bezier splines of higher degree can be constructed. The figure to the right below shows the construction of a Bezier spline of third degree. (→ Animation)

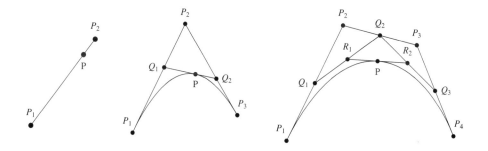

Various kinds of splines can be made in Max/MSP using the external icst.spline.
(→ *icst.spline.maxhelp*)

See: Gary D. Knott, Interpolating Cubic Splines *[81]*.

5.1.3.3 Interpolation Filters

The next illustration shows to the left a control signal consisting of only a few non-zero values. We will modify the signal by interpolation so that a continuous function passes through all the non-zero points.

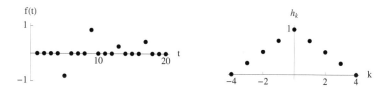

We interpolate linearly between the points by convoluting the control signal with the impulse response shown to the right. Convolution results in a non-causal FIR filter. A causal filter would shift the entire signal in time. Since the filter uses three previous samples and three future samples to compute the response, the interpolation only begins after three samples and ends three samples before the actual end of the signal.

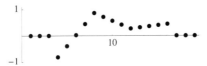

We get a smoother connection of the points by using an impulse response that corresponds approximately to the usual windowing functions instead of the triangle shown (3.3.2.5). This technique is also used in so-called oversampling.

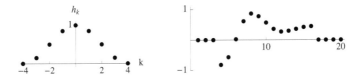

The following illustration shows, highly simplified, the spectrum of the signal resulting from the values at the interpolation points. It is, like all discrete signals, periodic in $\Omega = 2\pi$ (upper scale). Introducing the zeroes shown in the figure above does not change the form of the spectrum, but it does change the relationship between the periodic frequency and the sampling rate (lower scale). Ideally, an interpolation filter should suppress all the higher components caused by aliasing. Here we need a low-pass filter whose cutoff frequency is at $\Omega' = \pi/4$ (dashed line).

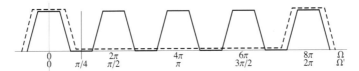

Both FIR filters above satisfy this requirement, as the illustration of their amplitude responses below shows. The second filter (right) suppresses the high frequencies markedly better than the first (3.3.2.5).

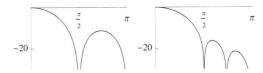

5.1.3.4 Variable Control Signals

Frequently, control signals are not set at the beginning of a sound, but have to take account of unexpected developments. Consider the problem of determining a signal's instantaneous amplitude. Often the amplitude cannot be predicted from the parameters and control signals of an instrument. Especially in systems with feedback, the amplitude can vanish or increase exponentially very quickly. In Csound the unit generator *rms* (root mean square) can be used to ascertain a signal's energy. This measurement gives a control signal that can be used, for example, to restore a strongly filtered signal to its original amplitude.

5.1 Fundamental Techniques of Sound Synthesis

To measure the varying amplitude of a sine tone, one could measure the curve's excursion directly at each turning point. However, this method is laborious and fails for complex signals. Therefore one often measures not the amplitude but the power, that is, the energy over time. To determine the energy, one squares and adds together all the (positive and negative) excursions:

$$E = \sum_{k=0}^{n} x(k)^2.$$

From this, we calculate the power as

$$L = \frac{1}{n} \sum_{k=0}^{n} x(k)^2.$$

This result is the mean squared value *msv*. Because energy and power are proportional to the square of the amplitude, we introduce a new measurement: the square root of the mean squared value *rms*. The following figures show a signal whose amplitude increases over time and the signals *msv* and *rms*.

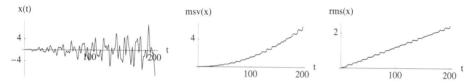

One can see that the resulting control signal reacts very slowly to changes: the original signal to the left breaks off suddenly at $t = 200$, but the control signals only decrease very slowly.

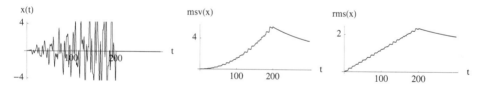

In the following example, we measure the power of a signal as a piecewise function. We perform the windowing and the simultaneous taking of the quadratic mean by running the signal through a FIR low-pass filter and scaling. The illustration below shows a signal with both high frequency and low frequency components (upper left). To the upper right is the filter's impulse response and below the computed control functions for the signal.

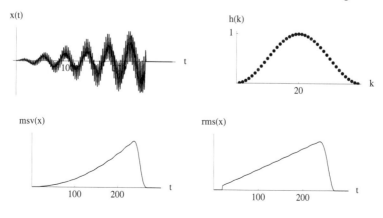

By using a small window we get a control signal that reacts quickly to changes. But if the frequency of a component of the signal falls below the filter's cutoff frequency, that component appears as ripple in the amplitude envelope (the *msv* and *rms* signals middle and right below).

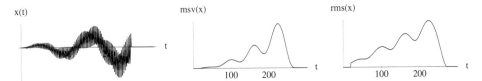

The control signal can be used to end a note when its amplitude gets below a certain threshold, or a tone's amplitude can be lowered or raised to a given value (for an example see 5.1.4.4). If we use a rectangular window, the FIR filter can be replaced by a moving average filter, which is a simple recursive filter (3.3.3.5).

To realize so-called envelope following, nonlinear filters are often used that react more quickly to increasing amplitude than to decreasing amplitude. In the system equation

$$env(k) = (1 - b) \cdot |x(k)| + b \cdot env(k - 1),$$

the factor b is chosen to be somewhat smaller when the absolute value of the input $|x(k)|$ is greater than the current value of the envelope $env(k)$. (→ Animation)

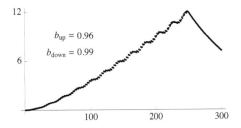

5.1.3.5 Tempo Functions

A particular difficulty arises when control functions influence themselves or the speed of a signal. Consider resampling a sound file. If one uses a control function whose control points are in a specific temporal relationship to the sound file, the control points will be shifted in the resampled file. For example, if we transpose a sound lasting one minute an octave down, we will only obtain the first half of the signal. One can see how to compute the function for the speed of reading out a sound file for a given transposition function in Chapter 5.1.4.3. At the end of that chapter, it is shown how the transposition function can be adapted so that the control points are in the same positions as they are relative to the original sound (5.1.4.3).
In this chapter we describe how the duration of time units (e.g. beats in traditional music) can be changed using functions of tempo.

Musical tempo T is defined as beats b per time unit t: $T = b/t$. The duration of a certain number of beats is $t = b/T$, and the number of beats occurring within a specific time is $b = Tt$. If the tempo is a function of time, it is defined as the derivative of the beat function

$$T(t) = \frac{d}{dt} b(t)$$

5.1 Fundamental Techniques of Sound Synthesis

and the beat function is the integral of the tempo function, i.e., the area under the curve $T(t)$ (see the following figure)

$$b(t) = \int_0^t T(\tau)\,d\tau.$$

The area under the curve can be calculated approximately as the discrete sum

$$b(t) = \sum_{k=0}^{t \cdot sr} f(k \cdot sp) \cdot sp = \frac{1}{sr} \sum_{k=0}^{t \cdot sr} f\left(\frac{k}{sr}\right)$$

where: sr = sample rate, $sp = \frac{1}{sr}$ = sample period of the discretization.

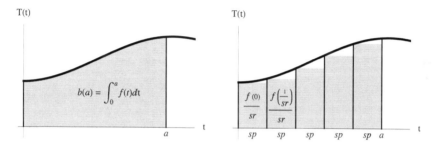

The beat function increases monotonically (if the tempo function is positive).

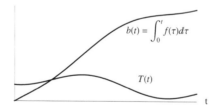

(→ Example for numeric integration)

The tempo function $T(t)$, defined as number of beats per time, is the derivative of the beat function, i.e. the slope of $b(t)$ at time t. The slope of the tangent to the function can be approximated by the slope of the line through two adjacent points of the function:

$$T(t) = \frac{\Delta b}{\Delta t} = \frac{b(t+sp) - b(t)}{sp} = (b(t+sp) - b(t)) \cdot sr \quad (\rightarrow \text{Figure})$$

The beat function $b(t)$ gives the number of beats up to time t. The inverse function to the beat function $b(t)$ is the time function $t(b)$ which gives the time at which beat b occurs. Conversely the beat function is the inverse of the time function. The graphs of inverse functions are symmetric to the main diagonal (A4.2).

$$t(b) = b^{-1}(t) \quad \text{and} \quad b(t) = t^{-1}(b)$$

If we denote the tempo at time 0 with T_0 and at time a with T_a, a linear tempo function is

$$T(t) = ct + T_0, \text{ with } c = \frac{T_a - T_0}{a}.$$

A linear tempo function is defined if three of the for variables T_0, T_a, a and $b(a)$ are defined. Given T_0, a and T_a the beat function $b(t)$ is

$$b(t) = \frac{ct^2}{2} + T_0 t.$$

If we equate the beat function to the number of beats i and solve the equation with respect to i, we get the end time t_i of beat i (starting time of beat $i + 1$)

$$t(i) = \frac{-T_0 + \sqrt{2ci + T_0^2}}{c}.$$

Example: $T_0 = 30$ beats/min $= 1/2$ b/s, $a = 6$, $T_a = 120$ beats/min $= 2$ b/s. The following diagram shows the tempo function and the starting times of the beats. The areas of each trapezoid is 1.

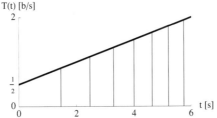

($\rightarrow$ Given: T_0, T_a and $b(a)$) ($\rightarrow$ Given: T_0, number of beats n, a)

An exponential tempo function is of the form $f(t) = T_0 e^{ct}$, with

$$c = \ln(T_a/T_0)/a \quad \text{and} \quad a = \ln(T_a/T_0)/c.$$

Given T_0, T_a and a, the beat function is:

$$b(t) = \int_0^t T_0 e^{c\tau} d\tau = \frac{T_0}{c}(e^{ct} - 1)$$

The starting time of the beat b is

$$t(b) = \frac{1}{c}\log(1 + \frac{ci}{T_0})$$

Example: $T_0 = 30, T_a = 90, a = 6$

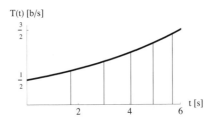

($\rightarrow$ Given: T_0, T_a, n. Given: T_0, n, a.)

A simple tempo function where the four variables T_0, T_a, a and n can be adjusted at will is the quadratic function. The function has the form $T(t) = c_2 t^2 + c_1 t + c_0$. The beat function is:

5.1 Fundamental Techniques of Sound Synthesis

$$b(t) = \int_0^t T(\tau)\,d\tau = \frac{c_2}{3}t^3 + \frac{c_1}{2}t^2 + c_0 t$$

Example: $T_0 = 90$, $T_a = 30$, $a = 5$, $n = 4$.

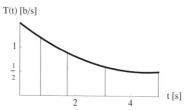

(→ Manipulate)

For more complex tempo function the beat function cannot usually be calculated algebraically. For numerical calculations, we discretize the time, defining the sampling rate as *sr*. The sampled time then is k/sr with $k = 0, 1, 2, \ldots$ (→ Example 1: Exponential tempo function)

Example 2: The illustration below shows a more complex tempo function.

```
f1[t_] := .7 * (Sin[.46 * 2 * π * t + 1.9] + t + 1);
```

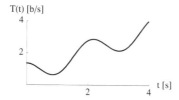

Computing the number of beats at $t = 4$:

```
sr = 1000; a = 4;    1   a*sr
                    ─── Σ  f1[k/sr]           8.17285
                    sr  k=0
```

Listing the starting times of the beats:

```
                                        f1[k/sr]
sum = 0; is = 1; nb = {}; sr = 1000; Do[sum += ────────;
                                          sr
  If[sum > is, is += 1; nb = Append[nb, 1. * k / sr]], {k, 0, a * sr}]
```

{1.063, 1.737, 2.113, 2.466, 2.875, 3.325, 3.681, 3.957}

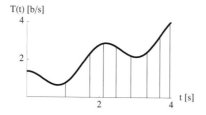

(→ See how to get an integral number of beats within a given time)

Periodic tempo functions in general do not result in periodic series of beats.

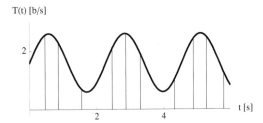

The simplest way to get a periodic series of beats is to add a constant to the tempo function or to multiply the function by a constant. The following algorithm computes such a constant c. The series of beats can repeat after one or any other integral number of periods of the tempo function.

```
sr = 100; a = 2 / fr; c = 0; dsum1 = dsum = .1;
dc = .5; bs = 7;
```

$$\text{While}\left[\text{Abs}[dc] > .005, \text{ sum} = 0; \text{Do}\left[\text{sum} += \frac{\text{f3}[k/sr] + c}{sr}, \{k, 0, a*sr\}\right];\right.$$
$$\left.\text{dsum} = bs - \text{sum}; \text{If}[dsum*dsum1 < 0, dc = -.5*dc]; c += dc; dsum1 = dsum;\right]; c$$

-0.0585938

Now we add the constant c to the tempo function

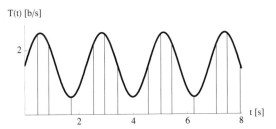

The above algorithm can be used for arbitrary periodic tempo functions. For the tempo function $T(t) = \sin(2\pi ft) + c$ the constant c can be calculated algebraically. (→ Calculation and Application)

In the following example we create tempo functions where the beat lengths are in geometric series. We begin with the time function $t(b)$ and derive the beat function $b(t) = t^{-1}(b)$ and the tempo function. To generate a geometric series of beat lengths, we need a time function $t(b)$ with a factor

$$q = \frac{t(b+1) - t(b)}{t(b) - t(b-1)}$$

and $t(0) = 0$. Every function of the form $t(b) = c_1(e^{c_2 b} - 1)$ fulfills this constraint. We find $q = e^{c_2}$ and $c_2 = \ln(q)$. In order to get positive values $t(b)$ for $b > 0$ the constants c_1 and c_2 must both be either positive or negative. (→ Animation)

5.1 Fundamental Techniques of Sound Synthesis

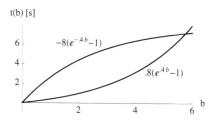

The length of the first beat is $d_1 = t(1) = c_1(e^{c_2 \cdot 1} - 1) = c_1(q - 1)$ and hence $c_1 = \frac{d_1}{q-1}$. The beat function is

$$b(t) = \frac{1}{c_2} \log(1 + \frac{t}{c_1})$$

and the tempo function

$$T(t) = \frac{d}{dt} b(t) = \frac{1}{c_2(c_1 + t)}.$$

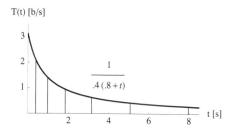
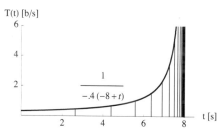

($\rightarrow$ Application)

The above tempo function is defined by the parameters c_1 and c_2 which can be calculated from d_1 and q. Since the durations form a geometric series the duration a of n beats can be calculated as follows

$$a = d_1(1 + q + q^2 + \ldots + q^n) = d_1 \frac{1 - q^{n+1}}{1 - q}.$$

When the tempo increases, that is for $0 < q < 1$, the greatest possible durations is

$$a_{max} = d_1(1 + q + q^2 + q^3 + \ldots) = \frac{d_1}{1-q}$$

and hence

$$q = 1 - \frac{d_1}{a_{max}}$$

Starting tempo T_0 and tempo at time a T_a are

$$T_0 = \frac{1}{c_2 c_1} \quad \text{and} \quad T_a = \frac{1}{c_2(c_1 + a)}$$

Given T_0 and q we get c_1 and c_2

$$c_2 = \ln(q) \quad \text{and} \quad c_1 = \frac{1}{c_2 T_0}.$$

Let us consider tempo functions for multiple voices which generate synchronous beats. Simple examples are: voices with constant tempi in a rational proportion (polyrhythms), voices with variable tempi in rational proportions and voices with beat functions having common periods. Below is an example of the second case. The beat function for a tempo function that generates a geometric series of beat lengths (see above) is a logarithmic function. In order to get beats at multiples of $t(b) = e^{b/r}$ we set

$$b_1(t) = r \cdot \log(t+1), \; b_2(t) = \frac{r}{2}\log(t+1), \; b_3(t) = \frac{r}{3}\log(t+1)$$

The common beats are at the times

```
r = 20; gr = Table[e^(6*x/r) - 1., {x, 0, 6}]
```
{0., 0.349859, 0.822119, 1.4596, 2.32012, 3.48169, 5.04965}

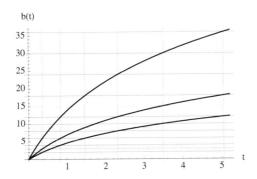

It is usually impossible to find two or more non trivial elementary tempo functions that provide more than two or three common beats. We therefore describe a generic procedure to construct such functions with the help of splines. There are splines whose interpolation points do not lie on the curve and splines whose interpolation points do lie on the curve. With the latter we can calculate beat functions by defining the times where beats coincide (list *tb*) and the numbers of the beats at these times in the different voices (lists *b1* and *b2*).

```
tb = {0, .9, 1.56, 3.67, 5.13}; b1 = {0, 3, 5, 7, 11}; b2 = {0, 5, 8, 11, 13};
```

If we connect the points with straight lines, we get a continuous beat function, but the tempo function (which is the derivative of the beat function) is not continuous.

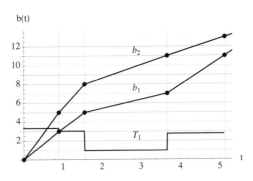

5.1 Fundamental Techniques of Sound Synthesis

If we then pass an interpolating function through the given points, we get a smooth beat function and a smooth tempo function.

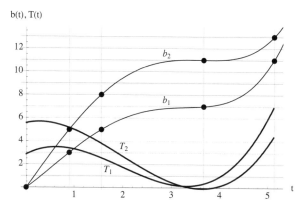

Let us now consider tempo functions which generate common beats for canonic voices. Trivial examples are polyrhythms without tempo changes (entry of the second voice after any number of beats) and periodic tempo functions with periodic beats (entry of the second voice after a period of beats). In the following example we generate exponential tempo functions which slow down in such a way that the duration of n beats of the leading voice always is equal to the duration of m beats of the following voice. We define the exponentially decaying tempo function $T(t) = be^{-ct}$. The beat function is

$$b(t) = \int_0^t b\, e^{-c\tau}\, d\tau = \frac{b}{c}(1 - e^{-ct}).$$

We now create a canon with an offset of the second voice of dt seconds. The slowing down shall be such that the first 3 beats of the second voice shall last as long as 2 beats of the preceding voice. We calculate c and b as follows:

$$c =.; \; dt =.; \; \text{Solve}\left[\frac{1. - e^{2*c*dt}}{1 - e^{c*dt}} == \frac{5}{3},\; c\right] \qquad \left\{\left\{c \to -\frac{0.405465}{dt}\right\}\right\}$$

$$b =.; \; dt =.; \; c =.; \; \text{Solve}\left[\frac{b}{c}\left(1 - e^{-c*dt}\right) == 3,\; b\right] \qquad \left\{\left\{b \to \frac{3\, c\, e^{c\, dt}}{-1 + e^{c\, dt}}\right\}\right\}$$

If we set $dt = 2$ we get

$$c = 0.4054651/2;\; b = \frac{3\, c\, e^{c*2}}{-1 + e^{c*2}};$$

The decellerando happens so fast that, although the tempo does not become zero, only a finite number of beats occur. We get the maximal number of beats by putting $t = \infty$ in the formula for the beat function

$$b(\infty) = \frac{b}{c}(1 - e^{-\infty}) = \frac{b}{c}$$

b/c   9.

The formula for the starting times of the beats is

b =.; c =.; i =.; Simplify[Solve[$\frac{b}{c}(1 - e^{-c*t}) == i, t$]]

$$\left\{\left\{t \to -\frac{\text{Log}\left[1 - \frac{c\,i}{b}\right]}{c}\right\}\right\}$$

The starting times in our example are

Table[$-\frac{\text{Log}\left[1 - \frac{c\,i}{b}\right]}{c}$, {i, 0, 8}]

The following figures show the tempo (upper figure below) and the beat function (lower figure below) with the beats of three canonic voices.

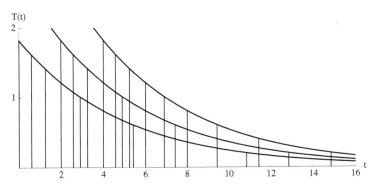

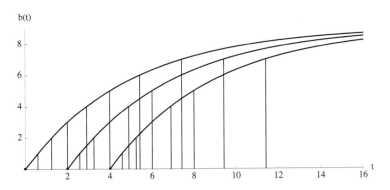

5.1 Fundamental Techniques of Sound Synthesis

We can subdivide the beats keeping the same tempo ratio.

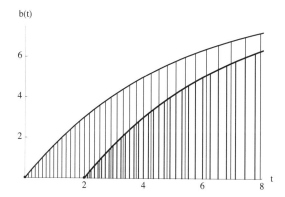

5.1.3.6 Synchronization

Temporally coordinated processes are said to be synchronized. If the processes are periodic, their frequencies and phases can be coordinated. Nature provides an impressive example of synchronization in Southeast Asian fireflies. Although the flashing of the fireflies in the same tree or in a given area is at first quite uncoordinated, after a certain time all the fireflies flash in synchrony. The spontaneous synchronization of machines was first described by Christiaan Huyghens (1629–1965), who observed that clocks affixed to the same beam can synchronize themselves. Organ pipes that do not have quite the same frequency can get in tune by a weak coupling through the air. Synchronization also plays an important role in laser technology, in neuronal nets, in chemical reactions, etc. Synchronization can be forced by any of a variety of impulse generators. In sound synthesizers, so-called *hard sync* uses a master oscillator to trigger a slave oscillator so that the slave interrupts its signal in the midst of a period and jumps back to the beginning of the current waveform. More interesting is the synchronization of several non-hierarchically organized systems, which can happen only with nonlinear systems (see 3.4.5).

To get the variable parameters of several voices to converge, it suffices to take the average of the parameters and to let the parameters of the individual voices approach this average. It is more difficult to synchronize periodic processes. In what follows we consider periodic control signals and show ways to synchronize them without using forced synchronization. The frequencies of oscillators can be synchronized easily by successive approximation. In the following example, each frequency moves closer to the middle frequency by 0.5 % at each sample time.

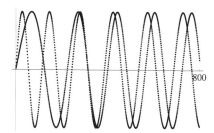

The illustration above shows that coordinating the frequencies of two oscillators does not necessarily coordinate their phases. However, if we compute the waveform using the formula

$x = amp \cdot \sin(\omega t)$ and then adjust the frequencies, the phases will also become synchronized (providing the initial phases were the same).

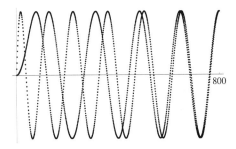

The phases of several oscillators can be synchronized by coordinating their unwrapped phases. If the phases are more than 2π apart, the waveforms will momentarily take on the same relative phase more than once during the process of synchronization. In the following example, the sine waves have the same frequency but the phases differ by 4π (this cannot be seen from the waveform). The synchronization begins at time $t = 100$. The waveform having the greater phase slows down rapidly and soon the relative phases are the same for a moment. However, the synchronization continues until the absolute phases are the same.

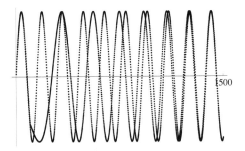

In the example which follows, we synchronize the relative (i.e. wrapped) phases of three oscillators having the same frequency. We use coupled form oscillators (5.1.2.1) and show the state of the oscillation in vector diagrams. The three dashed vectors show the instantaneous phases of the three waveforms, the solid vector shows their normalized sum. At each successive step of the computation, the phases of the three oscillators approach more closely the phase of the sum vector. (→ Animation)

The Max patch *Phase_Sync.maxpat* uses three oscillators to produce vibrato on three partials of a tone. When the phases of the vibratos are not in synchrony, one hears the partials separately. When the phases are in synchrony, they fuse to make a single sound (cf. 5.2.1.3). In the patch, the degree of synchronization can be varied. If this value is large, oscillators can be synchronized that have different frequencies when not coupled.

5.1.4 Delay Lines

We shall speak about delay lines in this introductory chapter because they are used in various synthesis techniques. In addition, they are of importance in realizing filters and spatial simulation.

5.1.4.1 Definition and Direct Implementation

A signal is delayed by N time units by reading the samples x_i into a buffer of length N and moving them one place farther in the buffer at each sample period.

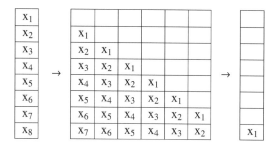

(→ Example)

To create delays that are not a multiple of the sample period, one must interpolate between two successive values. However, this changes the sound's spectrum (3.3.1.4).

5.1.4.2 The Circular Buffer

Instead of moving all the values through a buffer, only the reading and writing positions need to be moved. The following figure shows a circular buffer with the current read and write positions noted. At each sample period, the position of the arrows is moved one position and only the value x_{-9} is replaced by a newly read in value x_0.

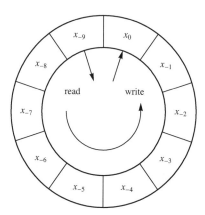

Let us make a circular buffer of length $N = 6$ using Mathematica. Read and write positions can coincide if we first read out the old value and then overwrite it with the next value to be read in. With this technique, we need N memory cells for a delay line of N sample periods.

    `n = 6; rb = Table[0, {i, 1, n}]`   {0, 0, 0, 0, 0, 0}

We first define a signal x and an empty list *xdel* into which we write the delayed values. We make the pointer for the read position using the modulo function Mod[t, n], which gives the remainder of the integer division t / n.

    `x = {1, -9, -10, -10, 6, -9, -3, 2, -4, 0, 6, -1, 5, 0, 0,`
        `-10, 9, -4, -9, 5};`

    `xdel = {}; tt = 17; Do[xdel = Append[xdel, rb[[Mod[t, n] + 1]]];`
    `rb[[Mod[t, n] + 1]] = x[[t]], {t, 1, tt}]`

From the list x above we get

    `xdel`   {0, 0, 0, 0, 0, 0, 1, -9, -10, -10, 6, -9, -3, 2, -4, 0, 6}

After 17 steps, the circular buffer contains the values x_{12} to x_{17}:

    `rb`   {-1, 5, 0, 0, -10, 9}

Programming languages for sound synthesis provide functions to generate delays which require only the input signal and the length of the delay.

5.1.4.3 Delay Lines With Variable Delay

Since all the samples between x_0 and x_{-N} are present in a delay line of length N, arbitrarily many delays of arbitrary length ($< N$) can be realized with the same buffer. In Csound such delays are made using several statements. The statement *delayr* reserves a buffer of the required length, *delayw* indicates the signal to be read into the buffer and the delays themselves are indicated by arbitrarily many *deltap* statements.

x_0	x_{-1}	x_{-2}	x_{-3}	x_{-4}	x_{-5}	...	x_{-i}	...	x_{-k}	...	x_{-l}	...	x_{-N}
↓							↓		↓				

The length of the delay can be made variable by changing the buffer's read position. If the delay becomes shorter while the signal is being read out, some values will be skipped and the signal shortened, which raises its pitch. If the delay is lengthened, some values will be read out twice, the signal will be lengthened and its pitch will drop (compare the Doppler effect in Chapter 9.1.3.1). Before we show how to calculate for the general case the transposition resulting from variable delay, let us illustrate the effect in a few special cases. If we shorten the delay from the entire buffer length $N = t \cdot sr$ to zero over the total time t of the read-in and read-out process, at every sample time when the signal moves towards the buffer's end the pointer for the read-out position moves one position towards the beginning of the buffer, thus skipping every other value of the signal. The signal is shortened to half its length and its pitch rises by an octave. The following illustrations show two moments of an animation in which a signal is read in and out of a delay buffer, shown by a rectangle. The lower arrow marks the read-out position, which is moved to the right, shortening the delay. In the time in which five periods of the input are read in, 10 periods are read out. (→ Animation) (→ *Var_Resampling.-maxpat*)

5.1 Fundamental Techniques of Sound Synthesis

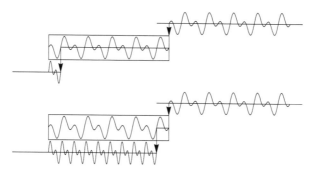

In the second example the delay is at first zero and increases during the reading in of the next five periods to 4/5 of the buffer length. The pointer moves with the signal, and only one period is read out. (→ Animation)

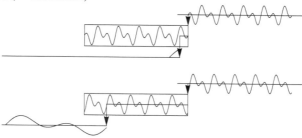

If the read-out pointer moves at the same speed as the signal, the same value is always read out, and the frequency goes to zero. If the pointer moves faster than the signal, the signal is read backwards. If it moves at twice the speed of the signal, we get the reverse of the original signal. Let us call the frequency of the original signal f_0, the frequency of the transposed signal f and the velocity of the read-out pointer v. The following table shows various values of the transposition interval $Tr = f/f_0$ and their relation to the velocity of the pointer.

v	-1	$-1/2$	0	$1/2$	$4/5$	1	2	...	v
f	$2f_0$	$3f_0/2$	f_0	$f_0/2$	$f_0/5$	$f_0\,0$	$f_0(-1)$	...	$f_0(1-v)$
f/f_0	2	$3/2$	1	$1/2$	$1/5$	0	-1	...	$1-v$
$Tr = \log_2\left(\left|\frac{f}{f_0}\right|\right)$	1	0.58 ..	0	-1	-2.32 ...	$-\infty$	-1	...	

The following figure shows the ratio f/f_0 and the transposition in octaves $\log_2|f/f_0|$ in relation to the pointer velocity v. Pointer velocities greater than one lead to time reversal of the signal as well as to transposition.

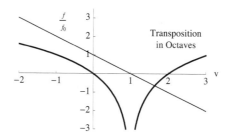

To obtain a transposition of Tr octaves, we set $v = 1 - 2^{Tr}$. Thus we get a transposition two octaves up with $v = 1 - 2^2 = -3$, a transposition a fifth down with $v = 1 - 2^{-7/12} = .33258$.

For a variable transposition $Tr(t)$, we calculate the change of position of the index ind caused by the variable pointer velocity $v(t)$ either by integrating the velocity with respect to time:

$$ind = \int v dt$$

or by summing it over T sample periods:

$$ind = \sum v\, T.$$

So, for example, to have a smooth glissando which falls an octave in two seconds, we set $Tr(t) = -.5t$ and get

$$ind = \int v dt = \int (1 - 2^{Tr}) dt = \int (1 - 2^{-.5t}) dt = t + 2^{1-.5t}/\ln(2) + C.$$

The constant C can take any value. We choose it so that the index is zero at the beginning:

$$C = - 2^{1-.5 \cdot 0}/\ln(2) = -2.8854.$$

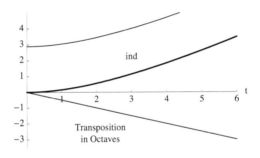

(→ Another Example)

In Csound, we can make an instrument that transposes according to the function $Tr(t)$ as follows. We have for the index

$$ind = \sum v \cdot T = \sum \left(1 - 2^{Tr(t)}\right) T = t - \sum 2^{Tr(t)}\, T.$$

For the second addend we define the sum $ksum$ (line 2 below). It is calculated by the formula above; for $Tr(t)$ we use ktr (line 5), which is read from an arbitrary function table (line 4). We generate the variable t for time with the linear function kt (line 3). In the last three program lines a delay buffer is defined (lines 7–9), into which a sine wave of frequency 440 Hz is read (line 10) and read out at the position $ind = kt - ksum$. The total length of the delay is given by $p4$, which must be estimated or calculated outside the program (line 7).

```
;5-1-4b
instr 1                                    ;1
ksum    init    0
kt      line    0,p3,p3
ktr     oscili  1,1/p3,2
ksum=           ksum+exp(log(2)*ktr)/kr    ;5
```

5.1 Fundamental Techniques of Sound Synthesis

```
asnd    oscil    30000,440,1
a0      delayr   p4
a1      deltapi  kt-ksum
        delayw   asnd
        out a1
endin
```

The following score transposes the sine tone corresponding to *f2* over the duration of the note (Sound Example 5-1-4b).

```
f1 0 32768 10 1
f2 0 32768 -7 0 10000 -1 10000 1 12768 -3
i1 0 8 1
```

The first of the following figures shows the control function *f2* for the transposition and the resulting function for the index *ksum*. The other two figures show that for slightly different control functions the index can become negative (middle) or greater than the length of the delay (right). In the first case, a constant has to be added to the index, in the second *p4* must be lengthened.

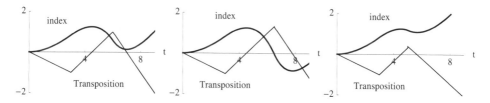

(→ Example with integration constant and corrected rounding errors)

Since not only the pitch of sounds is changed using such techniques, but also durations are shortened or lengthened, the values of the transposition function *Tr(t)* are no longer at the same positions after the transposition of the sound. In the following example, we read out the function *Tr(t)* from table 5 with the velocity 1000(*kt* + *ksum*). *kt* is the time in seconds, 1000 entries in the table are one second of the original sound. The addend *ksum* causes the table to be read at a speed corresponding to the transposition. The amplitude envelope *kamp* of the original sound *asnd* has the period 0.5 second. Because of the correction described above, the first six periods and the descending glissando no longer last 3 seconds but 4.3 seconds, the next six periods and the ascending glissando 3.1 seconds.

```
;5-1-4b
instr 3
ksum    init     0.0
kt      line     0,p3,p3
ktr     tablei   (kt+ksum)*1000.0,5,0,0,0
ksum =           ksum + (exp(log(2)*ktr)-1)/kr
kamp    oscil    20000,2,4
asnd    oscil    kamp,440,1
a0      delayr   p4
a1      deltapi  -ksum
        delayw   asnd
        outs     a1
endin
```

```
f 1 0 32768 10 1
f 4 0 32768 5 0.001 768 1 32000 0.0001
f 5 0 32768 -7 0. 3000 -1 3000 1 3000 0 19268 0
i 3 0 11 2
```

5.1.4.4 Delay Lines With Feedback

If one creates feedback by re-injecting the output signal into the delay line, one gets an IIR filter (3.3.3). A special application of such a filter is the *wave guide*, which is used in the simulation of wave propagation through a linear medium like a string or an air column (8.2). Because the filter becomes nonlinear if the delay is not constant, the result of its use is unpredictable. For this reason, the filter is rarely described or employed, although interesting results can be had with even the simplest nonlinear systems. In the following example we simulate feedback between an amplifier and a moving microphone. The delay corresponds to the time it takes the sound wave to go from the loudspeaker to the microphone. In the feedback between electronic devices, the amplitude of the sound picked up by the microphone depends on the distance of the microphone from the loudspeaker: the greater the distance, the weaker the feedback. If the distance is smaller than a given amount, the signal is constantly amplified and its loudness increases exponentially. The instrument below realizes a delay line whose length changes during the duration of the note linearly from *p7* to *p8* (line 3).

```
     instr 1                                          ;1
     kamp    linseg  0,p5/2,p4,p5/2,0,p3-p5,0
     kdel    line    p7,p3,p8
     ;asnd   rand    kamp
     asnd    oscil   kamp,440,1                       ;5
     a0      delayr  p6
     a1      deltapi kdel
             delayw  asnd+a1
             out     a1
     endin
```

In the first note (Sound Example 5-1-4c1) a short sine wave of 440 Hz *asnd* is read into the delay line together with the delayed output *a1* (line 8). In this example, the tone is transposed a minor third down, read in again and transposed, etc. The result is the arpeggio of a chord of a diminished seventh, each of whose tones enters later than the preceding tone because of the lengthening delay. In the second note (Sound Example 5-1-4c2) white noise is read in during the entire note (line 4). The delay is shortened from .2 to .01 second. First one hears an accelerating pulsation, which is caused by the amplification of small irregularities in the noise. When the delay is shorter than about 1/20 second, one hears the repetition of the nearly periodic signal as a pitched sound (compare the comb filter 3.3.3.5).

```
     f1 0 32768 10 1
;5-1-4c1      i1 0 5 18000 1 1 .2 1
;5-1-4c2      i1 0 9 3000 1 1 .2 .01
```

In this way, a sound of 100 Hz appears in acoustic feedback when the delay is $d = 1/440$ s. If the speed of sound is $c = 340$ m/s, this corresponds to a distance between loudspeaker and microphone of $cd = .773$ m. The much higher pitched sounds in acoustic feedback are independent of the distance between loudspeaker and microphone and are due to the fact that amplifier, loudspeaker and microphone are not strictly linear and thus amplify some frequency ranges more than others. We can simulate this behavior in instrument 2 below by filtering the signal before re-injecting it (line 5). After only a few periods, we get a high, whistling tone with the frequency of the filter. With a feedback factor of .2, the amplitude increases exponentially (Sound Example 5-1-4c3). After a few seconds the signal is so strongly overmodulated that it resembles a square wave.

```
instr 2
a2      init    0
asnd    rand    p4
a1      delay   asnd+p6*a2,p5
a2      reson   a1,p7,p8,2                  ;5
        out     a1
endin

i2 0 5 100 .3 .2 1000 100
```

When the feedback factor is small, the effects are weak, but if it passes a certain threshold, the amplitude begins to increase exponentially. To avoid this, it is necessary to find a control function calculated from the amplitude itself. Simple methods from control theory, like switching on or off when a signal falls short of or exceeds a given setpoint, lead to discontinuities and to an oscillatory behavior (3.4.4.4). A simple, continuous control signal can be made by defining the feedback factor as a function of the signal's energy. In the following Csound example, we measure the energy $krms$ of the signal using the unit generator rms and choose as feedback factor $c_1 e^{-c_2 \, krms}$.

```
......
krms    rms     a2
a1      delay   asnd+p7*exp(-p6*krms)*a2,p5
......
```

The amplification when $Amp = 0$ and the speed with which the feedback factor decreases as the amplitude increases can be regulated by the factors c_1 and c_2. The figure below shows the function for various constant values. The heavy line corresponds to the function used for the Sound Example 5-1-4c4.

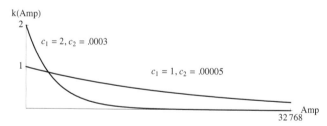

```
instr 3
......
krms    rms a2
a1      delay   asnd+exp(-p6*krms)*a2,p5
......

i3 0 9 20 .2 .00005 1500 400
```

The amplitude of the sound generated increases rapidly at first because the factor is 1, but then it quickly becomes stable. (→ Waveform) (→ Interacting Feedback Using Multiple Variable Delay Lines)

5.1.4.5 Applications

All applications that use delay lines to generate new signals are systems as defined in Chapter 3. When one is particularly interested in how such systems affect frequencies, one speaks of filters. All the examples discussed here are digital filters in the broadest sense, and their behavior can be described using the methods introduced in Chapter 3, at least as long as they are linear.

When one digitally simulates physical vibration, the current value of the new signal is computed from previous values, more precisely from values delayed by one or two sample periods (8.1.1). Systems conceived to model physical processes are nothing but complicated filters with feedback. The simulation of wave propagation, as shown in Chapter 8.2 for the synthesis of sounds from vibrating strings and air columns, uses delays corresponding to the time it takes a sound wave to move from one end of a string or a tube to the other. Such systems are comb filters (3.3.3.5).

For the simulation of sound waves in closed spaces, delay lines are used whose delays correspond to the sound waves' propagation times. Using delay lines, one can simulate acoustical phenomena like echo, reverberation, the Doppler effect and interaural time differences (9.1.3.1).

Special effects, familiar from analog electronics, can also be realized with delay lines. *Flanging* is made by adding a delayed signal to its non-delayed original. The resulting elimination of some partials and strengthening of others creates timbral change. If the delay changes periodically, one gets the characteristic rotating flanger effect, where the sound seems revolve around the listener's head. Usually durations of between 2 and 20 milliseconds are used, the frequency of the periodic variation is usually between .1 and 10 Hz. If one uses larger delay times, for example 10 to 30 ms, the delayed sound and its original no longer fuse and will be perceived as a doubling of the original. This is the so-called *chorus effect*.

5.2 Additive Synthesis

In *additive synthesis*, complex sounds are synthesized by combining simple sine waves. The technique has the advantage that the individual components, the sine waves, and their addition to form sounds can be described in a straightforward way. In addition, breaking down a complex sound into sine waves corresponds closely to the analysis the auditory perception performs on sounds. The disadvantage of this technique is the vast amount of information and computation required to generate differentiated and musically interesting sounds.

5.2.1 The Synthesis of Periodic Waveforms

5.2.1.1 Basic Techniques

The *Theorem of Fourier* says that every periodic oscillation of frequency f can be described as a sum of sinusoidal oscillations having the frequencies f, $2f$, $3f$, ... with the appropriate amplitudes a_1, a_2, a_3, ... and initial phases ϕ_1, ϕ_2, ϕ_3 ... (3.1.1.1). In additive sound synthesis, periodic waveforms are usually generated using a series of oscillators. The initial phase is often not given, because the phase has little or no influence on the resulting timbre.

5.2 Additive Synthesis

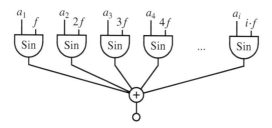

In Csound, one only needs to define a single oscillator and then specify the component sine waves in the Csound score.

```
instr 1
a1      oscil   p4,p5,1
        out a1
endin

f1 0 32768 10 1
i1 0 5 10000 100
i1 0 5 5000 200
i1 0 5 3000 300
```

A sum of sine waves having the amplitudes a_i and the frequencies f_i can be generated directly using the formula:

$$x(t) = a_1\sin(f_1 2\pi t) + a_2\sin(f_2 2\pi t) + \ldots + a_n\sin(f_n 2\pi t)$$

In Mathematica, the formula can be passed directly to the Play[ ] command. The default sample rate is $sr = 8192$ and the resolution is 8 bits. The following commands generate the sum of two sine waves of 100 and 200 Hz twice, the first time using the default parameters, the second time with a sample rate of 44,100 and quantization of 16 bits.

```
Play[Sin[100*2 π*t] + .5*Sin[200*2 π*t], {t, 0, 1}]

Play[Sin[100*2 π*t] + .5*Sin[200*2 π*t],
    {t, 0, 1}, SampleRate -> 44 100, SampleDepth -> 16]
```

We generate a digital signal by determining a sample rate sr or a sample period T and substituting kT or k/sr for the continuous time t. In C, we can write N sound samples of our example above into a buffer snd with this line:

```
for (k = 0; k < N; k++)
    snd[k] = 20 000*sin (100*2*pi*k/sr) + 10 000*sin (200*2*pi*k/sr);
```

5.2.1.2 Variable Parameters

The richness of most natural sounds depends less on the number of partials than on their constantly changing mixture. In synthesis, we use control functions for the amplitudes of the individual components to create the mixture. The changes of frequency in the upper partials normally correspond to the changes of frequency in the fundamental, and so we can generally use a single control function for the frequencies of all the partials. The Csound instrument below generates a sine tone whose amplitude is controlled by a function with four exponential segments. The segments correspond to the attack (from $p5$ in $p6$ seconds to $p7$), the release (in $p8$ seconds to $p9$), the steady-state phase (in $p3 - (p6 + p8 + p11)$ seconds to $p10$) and the decay (in *p11* seconds to *p12*) (5.1.3.1). The frequency varies periodically by the fraction *p13* at the frequency *p14*, which results in a simple vibrato. (Sound example 5-2-1a)

```
;5-2-1a.orc
instr 1
kamp    expseg  p5,p6,p7,p8,p9,p3-p6-p8-p11,p10,p11,p12
kfr     oscil   p13,p14,1
a1      oscil   kamp,p4*(1+kfr),1
        out     a1
endin

f1 0 32768 10 1
i1 0 2 440  1 .1 8000 .1 5000 4000 .1 1 .012 7.5
i1 0 2 880  1 .1 7500 .1 3400 3100 .1 1 .012 7.5
i1 0 2 1320 1 .1 9700 .1 4400 4000 .1 1 .012 7.5
i1 0 2 1760 1 .1 3300 .1 1500 1100 .1 1 .012 7.5
```

One can see how much calculation is required to produce differentiated sounds this way. The number of partials should be increased, at least for low sounds, and each amplitude envelope should have many points of articulation, particularly then when timbre changes quickly. In addition, in natural sounds the changing parameters interact with each other. For example, the relationship of the partials to each other changes as a function of the fundamental frequency and the total amplitude. The method of computing the partials individually shown above is principally used when the data required is available from spectral analysis (5.2.2). In order to produce differentiated envelopes economically, synthesis languages generally provide specialized generators.

5.2.1.3 Fusion

Whether several sine tones fuse into a single sound depends on many factors. Integer proportions between the frequencies of the partials is an important criterion, but by no means the only one. Sine tones reaching the ear from different directions are perceived separately, despite the simple relationship of their frequencies (in Sound example 5-2-1b, 440 Hz comes from the left, 880 Hz from the right). Two sine waves whose frequency ratio exceeds a certain size (for example, 1700/100 Hz, in Sound example 5-2-1c) are also heard as separate. When the tones begin one after the other, it takes a certain time before they fuse (200, 400, 600, 800, 1000 Hz at intervals of two seconds, Sound example 5-2-1c2). On the other hand, we perceive tones coming from the same direction or evolving in the same fashion as components of the same sound, even when their frequencies are not harmonic. In Sound example 5-2-1c3, all amplitudes decrease exponentially (frequency ratio 1:1.913:2.112:2.933:4.093:5.376), in Sound example 5-2-1c4 the tones all have the same strong vibrato (frequency ratio 1:1.937:2.134:3.145:3.941). Natural tones are not made up of sine tones, rather the way they are physically produced determines their waveforms. Hence, in nature it is usually easy to relate individual tones to a sound, because the criteria for fusion are always fulfilled. Even when instrumentalists play the same tones on the same instrument, one can distinguish the individual instruments. But with synthetic sounds, it can happen that tones not meant to be part of the same sonic event fuse or, on the the other hand, that individual partials of the same sonic event stick out. Individual partials can be imperceptibly exchanged between sounds, as Sound examples 5-2-1b2 and 5-2-1c5 show. In Sound example 5-2-1b2, 13 sine waves sound with the frequency ratio 1:2:3:... At the beginning, the partials are randomly distributed in the two stereo channels. After two seconds, the tones all move towards the middle for two seconds, then for another two seconds the even partials move to the right channel, the odd

5.2 Additive Synthesis

partials to the left channel. The most obvious difference between the beginning and the end of the example is that at the end the sound in the right channel is an octave higher than at the beginning, while the sound in the left channel has become somewhat nasal. Once again, in example 5-2-1c5 13 sine waves sound, six of which begin an octave above a middle frequency and seven of which begin an octave below that frequency. We hear a glissando in which the high tones move down an octave and the low tones up an octave. The two sounds meet after four seconds, that is, the tones are now all partials of a single fundamental. Then the even partials move down and the odd partials move up, giving the rising sound a nasal timbre and causing the fundamental of the descending sound suddenly to leap an octave. Interestingly enough, one does not immediately hear the octave leap, although it is instantaneous, but one does notice that the descending sound is only an octave lower than at the beginning, despite the glissando over two octaves. (→ Csound scores and orchestras)

5.2.1.4 Data Reduction

It is not always necessary to use separate amplitude and frequency functions for all the partials of a sound. For instance, if the timbre remains the same, all partials can use the same amplitude envelope (2.3.4.4). In this case, it is worthwhile to make a function table containing the sum of the partials' amplitudes (5.1.2.1). Tones within the same critical band (2.3.4.1) can hardly be distinguished from each other, and under certain circumstances the stronger of two can completely mask the weaker. In synthesis, it is often possible to leave out partials because of the masking effect, particularly since from about the eighth partial on, neighboring partials always fall within the same critical band. In example 5-2-1d1, we hear successive sounds whose first eight partials have the same amplitudes. The first sound also has the partials 9 to 13 with amplitudes 1000, 3000, 9000, 3000 and 1000. The second sound has only the additional partial 11 with amplitude 9000, and the third sound has only partial 11 with amplitude 13000. The fourth sound has no additional partials. The differences between the first three sound are so slight that the first spectrum can be replaced by the second or the third.

5.2.1.5 Acoustic Illusions

Our perception attempts to recognize form even when the available sensory data is incomplete. This mechanism is extremely important for the acoustic and visual perception of everyday objects and usually performs without our being aware of it. In synthetic images and sounds, however, it can sometimes lead to a misinterpretation of the sensory data. The best-known acoustic illusion is named after its discoverer, the psychologist Roger Shepard, or in connection with glissandi after Jean-Claude Risset. Consider a sound consisting of partials an octave apart (solid lines in the figure below) that moves up or down while the spectral envelope remains constant (dotted lines below). After the sound has moved an octave, it has its original spectrum again. This process can be repeated as often as one wants, resulting in an arbitrarily long glissando. Risset uses a bell-shaped amplitude envelope, so that new tones enter and old tones disappear unnoticed. (For Risset's original instrument, see [2] p. 106)

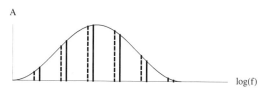

In Csound, Risset's glissando can be realized as follows. A first instrument produces a glissando (line 3) using a sine wave and the amplitude envelope shown above (the square of a half period of a sine) (lines 2 and 4). The second instrument repeats the sound of the first instrument (*ga1*) with a delay of *p4* seconds.

```
;5-2-1e
instr 1                                                      ;1
kamp    oscili   sqrt(log(p4)),.5/p3,1
kfr     expon    p5,p3,p6
ga1     oscil    exp(kamp*kamp),kfr,1
        out      ga1                                         ;5
endin
instr 2
a2      init     0
a2      delay    ga1+a2,p4
        out      a2                                          ;10
endin
```

The glissando produced by the first note of the score falls 10 octaves in 20 seconds. If instrument 2 repeats the tone every two seconds, we obtain partials in octaves (Sound example 5-2-1e1).

```
f1  0   32768      10 1
;       duration   amplitude   frequency1   frequency2
i1  0   20         14000       20480        20
i2  0   40         2
```

The effect can be varied in many ways. The next example uses narrowly filtered noise. The delay is chosen so that three sounds per octave are generated, giving an augmented triad. Example 5-2-1f1 is an excerpt from the Csound synthesis. Example 5-2-1f2 consists of augmented triads simultaneously ascending and descending.

```
;5-2-1f
instr 1
kamp    expseg   1,p3/2,p4,p3/2,1
kfr     expon    p5,p3,p6
arnd    rand     kamp
ga1     reson    arnd,kfr,3+p7*kfr,2
        out      ga1
endin

;          duration    f1      f2     BW
i1 0 50    15000       2560    80     .001
i2 0 80    3.333
```

In the next example, we move several sine tones with a constant frequency difference of 200 Hz. The tones are faded in at 2000 Hz and faded out at 1000 Hz. Although only five tones are sounding at any time, the fading in and fading out of the tones is almost imperceptible, and the impression of the glissando dominates. At high volumes, one hears the constant difference tone at 200 Hz.

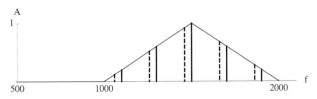

5.2 Additive Synthesis

With the first Csound instrument, we produce a single tone with the linear decreasing frequency *kfr* and the amplitude envelope *kamp*. The second instrument reproduces the tone using a delay line. The frequency space between 1000 and 2000 Hz is filled after 20 seconds, and the sound repeats every four seconds.

```
;5-2-1g
instr 1
kamp      linseg      0,p3/2,p4,p3/2,0
kfr       line        p5,p3,p6
ga1       oscil       kamp,kfr,1
endin

instr 2
a2   init        0
a2   delay       ga1+a2,p4
     out    a2
endin

f1 0 32768 10 1
;         dur         amp         f1          f2
i1 0      20          5000        2000        1000
;         dur         del
i2 0      40          4
```

In the next example, the envelope of a spectrum is transposed higher and higher under a second envelope. This results in a sound with constant fundamental and ascending formants (2.3.2.2).

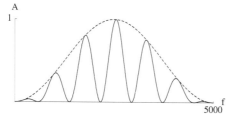

(→ Animation) (→ Realization in Csound)

5.2.2 Analysis-Resynthesis

In order to generate natural sounds using additive synthesis, one needs detailed knowledge of the composition and variation of their spectra. It is instructive to improve the quality of the synthetic sounds by comparing them with natural sounds. One not only learns how the natural sounds are put together; one also learns which elements of a sound are particularly important for our perception. Ideally, an acoustical analysis can provide data that permit a convincing resynthesis of the analyzed sound. Moreover, one can change the data, and hence the resulting sound, before the resynthesis.

Further Reading: Computer Sound Synthesis for the Electronic Musician *by Eduardo Reck Miranda [15], pp. 133-143,* Computer Music *by Charles Dodge and Thomas A. Jerse [2], pp. 220-261 and* The Theory and Techniques of Electronic Music *by Miller S. Puckette [87], pp. 267-300.*

5.2.2.1 Introduction

The block diagram below shows the principles of the analysis and resynthesis of a signal. The signal is put through N band-pass filters in parallel, F_0 to F_{N-1}, each of which filters the signal through a different frequency band. Originally this technique was used in the analog domain in telecommunications to reduce the data necessary for the transmission of speech by recording and transmitting only those channels relevant for the speech signal. As computing power increased, however, it became feasible to do both the analysis and the resynthesis digitally, even if that sometimes meant that the analysis produced more data than in the original signal. The precision of the analysis is essentially dependent on the number of filters, also called channels. Increasing the number of filters has the disadvantage, apart from the greater amount of data, that the filters react more slowly as the passband becomes smaller.

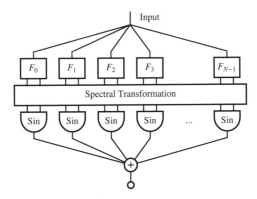

Analyzing and resynthesizing digital signals corresponds to the process shown above. One obtains the spectrum using the discrete Fourier transform DFT (3.2.2). A sound N samples long is analyzed into $N/2$ spectral components, called *bins*, equally distributed over the frequency range $[0, sr/2]$. For example, a sound 32,768 samples long at sampling rate 44,100 Hz will be analyzed into 16,384 bins of $44,100/32,768 = 1.34583$ Hz width. When analyzing time-varying sounds, one must calculate the spectrum in short time intervals. If $N = 2048$, one gets a new spectrum every $2048/44,100 = 0.04643$ seconds, each having 1024 bins $44,100/2048 = 21.5332$ Hz wide. If the frequency of an analyzed sine wave does not correspond exactly to the center frequency of a bin, there will be *spectral leakage* (3.2.2.3). The illustration below shows the analysis of two sine waves. The frequency of the first (figure left below) is at the edge of the third bin. Besides a strong component in the third bin, there are also smaller components in all the other bins. The frequency of the second sine wave (figure right below) lies exactly in the middle of the third bin, making a precise analysis possible. The quality of the analysis of a harmonic sound depends essentially on whether the partials fall on the middle of their respective bins. When this is the case, one speaks of pitch-synchronous analysis. Using the Inverse Fourier Transform, the original sound can be resynthesized from the analysis spectrum. If the spectrum is changed before resynthesis, all kinds of side-effects can arise (5.2.2.3).

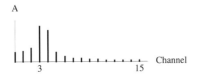
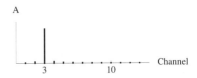

5.2 Additive Synthesis   251

5.2.2.2 The Discrete Fourier Transform DFT

In this chapter we shall try to illustrate the discrete Fourier transform and the concepts used in synthesis programs, without going into technical details. One can find a simple description of spectra in Chapter 2.3.2. Technical details of the Fourier analysis of analog signals are given in Chapters 3.1.1 and 3.1.3, and details of Fourier analysis of digital signals can be found in Chapters 3.1.2 and 3.2.2.

The discrete Fourier transform DFT is an invertible mapping: the spectrum of a sound of N samples is computed and stored in N so-called bins. Each bin contains information about both the amplitude and the phase of one of the frequencies 0, $1/sr$, $2/sr$, ..., $(N-1)/sr$. The sound samples can be reconstructed from this spectrum using the inverse discrete Fourier transform IDFT. That means that the entire temporal information about an arbitrarily long signal is also contained in the spectrum. In order to demonstrate the amplitudes and phases of the bins, we will use a *phasor representation* of the signal (2.2.1.1 and 3.1.2.5). The illustration below shows to the right a cosine oscillation with amplitude r and initial phase ϕ (solid line) and its decomposition into cosine and sine components (dashed lines). In the phasor diagram, the position of the pointer X corresponds to the cosine oscillation at time 0. The vectorial decomposition gives the amplitudes a and b for the cosine and sine components respectively. The frequency of the bin corresponds to the number of revolutions per second of the phasor.

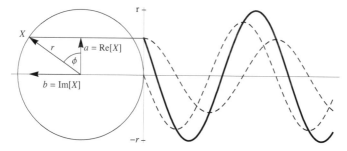

The following illustration shows the phasors, or rather the bins, of a spectrum with seven frequencies ω_i.

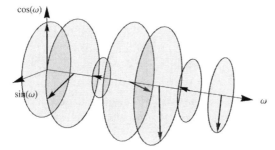

The DFT (and the FFT) of a sound consisting of N real-value samples $x(k)$ ($k = 0$ to $N - 1$) gives N complex values $X(n)$ ($n = 0$ to $N - 1$). The second half of the DFT consists of the complex conjugates of the first half (for $a + bi$ set $a - bi$) and is thus symmetrical to it. A complex number $a + bi$ is the sum of a real number a and an imaginary number bi (3.1.2.1). In the DFT spectrum the real part is the cosine amplitude component of the signal and the imaginary part is the sine component. Usually, the real part of a complex number is shown on the *x*-axis and the imaginary part on the *y*-axis. In the illustration above, the real axis is pointing up and the imaginary axis is pointing left so that the oscillation can be taken from the position of the phasor.

```
Fourier[{1, -2, 0, 3, -2, 0, 1}]
```

{0.38, -0.2 - 0.07 i, 0.7 - 2.58 i,
0.63+ 1.35 i, 0.63 - 1.35 i, 0.7+ 2.58 i, -0.2 + 0.07 i}

The values can be understood as follows. The N values $X(n)$ correspond to the positions of the phasor at time zero for each of the frequencies ω_n: 0, sr/N, $2sr/N$, ..., $n \cdot sr/N$, ..., $(N-1)sr/N$. The angle $\phi_n = \arg(X(n))$ is the initial phase of the cosine component, the radius $r_n = |X(n)|$ is the amplitude of the frequency $n \cdot sr/N$. The phasor of the *n*th component rotates counterclockwise with the frequency ω_n. Since a phasor with a frequency greater than half the sampling rate rotates through more than half a cycle per sample, its frequency is $\omega_{N-n} = -\omega_n$. The phasors $X(n)$ and $X(N-n)$ rotate symmetrically to the *x*-axis. Hence their sum is real and is equal to $2r_n\cos(\omega_n t + \phi_n)$. We get r_n and ϕ_n from the components $X(n)$ of the DFT:

$$r_n = \sqrt{a^2 + b^2}, \phi_n = \arctan(b/a).$$

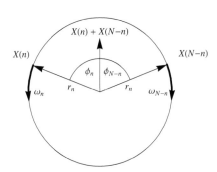

Let us compute a DFT and examine the results. We generate a sound 12 samples long, consisting of a DC component of amplitude 0.3, one period of a cosine, initial phase 0 and amplitude 1, two periods of a cosine of double the original frequency, initial phase $\pi/4$ and amplitude 2 and four periods of a cosine of four times the original frequency, initial phase π and amplitude 0.4. When we take the DFT, we compensate the scaling factor Mathematica uses with the factors $1/\sqrt{n}$ and $\sqrt{n}$ (3.2.2.4).

```
n = 12; snd = Table[.3 + Cos[2*π*i/n] + 2 Cos[2*2*π*i/n - π/4]
              + 0.4 Cos[4*2*π*i/n - π], {i, 0, n - 1}];
```

$$\mathtt{fft} = \frac{1}{\sqrt{n}} \mathtt{Fourier[snd]}$$

{0.3, 0.5, 0.707+ 0.707 i, 0, -0.2, 0, 0, 0, -0.2, 0, 0.707- 0.707 i, 0.5}

5.2 Additive Synthesis

The first value of the DFT is the amplitude of the DC component, that is, the amplitude at 0 Hz. The second and the last values give together the amplitude $0.5 + 0.5 = 1$ and the initial phase 0 for the component at frequency sr/N. The third and second-last values give the amplitude $2\sqrt{0.5 + 0.5} = 2$ and the initial phase $\arctan(1) = \pi/4$ of the component at the frequency $2sr/N$. The fifth and fourth-last values give the amplitude 0.4 and the initial phase π (negative values) of the component at the frequency $4sr/N$. ($\rightarrow$ IFFT)

The characteristics of the DFT were discussed in Chapter 3.2.2.4 and the DFT was processed using complex operators. Usually in applications for the treatment of sound, only the real amplitude and phase values up to half the sampling rate are converted and used. After the spectrum is treated, the resulting values are converted back into complex values and transformed into sound using the IDFT.

The figure on the left in the illustration below shows the two subpatches (or abstractions) *fft_bp~* und *fft_phase_rand~* a *fft_phase_rand~* used in the Max patch *FFT_1.maxpat*. The middle figure shows the contents of *fft_bp~*, the right figure shows the contents of *fft_phase_rand~*. Both abstractions have an audio input and an audio output. In addition, *fft_bp~* has two control inputs. The first number in each abstraction is the length in samples N of the signal block to be analyzed, the second the number of overlaps of the blocks (5.2.2.4). The Max object *fftin~* transforms the N samples of the input block into the N complex values of the DFT. The three outputs of *fftin~* give the real part, the imaginary part and the bin number of the value. Hence, for every data block the values at the third output count up from 0 to $N - 1$. The object *cartopol~* transforms the Cartesian coordinates a and b of the complex value into the polar coordinates r and ϕ. The polar coordinates are treated and then converted back into cartesian coordinates with *poltocar~* and transformed again into an audio signal using *fftout~*. The patch *fft_bp~* (middle figure) is a band-pass filter with control inputs for the upper and lower cut-off frequencies. The logical operators $>\sim$ and $<\sim$ test whether the bin number is between the upper and lower cut-off frequencies given. The bin number for a given frequency f is fN/sr (here $f \cdot 2048/44,100 = f \cdot 0.04644$). The patch *fft_phase_rand~* (to the right below) randomizes the phase, destroying the temporal information and causing all the spectral components to sound throughout the entire block (32,768 samples).

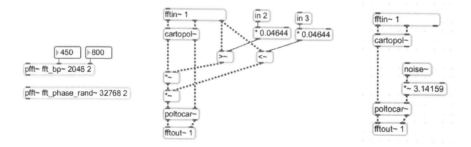

It is more intuitive to use polar coordinates than to use cartesian coordinates, although treatment of the spectrum is equally feasible in either representation. The illustrations below show realizations of the patches above without conversion of the coordinates. The amplitude can be multiplied by a factor by multiplying the components separately by that factor (figure left). One can rotate a phasor by the angle ϕ without changing the amplitude (figure right) with the formulas $a' = a \cdot \cos(\phi) - b \cdot \sin(\phi)$ and $b' = a \cdot \sin(\phi) + b \cdot \cos(\phi)$ (see A3.4).

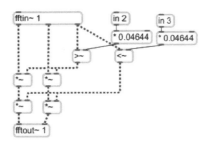 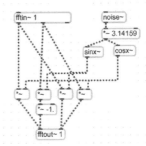

5.2.2.3 Long-Term Fourier Transform LTFT

For the treatment of a sound with a variable spectrum or of sounds in real time, one generally cuts the sound into blocks (usually 512 or 1024 samples) (STFT 5.2.2.4). Any phase changes are imperceptible over such a short time. In this chapter, we show an unusual application of the DFT over long blocks or an entire sound file where clearly audible timbre changes and changes in the timing of sound events are caused by manipulating phase information. The technique was developed by the Swedish composer Paul Pignon [3-3] and is called *mammut* FFT or, in contrast to the Short-Term Fourier Transform, the *Long-Term Fourier Transform* (LTFT). The effect described here can be tried out using the Max patch *fft_2.maxpat*. The maximum block size of the Max object *pfft~* is 131,072 samples. Longer blocks can be treated using the program mammut ([I-5]).

First, let us try to understand how a sound can be shifted in time by manipulating the DFT. The phasor of an individual bin rotates at the bin's frequency f_i times per second. The phasor of bin 0 does not rotate, the phasor of the first bin rotates at the frequency sr/N, that is, at $1/N$ per sample. The phasor of the ith bin rotates with the frequency $f_i = i \cdot sr/N$, that is at i/N per sample. Therefore, to shift a sound by m samples, we must rotate all the phasors by $2\pi mi/N$. In the example that follows, we shift an impulse from the first to the sixth sample. First we compute the amplitudes and phases of the bins.

```
snd = {1, 0, 0, 0, 0, 0, 0, 0};

fft = Fourier[snd]; amp = Abs[fft]
phase = Arg[fft]

{0.353553, 0.353553, 0.353553,
 0.353553, 0.353553, 0.353553, 0.353553, 0.353553}

{0., 0., 0., 0., 0., 0., 0., 0.}
```

Then we compute the bins' rotation by $2\pi mi/N$ and add this value to their phases.

```
m = 5; ph = Table[2.*π*(i - 1)*m/8, {i, 1, 8}]

{0., 3.92699, 7.85398, 11.781, 15.708, 19.635, 23.5619, 27.4889}

phase2 = phase + ph;
```

We obtain the complex spectrum using the formula $amp_i(\cos(\phi_i) + i \cdot \sin(\phi_i))$

```
fft2 = Table[amp[[i]] * (Cos[phase2[[i]]] + i*Sin[phase2[[i]]]),
    {i, 1, 8}]; Chop[fft2]
```

5.2 Additive Synthesis

```
{0.353553, -0.25 - 0.25 i, 0. + 0.353553 i, 0.25 - 0.25 i,
 -0.353553, 0.25 + 0.25 i, 0. - 0.353553 i, -0.25 + 0.25 i}
```

amp = Abs[fft2]
phase = Arg[fft2]

```
{0.353553, 0.353553, 0.353553,
 0.353553, 0.353553, 0.353553, 0.353553, 0.353553}
```

```
{0., -2.35619, 1.5708, -0.785398, 3.14159, 0.785398, -1.5708, 2.35619}
```

The inverse Fourier transform then gives the time-shifted pulse.

snd2 = InverseFourier[fft2]; Chop[snd2]

```
{0, 0, 0, 0, 0, 1., 0, 0}
```

The illustration below shows the bins' rotations proportional to their frequencies.

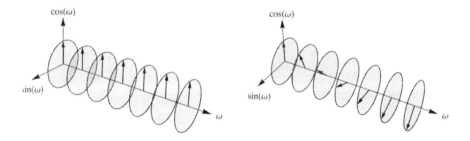

If we rotate the bins disproportionately, for example the ith bin by $2\pi i^p/N$, the spectral elements will be shifted in time according to their frequency. In the next example, an impulse is transformed into a frequency sweep. One can experiment with this effect in the Max patch *FFT_2.maxpat* with the object *fft_freq_drift~* using either an impulse or a sound file.

n = 10 000; snd = Table[0, {n}]; snd[[1]] = 1;

fft = Fourier[snd]; amp = Abs[fft]; phase = Arg[fft];

First we compute the rotation up to $n/2$.

p = 2.; ph1 = Table[2 * π * (i - 1)^p/n, {i, 2, n/2}];

Bin 0 and bin $n/2 + 1$ are pure cosine components of phase 0, the other bins are symmetrical to bin 1 to $n/2$, that is, they have the same amplitudes but inverted phases.

ph = Join[{0}, ph1, {0}, -Reverse[ph1]];

phase2 = phase + ph;

fft2 =
  Table[amp[[i]] * (Cos[phase2[[i]]] + i * Sin[phase2[[i]]]), {i, 1, n}];

```
snd2 = InverseFourier[fft2];

ListPlay[snd2]
```

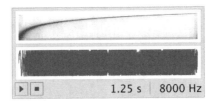

The illustration below shows a disproportionate rotation of the bins.

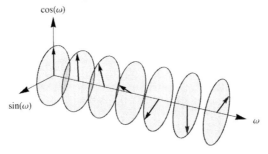

(→ Example in Mathematica)

The illustration below shows two cosine impulses that have been temporally shifted by different amounts by manipulating their phases in Mathematica. (→ Manipulate)

Temporal blurring can also occur when only the amplitudes are treated. If one sets to zero all those bins whose amplitudes are below a given threshold, bins can be zeroed out that belong to a strong partial that has been blurred by spectral leakage. In this case, the partial will sound, but the phase information of the zeroed-out bins will be missing (*fft_2.maxpat*, *fft_threshold~*).

5.2.2.4 Short-Term Fourier Transform STFT

In the Short-Term Fourier transform, the sound to be examined is taken in short segments which are analyzed in sequence. To avoid discontinuities caused by the abrupt ends of the segments, each segment is multiplied by a window function (3.3.2.5).

5.2 Additive Synthesis

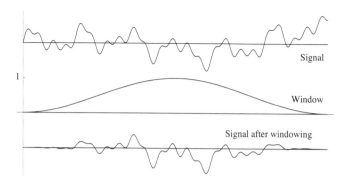

The windowing causes spectral leakage, even of components exactly in the middle of a channel. To compensate for this, two or more windows can overlap. Hence one must pass the size of the window and the number of overlaps to the Max object *pfft~*. The shape of the window can be given in the objects *fftin~* and *fftout~*.

The illustrations below show Max objects for the treatment of spectra used in the Max patch *FFT_3*. Since the *object fftin~* outputs the spectrum as an audio signal, the spectrum can be shifted using a delay (*fft_shift~*, middle figure). The object *delay~* expects in inlet 2 the delay in samples. If this value is negative, the length of the window *N* is added. One can give the upper frequency limit of the bins used in inlet 3. That way, the frequencies that disappear at the lower end of the spectrum because of the shift and reappear at the upper end can be suppressed. Small shifts result in transposition of the sound and make the partials inharmonic (for example, the frequencies 100, 200, 300, ... Hz might become 120, 220, 320, ... Hz). If one treats a melody using a larger shift, the individual partials become audible, and if they are stronger than the fundamental, or if the fundamental disappears, one hears a new sequence of tones made of the melody's overtones.

In order to transpose a sound (*fft_transpose~*, figure at the right below), the spectrum must be stretched or compressed. This is done by multiplying the bin number in the third output of *fftin~* by the transposition factor in inlet 2.

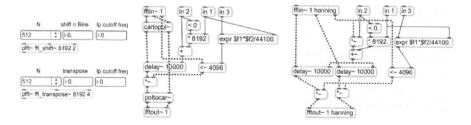

In the example *fft_threshold~*, all bins whose amplitudes are below the threshold given at inlet 2 are zeroed out. Alternatively, the spectrum can be distorted by any mathematical operations the user chooses. An example is *fft_shape~*, where the amplitudes are raised to the power in inlet 3 and then scaled by the factor in inlet 2. This increases the differences in the partials' amplitudes when the exponent is greater than 1 and lessens the differences when the exponent is between 0 and 1.

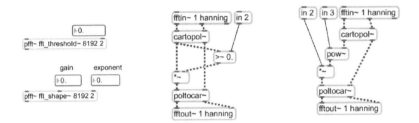

Since in Max the spectrum is treated as an audio signal, it can be processed with audio objects (*fft_reson~*, middle figure below). The example *fft_filter~* (right figure below) shows how to make a simple audio equalizer. Using the object table, one draws the amplitude response of the filter with the mouse. These values are written into an audio buffer (*amp_resp*) that is then used as a lookup table for the amplification factors for the bins.

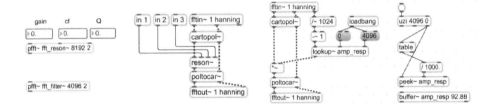

5.2.2.5 The Phase Vocoder

The *phase vocoder* is a device or a program for the analysis and resynthesis of audio signals in the frequency domain. It typically makes simultaneous pitch shifting and time compression or expansion possible. The phase vocoder measures the components of a signal and in addition determines the deviation of a component's frequency from the center frequency of its channel. The figure below shows the procedure for the analysis of a channel k. The left part of the figure corresponds to the Fourier analysis of the windowed signal described above. The amplitude and the instantaneous phase can be computed from the sine and cosine components a_k and b_k, and from two successive phase values the phase change per time unit can be calculated. The phase change is used to calculate the deviation from the center frequency of the channel, because the greater the deviation of the input frequency is from the nominal frequency, the faster its phase is shifting compared with the nominal phase of that channel. (For a detailed technical discussion of the phase vocoder, see [8] pp. 241-264, [70] pp. 78-82, [86] pp. 557-580, [88] pp. 31-90 and [64].)

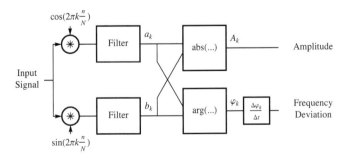

5.3 Subtractive Synthesis

Subtractive synthesis is in some sense the opposite of additive synthesis. The original material of subtractive synthesis is sound having a rich spectrum, usually white noise or a pulse train, which is filtered to obtain the desired result. One advantage of the technique is a certain naturalness that can be attributed to the fine irregularities coming from the noise. In addition, the process essentially resembles the production of many natural sounds that come from the transformation of an external excitement by a resonating body. The theory explained in Chapter 3 and the synthesis by physical modeling discussed in Chapter 8 show that a resonator can be understood as a special kind of filter.

5.3.1 Filters

A filter is a system that suppresses oscillations in a certain frequency range while passing other frequency ranges unchanged or even amplified. All objects are mechanical filters for oscillation, because they react differently to different frequencies. Objects with clear resonances or natural frequencies, like plates, membranes and strings, amplify exciting oscillations whose frequencies correspond to their natural frequencies, and suppress all others. Both technical and natural systems often maintain their stability by damping all oscillation thanks to rigid materials and irregular structure. In sound-producing systems, on the contrary, the intention is often to obtain specific tones by amplifying individual frequencies. In resonant bodies, the intention is usually to propagate as large a frequency range as possible by using elastic materials and simple structures. The filters used in electronic devices are resonant electric circuits. We speak about the design of digital filters and how they work in Chapter 3. Here we just give a summary of the most important notions in connection with filters, referring the reader to the appropriate chapters for technical details.

5.3.1.1 Characteristics of Filters in the Frequency Domain

The characteristics of filters are described by their response to sine waves of various frequency, known as the *frequency response* of the filter (3.3.1.4). *Linear filters* modify only the amplitude and the phase of an oscillation. The modification of amplitude as a function of frequency is called the *amplitude response* of the filter. The figure a) below shows the amplitude response of a so-called *band-pass filter*, which passes a given frequency, the *center frequency* f_m, and suppresses other frequencies more and more strongly the greater their distance is from the center frequency. The *cut-off frequencies*, f_{c1} and f_{c2}, are defined as the frequencies at which the signal's amplitude has decreased by 3 dB. The range between the two cut-off frequencies is called the filter's *bandwidth*. The quality of a filter, its so-called Q factor, is given by the ratio of the center frequency to the bandwidth. A filter that suppresses frequencies in a certain range but lets through lower and higher frequencies is called a *band-stop* or *band-reject* filter (figure b) below).

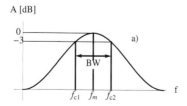
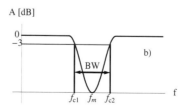

A filter which passes low frequencies and attenuates high frequencies is called a *low-pass filter* (figure a) below). One which passes high frequencies and attenuates low frequencies is called a *high-pass filter* (figure b) below).

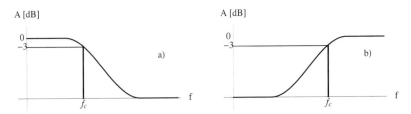

The transition between the passband and the stop band is continuous. The region of transition cannot be arbitrarily narrow, nor can the shape of the amplitude response curve be realized with arbitrary precision. In many applications, one has no control over the details of the amplitude response (3.3.1).

5.3.1.2 Types of Filters

We distinguish between linear filters and nonlinear filters (3.3.1.2). A *nonlinear filter* can be unpredictable and is often difficult to control. Consequently, one usually tries to avoid using nonlinear filters in technical applications. Most computer music applications and sound synthesis languages do not provide nonlinear filters. Using nonlinear filters can, however, lead to interesting results (6.2, 3.4.1.4, 3.4.4.4, 8.1.1.5).

Another way to distinguish between filters is to observe how they are implemented. If the output signal is computed only from values of the input signal, one speaks of a *non-recursive* filter. If both the input signal and the (past) output signal are used to compute the (current) output signal, one speaks of a *recursive* filter (3.3.1.6). The signal in all non-recursive and in certain recursive filters eventually dies out. These filters are called *Finite Impulse Response* (FIR) filters. The signal in most recursive filters (theoretically) never dies out. These filters are called *Infinite Impulse Response* (IIR) filters.

A third distinguishing feature of filters is their temporal behavior. A *time-invariant* filter does not change during use. A *time-variant* filter can have predetermined changes or it can change during use on the basis of characteristics of the signal being treated. In this case, we speak of an *adaptive filter* (5.3.1.6).

In a broad sense, a filter can be considered any technique producing an output signal from an input signal. Accordingly, a filter is realized using algorithms that generate new output values on the basis of a certain number of input values and earlier output values. A system producing sound by nonlinear distortion of sine waves (6.1.3) is a non-recursive, nonlinear filter. Here we see that even a simple nonlinear filter can change not only the amplitude and phase of an input signal, but also its frequency components. Interpolation between various values, a common technique used with both control and audio signals, corresponds to treating a signal with an interpolation filter (5.1.3.2). The basic element of the spring-mass model in Chapter 8.1.1 corresponds to a simple bandpass filter, and the basic element of the wave guide (8.2) corresponds to a comb filter (5.3.1.3). An amplitude envelope is a very simple time-variant filter that typically influences the spectrum of a sound, as amplitude modulation shows. An amplitude control function that reacts to the energy of its input is an adaptive filter (5.1.3.3, 5.1.4.4). The phase vocoder (5.2.2.1, 5.2.2.3) can be considered a complex FIR filter because the output signal is produced from the input signal (the analyzed sound).

5.3.1.3 Special Filters

A filter that only blocks a single frequency and lets pass almost all other frequencies unchanged is called a *notch filter* (3.3.3.5). A filter with a narrow passband is called a *resonator* (reson 3.3.3.5, applications 8.1.1). A filter having several regularly spaced passbands is called a *comb filter*. The center frequencies of the passbands of a comb filter form an harmonic series. If the first passband is centered at 0 Hz, all the partials of the series will be present, if the first stopband is at 0 Hz, only the odd partials of the series will be present (3.3.3.5). The *inverse comb* filter attenuates frequencies at regularly spaced intervals.

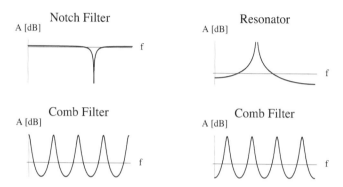

A filter which passes all frequencies with unchanged amplitudes and only changes the phases of the components is called an *all-pass filter* (3.3.3.5).

5.3.1.4 Combining Filters

It is one of the great advantages of linear filters that an arbitrarily complex linear filter can be decomposed into a sum of simple high- and low-pass, band-pass and band-reject filters, and any pattern of amplitude response can be approximated by such a sum. Filters can be combined in two ways. The illustration below on the left shows a parallel circuit where the outputs of two band-pass filters are added together. If the filters are of different types, differences in their phase responses can lead to problems when they are summed. In a serial circuit (illustration below on the right), the input signal passes two filters in succession. The amplitude response of the sum of two filters (here two band-pass filters) is the sum of the individual amplitude responses of each filter. When the passbands of the two filters do not overlap, the decibel (i.e. logarithmic) representation of the amplitude response of the sum is the envelope over the two individual amplitude responses. The amplitude response of two filters in series is the product of the individual amplitude responses. Its logarithmic representation corresponds to the sum of the two individual responses.

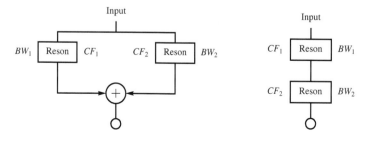

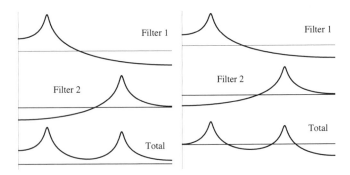

To construct a band-pass filter having steep falloff but nonetheless a broad resonance, one could combine two band-pass filters of similar center frequency (left figure below). The figure to the right shows the amplitude response of low-pass filters in series.

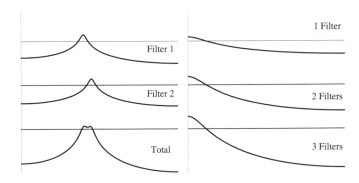

5.3.1.5 Effects in the Time Domain

In general, a filter changes a sound's spectrum. But in special situations, changes in the time domain can be predominant. A causal filter uses the current and the last input values as well as the last output values to compute the new output value. This means that certain events only take effect after some time. In particular, phase shifts result (3.3.1.4). In a recursive filter, the feedback results in a delayed reaction of the filter, but also in oscillation after the input has died out. The following illustration shows how the high-pass filter ($y(t) = x(t) - cy(t-1)$) reacts to a sine wave switched on and off when its value is zero (left figure below) and when its value is at a maximum (right figure below).

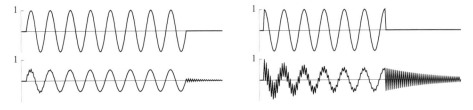

By choosing extreme parameter values, a filter can be used in this way to generate sound.

5.3.1.6 Variable Filters

When the parameters of a filter are suddenly changed, the amplitude and phase of the output change abruptly and kinks and discontinuities may appear in the output wave. The illustration below shows the result of sudden parameter changes in the non-recursive comb filter $y(t) = cx(t) + (1 - c)x(t - d)$. In the figure to the left, the time delay is changed from 1 to 6 while c remains constant. In the figure to the right, c changes from $c = 1$ to $c = .5$.

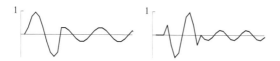

Therefore, the parameters of a variable filter must be computed using the highest possible control rate. The following example shows that new values cannot be generated by interpolation but must be recomputed. Let us design a low-pass filter whose cut-off frequency goes continuously from $\pi/3$ (left figure below) to $2\pi/3$ (center figure) (the unit is radial frequency per sample rate 3.3.1.4). We will use the 24th order FIR filter described in Chapter 3.3.2.4. If we replace the recomputation of the filter coefficients by interpolation between their initial and final values over a certain duration, after half this duration we do not get, as we expect, a low-pass filter with cut-off frequency $\pi/2$ but rather a filter that attenuates the high frequencies in two steps with cut-off frequencies at $\pi/3$ and $2\pi/3$ (right figure below).

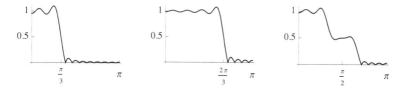

5.3.2 Applications

5.3.2.1 Sound Sources

For subtractive synthesis one generally uses sounds with rich spectra, because linear filters cannot generate any new frequency components. The most common sound source is white noise, where the energy is uniformly distributed over the entire spectral range (Sound 5-3-2b, 5.1.2.3). Filtered white noise is called *colored noise* (2.2.3.2), of which *pink noise* is a special case. Pink noise is generated from white noise by damping the frequencies above about 500 Hz by 3 dB per octave. This assures that pink noise has the same intensity in every octave. It sounds smoother and more consistent than white noise (Sound 5-3-2a). (→ Filter)

White noise made with uniformly distributed random numbers sounds even and regular. We can generate noise with temporal irregularities, like rustling, crackling, etc. by using other random distributions (10.1.6.1).

If the random numbers are not generated at the sampling rate but at a slower rate, the noise generated has fewer high frequencies than white noise; the lower the generating rate, the fewer high frequencies are produced. In sound example 5-3-2a2, the random values are generated at the constant rate $sr/10 = 4410$ Hz (see 5.1.2.3 for the spectrum). In the sound example 5-3-2a3 the generating rate begins at 20 kHz and falls to 10 Hz at the end of the sound.

```
k1        expon    p5,p3,p6
arnd      randh    p4,k1

i3 0 3 30000 4410 4410                              ;5-3-2a2
i3 0 15 30000 20000 10                              ;5-3-2a3
```

Like white noise, the energy of a pulse train is also distributed over the entire frequency spectrum. If the pulses are regular at the rate T, a harmonic sound results of frequency $1/T$. Using an appropriate generator (5.1.2.2), a pulse train with a limited spectrum (left figure below) or with exponentially decreasing partials (right figure below) can be produced.

A single pulse has the same spectrum as white noise (2.2.3.3). The reaction of a filter to a single pulse, known as its impulse response, plays a decisive role in the description and calculation of a linear filter (3.3.1.3). In sound synthesis using physical modeling, an impulse can simulate a short excitation, for example striking a string or a plate with a solid object.

5.3.2.2 Resonators and Formants

A resonator with a large bandwidth transforms white noise into colored noise. As the bandwidth diminishes, one can hear more and more clearly a noisy whistle at the center frequency. When the bandwidth is very small, the sound resembles a sine tone fluctuating in amplitude and frequency. If the damping at the center frequency is 0 dB, all the other frequencies are damped, and the signal's energy decreases with the bandwidth (left figure below). To obtain an output signal whose total energy corresponds to the energy of the input signal, the amplitude response of the filter at the center frequency must increase with diminishing bandwidth (right figure below). In the unit generator *reson* in Csound one can indicate whether the level of the center frequency is normalized at 0 dB or is increased so that the total RMS energy (4.1.1.7) becomes equal to one.

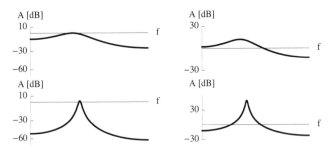

5.3 Subtractive Synthesis

In the Csound instruments below, the center frequency (*kcf*) and the bandwidth (*kbw*) can be varied during the duration of a note. The first instrument has the amplitude response shown at the left above, the second the amplitude response shown at the right above.

```
instr 1
kcf      expon    p5,p3,p6
kbw      expon    p7,p3,p8
arnd     rand     p4
ares     reson    arnd,kcf,kbw,1
         out      ares
endin
instr 2
.....
ares     reson    arnd,kcf,kbw,2
.....
```

Although the amplitude of the white noise (*p4*) in the first note is larger than that in the second note, the second note is always louder than the first. The loudness of the first note decreases rapidly with decreasing bandwidth. The loudness of the second note remains approximately constant (Sound example 5-3-2a1).

;instr	t	dur	amp	cf1	cf2	bw1	bw2	;5-3-2a1
i 1	0	8	30000	1000	1000	1000	2	
i 2	9	8	12000	1000	1000	1000	1	

Correcting the sound level after filtering with variable filters or with filters in series is often difficult. When the bandwidth is very small, an amplifying filter will overmodulate. A non-amplifying filter is likely to have such a weak signal that it cannot be recorded accurately and noise results, as at the end of the first tone in Sound example 5.3.2a1. One must be particularly careful when amplifying filters connected in series. One must also be careful when the input is not white noise, because the amplification is calculated to give an output level equal to the input level of white noise. It is often safest to measure the output level and correct it in a second step.

Resonators can be used to simulate the resonance of physical systems (8.1.1) like the sounding bodies of instruments or the resonances of the human voice. It is not the relative strength of the partials that is decisive for vowel recognition or for distinguishing individual voices, but rather the location and relative strength of formants (2.3.2.2). To synthesize vowel sounds, one defines a resonator with appropriate center frequency, amplitude and bandwidth for each formant to be simulated. The resonators are excited by a pulse train. The illustration below shows the spectral envelope when four resonators are used.

The first instrument in the following Csound program generates a pulse train *ga1* (Sound example 5-3-2b1) that is passed to the resonator in instrument 3.

```
instr 1
ga1      buzz     p4,p5,p6,1
;        out      ga1
endin
```

```
instr 3
a2      reson   p4*ga1,p5,p6,2
        out     a2
endin
```

By calling instrument 1 once and instrument 3 once for every resonance required, we can generate sounds with vowel-like timbres. The score below first approximates the sound |a| using four formants and then the sound |i| using six formants (5-3-2b2).

```
f1 0 32768 10 1
i1 0 3 30000 180 30            i1 4 3 20000 180 40
i3 0 3 1 660 150               i3 4 3 1 220 30
i3 0 3 .55 1100 170            i3 4 3 .5 2300 130
i3 0 3 .42 2700 230            i3 4 3 .5 3000 170
i3 0 3 .21 5500 330            i3 4 3 .1 4200 320
                               i3 4 3 .2 5500 430
                               i3 4 3 .2 6400 460
                               i3 4 3 .2 7200 490
```

5.3.2.3 Linear Prediction

The Phase Vocoder (5.2.2.5) works by dividing the frequency spectrum into channels and measuring the energy falling into each channel at regular time intervals. We are often not so interested in the exact frequencies of the components of a sound as we are in its spectral envelope, which generally changes over time. The technique of *linear prediction* generates a sequence of spectral envelopes of a given sound. These envelopes correspond to a continuously varying filter from which the original sound (or sounds derived from the original sound) can be reconstituted by subtractive synthesis. The next value of a signal $y(n)$ can be predicted with some accuracy from previous values $y(n-1), y(n-2), ..., y(n-N)$ by taking the sum

$$\tilde{y}(n) = b_1 y(n-1) + b_2 y(n-2) + ... + b_N y(n-N)$$

with appropriate coefficients $b_1, b_2, ..., b_N$. For example, it takes only three values of a sine wave to predict the next value precisely. If the original signal is not constant, the coefficients b_i change from moment to moment, and the sum differs from the real value by a certain error $e(n)$, called the *residue*. If we write the value $y(n)$ as the sum

$$y(n) = e(n) + \tilde{y}(n) = e(n) + b_1 y(n-1) + b_2 y(n-2) + ... + b_N y(n-N),$$

we see that equation represents a recursive filter with the coefficients b_i and the excitation $e(n)$. In speech analysis, where this technique was originally developed, the residue of a voiced signal is a pulse train whose frequency is the fundamental of the signal, while the residue of an unvoiced signal is essentially white noise. The following illustration shows schematically how the original sound is reconstituted from the residue using a filter.

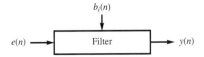

See [8] p. 294, [2] p. 242 and [18] p. 352.

6 Nonlinear Techniques

In linear processes, the superposition principle holds, which says that several simultaneously occurring events will not disturb each other. For instance, waves on the surface of water coming from different directions will pass through each other without disturbing. The linear superposition of sound waves is a prerequisite for our being able to hear individual sounds out of a complex sound mixture. Linear processes are much simpler in their behavior and easier to calculate than nonlinear processes. Most physical systems are only approximately or only in a limited domain linear. Waves in water that reach a certain size, but also simple systems like a pendulum or a spring at extreme displacements, show nonlinear behavior. Nonlinear superposition of sound waves, or rather of their sensory stimuli, gives rise to combination tones. The drawback of nonlinear systems, their complex behavior, is simultaneously their advantage when it comes to generating differentiated processes or sounds.

6.1 Modulation Techniques and Distortion

It takes many oscillators and control functions to produce differentiated sounds by additive synthesis. Because the tools of analog electronic music and of early computer music were severely limited, one employed nonlinear techniques to produce overtone-rich sounds with only a few oscillators. Despite their nonlinearity, the spectra of these sounds can sometimes be calculated, although both the prediction of the spectrum for given parameter values and above all the calculation of the parameter values required for a given spectrum are extremely tedious. Fortunately, after some practice the parameters can be determined quite accurately by intuition.

Further Reading: Elements of Computer Music *by F. Richard Moore [8], pp. 315-337,* Computer Music *by Charles Dodge and Thomas A. Jerse [2], pp. 115-156,* The Theory and Techniques of Electronic Music *by Miller S. Puckette [87], pp. 119-147 and* Computer Sound Synthesis for the Electronic Musician *by Eduardo Reck Miranda [15], pp. 57-81.*

6.1.1 Amplitude Modulation and Ring Modulation

Further Reading: Computer Music *by Charles Dodge and Thomas A. Jerse [2], pp. 90-94,* Elements of Computer Music *by F. Richard Moore [8], pp. 185-190 and* Computer Sound Synthesis for the Electronic Musician *by Eduardo Reck Miranda [15], pp. 58-61.*

6.1.1.1 Introductory Example

When the amplitude of a signal changes slowly, the signal's spectrum does not change noticeably. But if the amplitude changes rapidly, one hears a change of spectrum, as the following example shows.

Beating arises when two sine waves of similar frequency f_1 and f_2 are added together. The beats can be described as the product of a sine wave of frequency $(f_1 + f_2)/2$ and a cosine function of frequency $f_m = (f_1 - f_2)/2$ (2.2.1.4). Inversely, by multiplying two sine waves of frequency 440 Hz and 40 Hz (drawn with thin lines below), we can produce "beating" (drawn with a thicker line) that is so rapid that we cannot hear the amplitude changes but instead hear two tones of frequency 440 Hz + 40 Hz = 480 Hz and 440 Hz − 40 Hz = 400 Hz. (If, however, the illustration is meant to show two sine waves with frequency 77 and 7 Hz, then we hear a tone of 77 Hz with periodically increasing and decreasing amplitude. The beat frequency is 2·7 Hz = 14 Hz.)

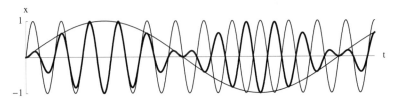

In the sound example that follows, a sine wave of 440 Hz was multiplied by a second sine wave of increasing frequency f_m. (→ Sound Example)

6.1.1.2 Basic Techniques

In amplitude modulation the spectrum of a waveform is modified by changing the waveform's amplitude periodically. The lower oscillator in the block diagram below generates an audio signal, the so-called *carrier wave*, with a constant frequency f_c. Its amplitude is composed of a constant amount A and a variable amount, the modulation signal generated by the upper oscillator. The frequency f_m of this oscillator is the *modulation frequency*. Its amplitude, which is usually expressed as a multiple of the constant part of the amplitude $m \cdot A$, determines the maximal increase and decrease of the carrier's amplitude. The factor m is called the *modulation index*.

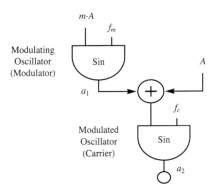

In Csound we can write an amplitude modulation instrument by sending the sum of the constant amplitude *Amp* and the signal *a1* of the first oscillator to the amplitude input of the second oscillator.

```
a1    oscil      m*Amp, fm, 1
a2    oscil      Amp+a1, fc, 1
      out        a2
```

The following illustration shows four amplitude modulated waveforms with increasing values for the modulation index ($0 \leq m \leq 3$). The waveforms were generated using the function $x(t) = (A + mA \cdot \sin(\omega_m t)) \cdot \sin(\omega_c t)$. (→ Animation)

6.1 Modulation Techniques and Distortion

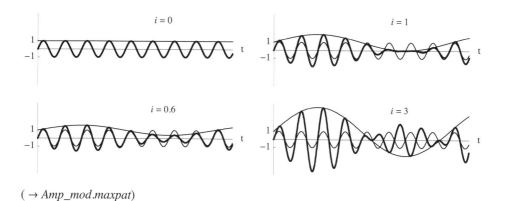

(→ Amp_mod.maxpat)

6.1.1.3 The Spectrum of Amplitude Modulated Waveforms

To calculate the spectrum of amplitude modulated sounds we begin with the sine functions for carrier and modulator. From

$$x(t) = (A + mA \cdot \sin(\omega_m t)) \cdot \sin(\omega_c t)$$

we get

$$x(t) = A \cdot \sin(\omega_c t) + mA \cdot \sin(\omega_m t) \cdot \sin(\omega_c t).$$

The first term shows that the carrier frequency appears with amplitude A independently of the factor m. From the second term, using the formula $\sin(\alpha)\sin(\beta) = 1/2(\cos(\alpha - \beta) - \cos(\alpha + \beta))$, we derive

$$mA \cdot \sin(\omega_m t) \cdot \sin(\omega_c t) = \frac{1}{2} mA \cdot \cos((\omega_c - \omega_m)t) - \frac{1}{2} mA \cdot \cos((\omega_c + \omega_m)t).$$

This means that two additional frequencies, so-called *side bands*, appear in the spectrum. The side bands' frequencies are $f_c + f_m$ and $f_c - f_m$, their amplitudes $\frac{1}{2} mA$.

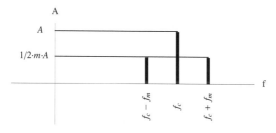

If we use an arbitrary waveform $f(t)$ in place of a simple sine wave for the modulator, we have to decompose the waveform into its spectrum, that is, a sum of sine and cosine functions

$$B_0 + \sum_{i=1}^{\infty} A_i \sin(i\omega t) + B_i \cos(i\omega t),$$

substitute this sum for the simple sine function in the equation above and get for each term the result above. The entire spectrum of the modulator is shifted upward by the amount of the carrier frequency f_c to form the right-hand side bands. The left-hand side bands are symmetrical to them about the carrier.

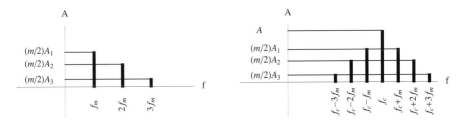

Left-hand side bands of negative frequencies can be represented by positive frequencies of the same magnitude, as long as they do not coincide with already present frequencies (upper figures in the illustration below; $f_c = 440$ Hz, $f_m = 600$ Hz). If the negative frequency coincides with an already present positive frequency, their amplitudes cannot simply be summed because the two components may be of different phase (lower figures below), as the following example shows. We take as modulating frequency the double of the carrier, so $f_c = 440$ Hz, $f_m = 880$ Hz. We set the modulation index to 2 in order to have all the spectral components equally strong, and we set the amplitude to .5. This gives us a right-hand side band at $f_c + f_m = 1320$ Hz.

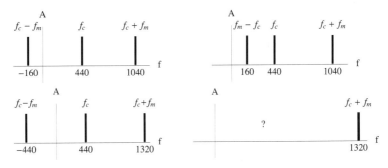

The left-hand side band $f_c - f_m = -440$ Hz reflects about 0 Hz to become positive, but the sum of it and the carrier depends on the phase of the two components. The illustration below shows the modulation of a sine wave by another sine wave (left) and the modulation of a sine wave by a cosine wave (right). At the left one can easily recognize the contributions of the carrier and the third partial, on the right the carrier is missing altogether.

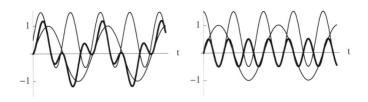

In both cases shown above, the magnitude of the amplitude at -440 Hz is the same as the magnitude of the amplitude at 440 Hz. If we use a cosine as modulator, the sign after reflection about 0 Hz remains the same as before reflection because of $\cos(-\alpha) = \cos(\alpha)$. If we use a sine wave, the sign changes on reflection because of $\sin(-\alpha) = -\sin(\alpha)$. (→ Computation and Sound Example)

6.1.1.4 Ring Modulation

If we set the constant amplitude of the carrier to zero, the block diagram of Chapter 6.1.1.2 can be simplified to the following diagram of so-called *ring modulation* (a below). (Ring modulation takes its name from the analog circuit originally used to produce it, which contains four diodes connected in a ring.) If both oscillators in figure a) use a sine or cosine wave, they generate a spectrum containing only the side bands with frequencies $f_1 + f_2$ and $f_1 - f_2$, each of amplitude $m/2$.

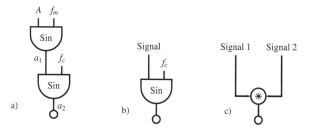

If we use an arbitrary signal as the amplitude input to the carrier (figure b above), the carrier's sine wave is multiplied by that signal. If we replace the carrier oscillator by a second arbitrary signal, we get the product of the two signals (figure c). The spectrum of the product contains all the sums and differences of the components of both signals. If one signal has p, the other q components, the product will contain at most $2pq$ components. To avoid aliasing, the sum of the highest frequencies should be less than the Nyquist frequency. The following illustration shows at the left and in the middle the spectra of two signals of two and three components respectively. At the right the spectrum of the product of the two signals is shown.

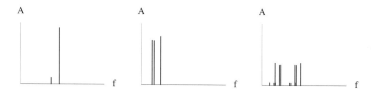

Using the terminology and results developed in Chapter 3, we can say: Multiplication of two signals $x(k)$ and $y(k)$ in the time domain is equivalent to the convolution of the two signals in the frequency domain.

$$\text{DTFT}\{x(k)y(k)\} = \sum_{k=-\infty}^{\infty} x(k)\,y(k)e^{-i\Omega k} = \frac{1}{2\pi}\,X(e^{i\Omega}) * Y(e^{i\Omega})$$

To avoid aliasing, components near the Nyquist frequency should be filtered out of both signals before multiplication. After multiplication, the damping of the high frequencies can be compensated by a high-pass filter. A DC block can be used to remove any direct current produced by the multiplication. (→ Filters)

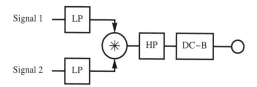

6.1.2 Frequency Modulation and Phase Modulation

Because frequency modulation is realized by modifying a signal's phase, the technique addressed in this chapter should actually be called phase modulation. We shall explain the relationship between the two in Chapter 6.1.2.2; subsequently we shall only speak of frequency modulation.

See [84] *Chowning J., Bristow D., FM Theory and Applications: By Musicians for Musicians [7] S.441-445, [2] S.115-138, [8] S.316-332.*

6.1.2.1 Introductory Example

Vibrato is produced by a periodic change of pitch. The maximal pitch difference in instrumental and vocal sounds is typically a fraction of a semitone, the frequency of the vibrato a few Hz. If we increase the pitch difference, the vibrato begins to sound unnatural, but we can still hear the variation of pitch. If we now increase the frequency of the pitch change to above 20 Hz, we no longer hear a vibrato but rather a different sound. If the vibrato's frequency is a simple fraction of the frequency of the original tone, we will hear an harmonic sound whose fundamental is determined by the frequency of the vibrato. The command below generates a waveform whose frequency varies 100 times per second between 400 + 400 = 800 Hz and 400 − 400 = 0 Hz (sr = 10000 Hz) (see 6.1.2.2 for an explanation of the formula).

```
snd = Table[Sin[400*2 π t + (400/100)*Sin[100*2 π t]], {t, 0, 1, .0001}];
```

In the illustration below one can see frequency maxima at $t = 0, 100, 200, \ldots$ and a period of somewhat more than 10 time units (800 Hz at a sampling rate of 10 kHz). There are frequency minima at $t = 50, 150$, etc. One can see from the horizontal tangents of the waveform at $t = 50$ and $t = 150$ that the frequency is momentarily 0 Hz at these points. These variations are too rapid to be perceived as such. Rather one hears an oscillation of period 100, or 100 Hz at a sampling rate of 10 kHz.

```
ListPlot[Take[snd, {1, 200}], Joined → True, AspectRatio → 0.2]
```

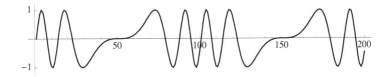

6.1.2.2 The Basic Method

The diagram below shows an instrument for generating frequency modulated signals. The lower oscillator generates the *carrier*, here with a constant amplitude *Amp*. The frequency of the resulting signal consists of a constant part f_c, the *carrier frequency*, and a variable part, the modulation signal generated by the upper oscillator, called the *modulation frequency* f_m. Its amplitude d determines the deviation of the carrier frequency from f_c. The ratio of this deviation and the modulation frequency is called the *modulation index* $I = d/f_m$.

(→ *Freq_mod.maxpat*)

6.1 Modulation Techniques and Distortion

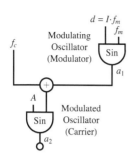
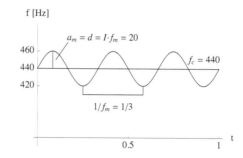

In Csound we realize the instrument by summing the constant carrier frequency *fc* and the output *a1* of the modulating oscillator and using this sum as the frequency input of the carrier oscillator. One can also use the unit generator *foscil* to produce frequency modulated sounds directly.

```
a1    oscil     d, fm, 1
a2    oscil     amp, fc+a1, 1
      out       a2
```

If we try to generate a frequency modulated signal by substituting for the frequency f in the equation for a sine wave $y(t) = A \cdot \sin(f \cdot 2\pi t)$ the sum of the carrier frequency and the sine wave $d \cdot \sin(f_m \cdot 2\pi t)$, that is $f = f_c + d \cdot \sin(f_m \cdot 2\pi t)$, we get the waveform

$$y(t) = A \cdot \sin((f_c + d \cdot \sin(f_m \cdot 2\pi t)) \cdot 2\pi t)$$

shown below.

Sin[(400 + 400 * Sin[100 * 2 π * t]) * 2 π * t]

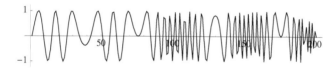

This is clearly not the frequency modulated signal we wanted; not only is it not periodic but the carrier frequency increases continuously. The problem is that in the frequency of the outer term $f_c + d \cdot \sin(f_m \cdot 2\pi t)$ the variable time t appears. The frequency is defined as the number of oscillations per time unit, but this definition is appropriate only for constant frequencies. In general, the so-called *instantaneous frequency* can be defined as the time derivative of the phase (2.2.3.5). The waveform above has the instantaneous angular frequency

D[(ω_c + d * Sin[ω_m * t]) * t, t]

d Sin[t ω_m] + ω_c + d t Cos[t ω_m] ω_m

To obtain a function with the *instantaneous angular frequency* $\omega_c + d \cdot \sin(\omega_m t)$, we integrate and get the instantaneous phase.

Integrate[ω_c + d Sin[t ω_m], t]

$$t \omega_c - \frac{d \cos[t \omega_m]}{\omega_m}$$

The required function is:

$$\sin(\omega_c t - \frac{d}{\omega_m}\cos(\omega_m t)) = \sin(f_c \cdot 2\pi t - \frac{d}{\omega_m}\cos(f_m \cdot 2\pi t))$$

or for arbitrary starting phases of carrier and modulator:

$$\sin\left(\omega_c t + \frac{d}{\omega_m}\sin(\omega_m t + \phi_m) + \phi_c\right)$$

This is the function used in the example in Chapter 6.1.2.1. The factor $\frac{d}{\omega_m}$ is equivalent to the modulation index I. (→ Phase_mod.maxpat)

6.1.2.3 The Spectrum of a Frequency Modulated Waveform

Frequency modulated sounds show a line spectrum, even when carrier and modulating frequencies are not harmonic. As the modulation index increases, more and more of the carrier's energy is distributed to the side bands. The frequencies and amplitudes of these side bands can be calculated using the formula:

$$\sin\left(\omega_c t + \frac{d}{\omega_m}\sin(\omega_m t + \phi_m) + \phi_c\right) = \sum_{n=-\infty}^{\infty} J_n(I) \sin((\omega_c + n\omega_m)t + n\phi_m + \phi_c)$$

The frequencies of the side bands are $f_c \pm n f_m$.

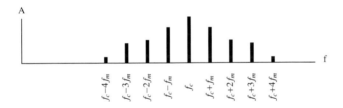

The amplitudes of the side bands can be calculated using *Bessel functions* $J_n(I)$ of the first kind. The values for specific n and I can either be computed by a program or taken from tables or graphic representations. The number of side bands on either side of f_c with significant amplitude is approximately $I + 1$. The following table shows how to calculate the amplitude of the nth side bands.

n	Frequency	Amplitude	Frequency	Amplitude
0	f_c	$J_0(I)$		
1	$f_c - f_m$	$-J_1(I)$	$f_c + f_m$	$J_1(I)$
2	$f_c - 2f_m$	$J_2(I)$	$f_c + 2f_m$	$J_2(I)$
3	$f_c - 3f_m$	$-J_3(I)$	$f_c + 3f_m$	$J_3(I)$
4	$f_c - 4f_m$	$J_4(I)$	$f_c + 4f_m$	$J_4(I)$
5	$f_c - 5f_m$	$-J_5(I)$	$f_c + 5f_m$	$J_5(I)$
:	:	:	:	:

6.1 Modulation Techniques and Distortion

For the example above (6.1.2.1) with $f_c = 400$ Hz, $f_m = 100$ Hz and $I = d/f_m = 400/100 = 4$, Mathematica gives the following values for $J_n(I)$:

```
Table[N[BesselJ[n, 4]], {n, 0, 8}]
```

{-0.39715,-0.0660433,0.364128,0.430171,0.281129,0.132087,0.0490876, 0.0151761,0.00402867}

We put the values into the table above and generate as a check one period of a waveform by adding the partials listed.

n	Frequency	Amplitude	Frequency	Amplitude
0	400	−.397		
1	300	.066	500	−.066
2	200	.364	600	.364
3	100	−.43	700	.43
4	0	.28	800	.28
5	−100	−.13	900	.13

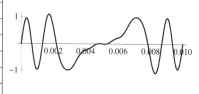

To illustrate how to deal with negative frequencies and amplitudes when computing a spectrum, let us double the modulation index $I = 8$ in the example above. We get for $J_n(8)$ the following values:

```
Table[N[BesselJ[n, 8]], {n, 0, 12}]
```

{0.171651, 0.234636, -0.112992, -0.291132, -0.105357, 0.185775, 0.337576, 0.320589, 0.223455, 0.126321, 0.060767, 0.0255967, 0.00962382}

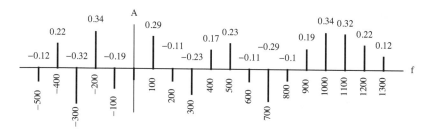

Because $\sin(-\alpha) = -\sin(\alpha)$, we can subtract the amplitudes of the negative frequencies from the amplitudes of the corresponding positive frequencies (if $\phi_{0,m} = \phi_{0,c} = 0$). (6.1.2.8).

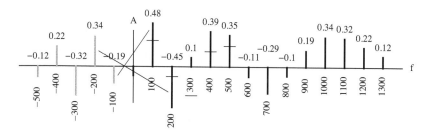

Now if we consider only the magnitudes of the amplitudes, we have the following spectrum:

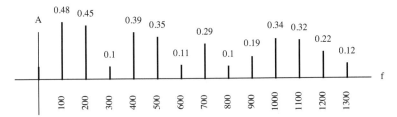

Bessel functions are defined as the solutions of certain differential equations. Bessel functions of the first kind, usually written as $J_n(x)$, can be developed into an infinite sum as follows:

$$J_n(x) = \sum_{k=0}^{\infty} \frac{(-1)^k (x/2)^{n+2k}}{k! \cdot \Gamma(n+k+1)}$$

Here $\Gamma(n + 1)$ is the *Gamma function*. If n is a natural number, $\Gamma(n + 1) = n!$. From this we derive for $n = 0$ and $n = 1$:

$$J_0(x) = 1 - \frac{x^2}{2^2} + \frac{x^4}{2^2 \, 4^2} - \frac{x^6}{2^2 \, 4^2 \, 6^2} + \ldots$$

$$J_1(x) = \frac{x}{2} - \frac{x^3}{2^2 \, 4} + \frac{x^5}{2^2 \, 4^2 \, 6} - \ldots$$

The next figures show the Bessel functions of the first kind $J_n(I)$ for $n = 0, 5, 12, 22$ and $0 \le I \le 40$. The animation shows the development from $n = 0$ to $n = 30$. The individual frames of the animation can be used to read out concrete values of the Bessel function. ($\rightarrow$ Animation)

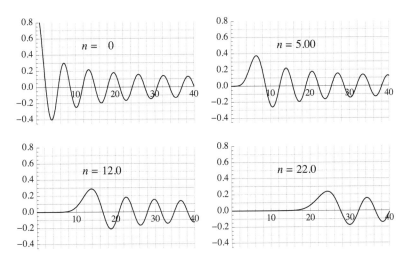

In certain contexts, it is useful to plot the function against the order n, keeping the index constant. The figures below show Bessel functions of the first kind $J_n(I)$ for $I = 1, 4, 8, 14, 19$ and 29, over a range of $n = 0$ to $n = 40$. The animation shows the development from $I = 0$ to $I = 30$. ($\rightarrow$ Animation)

6.1 Modulation Techniques and Distortion

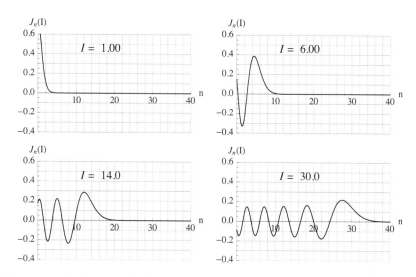

$J_n(I)$ with constant index as a function of $n = (f - f_c)/f_m$, here $J_{f/100-4}(8)$, corresponds to its own spectral envelope.

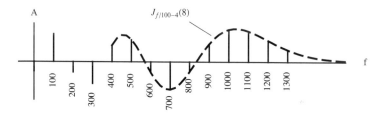

In the following illustration, the function $J_n(I)$ is shown three-dimensionally as a function of the variables n and I. The illustrations above are longitudinal sections and cross sections respectively of the three-dimensional surface. (B.6)

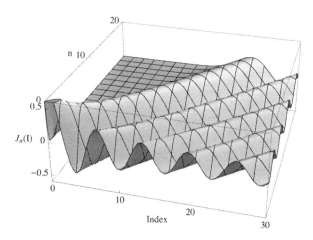

See [1] pp. 224-242, [9] pp. 440-444.

6.1.2.4 The Proportion $f_c : f_m$

In the previous examples the carrier frequency was a multiple of the modulation frequency. In this case, all frequencies nf_m appear, unless either the modulation index or the function $J_n(I)$ becomes exactly zero. If the proportion f_c/f_m is not an integer but a fraction n/m, the spectrum is harmonic, but not all partials appear. For example, if we generate a frequency modulated sound with $f_c = 300$ Hz and $f_m = 200$ Hz, the frequencies ... –300, –100, 100, 300, 500 Hz ... appear, and we get a sound with fundamental 100 Hz and only odd partials. If we choose $f_c = 500$ Hz and $f_m = 300$ Hz, the frequencies ... –400, –100, 200, 500, 800, 1100 Hz ... appear. The fundamental is again 100 Hz, but here the partials 3, 6, 9 ... are missing. In general, we have the fundamental $f_0 = f_c/n = f_m/m$ and the partials $n \pm km$. Because of the reflection of negative frequencies a second series of frequencies $i + km$ can appear (in bold type in the following table).

$n : m$	Partials Present	Missing Partials
5 : 1	1, 2, 3, 4, 5, 6, ..	
5 : 2	1, 3, 5, 7, 9, ..	2, 4, 6, ..
5 : 3	1, 2, **4**, 5, 7, 8, ..	3, 6, 9, ..
5 : 4	1, **3**, 5, **7**, 9, ..	2, 4, 6, ..
1 : 1	1, 2, 3, 4, ..	
5 : 6	**1**, 5, **7**, **11**, 13, ..	2, 3, 4, 6, 8, 9, 10, ..
5 : 7	**2**, 5, **9**, 12, **16**, ..	1, 3, 4, 6, 7, 8, ..
5 : 8	**3**, 5, **11**, 13, 19, ..	1, 2, 4, 6, 7, 8, ..
5 : 9	**4**, 5, **13**, 14, **22**, 23, ..	1, 2, 3, 6, 7, 8, ..

If the proportion f_c/f_m is irrational, the spectrum is not harmonic. In the following example the carrier frequency is 400 Hz, the modulation frequency $100\sqrt{3}$ Hz = 173.20508... Hz and the index 4. When the proportion f_c/f_m is irrational, the negative frequencies cannot reflect onto the positive frequencies.

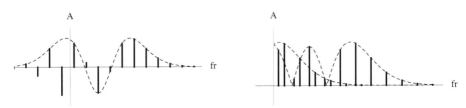

6.1.2.5 Variable Spectra

The three parameters of frequency modulation, f_c, f_m and I can be varied during a sound. Only the modulation index is relevant for the sound's timbre. As the examples above show, we cannot vary the amplitude of the partials individually because a change of the index value always affects all the partials. For this reason, frequency modulation is particularly well suited to synthesis of sounds whose spectra as a whole change. The figures below are snapshots from an animation showing the spectrum of a frequency modulated sound with $f_c = 400$ Hz and $f_m = 100$. The fundamental is 100 Hz and the index I goes from 0 to 10 (Sound Example 6-1-2a). ($\rightarrow$ Animation)

6.1 Modulation Techniques and Distortion 279

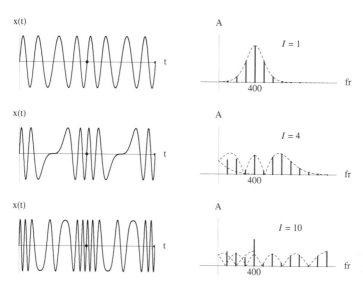

Because positive and negative frequencies are superposed in the example above, the spectral lines do not always coincide with the envelopes shown. When the proportion f_c/f_m is irrational, the spectral lines of positive and negative frequencies are all different (6.1.2.4). The figures below are snapshots from an animation showing the spectrum of a carrier sine wave of frequency $f_c = 400$ Hz modulated by a sine wave of frequency $f_m = \sqrt{3} \cdot 100$ Hz. The modulation index I goes from 0 to 8 (Sound Example 6-1-2b). ($\rightarrow$ Animation)

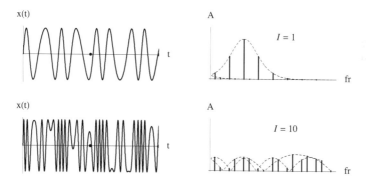

6.1.2.6 Synthesis Models and Examples

In this section we will discuss various synthesis examples and model them in Csound instruments. An instrument which can modify both loudness and timbre during the course of a sound has two envelopes (figure at the left below), one for the amplitude of the carrier and one for the modulation index. Because a spectrum's richness often depends on the sound's amplitude, it can be useful to use the same function for both envelopes. In the first publication discussing frequency modulation as an audio technique (it had long been used in radio broadcasting), its inventor, John Chowning, used an exponential function for both envelopes in an instrument to synthesize a bell-like tone ([2], pp. 124-126). The Sound Example 6-1-2c is made with the values given by Chowning: $Imax = 10$, $f_c = 200$ Hz, $f_m = 280$ Hz and a duration of 15 seconds.

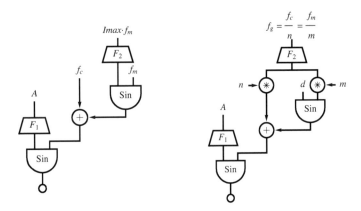

We can make an instrument that changes pitch during a sound by using the same envelope for both carrier and modulation frequency (figure to the right above). In addition, we make a control function F_2 for the fundamental and multiply it by n for the carrier and m for the modulation frequencies respectively. (→ 6-1-2d.csd)

Because the amplitudes of the individual partials cannot be controlled independently of one another, some imagination is often required to synthesize certain acoustic situations. For example, one can produce formants by synthesizing several sounds with the same fundamental but having different carrier frequencies. If one uses low index values, each carrier will be surrounded by partials spaced at intervals of the modulation frequency. In the following example, we generate a sound with fundamental 150 Hz and formants suggesting the vowel |a:| ([2], p. 230) by calling five times a simple frequency modulation instrument with a modulator of 150 Hz and a carrier frequency corresponding to the center frequency of the formant required. The formants can only be approximated, because the carriers have to be multiples of the fundamental. The higher the fundamental, the greater the discrepancy in the formants' center frequencies. The center frequencies given in [2] are: 609, 1000, 2450, 2700 and 3240 Hz. The index determines the bandwidth of the formants. (→ 6-1-2e.csd).
(→ Csound score)

6.1.2.7 Extensions of the Basic Method

In the examples until now both carrier and modulator signals have been sine waves. But of course any function can be used for either carrier or modulator or both. The following illustrations show two periods each of a frequency modulated sound where $f_c{:}f_m = 2$ for an index $I = 1$. In the upper figure the carrier is a square wave and the modulator a sine wave, in the lower figure the carrier is a sine wave and the modulator the square wave. Once again, if the proportion $f_c{:}f_m$ is rational, periodic waveforms are generated.

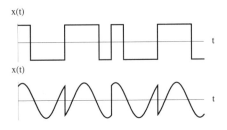

6.1 Modulation Techniques and Distortion

If the modulator is the sum of two sine waves with the frequencies f_{m1} and f_{m2} (left figure below), the spectrum of the sound generated will contain the frequencies $f_c \pm k_1 f_{m1} \pm k_2 f_{m2}$. The amplitudes of the components are $J_{k_1}(I_1) J_{k_2}(I_2)$. The modulator can itself be a frequency-modulated signal (right figure below).

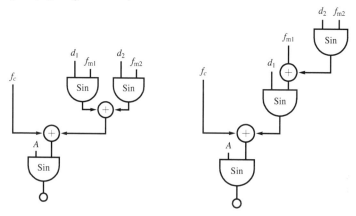

The modulating oscillator can also be replaced by feedback from the carrier. Instead of the modulation index, we use a feedback factor b. The advantage of this technique is that the amplitudes of the sidebands increase more uniformly than in classical frequency modulation, where the amplitudes increase and decrease according to the Bessel functions. Feedback usually modulates the phase of the carrier.

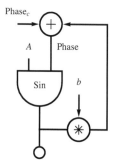

The system equation of digital modulation with feedback is: $y_k = A \cdot \sin(\omega_c t + b y_{k-1})$. We perform the modulation using the following command:

```
Do[snd1[[t]] = Sin[fc*2*π*t + b*snd1[[t - 1]]], {t, 2, 100}]
```

The figures below show the signals for the feedback factors $b = 0.4$ and $b = 1$. The feedback factor acts like a modulation index and determines the degree of modulation.
($\rightarrow$ Animation) ($\rightarrow$ Example in Csound)

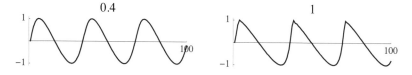

The spectrum can be computed using this formula [3-4].

$$\sum_{n=1}^{\infty} \frac{2}{nb} J_n(nb) \cdot \sin(n\omega_c t)$$

It is recommended to remove the DC component resulting from the modulation.
(→ *Freq_mod_feedback.maxpat*)

The situation gets more complicated if the output signal is added to the frequency of an oscillator instead of to its phase. Since the feedback depends on the amplitude of the signal generated, we divide the feedback factor by the amplitude, after adding $\delta = .00001$ to the amplitude to avoid its becoming zero. (→ Example in Csound)

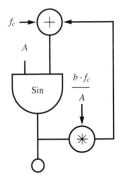

The waveform below was generated using the values $A = 1$, $f_c = .04$ and $b = .17$

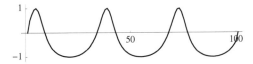

The unusual shape of the waveform is a consequence of the feedback. When the displacement is positive, the oscillator's frequency increases, accelerating (and thus shortening) that part of the waveform. When the displacement is negative, the frequency drops and the waveform slows down. As the feedback factor increases, the part of the signal with negative displacement gets longer and the frequency falls. In the example above, the period is about 40 time units long, which corresponds to a frequency of $f_c = .025$ instead of $f_c = .04$. The following snapshots from an animation with increasing feedback factor show the signal slowing down until it comes to a standstill when the feedback factor and the carrier frequency take on the same value. (→ Animation) (→ *Freq_mod_feedback.maxpat*)

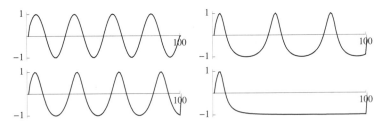

6.1 Modulation Techniques and Distortion

Before we try to make the frequency independent of the feedback factor, let us consider how to represent as a function the modulation with feedback. To get as output a function of time $f(t)$ having the required instantaneous angular frequency $\omega_c + b \cdot f(t)$, we integrate the instantaneous frequency and get the instantaneous phase (2.2.3.5):

$$\int (\omega_c + b \cdot f(t)) dt = \omega_c t + b \cdot \int f(t) dt$$

In discrete functions, the integral of $f(t)$ is replaced by the sum of the discrete values $f(k)$. The figures above were made using the following commands:

```
sum = 0; fc = .04; cfb = .17;
Do[sum = sum + snd[[t - 1]];
   snd[[t]] = Sin[fc*2*π*t + b*sum], {t, 2, 100}];
```

After declaring constants, we loop for the duration of sound and take the sum of the previous displacements by adding at each step the last computed signal value to the variable *sum* and then passing the instantaneous frequency to the sine oscillator.

To make the frequency in our example independent of the feedback factor, we begin by determining the frequency f_r of an uncorrected waveform and then correcting it by f_c/f_r. It is not difficult to determine the frequency of the waveform. It cannot ever be negative, since the oscillator ceases to oscillate if the frequency reaches $f = 0$. Thus, a period is over at every second zero crossing. The following animation, showing somewhat more than three periods of a modulated sound with various feedback factors but constant frequency, was made using the commands below. First we define the carrier frequency *fc* and its period *tp* (line 1). We call the uncorrected period we want to measure *tup* and set *tup* = 2, since we need the previous value *tup* − 1 and the indices start at 1. Then we initialize the sum *sum* and a counter *k* for the zero crossings (line 2). The uncorrected waveform is generated in the while-loop until the second zero-crossing is reached (lines 3-6). We determine the zero-crossings by multiplying two consecutive signal values. When the product becomes negative, one of the terms is before, the other after the zero-crossing (line 6). The final command computes the sound, correcting the frequency by *tup/tp* (lines 8-10).

```
fc = .03; tp = 1 / fc;                                              (1)
tup = 2; sum = 0; k = 0;
While[tup < 100 && k < 2,
    {sum = sum + snd[[tup - 1]];
     snd[[tup]] = Sin[fc*2*π*tup + .01*i*sum];                      (5)
     If[snd[[tup]] * snd[[tup - 1]] < 0, k = k + 1; tup = tup + 1;}];
sum = 0;
Do[ sum = sum + snd[[t - 1]];
    snd[[t]] = Sin[(fc*2*π*t + .01*i*sum) *tup/tp],
{t, 2, 100}];
```

The waveform now has a constant frequency. In the snapshots below taken from an animation, the numbers above the waveform indicate the lengths of the uncorrected periods.
(→ Animation)

6.1.2.8 The Influence of the Phase

The relative phase of the two waveforms used for frequency modulation is usually not taken into account. As the following illustration shows, however, different initial phase values give different waveforms. An initial phase difference of $\varphi = \pi/2$ gives the same waveform as $\varphi = 0$. The Sound Example 6-1-2f shows that the spectra of the waveforms also differ from each other.

`fm = 100; fc = 400; ind = 8; Sin[fc*2*π*t + ind*Sin[fm*2*π*t + φ]]`

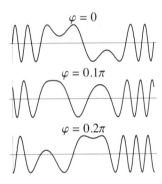
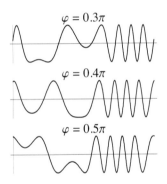

In order to understand the influence of the initial phase, let us go through the discussion of the basic method again (6.1.2.2), adding to the phase of the modulator a phase constant φ_m. We integrate again the instantaneous frequency

$$\int (\omega_c + d \cdot \sin(\omega_m t + \varphi_m)) dt$$

and get the phase

$$\theta(t) = \omega_c t - \frac{d}{\omega_m} \cos(\omega_m t + \varphi_m) + C$$

To determine the constant of integration C, we set $t = 0$ and note that the phase $\theta(t)$ at time $t = 0$ is equivalent to the initial phase of the carrier φ_c.

$$\theta(0) = \varphi_c = \omega_c \cdot 0 - I \cdot \cos(\omega_m \cdot 0 + \varphi_m) + C$$

$$C = \varphi_c + I \cdot \cos(\varphi_m)$$

Therefore, the general function $a(t)$ for a frequency modulated signal with carrier frequency ω_c, modulating frequency ω_m, index I and initial phases φ_c and φ_m is:

$$\boxed{a(t) = \sin(\omega_c t - I \cdot \cos(\omega_m t) + C) \quad \text{with } C = \varphi_c + I \cdot \cos(\varphi_m)}$$

6.1 Modulation Techniques and Distortion

We obtain the example discussed in Chapter 6.1.2.2 when we set the initial phases to $\varphi_c = 0$ and $\varphi_m = \pi/2$.

The waveform $a(t)$ can be decomposed into a sum of sine wave (see [1-11] pp. 44-45):

$$a(t) = \sum_{i=-\infty}^{\infty} J_{|i|}(I) \cdot \sin((\omega_c - i\omega_m)t + \varphi_i) \quad \text{with} \quad \varphi_i = C + i\varphi_m - |i| \cdot \frac{\pi}{2}$$

Once again, we have a spectrum with the frequencies $\omega_c \pm i\omega_m$ and the amplitudes $J_i(I)$. But because in a harmonic spectrum, that is where the proportion ω_c/ω_m is rational, positive and negative frequencies of the same magnitude can be of different phase, their amplitudes cannot simply be added together as above. Instead, the total amplitude, taking into account the phase difference ϕ, is calculated with the formula

$$A = \sqrt{\left(A_1^2 + A_2^2 + 2 A_1 A_2 \cos(\phi)\right)}.$$

The effect of the superposition of positive and negative sidebands is only significant when sufficiently strong side bands with negative frequency are generated, that is, when both the modulation index and the proportion f_m/f_c are large and when the negative side bands reflect onto corresponding positive side bands.

To be able to compare the differences in timbre, we increase the phase difference between carrier and modulator over time by introducing a time-dependent function $\varphi(t) = 2t$ for the variable φ.

```
fm = 100; fc = 400; φ[t_] = 2 * t; ind = 8;
Sin[fc * 2 * π * t + ind * Sin[fm * 2 * π * t + φ[t] ]]
```

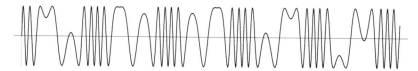

We obtain the same result by raising the modulation frequency slightly, because we can write the phase of the carrier $f_m 2\pi t + ct$ as $(f_m 2\pi + c)t$. This means that the waveform generated is the same as a frequency modulated waveform with a slightly inharmonic relationship of carrier frequency to modulation frequency f_c/f_m. From this point of view, the slow periodic variations of timbre can be interpreted as beats arising between neighboring sidebands.

(→ 6-1-2f.csd)

[1-11] pp. 38-45

6.1.3 Nonlinear Distortion – Waveshaping

As we saw in Chapter 3, linear systems only change the amplitudes and phases of a sound's components. Even simple nonlinear operations give rise to new frequencies or change the frequencies of individual components of a sound. The amplitude and frequency modulation discussed in previous chapters are typical examples of nonlinear systems. In both cases two input signals are used to compute the output signal. In this chapter we will consider nonlinear systems in which only the instantaneous value of the input signal is used to compute the instantaneous value of the output. This technique is called nonlinear distortion or waveshaping, because the original waveform is distorted. Complex spectra are generated from simple spectra. Despite the straightforward procedures, the computations involved can be quite elaborate.

See [2] pp. 139-158, [15] pp. 78-80.

6.1.3.1 Introductory Examples

Multiplying a signal by itself is a special case of ring modulation. The theory leads us to expect that ring modulating a sine wave of frequency f with itself will produce two new components, one of frequency $f + f = 2f$ and one of frequency $f - f = 0$. The following illustration shows that the waveform of frequency f disappears and that a waveform of frequency $2f$ and a DC component appear, each of amplitude 1/2. The trigonometric identity

$$\sin^2(\alpha) = \frac{1}{2}(1 - \cos(2\alpha))$$

confirms that no other partials are generated.

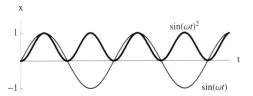

The response of a linear system to a composite signal can be computed as the sum of the responses to the individual components. The figure below, illustrating the squaring of the signal $x(t) = \sin(\omega t) + 1$, shows that this is impossible even for the simplest nonlinear systems. The signal differs from the sine wave shown above only by the addend 1. But the signal $x(t)^2$ is not the sum of a sine wave of twice the original frequency and a constant. The resulting signal has the same period as the original, but since the waveform is not a sine wave, it must be composed of several partials.

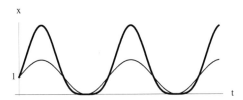

One can experiment with the examples in this Chapter using the Max-patch *Distortion.max-pat*.

The square wave of amplitude .5 (left above in the following figure) can be described by the function sign(sin(ωt)) and consists of the sum of infinitely many odd partials (3.1.1.3). Squaring this function gives the constant $x = .25$ (left below). All other partials disappear. But if we add .5 to the square wave, we shift it up by that amount (right above). Squaring this waveform gives the original signal again (right below). In this case the spectrum is not changed.

6.1 Modulation Techniques and Distortion

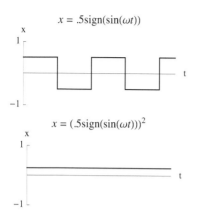

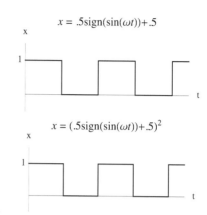

The phase of individual components is also relevant in nonlinear systems. In the examples below the input signal consists of a fundamental and a third partial. In the first example, both components have the initial phase 0, in the second the third partial has the initial phase π. After squaring, their spectra have the following components:

```
TrigReduce[TrigExpand[(.5*Sin[t] + .5*Sin[3*t + 0])^2]]
0.25+ 0.125 Cos[2 t] - 0.25 Cos[4 t] - 0.125 Cos[6 t]

TrigReduce[TrigExpand[(.5*Sin[t] + .5*Sin[3*t + π])^2]]
0.25- 0.375 Cos[2 t] + 0.25 Cos[4 t] - 0.125 Cos[6 t]
```

The amplitudes of the second partial differ from one another, and the signs of the second and fourth partials are also different, which means that these components are phase shifted by π with respect to each other.

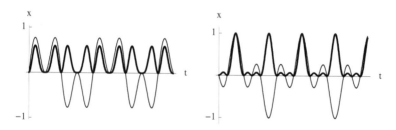

The function Sign($x(t)$) assigns the value 1 to $x(t)$ when $x(t)$ is greater than 0, -1 when $x(t)$ is less than 0 and 0 when $x(t)$ is exactly 0. This nonlinear, discontinuous function transforms a sine wave into a square wave.

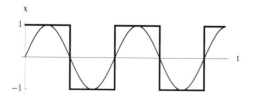

6.1.3.2 Waveshaping

Because nonlinear distortion changes the shape of the original signal, we speak of waveshaping. The following figure shows the basic block diagram for this technique. An oscillator generates a simple signal which passes through an element called a (nonlinear) *distortion module* or *waveshaper*. This element changes the original waveform and thus the original spectrum. The function describing the change of the waveform is called the *characteristic* or the *transfer function* of the distortion module.

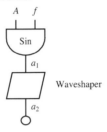

The distortion may be realized by a mathematical formula, as was the case in the introductory examples. Usually, however, a table is used in which the function is stored for all possible input values. An advantage of using look-up tables is that the mathematical formula need not be known. In fact, a table can contain arbitrary values. The figure below shows how a transfer function changes the instantaneous value $a_{in}(t)$ of the input signal into the instantaneous value $a_{out}(t)$ of the output.

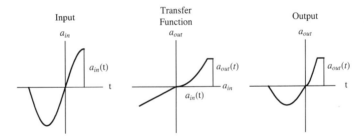

If we rotate the figure of the input signal by 90 degrees and superimpose it on the transfer function, the a_{in}-axes of input and transfer function coincide. The following figures show how a transfer function progressively changes an input signal into a distorted output signal.
(→ Animation)

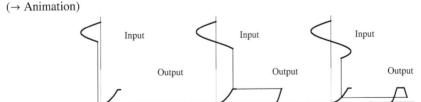

In Csound we realize waveshaping either by computing the output signal directly as in Instrument 1 below, or by generating a look-up table with the required values and then using the input signal as an index into the table (with the unit generator *table*) as in Instrument 2. In

6.1 Modulation Techniques and Distortion

both instruments we fade the sound in and out with the amplitude envelope *kamp*. In the first instrument, the sine wave's amplitude is 1. The unit generator *table* requires an index (*a1*), the function table number (2), the index data mode (1, i.e., index values between 0 and 1), and offset (.5). Since we use index values between −.5 and .5, we set the amplitude of the sine wave to .5. (→ *6-1-3a.csd*)

```
;6-1-3a.orc
instr 1
kamp     linseg   0,.5,1,p3-1,1,.5,0
a1       oscil    kamp,p5,1
a2=               a1*a1*p4
         out      a2
endin
instr 2
kamp     linseg   0,.5,.5,p3-1,.5,.5,0
a1       oscil    kamp,p5,1
a2       table    a1,2,1,.5
         out      a2*p4
endin


f1 0 32768 10 1
f2 0 32768 3 -1 1 0 0 1
;        Amp      Fr
i1 0 3   20000    440
i2 3 3   20000    440
```

(→ *Waveshaping.maxpat*)

6.1.3.3 The Modulation Index

The spectrum of natural sounds typically becomes richer as the sound gets louder. To simulate this, transfer functions are often used for waveshaping that are nearly linear close to zero and thus produce no new partials, hardly changing the input signal at low amplitudes. The greater the distance from zero however, the more complex the function becomes, which means that the greater the input amplitude, the more new partials will be generated. Since in waveshaping spectral richness can be controlled by the input amplitude, the input amplitude plays here the same role as the modulation index in amplitude and ring modulation and is often referred to as the index. In most applications, the index is limited to values between 0 and 1. (→ Animation)

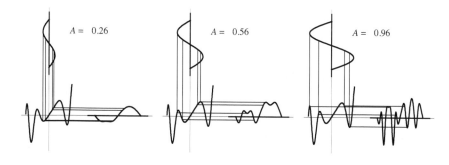

6.1.3.4 Polynomials as Transfer Functions

It is possible to compute the spectrum for certain transfer functions using a sine wave as input. In the first introductory example above, we showed that the transfer function $y = x^2$ transforms the input signal $x = \sin(\omega t)$ into an output signal with the components .5 and .5·cos(2ωt). We can carry out the same calculations for any powers and sums of sine functions and so compute the spectrum generated by any polynomial transfer function. With the command TrigReduce[] we compute the coefficients of the partials for the nth power of the sine function.

TrigReduce[Sin[t]³]

$$\frac{1}{4}(3\,\text{Sin}[t] - \text{Sin}[3\,t])$$

In the following table we indicate the coefficients in the same way as does Mathematica. One obtains the nth coefficient of the ith power by dividing the value a_n by the divisor of the ith row of the table.

	Div	a_0	a_1	a_2	a_3	a_4	a_5	a_6	a_7	a_8
x^0	.5	1								
x^1	1		1							
x^2	2	2		1						
x^3	4		3		1					
x^4	8	6		4		1				
x^5	16		10		5		1			
x^6	32	20		15		6		1		
x^7	64		35		21		7		1	
x^8	128	70		56		28		8		1

The values a_n form the right half of Pascal's triangle, a geometric arrangement of binomial coefficients where each row begins and ends with 1 and each of the other numbers corresponds to the sum of the two numbers above.

					1					
				1		1				
			1		2		1			
		1		3		3		1		
	1		4		6		4		1	
1		5		10		10		5		1
	6		15		20		15		6	

When the transfer function is an odd polynomial, only odd partials are generated, when it is an even polynomial, only even partials are generated. The following figures illustrate what happens, using the composite odd transfer function $y = 1.6x - x^7$ (on the left) and the even transfer function $y = 2x^2 - x^8$ (to the right). When only even partials are generated, we hear the second partial as the fundamental, the fourth as the second, the sixth as the third, etc. This can be seen in the figure on the right. Because the output waveform repeats within the period of the input signal, we hear a sound an octave higher than the input tone.

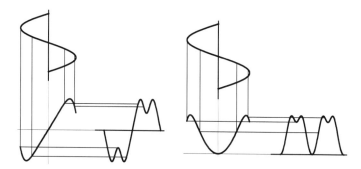

We can compute the spectrum generated by the transfer function $y = x + .7x^2 - .2x^5$ by using the table above as follows. The coefficient of the first partial is .875, this is, the sum of 1 for x^1 and $-.2 \cdot 10/16$ for $-.2x^5$. The coefficient of the second partial is $.7/2 = .35$ for x^2, the coefficient for the third partial is $-.2 \cdot 5/16 = -.0625$ for $-.2x^5$, etc. The direct calculation in Mathematica confirms these coefficients.

TrigReduce$\left[\text{Sin}[t] + .7*\text{Sin}[t]^2 - .2*\text{Sin}[t]^5\right]$

$0.35 - 0.35 \cos[2 t] + 0.875 \sin[t] + 0.0625 \sin[3 t] - 0.0125 \sin[5 t]$

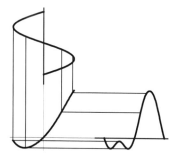

It is easy to show the influence of the modulation index when using polynomial transfer functions. If in the polynomial $y = c_0 + c_1 x + c_2 x^2 + ... + c_n x^n$ we replace the input signal x by ax, we get $y = c_0 + a c_1 x + a^2 c_2 x^2 + ... + a^n c_n x^n$. That means that the amplitude of the nth partial is multiplied by a^n. If we apply this to the transfer function $y = .5 \cdot x + .8x^2 + .3x^3 - .9x^5$, we get

TrigReduce$\big[$
$.5*a*\text{Sin}[t] + .8*(a*\text{Sin}[t])^2 + .3*(a*\text{Sin}[t])^3 - .9*(a*\text{Sin}[t])^5\big]$

$0.4 a^2 - 0.4 a^2 \cos[2 t] + 0.5 a \sin[t] + 0.225 a^3 \sin[t] - 0.5625 a^5 \sin[t] - 0.075 a^3 \sin[3 t] + 0.28125 a^5 \sin[3 t] - 0.05625 a^5 \sin[5 t]$

The following illustration shows the amplitude a_i of the ith partial plotted against the amplitude of the input signal (index). When the index is small, the lower partials predominate, when it is large, the high partials predominate. The term $.4a^2$ describes the DC component a_0.

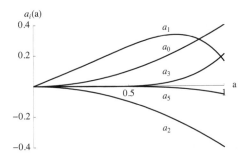

(→ Animation)

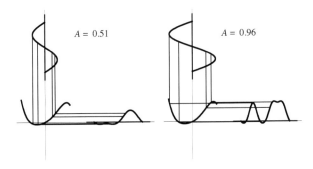

6.1.3.5 Chebyshev Polynomials as Transfer Functions

Usually, polynomials generate spectra containing several partials. By using special polynomials, however, spectra can be generated in which all components but one cancel each other out, leaving only one partial. For example, to generate the third partial we subtract the term $3x$ from $4x^3$, removing the first partial's coefficient $a_1 = 3$. Similar calculations for the nth partial lead to the so-called *Chebyshev polynomials* of the first kind T_n, defined by the equation $T_n(\cos(\alpha)) = \cos(n\alpha)$.

```
ChebyshevT[5, x]
```

$5\,x - 20\,x^3 + 16\,x^5$

$T_0 = 1$
$T_1 = x$
$T_2 = -1 + 2\,x^2$
$T_3 = -3\,x + 4\,x^3$
$T_4 = 1 - 8\,x^2 + 8\,x^4$
$T_5 = 5\,x - 20\,x^3 + 16\,x^5$
$T_6 = -1 + 18\,x^2 - 48\,x^4 + 32\,x^6$
$T_7 = -7\,x + 56\,x^3 - 112\,x^5 + 64\,x^7$
$T_8 = 1 - 32\,x^2 + 160\,x^4 - 256\,x^6 + 128\,x^8$
$T_9 = 9\,x - 120\,x^3 + 432\,x^5 - 576\,x^7 + 256\,x^9$
$T_{10} = -1 + 50\,x^2 - 400\,x^4 + 1120\,x^6 - 1280\,x^8 + 512\,x^{10}$

6.1 Modulation Techniques and Distortion

More Chebyshev polynomials can be derived using the recursive formula

$$T_{k+1}(x) = 2xT_k(x) - T_{k-1}(x). \qquad (B.7)$$

The illustration below shows the effect of the Chebyshev polynomials T_3 and T_6.

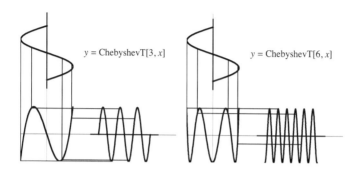

Simple spectra arise only when the amplitude of the input signal is 1. In the following example, we analyze the relationship between spectrum and index for the transfer function y = $T_2 - .3T_9$. The figure below shows that when the index is 1 only the second and the ninth partials appear in the proportion 1/.3.

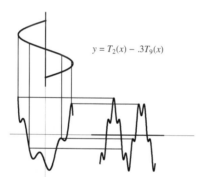

We convert the sum of the Chebyshev polynomials to an ordinary polynomial, substitute ax for x and show the computed functions of the coefficients a_i.

ChebyshevT[2, x] - .3 * ChebyshevT[9, x]

$-1 + 2 x^2 - 0.3 \left(9 x - 120 x^3 + 432 x^5 - 576 x^7 + 256 x^9\right)$

TrigReduce[-1 - .3 * 9 * a * Sin[t] + 2 * (a * Sin[t])² + .3 * 120 * (a * Sin[t])³ - .3 * 432 * (a * Sin[t])⁵ + .3 * 576 * (a * Sin[t])⁷ - .3 * 256 * (a * Sin[t])⁹]

$-1 + a^2 - a^2$ Cos[2 t] - 2.7 a Sin[t] + 27. a^3 Sin[t] - 81. a^5 Sin[t] + 94.5 a^7 Sin[t] - 37.8 a^9 Sin[t] - 9. a^3 Sin[3 t] + 40.5 a^5 Sin[3 t] - 56.7 a^7 Sin[3 t] + 25.2 a^9 Sin[3 t] - 8.1 a^5 Sin[5 t] + 18.9 a^7 Sin[5 t] - 10.8 a^9 Sin[5 t] - 2.7 a^7 Sin[7 t] + 2.7 a^9 Sin[7 t] - 0.3 a^9 Sin[9 t]

With $a = 1$, we obtain the same spectrum as before. When a is between 0 and 1, the partials 1, 3, 5 and 7 appear.

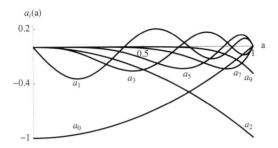

We notice the dominance of the 0th partial, that is of the DC component, when the index is small. The following figures show the distortion as the index increases. (→ Animation)

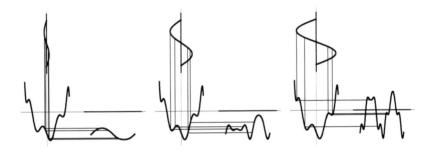

6.1.3.6 Limiters, Compressors and Expanders

Nonlinear distortion cannot only change a signal's timbre, it can also modify its amplitude. Here one tries to affect the timbre as little as possible. In recording and in sound treatment where it is difficult to predict the resulting amplitude, it is customary to use so-called *limiters* to keep an input signal from exceeding a given maximum amplitude. *Expanders* increase the differences in amplitude, *compressors* lessen them. The following illustrations show above transfer functions and below the amplitudes of the corresponding outputs plotted against the inputs. The graphical representations of these functions are often used in commercial software to symbolize the elements expander, compressor and limiter.

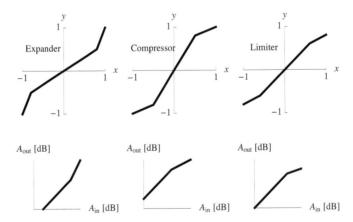

6.1 Modulation Techniques and Distortion

In contrast to the distortion techniques described above, compressors, limiters and expanders do not affect their input signals directly but only the signals' amplitude. The amplitude is measured over a certain period of time. The *attack time* is the time during which the input is corrected after having exceeded a given threshold. The *release time* is the time during which the correction returns to zero after the input has fallen below the threshold. An additional parameter can optionally round the corner of the transfer function.

Sometimes it is necessary to use a limiter in sound synthesis to avoid overmodulation. Depending on the program used, overmodulation is either clipped by setting the overflow to the maximum permissible value, or the overflow is reflected into the lower end of the amplitude range.

When overmodulation occurs, one usually either resynthesizes the sounds with different amplitudes or else uses floating point format and converts the sound afterward to the conventional format. But neither possibility works for real-time synthesis. Where the differences in amplitude are great, it can be useful to limit the amplitude so as not to have to keep the general level too low. What are reasonable limiting functions for various situations? All that is required to avoid the loud noise produced when amplitude overflow appears at the other end of the amplitude scale (right figure above) is a simple control function that leaves all values between a_{min} and a_{max} unchanged and replaces all values less than a_{min} by a_{min} and all values greater than a_{max} by a_{max}. This can be programmed in Mathematica with the command If[x < -1, -1, If[x < 1, x, 1]] (left figure below). That way, all sounds within the given amplitude limits remain undistorted. There is, however, still noise. To avoid it, we can define a function that leaves amplitudes unchanged only up to a certain limit $\pm a_{lim}$, then correcting more and more towards the extremes. To insure a smooth transition, the function should have the slope 1 at the points a_{lim} and $-a_{lim}$ and converge asymptotically to a_{min} and a_{max} respectively (right figure below). (→ *Limiter.maxpat*)

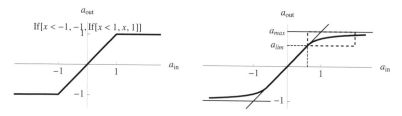

Functions having these characteristics can be derived, for example, from the functions $y = 1/x$ (left below) and $y = e^{-x}$ (right below), which for $x = 1$ and $x = 0$ respectively have a slope of -1 and for $x \to \infty$ converge asymptotically to 0.

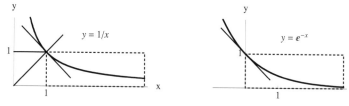

Let us now formulate the transformations within the ranges $[a_{lim}, a_{max}]$ and $[-a_{lim}, -a_{max}]$. We call the extent of the range d. In the first case, we set

$$y = -a_{max} - d/((x + a_{lim})/d - 1)$$

for values less than $-a_{lim}$ and

$$y = a_{max} - d/((x - a_{lim})/d + 1)$$

for values greater than a_{lim}. In the second case we set

$$y = -a_{max} + de^{(x+a_{lim})/d}$$

for values less than $-a_{lim}$ and

$$y = a_{max} - de^{-(x-a_{lim})/d}$$

for values greater than a_{lim}. The composite functions can then be described as follows:

```
If[x < -lim, -max - d/((x + lim)/d - 1),
        If[x < lim, x, max - d/((x - lim)/d + 1) ]]
If[x < -lim, -max + d*e^((x+lim)/d), If[x < lim, x, max - d*e^(-(x-lim)/d) ]]
```

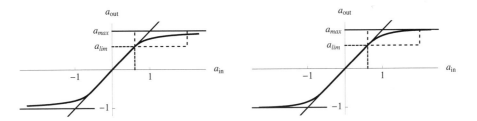

If the amplitude values exceed the limits only rarely, the additional computation time for the correction will be negligible, but if they are frequently out of range, it is worthwhile to use a table lookup procedure. If the magnitude of the overflow is very large, there can still be problems using a table: either the values can exceed those in the table, or the range of values, and hence the number of points in the table, has to be very large. In this case, it is useful to use a transfer function of five segments. When $x > a_{lim2}$ ($x < -a_{lim2}$), the values are replaced by a_{max} ($-a_{max}$). When the values are within the transition range $-a_{lim2} < x < a_{lim2}$, they are changed using a table. The illustration below shows only the positive half of the transfer function.

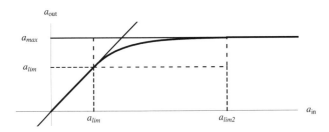

6.2 Nonlinear Systems

6.2.1 Non-Recursive Systems With a Single Input

6.2.1.1 Functions of a Single Input Value

The output of a system which computes the value $y(k)$ from only a single value of an input signal is called *distorted* (6.1.3). Such systems can be described by an equation of the form $y(k) = f(x(k))$. If the function f is linear, that is if $y(k) = c_1 x(k) + c_2$ holds, then the output is simply an amplification of the signal plus a constant. The proportions of the spectral components remain the same, except at 0 Hz. The component at 0 Hz corresponds to the constant part of the signal. An arbitrary function f can be stored as a look-up table and be read out using the input signal $x(k)$ as arguments (i.e. as index of the look-up table) (6.1.3.2). For simple functions f, for example polynomials and for sinusoids as input signal $x(t)$, the spectral changes can be calculated (6.1.3.4 and 6.1.3.5). If the function f is bounded, the system is stable.

6.2.1.2 Functions of More Than One Input Value

If we use more than one value of the input signal to compute the output signal $y(k)$, we get an extension of the non-recursive filter from Chapter 3.3.2. Such a system takes the form $y(k) = f(x(k), x(k-1), x(k-2) ...)$ for an arbitrary function f. In general, it is difficult to make predictions about the frequency behavior of such a system.

Let us first consider a system that takes two input values. To get an idea of such a system's behavior in the time domain, we can represent the input signal in a coordinate system with the axes $x(k)$ and $x(k-1)$ (see the illustrations below). A sine wave of frequency $sr/8$ corresponds to an ellipse (left). The lower the frequency, the narrower the ellipse (middle). A periodic oscillation, for example $.34 \cdot \sin(t) + .6 \cdot \sin(2t + 2.5)$, yields a closed orbit (right).

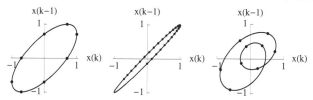

The function value $y(k) = f(x(k), x(k-1))$ can be drawn as the third axis in the representation. A nonlinear function is then represented as a curved surface. The following illustrations show the functions $y(k) = x(k-1)x(k)$ and $y(k) = x(k-1)\sin(5x(k))$. ($\rightarrow$ *sys0* in *NonLinSys_1.maxpat*)

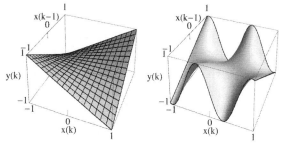

The function values $y(k) = f(x(k), x(k-1))$, belonging to an input signal x, lie vertically over the curve described by the oscillation in the $x(k)$-$x(k-1)$-plane.

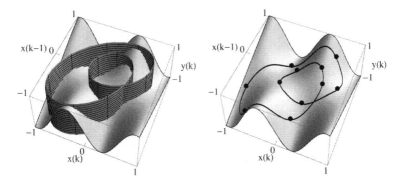

To get an idea of the behavior in the frequency domain, let us examine the linear filters $y(k) = ((x(k) + x(k-1))/2$ and $y(k) = (x(k) - x(k-1))/2$. The following figure shows the plane belonging to the high-pass filter $y(k) = (x(k) - x(k-1))/2$. The y-values over the first diagonal $x(k) = x(k-1)$ are zero. This means that the amplitude of low frequencies is small (figure right). The y-values over the second diagonal $x(k-1) = -x(k)$ vary between -1 and 1, hence the amplitude of higher frequencies is large (figure left).

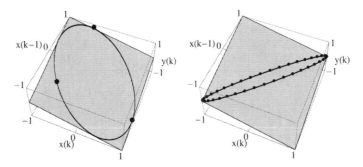

In a low-pass filter, the y-values over the second diagonal are zero, hence the amplitude of higher tones is smaller (left below). In the illustration to the right, the y-values over the first diagonal vary between -1 and 1, hence the amplitude of higher tones is greater.

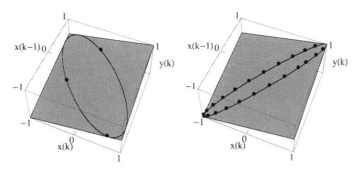

6.2 Nonlinear Systems

Nonlinear systems can be designed using such diagrams. The equation

$$f_1(x(k), x(k-1)) = -1.3 e^{-25\,|x(k)-x(k-1)|}\, e^{-25\,x(k)\,x(k-1)} \sin\left(\frac{1}{x(k)^2 + x(k-1)^2 + .1}\right)$$

gives the surface shown in the illustration below (to the left the entire domain, to the right a part of it).

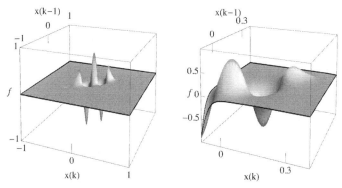

The exponential functions in the equation above bring about a decrease in the function values for large values of x. The argument of the sine function was chosen so that the oscillations increase in frequency with diminishing x. From the shape of the curve we can deduce the following characteristics of the system. Low frequencies are strongly distorted even at low amplitudes (their curves lie in the first diagonal). High frequencies correspond to wider ellipses and are strongly damped at high amplitudes but not distorted (their curves pass around the peaks). At low amplitudes, however, they are both amplified and distorted.
($\rightarrow$ *sys1* in *NonLinSys_1.maxpat*)

Now let us change our system so that the peak in the middle disappears, since it causes considerable amplification of very small amplitudes and strong distortion of high frequencies. To do so, we multiply the function by $x(k)x(k-1)$ and adjust the other parameters correspondingly. ($\rightarrow$ *sys2* in *NonLinSys_1.maxpat*).

$$f_2(x(k), x(k-1)) = -40\,x(k)\,x(k-1) e^{-12\,|x(k)-x(k-1)|}\, e^{-12\,x(k)\,x(k-1)} \sin\left(\frac{1}{x(k)^2 + x(k-1)^2 + .15}\right)$$

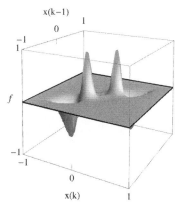

The following figures show how the waveform of the output signal depends on the frequency of the input signal, a sine wave with increasing frequency and constant amplitude.
(→ Animation)

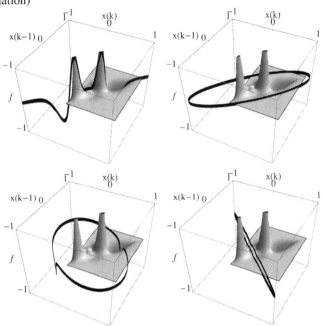

The following figures show how the waveform of the output signal depends on the amplitude of the input signal, here a sine wave of constant frequency and increasing amplitude.
(→ Animation)

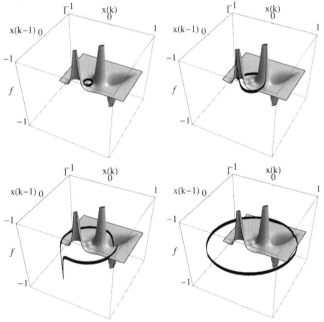

6.2 Nonlinear Systems

The distorted output of a sum of sine waves does not correspond to the sum of the distorted outputs of the individual sine waves. This can be seen in the following illustration of two sine waves in inharmonic relationship to each other. The curve representing the oscillation is not closed but wanders in the plane. The frequency proportion in the first figure is 1/5.315..., in the second figure 1/4.73...

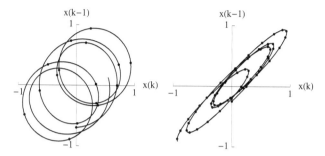

The following animation uses the first of the signals above and the function f_2 above. As the circles wander, ever new waveforms arise. ($\rightarrow$ Animation)

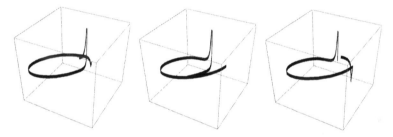

The program *NonLinSys1_1.maxpat* allows us to input the sum of two sine functions to the system *sys2*. The constant variation of timbre can best be followed when the frequency relationship is nearly harmonic (e.g. 100/200.037...), so that the changes arise very slowly. In the same program, values can be input with arbitrary delay.

Now let us consider the representation of a function of the values $x(k)$ and $x(k - n)$, rather than of $x(k)$ and $x(k - 1)$. The following illustration shows the curves made by a sine wave of frequency $sr/20$ in coordinate systems with $x(k - n)$ as the second axis for various delays n.
($\rightarrow$ Animation)

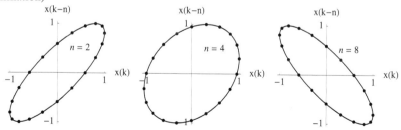

That the succession of these curves corresponds to the succession of sine waves with increasing frequencies means that we get the same variations of the output waveform for increasing delay as for increasing frequency. Systems using more than two input signals cannot be simply represented by graphical means.

6.2.2 Non-Recursive Systems With More Than One Input Signal

6.2.2.1 Functions Using a Single Value of Each Input

Ring modulation (6.1.1.4) and amplitude modulation (6.1.1.2) are examples of nonlinear systems whose output signals $y(k)$ are computed from one value each of several input signals $x_i(k)$. Ring modulation of the signals x_1 and x_2 corresponds to the system $y(k) = x_1(k)x_2(k)$, while amplitude modulation with the modulator x_1 and the carrier x_2 corresponds to the system $y(k) = x_2(k)(c + x_1(k))$. The systems of Chapter 6.2.1 can be considered special cases in which the input signals are identical or are made up simply of various delays of a single signal. We can describe these systems using an equation of the form

$$y(k) = f(x_1(k), x_2(k), x_3(k), ...).$$

If the function f is linear, that is, if $y(k) = c_1x_1(k) + c_2x_2(k) + ...$ holds, then the output is a weighted sum of the input signals corresponding to mixing the signals. Arbitrary functions of two input signals can be represented as surfaces above the plane with the axes $x_1(k)$ and $x_2(k)$, as shown in Chapter 6.2.1.2. The curve on the plane, that is, the succession of the points $(x_1(k), x_2(k))$, is easy to interpret for simple signals. Two sine waves of the same frequency give a straight line or an ellipse, depending on their relative phase.

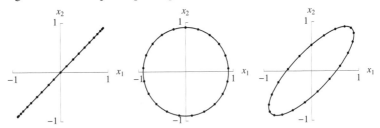

The following illustration shows the points $(x_1(k), x_2(k))$ for a frequency relation of 1/3, where the sine waves are of different phases $(x_1(k) = \sin(k), x_2(k) = \sin(3k + \varphi), \varphi = 0, .6, .18$. (→ Animation)

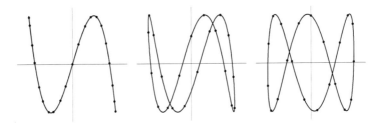

It is clear that the relative phase plays an essential role here. The first of the curves above, in the system defined by the function f_2 (6.2.1.2), passes around the peaks of the surface. The others do not. If one chooses as input signals in the program *NonLinSys_1.maxpat* two waveforms whose frequencies are in nearly harmonic relationship, one can hear the change of timbre resulting from the increasing phase shift.

In simple cases, we can visualize the behavior of a system with three input values by reckoning one input signal to the system and representing this extended system as we saw above.

6.2 Nonlinear Systems

The system $y(k) = \sin(x_1(k)\pi)x_2(k)x_3(k)$ gives for a constant signal $x_3(k) = 1$ the function

$y(k) = \sin(x_1(k)\pi)x_2(k)$ (left figure below).

If we use the input signal $x_3(k) = \sin(.7k)$, the surface varies with time k.

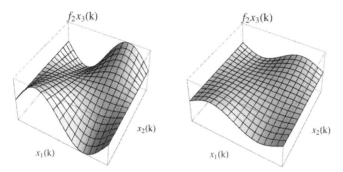

The following illustrations show the variable function and the path belonging to the input signals $x_1(k) = \sin(k)$ and $x_2(k) = \sin(3k)$ (see above). This method is particularly useful when an input signal is very simple and only changes slowly. (→ Animations)

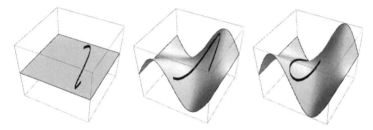

The behavior of a nonlinear system with many input signals cannot be visualized, but sometimes one can recognize certain characteristics of such a system by simplifying or re-grouping its equation. Ring modulation of the sum and difference of two signals $x_1(k)$ and $x_2(k)$ yields the system $y(k) = (x_1(k) + x_2(k))\cdot(x_1(k) - x_2(k))$, which corresponds to the difference of the squares of the signals: $y(k) = x_1(k)^2 - x_2(k)^2$.

One can get a certain idea about the behavior of the nonlinear system

$y(k) = \sin(x_1(k) + x_2(k) + x_3(k))$

in the time domain, but its behavior in the frequency domain can be seen more clearly in the following representation showing that the system corresponds to the sum of the ring-modulated signals $\sin(x_i)$ and $\cos(x_i)$.

```
TrigExpand[Sin[x₁ + x₂ + x₃]]

Cos[x₂] Cos[x₃] Sin[x₁] + Cos[x₁] Cos[x₃] Sin[x₂] +
    Cos[x₁] Cos[x₂] Sin[x₃] - Sin[x₁] Sin[x₂] Sin[x₃]
```

6.2.2.2 Special Functions of Two Input Signals

In the first example below we are looking for functions that can be controlled by a parameter $0 \le c \le 1$ so that the signals are summed when $c = 0$ and the mutual distortion increases with

c. Such systems can be most easily realized when a linear part $.5 \cdot (1 - c)(x_1 + x_2)$ and a nonlinear part $c \cdot f(x_1, x_2)$ are added together. For example, with the equation

$$y = f_0(x_1, x_2) = .5 \cdot (1 - c)(x_1 + x_2) + c x_1 x_2$$

simple mixing of two signals changes into ring modulation as c increases. The following illustration shows the function

$$y = f1(x_1, x_2) = .5 \cdot (1 - c)(x_1 + x_2) + c \cdot \sin(4x_1) \cdot \sin(4x_2)$$

for various values of the parameter c.

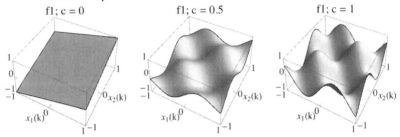

By looking at either the formula or the first of the following illustrations, we can see that for $c = 1$ and either $x_1 = 0$ or $x_2 = 0$, we get an output with the constant value 0. With input signals $x_1 = \sin(\omega t)$ and $x_2 = \sin(2\omega t)$ we get in the $x_1 x_2$-plane the curve shown in the second illustration below, as well as the function values for the output y shown in the third illustration (→ NonLinSys_2.maxpat, sys1).

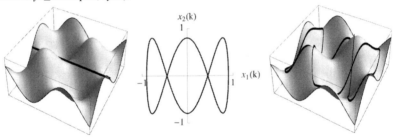

Instead of fading in the nonlinear part of the equation by using the factor c, we can multiply the argument of the sine function by c and get for the nonlinear part $\sin(4cx_1)\sin(4cx_2)$. The function is now

$$y = .5 \cdot (1 - c)(x_1 + x_2) + \sin(4cx_1)\sin(4cx_2). \quad (\rightarrow \text{Animation})$$

6.2 Nonlinear Systems

When c takes a maximum value of 1 and with sine waves as input signals, we get two sine waves whose frequencies are the sum and difference of the input frequencies.

TrigReduce[Sin[a] * Sin[b]]

$$\frac{1}{2}(\text{Cos}[a-b] - \text{Cos}[a+b])$$

In the system *sys2*, we introduce two further parameters d_1 and d_2 allowing us to change the factors in the arguments of the sine functions individually. The system now has the equation

$$y = f_2(x_1, x_2) = .5 \cdot (1-c)(x_1 + x_2) + \sin(cd_1 x_1) \cdot \sin(cd_2 x_2).$$

The function in *sys3* has the equation

$$y = f_3(x_1, x_2) = .5 \cdot (x_1 + x_2) \cdot \cos(cx_1 x_2).$$

This function cannot be divided into linear and nonlinear parts in a straightforward way. For $c = 0$ the system reduces to $y = .5 \cdot (x_1 + x_2)$. The following illustration shows the function for different values of c. (→ Animation)

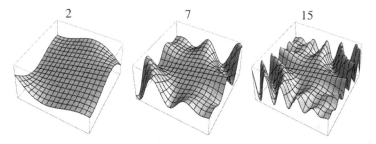

The function can be extended by adding a term that disappears when $c = 0$. The function in system *sys4* is

$$y = f_4(x_1, x_2) = .5 \cdot (x_1 + x_2) e^{-2.8 c}(c \cdot \sin(d_1 x_1) \sin(d_2 x_2) + \cos(ce\, x_1 x_2))^4$$

The additional factor $e^{-2.8c}$ assures that the function values do not become too large. Raising the sum to a power does not change the fact that the function becomes $y = .5(x_1 + x_2)$ when $c = 0$. The following illustrations show the function for the constant values $d_1 = d_2 = 15$, $e = 11$ and various values of c (→ Animation),

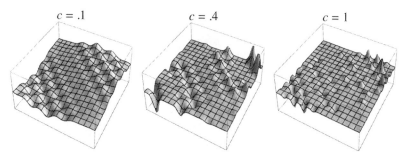

as well as for constant values $c = .5$ and $e = 15$ and various values of $d = d_1 = d_2$.

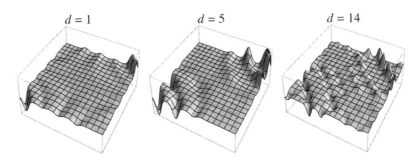

(→ Variation of e) (→ sys5 from *NonLinSys_2.maxpat*)

In the following examples we are looking for functions that can be controlled by a parameter $0 \leq c \leq 1$ so that when c is equal to zero we hear only the signal x_1 and as c increases we hear nonlinear distortion due to the signal x_2. The function in system sys7 (*NonLinSys_2.maxpat*) is

$$y = f_7 = \text{sign}(x_1)\text{abs}\,(x_1)^{1 + cd \cdot \sin(cex_2)^2}.$$

The number of folds in the surface depends on the parameter e, their curvature on the parameter d. (→ Animation)

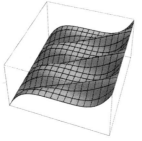

System *sys8* uses a somewhat more complicated function:

$$y = f_8 = \text{sign}(x_1)\sqrt{x_1^2 \cos(cdx_2 \cdot \sin(dx_1))^2}$$

(→ Animation)

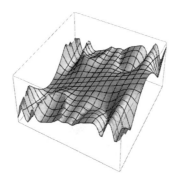

In a final group of functions, we want to control the system so that when $c = 0$, we hear only x_1, when $c = 1$ we hear x_2, and when $0 < c < 1$ we hear a nonlinear transition between the signals.

6.2 Nonlinear Systems

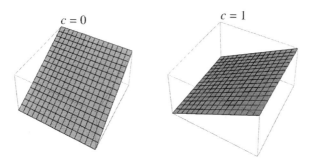

The function in system *sys9* (→ *NonLinSys_2.maxpat*) is

$$y = f_9 = cx_1 \cos((1-c)\,d\pi x_1\,x_2) + (1-c)\,x_2 \cos(cd\pi x_1\,x_2)$$

The animation shows how with increasing c the surface is distorted, tipped and then smoothed again. The number of folds depends on the parameter d. (→ Animation)

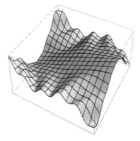

The function in *sys10* is

$$y = f_{10} = (cx_1 + (1-c)x_2 + \sin(c(1-c)dx_1x_2))e^{-2\sqrt[3]{c(1-c)}}$$

(→ Animation)

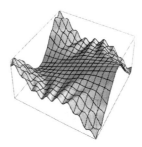

6.2.2.3 Functions of Several Values of Several Input Signals

Systems which treat several values of several input signals can be considered special cases of the systems $y(k) = f(x_1(k), x_2(k), x_3(k), ...)$ described in Chapter 6.2.2.1. The input signals here, however, are not independent; rather $x_i(k) = x_l(k - dt)$ holds for at least one signal $x_i(k)$. We saw in Chapter 6.2.2.1 that only systems with at most a few inputs can be visualized. Therefore, in this section we shall limit ourselves to general observations.

The more signals treated in a system, the more difficult it becomes to make predictions about the system's stability and boundedness. In addition, one must be careful not to divide by zero.

Since the output values can vary considerably, it may be wise to adjust the level, not to the largest output values, but to an optimum value and to limit the input values (6.1.3.6).
If an input signal contains no high frequencies, then its values only change slowly. The behavior of a system treating such a signal can be approximately described by a system treating all signals delayed by only a few samples from each other as equal. So, for example, we can simplify $x_1(k)x_1(k-1)x_1(k-2)$ to $x_1(k)^3$.
If there is only nonlinear interaction between differently delayed copies of the same signal, then we have a sum of distorted input signals

$$y(k) = f_1(x_1(k), x_1(k-dt_1), x_1(k-dt_2), ...) + f_2(x_2(k), x_2(k-dt_2), x_2(k-dt_2), ...$$
$$= \sum_i f_i(x_i).$$

6.2.3 Recursive Systems

Recursive systems use previous output values to compute the value of the current output. This feedback means that after a brief excitation recursive systems can resonate for an infinitely long time. Hence such systems are called *infinite impulse response* (IIR) filters (3.3.3.1). We can use these systems not only as filters but also as sound generators. The feedback can cause the output amplitude to increase indefinitely, so the Max programs illustrating this chapter should be used with caution.

6.2.3.1 Functions of One Value

In the simplest case, a recursive system uses only the last value to compute a new value. These systems have no input, and all the output values $y(k)$ are determined by the initial value $y(0)$. In the Max examples, however, the system variable $y(k)$ is initially zero, so that an output is only generated when a pulse is input. The formula is $y(k) = f(y(k-1))$. If the system is linear ($y(k) = ay(k-1)$), the output amplitude decreases exponentially when $0 < a < 1$ and increases exponentially when $a > 1$. When $a < 0$, the output oscillates with the frequency $sr/2$. Some nonlinear systems of this simple kind not only can produce signals of arbitrary frequency, but also can show all the typical aspects of chaotic behavior. We described the chaotic behavior of the logistic equation $y(k) = ay(k-1)(1-y(k-1))$ in detail in Chapter 3.4.3.3. Using the Max program *Logistic_Map.maxpat* we can either generate an audio signal or a control signal in real time. The parameter a can vary between 3 and 4. In order to generate an audio signal with a recursive system, the system's control rate must be equal to its sampling rate, that is, the vector size of the audio buffer must be 1. The Max program uses the external *icst.fexpr~*, where the system equation can be entered and which functions regardless of the size of the audio buffer.

When $a < 3$, the values generated converge rapidly towards a fixed value, independently of the starting value (3.4.3.3), which means that no oscillation is produced. As a increases, $y(t)$ oscillates between the same 2, then 4, then 8 etc. values, giving rise to frequencies of $sr/2$, $sr/4$, $sr/8$ etc. Because computer programs calculate with limited precision, periodic oscillation also can arise with parameter values which should cause chaotic behavior (3.4.3.3). In programs with 16-bit precision, the values repeat after 32,768 steps (or less). For certain values of the parameter a between 3 and 4, even an impulse of amplitude .0001 is enough to cause the system to switch between chaotic and periodic behavior or to jump from one period to another. The same is true of tiny changes in a. By adding very low-amplitude noise to a, the periodicity caused by imprecision of the computations can be avoided. Programs like

6.2 Nonlinear Systems

Mathematica, which work with greater precision, do not require such correction. (→ Example in Mathematica)

The system described above is stable as long as the values generated are between 0 and 1. Adding an input value however can rapidly lead to large numerical values. We can avoid this problem by setting $y(k) = c \cdot \sin(d\pi y(k-1))$. For $d = 1$ and $c < 1$ we get a system having a behavior similar to the logistic map (3.4.3.3) (below left), but which gives values in $[-c, c]$ for $y(k-1) > 1$ or $y(k-1) < 0$ (middle figure below). The third figure shows a sequence for a system having $d > 1$.

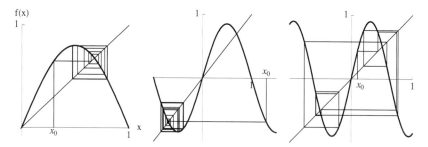

The system's attractor (3.4.3.3) for $d = 1$ corresponds very nearly to that for the logistic map.

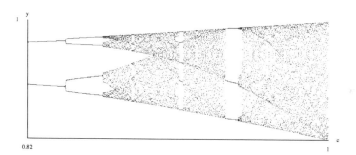

Now we can choose $d \neq 1$ and $c > 1$. The illustration below shows the attractor for $d = 1.2$ and $.5 < c < 2.5$.

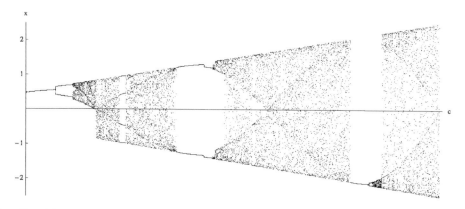

(→ Sine_Map.maxpat, sine_map.pde)

6.2.3.2 Functions of Two Values

If we use the last two output values $y(k-1)$ and $y(k-2)$ of a recursive system to calculate the new output, the formula is $y(k) = f(y(k-1), y(k-2))$. We can interpret such a system as a direct analogy to a first-order differential equation (3.4.1.4). Conversely, by using approximation procedures we can go from a differential equation describing a physical system to a discrete recursive system. We can draw conclusions about the behavior of a discrete system from the behavior of the corresponding physical system. The physical models of Chapter 8.1, describing the movement of a spring or a string without the addition of external energy, are stable systems. Linear systems produce sine waves whose frequencies we can determine using the approximations of Chapter 8.1 or compute exactly using the methods shown in Chapter 3. The discretization can however introduce inaccuracies and cause the systems to become unstable.

For the description of such systems, we refer the reader to Chapters 3.4 and 8.1. Here we shall only give a few Max examples with which one can get an idea of the systems' behaviors.

The system in the Max example *Pendulum_6-2-3-2.maxpat* is analogous to the pendulum described in Chapter 3.4.1.5. The frequency and the friction can be changed. Either a pulse or white noise can be input as excitation. Note that the frequency is dependent on the amplitude.

In the systems of the Max-patch *NonLinSys_6-2-3-2.maxpat*, the reactive force is a nonlinear function of the displacement (8.1.1.5). Note that the graphs of the functions must run from lower left to upper right; otherwise the systems are unstable.

To get an idea of the behavior of a recursive system, let us look at the slope field of the differential equation in phase space with the coordinates $y(k)$ and $y(k) - y(k-1)$ (3.4.1) and estimate where the trajectories are and which areas are stable. The following illustration show the slope field for the equation

$$y(k) = 1.9y(k-1) - .9y(k-2) - .1 \cdot \sin[3.5(y(k-1) + \sin[y(k-2)])],$$

together with a possible trajectory.

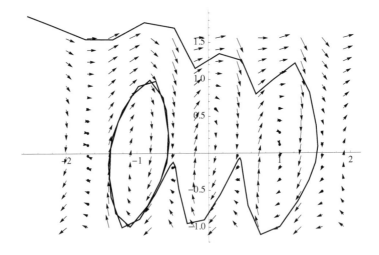

(→ Another, rather more complex slope field)

6.2.4 Time-Variant Systems

6.2.4.1 Delimiting Systems

Because for them the principle of superposition applies, linear time-invariant systems can be decomposed into systems with one input and one output. Nonlinear variable systems cannot usually be decomposed into subsystems having one input and one output. Failing to take into consideration even one single input can change the entire reaction of a system. When dealing with complex patterns of interaction, it is difficult and to some extent arbitrary to delineate systems and to differentiate between inputs and variables inherent to a system. Organisms and social systems are typically so complex that only individual parts can be approximately described mathematically. Sometimes complex nonlinear systems can be decomposed into roughly linear subsystems by progressive differentiation. There are also systems, however, for example control circuits (3.4.4.1), that react linearly, or even in a constant manner, compensating disturbance variables and generating a constant signal corresponding to a given set-point value. Such systems are often made up of complex nonlinear subsystems.

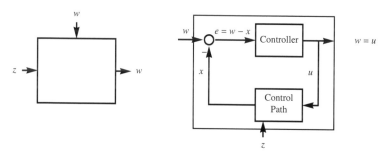

The following illustration shows the block diagram of frequency modulation from Chapter 6.1.2.2 with two of many possible descriptions. The system S_1 has one control input, one audio input and one audio output. Inside the system is an oscillator producing a waveform with constant amplitude A and a frequency computed from the input signals f_c and x. System S_2 has four control inputs and one audio output.

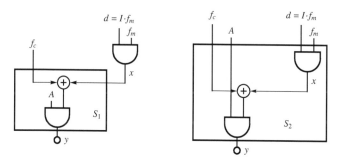

6.2.4.2 Non-Recursive Systems With Constant Delay

Non-recursive systems using constant delays to compute their outputs are described by equations having the form $y(k) = f(t, x_1(k), x_1(k-1), ... x_2(k), x_2(k-1), ...)$. When the function f varies slowly, we can approximate its value at various times by keeping the times in the function constant. The equation $y(k) = tx(k) + (1 - t)x(k - 1)$ approximately describes a linear filter. At time $t = .5$, it is the low-pass filter $y(k) = .5x(k) + .5x(k-1)$, at times $t = 0$ and $t = 1$, it is the all-pass filter $y(k) = x(k)$ or $y(k) = x(k-1)$ respectively. (→ Sound Example)

The equation $y(k) = \cos(2\pi ft)x(k) + \sin(2\pi ft)x(k-1)$ produces high-, low- and all-pass filters in regular succession: at time $t = 0$ the all-pass filter $y(k) = x(k)$, at time $t = 1/(8f)$ the low-pass filter $y(k) = .707x(k) + .707x(k-1)$, at time $t = 1/(4f)$ the all-pass filter $y(k) = x(k-1)$, at time $3/(8f)$ the high-pass filter $y(k) = -.707x(k) + .707x(k-1)$, at time $t = 1/(2f)$ the all-pass filter $y(k) = -x(k)$, at time $t = 5/(8f)$ the low-pass filter $y(k) = -.707x(k) - .707x(k-1)$, etc. One hears the variable effect of the filter at frequencies below 20 Hz, at frequencies around 20 Hz one hears a pulsating noise (→ *TimeVarFilter.maxpat*).

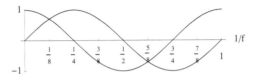

At high frequencies the nonlinearity of the system becomes quite obvious. At $f = sr/4$, the filter coefficients for successive samples are $\{0, -1, 0, 1, 0, -1, ...\}$ and $\{1, 0, -1, 0, 1, 0, ...\}$. The system transforms both the constant input $x_1 = \{1, 1, 1, 1, ...\}$ and the signal $x_2 = \{1, -1, 1, -1, ...\}$ with the frequency $sr/2$ into the signal $y_{1,2} = \{1, -1, -1, 1, 1, -1, ...\}$ with frequency $sr/4$. If this output is put back into the system, we get as output $y_3 = \{-1, 1, -1, 1, ...\}$. When the system is defined as a filter with variable coefficients $\cos(2\pi ft)$ and $\sin(2\pi ft)$, it can be interpreted as the sum of the ring-modulated signals $\cos(2\pi fk/sr)x(k)$ and $\sin(2\pi fk/sr)x(k-1)$. Hence its behavior can be described using the results from Chapter 6.1.1.

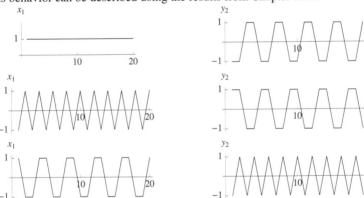

The coefficients of linear filters are usually variable. When the variations of the coefficients are rapid and/or discontinuous, nonlinear effects result (5.3.1.6). The coefficients vary discontinuously when either the control rate is low (5.1.2.3) or the control function only gives relatively few discrete values.

6.2.4.3 Non-Recursive Systems With Variable Delay

Non-recursive systems that use the variably delayed input signal to compute the output signal can be described by equations of the form $y(k) = f(t, x_1(k_1(t)), x_1(k_2(t)), ...)$. The system $y(k) = x(k - ct)$ transposes the input signal by resampling it. Because of this effect of transposition, systems with variable delay often cause audible signal changes even when the functions $k_i(t)$ change slowly (5.1.4.3).

6.2 Nonlinear Systems

In the usual explanations of frequency modulation (6.2.4.1, 6.1.2.2), the carrier oscillator is represented as the system generating the output signal. But the carrier signal can be generated without distortion and passed as the input to a system in which it is modulated by a variable delay. In the usual implementation of the oscillator, the output values are read from a table. In the following illustration an arbitrary signal $x(k)$ takes the place of such a table (figure left). The variable delay, in the simplest case controlled by a sine generator, can also be replaced by an arbitrary signal (figure right).

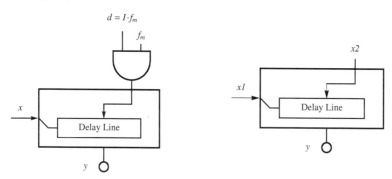

In the program *Var_Delay.maxpat* the input can be generated by an oscillator or come from a microphone. The signal determining the length of the delay can either be generated independently of the input signal by an oscillator or can be derived from the input by filtering. If this signal is generated by an oscillator, the frequency modulation described above results. If the delay is made by filtering the input, the distortion increases as the input amplitude increases. If one uses a band-pass filter to determine the delay, the degree of distortion depends on the spectrum of the input signal as well as on its amplitude. If the spectrum contains strong components in the passband, there will be strong distortion. If one uses a filter that measures the energy of the signal by taking its RMS, the delay will only change slowly and audible transpositions of the signal will result. The system transposes downward when the signal's energy increases, upward when the energy decreases. Virtually any sound can be used as the input signal and as the signal for the filter. This nonlinear Max program consists of linear subsystems like the amplitude control and the band-pass filter and of nonlinear subsystems like the variable delay and the RMS filter. Varying the parameters of the linear subsystems can also lead to nonlinear effects, particularly when the changes are rapid and/or discontinuous. (→ Example in Mathematica)

Further examples and details of the implementation can be found in Chapters 5.1.4.3.

6.2.4.4 Recursive Systems With Constant Delay

Even without an input signal, a recursive system can generate a signal that does not disappear (6.2.3). Variable coefficients can excite the system to oscillate. Time-variant recursive systems are always nonlinear. In Chapter 8.1.1.4 we shall derive a system that simulates the influence of gravity on the movement of a spring. The system's equation is

$$y(k) = ay(k-1) - y(k-2) - g.$$

The first two summands give a band-pass filter that simulates the resonance of the spring. The constant factor $-g$ simulates a force directed downward. In the following Mathematica program we set *y* for *y(k)*, *y1* for *y(k − 1)* and *y2* for *y(k − 2)*. A mass suspended from the spring would find its rest position at *y* = 0 without the influence of gravity. Under the influence of gravity, it oscillates about a position somewhat below this rest position when dropped.

```
a = 1.9; g = 0.2; y1 = y2 = 0; snd = {y1};
Do[y = a*y1 - y2 - g; y2 = y1; y1 = y; snd = Append[snd, y], {t, 0, 200}];
```

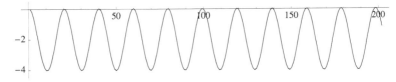

By changing the spring constant, we can excite the system. The change corresponds to varying the coefficient *a*. (→ Sound Example)

```
a = 1.9; g = 0.2; y1 = y2 = 0; snd = {y1};
Do[y = (a + RandomReal[{-0.1, 0.1}]) y1 - y2 - g;
   y2 = y1; y1 = y; snd = Append[snd, y], {t, 0, 600}];
```

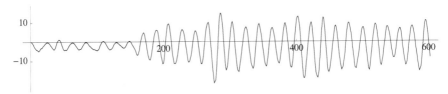

In program *Spring.maxpat* one can produce a sound by varying the frequency. Changing the "gravity" *g* also produces a sound. Since *g* is independent of the system variables, a varying *g* can be interpreted as an input signal. When *g* is zero, there is no output. This corresponds to the physical fact that a spring does not begin to oscillate when its elasticity is changed without the influence of an external force.

Another example of a recursive system is feedback control (3.4.4). A feedback loop is recursive and has a variable control value, known as the set-point (reference value). In adaptive feedback, certain values are adapted to the input signal or to the disturbance variables. In this way, system variables are modified by the system itself.

Stable recursive systems can become unstable when the coefficients vary rapidly. The system *y(k)* = *ay(k − 1)* − *y(k − 2)* is stable for 0 < *a* < 2. If the coefficient *a* changes quickly and hence discontinuously, the system becomes unstable.
In the first of the examples below, *a* varies periodically within the range [.3, 1.7], in the second randomly within the range [1.3, 1.9].

```
a = 1; b = 1; g = 0.2; y1 = y2 = 1; snd = {y1};
Do[y = (1 + .7*Cos[.306*π*t]) *y1 - b*y2;
   y2 = y1; y1 = y; snd = Append[snd, y], {t, 0, 100}];
```

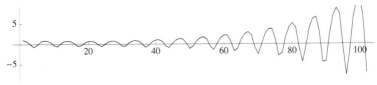

6.2 Nonlinear Systems

```
Do[y = (1.6 + 0.3 RandomReal[{-1, 1}]) y1 - b y2;
    y2 = y1; y1 = y; snd = Append[snd, y], {t, 0, 400}];
```

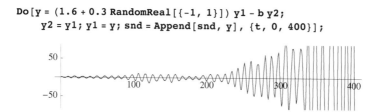

For further examples, see 8.1.6.1, 8.1.5.5.

6.2.4.5 Recursive Systems With Variable Delay

The recursive system $y(k) = ay(k - T) + x(k)$ is a comb filter (3.3.3.5). When a is only minimally less than 1, exciting the system briefly produces a tone that sounds similar to a plucked string. The period of the sound generated corresponds to the delay T. When the delay varies slowly, audible changes of frequency result. When the delay varies rapidly, the same effects are produced as in frequency modulation. In the program *Var_Rec_Comb.maxpat* one can define the delay by setting a constant part and a variable part. The system can be excited by constant noise, a click or by a signal from a microphone. The timbre change in the damping of the sound comes from the interpolation in the Max object when the period is not an integral multiple of the sample rate.

Recursive systems $y(k) = ay(k - T) + x(k)$ that use only one delayed output value to compute the new output are stable for $T = T(t)$ when $-1 < a < 1$. ($\rightarrow$ See example)

The systems $y(k) = a_1 y(k - T_1) + a_2 y(k - T_2) + \ldots x(k)$ that use more than one delayed output value to compute the new output are stable when $\sum_i |a_i| < 1$. In the following example, $\sum_i |a_i| = 1.1$.

```
a1 = -.505; a2 = .505; y1 = y2 = 1; snd = {y1}; Do[y = a1*y1 + a2*y2;
    If[Mod[t, 2] == 0, y2 = y1; y1 = y;]; snd = Append[snd, y], {t, 0, 200}];
```

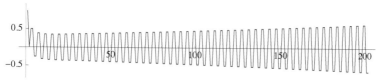

Systems that fulfill the above condition are strongly damped. The following illustration shows two signals generated by the system

$$y(k) = a_1 y(k - T_1) + a_2 y(k - T_2).$$

The delays $T_1(t)$ and $T_2(t)$ vary randomly between 1 and 5. Although the sum $|a_1| + |a_2| = 1.76$ ($a_1 = .88$ and $a_2 = -.88$) is obviously greater than 1, the output diminishes rapidly in amplitude for a temporally limited input (upper figure). Here the input is six real random numbers. For some successions of numbers, however, the output increases unchecked after a certain time (figure below).

```
a1 = 0.88; a2 = -0.88; snd = RandomReal[{-1, 1}, 6];
Do[y = a1 snd[[t - RandomInteger[{1, 5}]]] +
    a2 snd[[t - RandomInteger[{1, 5}]]]; snd = Append[snd, y], {t, 5, 100}];
```

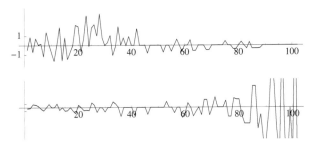

Despite these issues, a recursive system with several variable delays can have an infinitely long limited impulse response if the amplitude is controlled (3.4.4, 3.4.4.7, 5.1.4.4). In the program *Delay2.maxpat*, the outputs of two delay lines are added together and then added back into the (common) input of the two delay lines. To avoid both the rapid disappearance and the exponential increase of the signal, the amplitude of the feedback signal is multiplied by $\exp(a \cdot \mathrm{rms}(y(k)))$ (5.1.4.4). The delays are controlled by independent oscillators producing sine waves with a phase difference from each other of π, so that while one delay increases, the other decreases. In this way, the delayed sounds are transposed simultaneously up and down. The input can be noise, a click, or a sound picked up by microphone. Because the level of the feedback is controlled, even the output signal from the loudspeaker can be picked up by the microphone and fed back into the system without overmodulation.

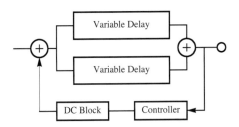

Because input signals, particularly impulses and recorded sounds, often have a DC component, the output signal can increase unlimitedly. Die following figure shows the reaction of the system $y(k) = .51y(k-1) + .51y(k-2)$ to a noise burst.

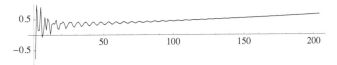

In the program *Delay2.maxpat*, the increase of the DC component is avoided by using a DC block.

The program *Delay4.maxpat* uses four delay lines each. All the delays increase linearly and jump back to their initial values. Their initial phases are 0, .25, .5 and .75. In order to avoid discontinuities when the delay pointer jumps back to the beginning of the delay line, the outputs of the delay lines are enveloped. Although the systems only transpose the feedback signal down, the sounds do not disappear thanks to the amplitude control. A DC block makes sure that the low frequencies do not dominate.

7 Other Techniques for Sound Analysis and Synthesis

7.1 Granular Synthesis

Many natural sounds and noises consist of countless short tones or noises that cannot be perceived singly. The sound of rain and the rustling of leaves are two such examples. Traffic noise, clapping or the murmur of a crowd are all made up of many sounds of varying loudness and length, each of which consists of many individual sound events. Even seemingly constant sounds, like a fountain or a motor, arise through regular sequences of short noises. In sound synthesis, repeating short noises even approximately regularly generates sounds reminding one of tearing or scratching. The repetition of sounds that can be heard either individually or globally generates sounds like humming or growling. Repeating several different noises irregularly generates sounds that can be perceived as crackling or rattling, depending on whether the listener's perception is directed to small-scale or large-scale events.

Granular synthesis includes many different techniques for generating complex sounds from single sonic events, so-called grains. Among these are algorithms to simulate the human voice and programs to chop up pre-existing sound material and reassemble it any way one chooses.

Further Reading: Microsound *by Curtis Roads [71],* Computer Music *by Charles Dodge and Thomas A. Jerse [2], pp. 262-276,* Computer Sound Synthesis for the Electronic Musician *by Eduardo Reck Miranda [15], pp. 107-114,* The Csound Book *edited by Richard Boulanger [40], pp. 281-292,* Real Sound Synthesis for Interactive Applications *by Perry R. Cook [70], pp. 15-18 and* Musical Signal Processing *by Curtis Roads [88], pp. 155-186.*

7.1.1 Fundamentals

7.1.1.1 Grains

The grains used in granular synthesis can either be synthesized or taken from recorded sounds. They can either be self-contained units or extremely short excerpts of continuous signals. The following illustration shows schematically simple grains being produced. In the first figure, a sine wave is synthesized whose amplitude envelope corresponds to a window function. In the second, a filter transforms an impulse into a short signal with exponentially decreasing amplitude.

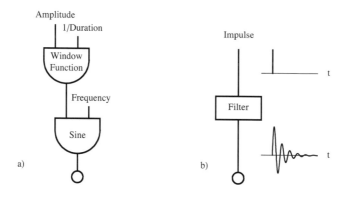

Many different window functions and virtually any waveform can be used to synthesize grains. A grain's spectrum is equivalent to the convolution of the spectra of its window function and the waveform used for its synthesis (3.2.2.4). The following figure shows grains made with four different window functions (above): a triangular window, a so-called Hann window, fade-in and -out lasting 1/10 of the grain's duration and fade-in and -out lasting 1/3 of the duration. Because the grains were synthesized using a sine wave, each grain's spectrum is the spectrum of its window function shifted to the frequency of the sine wave.

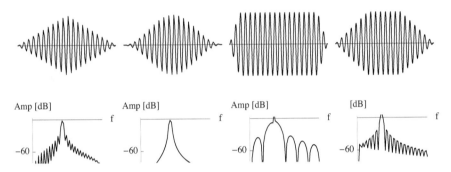

The spectrum of the grain produced by filtered impulses corresponds to the resonance curve of the filter. The impulse response of a resonator is an exponentially damped sine wave. The following figure shows two grains made with different resonators. The shorter the impulse response of a resonator, the larger the resonator's bandwidth and hence the broader the corresponding rise in the spectrum of the synthesized grains.

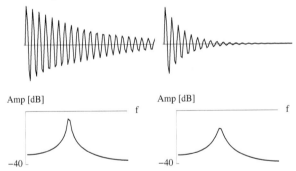

With more complex waveforms, the window function's spectrum is reproduced at the frequency of each component of the signal. The following figure shows to the left a grain made by windowing the sum of three sine waves with the frequency relationships 1:3:9 and amplitudes of .5, .3 and .2 with a Hann window. The spectrum at the right shows peaks at the frequencies of the components. The individual formants (resonances) have the same shape as the spectrum of the second of the four grain examples above.

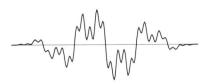 

7.1.1.2 Techniques for Making and Controlling Grains

There are many techniques for making grains and controlling their succession. If the grains change over time, they must always be calculated anew, but if they remain the same over the duration of a note, a grain can be stored in a table and read out at a rate corresponding to the desired frequency. The second grain in Chapter 7.1.1.1 was made with this command:

```
grain = Table[(.5 - .5*Cos[t*2 π])*Sin[20*t*2 π], {t, 0, 1, .01}];
```

To obtain a regular sequence, the grains can be written in alternation with a specific number of zeroes.

```
snd1 = {}; null = Table[0, {i, 1, 13}];
Do[snd1 = Join[snd1, grain, null], {t, 0, 1}];
```

The grains can also be inserted into an existing list of zeroes.

```
snd2 = Table[0, {i, 1, 250}];
Do[Do[snd2[[i + t*124]] = grain[[i]], {i, 1, 99}], {t, 0, 1}];
```

We can do this in Csound as follows. We make a table for the waveform (*f1*) and a table for the window function (*f2*). In order first to read sequentially from the two tables and then to wait until the beginning of the next grain, we make the function for the phase shown in the figure below. The phase *kph* increases during the duration of the grain (*il*) and then jumps to zero and remains there until the end of the period (*kp*) of the sound (instr 1, lines 11-15). In order to synthesize the sound at various frequencies, a new random frequency is chosen at the end of each (randomly long) period *kp* (line 6).

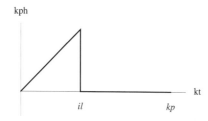

```
;7-1-1a
instr 1                                              ;1
il=             64
kt      init    0
kph     init    0
kprnd   randh   p6,p7                                ;5
kp=             p5+kprnd
        if      kt  >   il  goto    c1
kph=            kph+1
        goto                    c2
c1:                                                  ;10
```

```
         kph=              0
                  if      kt  <  kp  goto  c2
         kt=               0
         c2:
         kt=       kt+1                              ;15
         a1        table   kph,1
         a2        table   kph,2
                   out     p4*a1*a2
         endin
```

Sound Examples 7-1-1-a1 and 7-1-1-a2 were made with the following score. The grains contain components with frequency proportions 2:5:8 and 3:7 respectively. In the first example, the grains were windowed with a Hann window, in the second with a triangular window.

```
f1 0 64 10      0 1 0 0 .5 0 0 .3
f2 0 64 20 2
;      amp per rnd  frrnd
i1 0 5  30000       400 300 4

f1 0 64 10 0 0 1 0 0 0 .5
f2 0 64 20 3
;      amp per rnd  frrnd
i1 0 5  30000       120 55 13
```

The methods shown up to now can only generate sequences of grains that do not overlap. To generate overlapping sequences, real-time programs and sound synthesis languages that write sequentially into a file use several instruments or oscillators simultaneously. Overlapping grains can be produced more easily with programs that write to any particular place in a sound file. In Mathematica, we write the grains one after another into the sound file, but for each new grain, we jump back in the file by a specific duration. The new values must be added to the values already written in the file with the operator += (figure left), because simple assignment would overwrite the existing values (figure right).

```
Do[Do[snd2[[i + t * 65]] += grain[[i]], {i, 1, 99}], {t, 0, 3}];
```

When synthesizing with pulse trains and linear filters, the impulse responses can overlap. When simple filters are used, the computation time can be much shorter than with the methods described above. We make a simple IIR filter (3.3.3.5) using the command

$$a2 = a1 + (1-d)(.3a1 - a0).$$

We make a sequence of unit impulses with the command If[Mod[t,40]==1,1,0] (figure left below), and with the command If[Random[Integer, 15]==1,1,0] we make an irregular pulse train.

```
d = .2; snd = Table[a0 = a1; a1 = a2;
    a2 = a1 + (1 - d) * (.3 * a1 - a0) + If[Mod[t, 40] == 1, 1, 0], {t, 1, 100}];
```

7.1 Granular Synthesis

```
d = .1; snd2 = Table[a0 = a1; a1 = a2;
  a2 = a1 + (1 - d) * (.3 * a1 - a0) + If[Random[Integer, 15] == 1, 1, 0],
  {t, 1, 100}];
```

In the first case, we get a sequence of grains that hardly overlap, in the second a sequence of overlapping grains of varying density.

7.1.1.3 Synchronous Granular Synthesis

The regular repetition of a grain produces a periodic waveform, and we hear a harmonic sound. The frequency of the fundamental is equal to the reciprocal of the duration of the grain's duration and the spectral envelope corresponds to the spectrum of the grain itself. The figure below was made using the grain from Chapter 7.1.1.1. In the first example the grain is repeated without any interruption, giving a sound whose fundamental is a twelfth lower than the first partial of the grain. In the second example the grain is repeated at half the frequency of the first example. The fundamental is an octave lower than in the first example. The spectra of the sounds show one of the great advantages of granular synthesis, the invariability of the formants even when the fundamental changes.

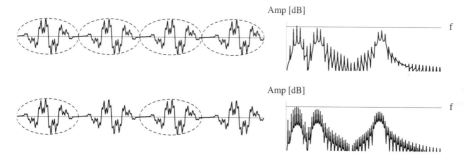

If the grains are repeated so fast that they overlap, we have a new waveform, but the spectral envelope remains the same. Within the envelope we can expect changes in the spectral detail, because the superposition of the grains acts like a filter. In the next example the repetitions overlap at half the grain's duration, causing the fundamental to be twice the frequency of the first example above.

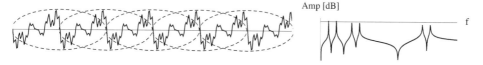

To change a sound's pitch during its duration, one only needs change the rate at which the grains are repeated. To change the timbre however, the waveform, that is, the spectral composition of the grain, must be changed. In granular synthesis, the amplitude depends on the frequency, since as the frequency gets lower, the segments of the signal with little or no energy get longer, and so the signal's total energy decreases. Inversely, the signal's energy increases with increasing frequency. When the grains overlap, the exact increase in energy is no longer predictable but depends on the form of the grains.

The spectral envelope resulting from filtering a pulse train is not influenced by the superposition of grains. The following figure show the waveforms and spectra of two filtered pulse trains (7.1.1.2).

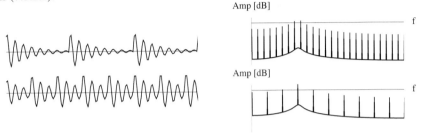

When nonlinear filters are used, the spectral envelope of the resulting sound depends on the sound's amplitude and frequency. The first two sounds below have the same frequency but different amplitudes (→ Sound 1 and Sound 2). The higher partials of the louder sound are clearly stronger than those of the softer sound. Since the individual impulse responses overlap more and more as the frequency increases, the higher partials are stronger at high frequencies even when the amplitude of the impulses remains the same (→ Sound 3).

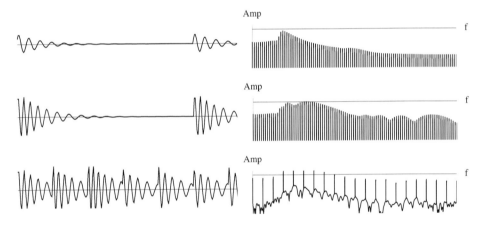

7.1.1.4 Asynchronous Granular Synthesis

When the grains are repeated at irregular intervals, noisy sounds are produced whose spectral envelopes correspond to those of a single grain. Depending on the number of grains per time unit and on their distribution, we get sounds of various roughness and graininess. The following figure shows the time-domain representation and the spectrum of a grain with two components.

```
grain = Table[(.7 - .5*Cos[t*2 π]) *
    (.5*Sin[3*t*2 π] + .3*Sin[10*t*2 π]), {t, 0, 1, .02}];
```

7.1 Granular Synthesis

In the first of the following examples, 100 grains are distributed randomly over 10,000 sample periods, in the second example 1000 grains are distributed randomly in the same time. The only difference in the spectra of the two sounds is that the level of the second spectrum is higher. In the time domain, on the other hand, one sees a crucial difference in the sounds, namely the looser distribution of the grains in the first example and the complete superposition of grains in the second (→ Sound 1 and 2). We can have a more regular distribution of the grains by adding a random factor to a constant time difference between the grains. If the differences to a mean value are small, we will have a slightly irregular sound with a clearly discernible pitch (→ Sound 3). If, for example, the time difference is 100 + random(10), that is, on an average 105 sample periods, then the perceived pitch is 8192/105 Hz = 78.019 Hz at a sampling rate of 8192 Hz. The spectrum shows clear lines at the fundamental and at the partial frequencies. The spectral envelope corresponds very closely to that of the single grain because the grains never overlap. Because there is no overlap, we can replace the operator += by the faster simple assignment.

```
Do[rn = Random[Integer, 9940];
  Do[snd1[[k + rn]] += grain[[k]], {k, 2, 50}], {i, 1, 100}]
Do[rn = Random[Integer, 9940];
  Do[snd2[[k + rn]] += grain[[k]], {k, 2, 50}], {i, 1, 1000}]
Do[rn = Random[Integer, 10];
  Do[snd3[[k + rn + 100*i]] = grain[[k]], {k, 2, 50}], {i, 1, 95}]
```

7.1.2 Applications

7.1.2.1 FOF

The so-called *FOF-generator* (*Fonction d'Onde Formantique*) generates sequences of exponentially damped sine wave bursts, which correspond to the impulse response of simple resonators. In contrast to Chapter 7.1.1.1, impulses are not filtered. Rather the frequency and the amplitude envelope for each grain are calculated from these parameters: fundamental frequency, center frequency and bandwidth of the formant and maximum amplitude of the grain. By adding the outputs of several FOF-generators sounds of several formants can be generated. The technique was developed for the synthesis of the human voice (see [2] p. 266).

Csound provides an FOF-generator which can be called by the command *fof*. Since the impulse response of a resonator is unlimited, the individual grains are windowed. The unit generator *fof* requires these parameters: amplitude, fundamental frequency, center frequency of the formant, octaviation index (normally zero; if greater than zero, lowers the effective fundamental frequency by attenuating odd-numbered sine bursts), bandwidth of the formant, rise time, grain duration, decay time, number of overlapping grains expected, table number of the waveform, table number for the window function, total duration. (→ Illustration of two grains from Sound 7-1-2a1)

```
;7-1-2a
instr 1
a1       fof      p4,p5,p6,0,p7,p8,p9,p10,p11,p12,p13,p3
         out      a1
endin

f1 0 32768 10 1                                    ;7-1-2a1
f2 0 1024 7 0 1024 1
;        p4       p5   p6       p7  p8   p9   p10  p11   p12 p13
;        amp      fr   fr-form  bw  ris  dur  dec  olaps fn1 fn2
i1 0 2   30000    100  1600     20  .001 .008 .003 10    1   2
```

Sound 7-1-2a2 was synthesized using four FOF-generators with the same fundamental, each producing a different formant. (→ Illustration of two grains from Sound 7-1-2a2)

```
;7-1-2a2
i1 0 2   15000 80 600    90  .001 .01 .003 10  1  2
i1 0 2    9000 80 1030   148 .001 .01 .003 10  1  2
i1 0 2    6000 80 2710   230 .001 .01 .003 10  1  2
i1 0 2    3000 80 3240   430 .001 .01 .003 10  1  2
```

See [2] pp. 265-266, [15] pp. 144-145.

7.1.2.2 VOSIM

VOSIM (VOice SIMulation) was developed for speech synthesis (see [2] p. 267). The grains consist of a sequence of oscillations with exponential decay. The individual oscillations correspond to half a sine wave squared, which gives a sine wave with a DC component. The grain is defined by five parameters: the number of oscillations N, duration of a single oscillation T, time remaining until the next grain D, amplitude A of the first oscillation and the factor c by which the amplitude is diminished at each oscillation. The total duration of a VOSIM grain, and hence the period of the generated sound, is $NT + D$ and the frequency of the fundamental is $1/(NT + D)$ Hz.

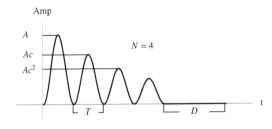

7.1 Granular Synthesis

To generate a grain we make a single oscillation (left below):

```
sinq = Table[Sin[t*π]², {t, 0, .9, .1}];
```

We join the required number of oscillations with decreasing amplitude and add the necessary zeroes (middle below):

```
Do[grain2 = Join[grain2, sinq*.89^i], {i, 1, 8}];
grain2 = Join[grain2,{0,0,0,0,0,0,0,0,0,0,0,0,0,0,0,0,0,0,0,0,0}];
```

Because of the DC component and the frequency of the oscillations, the spectrum of the grain shows peaks at 0 Hz and $1/T$ (right below).

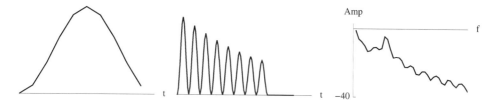

In the following example we reduce the amplitude of the oscillation continuously using an exponential envelope and generate the required number of oscillations in one step (left below).

```
grain4 = Table[((If[t < .666, e^-3*t, 0])*Sin[12*t*π]², {t, 0, 1.1, .01}];
```

In this way we get a grain with a flatter spectrum (middle below). By repeating the grain we get a sound with spectrum shown on the right below. (→ Sound)

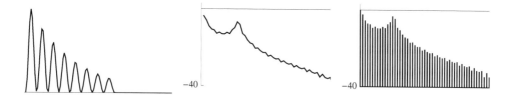

The choice of the parameters N, T, D and c determines the fine structure of the spectral envelope (see [2] p. 268). By changing the parameters one gets sounds with continuous transitions between pitches and timbres, which is useful for speech synthesis, particularly for generating diphthongs. By increasing the duration D one can lower the fundamental and decrease the amplitude. Changing the period T changes the center frequency of the formant. To keep the fundamental frequency constant while changing the formant frequency, $NT + D$ must be kept constant. By varying the duration D periodically vibrato can be produced. Random changes of D produce a certain roughness, which can be useful for the simulation of transients, fricatives and percussive instrumental sounds. To generate sounds with several formants, several VOSIM instruments with the same fundamental frequency producing different formants can be summed.

See [1-2], [7] p. 432, [2] pp. 267-269.

7.1.2.3 Granulating Sampled Sounds

The technique of granulation, that is, the decomposition of continuous sounds into successive or overlapping discontinuous elements, can be used to process existing sounds. If the individual grain is longer than a few tens of milliseconds, its "contents" can be recognized. If the grains are very short, new sounds and sound sequences can be generated. In both cases, the grains can be isolated or can overlap. Their duration and the point in time in the original sound from which they are taken can vary or remain constant. We shall describe a few simple applications from the many processes possible.

If a section of constant length T is taken at a specific point in time from a given sound and then windowed and repeated, a new sound is generated with a constant fundamental of frequency $1/T$ Hz and a spectral envelope corresponding to that of the original sound at the corresponding point in time.

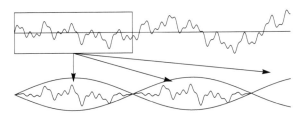

Sounds can be lengthened or shortened without changing the fundamental pitch using granulation, particularly if one chooses the sections carefully. Such time stretching involves either overlapping sections (left below) or skipping certain sections (right below).

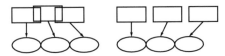

It is often necessary that sections overlap in the generated sound as well.

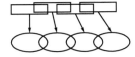

Granulation can also be used to transpose a sound without changing its duration (pitch shifting). To transpose a sound down, the individual sections are stretched by resampling (6.2.4.3) them and then re-assembled using correspondingly greater overlapping than in the original sound.)

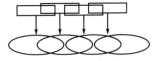

The mxj~ external *cm_granulator* inputs two seconds of sound at inlet 1, then stores and granulates the sound. The user can set the number of grains per second, their length and a transposition factor. The number of grains and their amplitudes can be varied randomly (inlet 4 and 5), and the grains can be delayed by a random amount (inlet 6). One can choose from

7.1 Granular Synthesis

three window functions: a Hann window, an exponentially decreasing envelope, and an exponentially decreasing envelope with adjustable rise time (inlet 8). The code below shows the inlet function and the external's DSP routine. There are three arrays: *sndin*[ ] to store the input sound, *sndout*[ ] to write and output the grains and *window*[ ] for the window function. The variables used are: *n* the length of the arrays *sndin* and *sndout*; *glen* the length of the grains; *wlen* the window length; *dt* the duration until the next grain; *dt_count* a counter for this duration; *amp* amplitude; *transp* the transposition function; *at* the rise time for the third window function; *flag1* a flag for the current window function; *ddt* and *damp* the amount of random scatter for duration between grains and the amplitude respectively; *dread* the amount of randomness of the reading position in the stored sound; *smp* and *sndin_del1* the current sample and the last sample read in respectively. In addition, there are several pointers for the arrays: *write_read* for the reading and writing position in *sndin* and *sndout*; *write_out* for write position in *sndout*; *read* for the reading position in *sndin*. For the interpolation when transposing, *read_int* and *read_frac* are the integer and fractional parts of the pointer read.

The inlet function inputs the variables, calculates the durations of the grains in samples and computes the window functions 1 and 2. If window function 3 is required, a flag is set that invokes the computation of the envelope during the perform routine. The perform routine checks whether a grain should be generated (line 33). If so, it is computed and written to the array *sndout*[ ]. Then the input buffer *in*[ ] is written to the array *sndin*[ ] and the output buffer is filled with values from *sndout*[ ] (lines 57-61). (→ *Granulator.maxpat*)

...

```
  public void inlet(float f){                                        //1
    if (getInlet() == 0) dt = (int)(44100.f/f);
    if (getInlet() == 1) glen = (int) (44.1f*f);
    if (getInlet() == 2) transp = f;
    if (getInlet() == 3) ddt = f;
    if (getInlet() == 4) damp = (int)f;
    if (getInlet() == 5) dread = (int)(44100.0f*f);
    if (getInlet() == 6){
      if((int)f == 0) {
        flag1 = 0;                                                   //10
        for(int i = 0; i < wlen; i++){
          window[i] = (float)(0.5*(1 - Math.cos(i*6.2831853f/wlen)));
        }
      }
      if((int)f == 1){
        flag1 = 0;
        for(int i = 0; i < wlen; i++){
          window[i] = (float) (Math.exp(-6.0f*i/wlen)*
                      (1.0f - Math.exp(-12.0f*i/wlen)));
        }                                                            //20
      }
      if((int)f == 2) flag1 = 1;
    }
    if (getInlet() == 7) at = f;
  }
  public void perform(MSPSignal[] ins, MSPSignal[] outs){

    float[] in = ins[0].vec;
    float[] out = outs[0].vec;
                                                                     //30
    for(int i = 0; i < in.length; i++){
```

```
      dt_count -= 1;
      if(dt_count < 0){
        dt_count = dt + (int) (dt*(2*(0.01*ddt*(Math.random()-0.5f))));
        amp = 1 + (2.f*(0.01f*damp*((float)Math.random()-0.5f)));
        write_out = write_read + 1024;
        read = write_read-2*glen-(int)(dread*((float)Math.random()-0.5f));
        if(read < 0) read += n;

        for(int k = 0; k < glen; k++){                              //40
          write_out += 1;
          if(write_out >= n) write_out -= n;
          read += transp;
          if(read >= n) read -= n;
          read_int = (int)Math.floor(read);
          read_frac = read - read_int;
          if(read_int == 0) sndin_dell = sndin[n-1];
          else sndin_dell = sndin[read_int - 1];
           smp = read_frac*sndin[read_int] + (1.0f - read_frac)*sndin_dell;
          if(flag1 == 1)                                            //50
             sndout[write_out] += (float)amp*smp*Math.exp(-7.0f*k/glen)*
                                  (1 - Math.exp(-at*k/glen));
          else
             sndout[write_out]+=amp*smp*window[(int)(k*(float)wlen/ glen)];
        }
      }
      sndin[write_read] = in[i];
      out[i] = sndout[write_read];
      sndout[write_read] = 0;
      write_read += 1;                                              //60
      if(write_read >= n) write_read = 0;
    }
  }
```

See [2] p. 271, [15] pp. 109-117.

7.2 Special Analysis Methods

7.2.1 Walsh Synthesis

In 1923 the mathematician Joseph Walsh described a complete, orthonormal set of functions from which any arbitrary periodic function can be generated. When summing the functions, the same problem arises as in analysis and synthesis of periodic functions using the sine function, namely the calculation of the coefficients. The Walsh functions can only have two values (−1 and 1) and can therefore be easily realized digitally. The great disadvantage of analysis and synthesis with Walsh functions is that the coefficients indicate nothing about the Fourier spectrum of the sound generated. Transformations for analysis and re-synthesis were developed in the 1970's, and a few techniques for sound synthesis with Walsh functions have been worked out (see [11] p. 155).

See [11] pp. 153-156, [15] pp. 81-82, [35] pp. 205-206.

7.2 Special Analysis Methods

7.2.1.1 Walsh Functions

The figure below shows the eight Walsh functions with which eight-point analysis or synthesis can be performed. The first is always equal to 1, the second changes from 1 to –1 in the middle, etc. Every change of sign is equivalent to a zero crossing, and the frequency is given correspondingly in zero crossings per second (zps) rather than in Hz or cycles per second (cps).

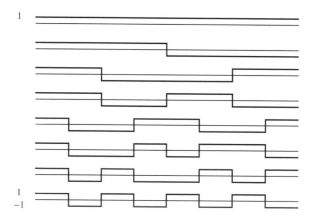

7.2.1.2 Examples

The following figure shows the Walsh functions $w1$ and $w2$ for two-point analysis (left) and a function f to be generated from the Walsh functions (right).

The sum of the functions w_1 and w_2 multiplied by the coefficients a_1 and a_2 gives the function $f = a_1 w_1 + a_2 w_2$. Then from the two equations $a_1 + a_2 = -1$ and $a_1 - a_2 = 3$ we calculate the coefficients $a_1 = 1, a_1 = -2$.

To do the eight-point analysis of a sine function, we derive eight equations from the eight Walsh functions and the eight values $\sin(2n\pi/8)$ and compute from them the coefficients.

```
sin = Table[N[Sin[i*2*π/8]], {i, 0, 7}]
```

{0., 0.707107, 1., 0.707107, 0., -0.707107, -1., -0.707107}

```
w1 = {1, 1, 1, 1, 1, 1, 1, 1}; w2 = {1, 1, 1, 1, -1, -1, -1, -1};
w3 = {1, 1, -1, -1, -1, -1, 1, 1}; w4 = {1, 1, -1, -1, 1, 1, -1, -1};
w5 = {1, -1, -1, 1, 1, -1, -1, 1}; w6 = {1, -1, -1, 1, -1, 1, 1, -1};
w7 = {1, -1, 1, -1, -1, 1, -1, 1}; w8 = {1, -1, 1, -1, 1, -1, 1, -1};
```

```
Solve[{a1 + a2 + a3 + a4 + a5 + a6 + a7 + a8 == sin[[1]],
    a1 + a2 + a3 + a4 - a5 - a6 - a7 - a8 == sin[[2]],
    a1 + a2 - a3 - a4 - a5 - a6 + a7 + a8 == sin[[3]],
    a1 + a2 - a3 - a4 + a5 + a6 - a7 - a8 == sin[[4]],
    a1 - a2 - a3 + a4 + a5 - a6 - a7 + a8 == sin[[5]],
    a1 - a2 - a3 + a4 - a5 + a6 + a7 - a8 == sin[[6]],
    a1 - a2 + a3 - a4 - a5 + a6 - a7 + a8 == sin[[7]],
    a1 - a2 + a3 - a4 + a5 - a6 + a7 - a8 == sin[[8]]},
    {a1, a2, a3, a4, a5, a6, a7, a8}]

{{a1 → 0., a2 → 0.603553, a3 → -0.25,
    a4 → 0., a5 → 0., a6 → -0.25, a7 → -0.103553, a8 → 0.}}
```

As a check we sum the component functions multiplied by the calculated coefficients.

```
.604*w2 - .25*w3 - .25*w6 - .104*w7

{0, 0.708, 1., 0.708, 0, -0.708, -1., -0.708}
```

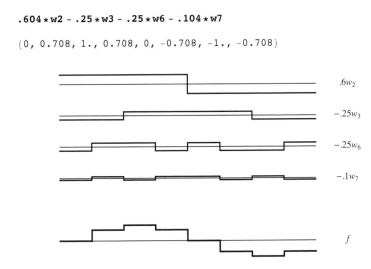

7.2.2 The Logarithmic Frequency Range in Spectral Analysis

The frequency resolution of the discrete Fourier Transform depends on the number of values used for the transformation (3.2.2). If 1024 values with a sampling rate of 44,100 Hz are used, the range from 0 to 22,050 Hz is divided into channels of 44,100/1024 = 43.0664 Hz width each. This means that the octave between 10,000 and 20,000 Hz is divided into 232.2 channels, corresponding to virtually indistinguishable micro-intervals, while the octave between 100 and 200 Hz is divided into only 2.322 channels each larger than a fourth.

Techniques have been developed to keep the (perceptual) frequency resolution constant throughout the required range by increasing the bandwidth of the channels proportionally to their increasing center frequencies. That way, the frequency range is subdivided logarithmically and the relation between the bandwidth and the center frequency of the channels, the so-called *quality factor Q* (5.3.1.1), is the same for all channels. For this reason, analysis using these techniques is called *Constant Q Filter Bank Analysis*. The following illustration shows schematically the subdivision of the frequency range 0 – 12000 Hz in channels of equal bandwidth as in the FFT (above) and in equal intervals as in the Constant Q Filter Bank Analysis (below).

7.2 Special Analysis Methods

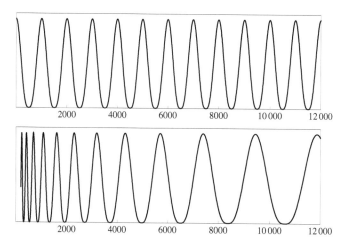

Constant Q Filter Bank Analysis has for most musical applications the further advantage that the channels, which correspond to bandpass filters, react faster in the higher frequencies as the bandwidth increases. The disadvantage of the method is that the computation time cannot be reduced to the same degree as for the FFT.

See [11] pp. 578-581.

7.2.3 Wavelets

Fourier analysis of sounds has two essential weaknesses. First, the frequency range to be analyzed is subdivided into equal bands, causing the bands to be unnecessarily narrow for high frequencies and too broad for low frequencies (7.2.2). Second, the frequencies present are identified, but the analysis does not show at what point in time they appear. The wavelet transformation offers greater flexibility by using short wavelets rather than the (theoretically) infinitely long sine and cosine functions of FFT analysis. Wavelets can be virtually any function, which is displaced and compressed or stretched during signal analysis.

See [52], [53], [55], [71] p. 282, [88] pp. 127-154.

7.2.3.1 Wavelets

Although theoretically virtually any function can be used for wavelet analysis, in general one will choose simple bounded functions. The function's mean value must be zero. The following illustration shows on the left the *Haar wavelet*, which is equal to zero outside the interval [0, 1] and whose mean value is obviously also zero. On the right is the *Mexican Hat wavelet*, defined as

$$f(t) = (1 - t^2)\exp(-t^2/2).$$

Although the function never goes to zero, its value decreases with increasing $|t|$ so rapidly that one can limit it to the interval [−4, 4]. We obtain the function's mean value by integrating:

$$\int_{-\infty}^{\infty} f(t)\,dt = 0.$$

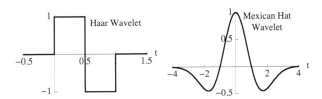

The wavelets are temporally displaced about *d* by subtracting *d* from the argument *t*, and they are compressed or stretched by the factor *s* by dividing *t* − *d* by *s*.

```
d = 1; p1 = Plot[hatw[t - d], {t, -4, 4}]
d = 1; s = .3; p2 = Plot[hatw[(t - d) / s], {t, -4, 4}]
```

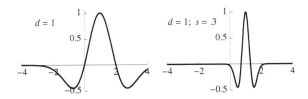

7.2.3.2 The Continuous Wavelet Transform

To illustrate the wavelet transform, we will use a test signal *test(t)* consisting of a low-frequency sine wave with an instantaneous attack and exponentially decreasing amplitude and a high-frequency sine wave with a continuous envelope.

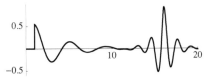

The wavelet transform using the wavelet *w(t)* gives a value for every point in time *d* and every factor $1/s$.

$$w\_t(d, s) = \frac{1}{\sqrt{s}} \int f(t) w\left(\frac{t-d}{s}\right) dt$$

The wavelet transform $w\_t(d, s)$ is a function of two variables, the displacement *d* and the stretch factor *s*. The value at the point (d, s) corresponds to the integral of the product of signal and wavelet, that is, the area under the function $f(t)w((t − d)/s)$. The factor $1/\sqrt{s}$ is used for normalization. The following illustration shows the test signal $f(t) = test(t)$ and the wavelet $w = hatw((t − d)/s)$ above and the product of the two below.

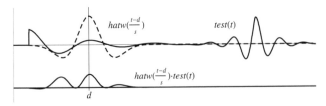

7.2 Special Analysis Methods

We run the wavelet over the entire signal using various stretch factors and compute the value $w\_t(d, s)$ for every point d. The following illustrations show at the top the test function $test(t)$ and the wavelet $w = hatw((t - d)/s)$ at various times d (s is held constant $s = 1$), in the middle their product and at the bottom the wavelet transform $w\_t(d, s)$, where the value $w\_t(d, s)$ at time d is shown by the large dot. (→ Animation)

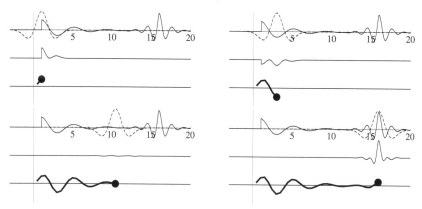

With a stretch factor $s = .4$ we get the following results. (→ Animation)

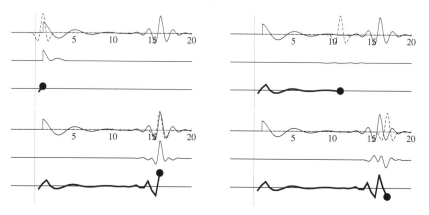

If we combine these functions, we obtain the function $w\_t(d, s)$.

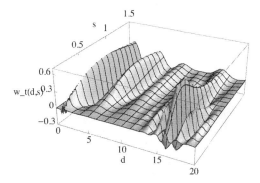

The wavelet transform $w\_t(d, s)$ is not usually represented as a three-dimensional graph, but rather as a contour or density diagram. Dark areas represent the low values of the function, lighter areas its high values. The figure on the left below shows the direct translation of the function above into such a diagram. The smaller the stretch factor s, the higher the frequencies represented, and so the figure is usually reflected to place the high frequencies at the top.

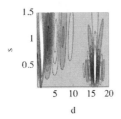 

The following figure shows the analysis of the test signal using the Haar wavelet. (→ Animation: Analysis using the Haar wavelet)

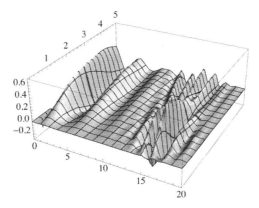

The wavelet used for the analysis and the values for s and d can be adapted to the signal at hand.

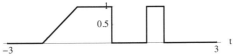

The illustration below shows the wavelet transform of the signal above (on the left with the Haar wavelet, on the right with the Mexican Hat wavelet).

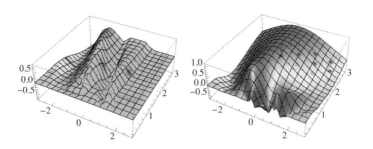

7.2.3.3 The Discrete Wavelet Transform

The wavelet transform of a signal can only be calculated algebraically for simple cases. More complex signals require numerical computation. In this case, both signal and wavelet are sampled for time and amplitude. We shall only consider a special case of a discrete wavelet transformation, the so-called Fast Wavelet Transform. We examine first how a signal consisting of two values $\{a_1 = 5, a_2 = -1\}$ can be decomposed (7.2.1.1). If we take the mean of the two values $(a_1 + a_2)/2 = 2$, we obtain the DC component (upper right in the illustration below). Half of the difference $(a_1 - a_2)/2 = 3$ gives the magnitude of the Haar wavelet (lower right in the illustration below).

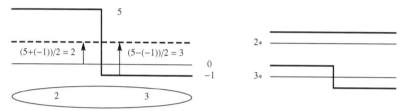

If a signal consists of four values $\{9, 1, 4, -6\}$, we decompose it into two signals with two values each $\{9, 1\}$ and $\{4, -6\}$. Taking the mean and half the difference for each signal, we get the numerical sequence $\{5, 4, -1, 5\}$. The mean values are shown as a dashed line and correspond to the sequence above $\{5, -1\}$. The mean of these two values gives the DC component of the entire signal (above right in the illustration below). Halving the difference gives the magnitude of the Haar wavelet $\{1, 1, -1, -1\}$. Halving the differences of the two partial sequences, that is, 4 for the first part and -1 for the second part, gives the magnitudes of the Haar wavelets $\{1, -1, 0, 0\}$ and $\{0, 0, 1, -1\}$.

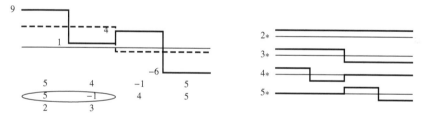

As a check, we sum the Haar wavelets weighted by the computed coefficients.

2 {1, 1, 1, 1} + 3 {1, 1, -1, -1} + 4 {1, -1, 0, 0} + 5 {0, 0, 1, -1}

{9, 1, 4, -6}

For a signal of length 2^N, the above procedure is repeated N times. The general algorithm can be described as follows. For groups of two successive values a_1 and a_2 (first row below), write $(a_1 + a_2)/2$ and $(a_1 - a_2)/2$ (second row). The values $(a_1 - a_2)/2$ are the coefficients of the wavelet $\{1, -1, 0, 0\}$ and of its displacement $\{0, 0, 1, -1, 0, ...\}$, etc. The first half of the values is the input signal for the next step, where we obtain the coefficients of the enlarged wavelet $\{1, 1, -1, -1, 0, 0, 0, 0\}$ and its displacement $\{0, 0, 0, 0, 1, 1, -1, -1\}$ as well as another signal $\{5, -1\}$, only half as long as before and once again the input signal for the following step. In the final row is the DC component and the coefficient for the wavelet having the same length as the original signal.

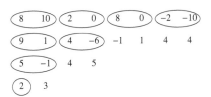

The computations correspond to both high- and low-pass filtering the signal and then down-sampling the result. The high-pass filter $(a_1 - a_2)/2$ first determines the magnitude of the highest frequencies. Then the complementary low frequencies are filtered out with the low-pass filter $(a_1 + a_2)/2$. Now the signal can be reduced to half its previous size without loss of information. Then the same procedure is applied to the reduced signal, etc. Let us observe this technique using the following signal:

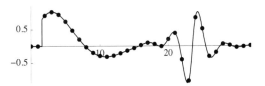

The following illustration shows the first four steps of the analysis. We see the signal after the low-pass filtering and data reduction and the Haar wavelet with which the coefficients are determined.

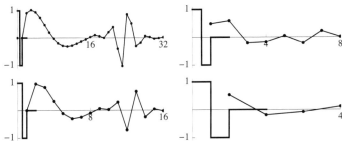

The magnitude of the highest frequency $(f_s/2)$ is measured $N/2$ times, the magnitude of the lowest frequency (0 Hz), the DC component, is only measured once over the entire signal. The filters halve the frequency range. This means that the analysis takes place in octave bands. Hence, the frequency-time plane is subdivided as shown below.

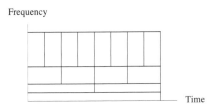

Only signals consisting of sine waves of frequencies $f_s/2^n$ and of constant amplitude, like $2 \cdot \sin((f_s/2)2\pi k) - 3.3 \cdot \sin((f_s/8)2\pi k)$, can be analyzed into constant values limited to single bands. Signals with spectra varying over time and signals whose frequencies cannot be represented by the term $f_s/2^n$, like our test signal, will result in oscillating values in all bands (compare Spectral Leakage 3.2.2.3). (→ Illustration)

8 Physical Modeling

The techniques subsumed under the notion of physical modeling generate sound by simulating physical processes. Spectra, dynamic behavior, transient effects, etc. are not produced, as in other techniques, by replicating their acoustic characteristics. On the contrary, all the acoustic characteristics of a sound ideally come from the model of an instrument and from the way its oscillation is excited. In all computer simulations, whether of an avalanche or of the deformation of heated metal, objects are assembled out of individual elements. These elements and their interactions with each other have to be so simple that they can be computed quickly and efficiently.

We come to a better understanding of natural processes by comparing them to simulations. But one can also employ these techniques to obtain natural sounding synthesis. This does not necessarily mean that the synthesized sounds mimic those of traditional instruments, but rather that they show the richness, the constant variation and above all the natural relationships between the components that characterize all real sounds.

In this chapter, we will discuss two ways to analyze or segment three-dimensional bodies. The first can be used to model any body whatsoever by resolving the object into many single points. The interactions result from the forces obtaining between neighboring points. Such models give very good results when used to simulate systems that are in fact composed of individual elements, like crystals in which a few kinds of atoms are regularly ordered. Such models are often referred to as *mass-spring models* or, when their order is linear, as *chains*. The second way to analyze a body is to segment it into linear sections. The essential oscillation of many elements important for sound production, like strings and air columns, comes from waves in or on these elements and it can be simulated using a delay line. This technique is called *wave guide synthesis* and is the oldest of the approaches described here which led to musically satisfying results. There is a third way to segment three-dimensional objects which we will not discuss in detail here, because its computational complexity makes it impractical for sound synthesis. The idea is to use a small number of basic spatial elements that can be assembled into any arbitrary form or shape. This technique is called the *finite element method* (FEM) and is used in the design of structural elements to calculate load, pressure distribution and resonance. There are programs which analyze bodies drawn in a *computer-aided design* (CAD) environment into finite elements and calculate the physical characteristics of these elements. These calculations are very complex, even for static situations. Therefore, dynamic processes, particularly oscillation, are not constantly recalculated as in other techniques. Instead, only the natural resonances are computed and then used to generate sounds corresponding to the body analyzed.

Sound synthesis using physical modeling has other advantages besides the naturalness of the sounds. It is possible to let various instruments interact with each other, and it is also possible to define instruments that would be physically unlikely or impossible in the real world. Designing a model and implementing it in a synthesis program is complicated and often time-consuming, but the programs themselves are frequently short and simple. The biggest problems and drawbacks of physical modeling synthesis are due to the simplifications that must be made at all levels. The subdivision of time into discrete points (corresponding to the sampling interval), the decomposition of continuous three-dimensional bodies into individual elements and the simplification nearly always necessary to define the interactions between elements will occupy us again and again in the chapters that follow.

Further Reading: Numerical Sound Synthesis: Finite Difference Schemes and Simulation in Musical Acoustics *by Stefan D. Bilbao [85],* Wave and Scattering Methods for Numerical Simulation *also by Stefan D. Bilbao,* Digital Sound Synthesis by Physical Modeling Using the Functional Transformation Method *by Lutz Trautman and Rudolf Rabenstein [82].*

8.1 Mass-Spring Models

In this modeling technique, oscillating bodies are decomposed into individual mass points that are connected among themselves by forces represented in the model as springs. We first consider in detail the case of a single mass. Here the task is to get an idea of how the oscillations arise, how frequency and amplitude can be calculated and how to program the model. We will see that this simple model corresponds to a digital filter, and so this chapter is also a good preparation for the chapter about digital filters (3.3), which might otherwise seem rather abstract and mathematical. Despite the model's simplicity, it has some interesting applications. Resolving the continuous oscillation into discrete steps has the advantage that the calculations that would require higher mathematics to be treated precisely are reduced to simple addition and multiplication. We will, however, use the exact computations as approximations in order to calculate the frequencies and amplitudes of the digitalized wave forms. In the second chapter we will explain how coupled masses oscillate and how the characteristics of the oscillations can be calculated. The third chapter deals with the calculation of frequency, or rather with the calculation of the ratio between the frequencies of a fundamental and its overtones. We will simulate the behavior of a string using a model with many linearly arranged masses. In the fourth chapter, we examine the disposition of coupled masses in two dimensions, in chapter five the disposition in three dimensions. We present various geometric configurations and show how to realize them in Mathematica and in C. In these chapters, we will see that the graphic representation of the simulations not only clarifies the functioning of the simulation, but also helps verify the models and programs themselves, since one often cannot judge by the generated sounds alone whether the model works as intended and whether the program is a precise implementation of it. In the final chapter, we will see ways to simulate the oscillation of bodies that are impossible in the real world. The calculations in the programs are not optimized for speed or efficiency but are left as they were derived or as they are easiest to understand.

Further Reading: Any physics textbook will discuss the behavior of springs in the chapter devoted to Mechanics. One can find examples of simple spring-mass models in mathematic texts, often as applications of systems of differential equations, but also in electronics texts as models for electronic oscillating circuits, which can sometimes be more accurately described using these models than can mechanical systems. Volume 3 of the Berkeley Physics Course, Waves, by Frank S. Crawford [6] can be especially recommended for the way it presents clearly and with few prerequisites the most important characteristics of vibration and waves. There are brief introductions in Real Sound Synthesis for Interactive Applications *by Perry R. Cook [70], in* Computer Music *by Charles Dodge and Thomas A. Jerse [2], pp. 283-286 and in* Computer Sound Synthesis for the Electronic Musician *by Eduard Reck Miranda [15], pp. 90-93. The Cordis-System is a computer application for generating sound with a spring-mass model [1-1] CMJ, Vol. 8, No. 3.*

8.1.1 Systems With One Mass

8.1.1.1 Harmonic Oscillation

The oscillation of springs, pendula or electrical circuits is based on two properties of the respective physical system that are in opposition to each other: the *restoring force* and *inertia*. The restoring force attempts to bring the system into equilibrium, while inertia causes every system uninfluenced by external forces to maintain its movement. In the systems mentioned

8.1 Mass-Spring Models

above, the restoring force is proportional to the displacement from equilibrium. The systems below show oscillation: a) a mass M connected by springs or rubber bands to two fixed points ($F1$ and $F2$), b) a mass hanging on a spring, and c) an elastic rod fixed at one end.
(→ *mass_spring_1.pde*)

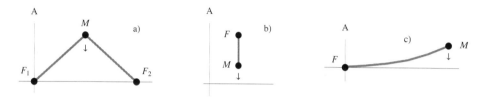

We can represent the system's oscillation over time by tracing the movement of the mass point in a diagram where the axes represent time and displacement. We imagine the oscillation as vertical motion about a constant middle position. The diagrams below show the change in displacement as a function of time. Hence the path of a point during one sampling period does not correspond to the distance between two successive points in the diagram, but is rather the difference between two successive displacements.

Let us generate oscillatory motion without going into physical details. To determine the velocity of the mass point we need two successive values for the displacement. In the figure below these values are $A(0) = a_0 = 1$ and $A(1) = a_1 = 4$. We can determine the displacement at the next moment in time from these values.

The velocity corresponds to the distance traveled per time unit, that is $a_1 - a_0$ per sampling period. If we define the sampling period as 1 time unit, the velocity is $a_1 - a_0$. Without the restoring force, the velocity would remain constant ($a_1 - a_0 = 3$), and the new value a_2 would be:

$$a_2 = a_1 + (a_1 - a_0) = 2a_1 - a_0 = 2 \cdot 4 - 1 = 7$$

The restoring force k is proportional to the mass's displacement. The factor of proportionality is called the *spring constant*. Let us call it c_s. Then we can write: $k = c_s a_1$. Using Newton's law "Force = Mass·Acceleration", we write for the acceleration $A = k/m$ or $A = a_1 c_s/m$. To simplify in what follows, we will write c for the constant c_s/m. Since the restoring force acts in the direction of the equilibrium position, we subtract $c a_1$ in our equation to account for the acceleration. For the example we choose $c = .4$ and obtain:

$$a_2 = a_1 + (a_1 - a_0) - ca_1 = (2 - c)a_1 - a_0 = (2 - .4) \cdot 4 - 1 = 5.4$$

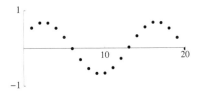

If at each sampling period we use the results of the previous sampling period a_1 and a_2 as new initial values a_0 and a_1 and perform the calculation above, we obtain a sine wave. The initial values are $a_0 = 0$, $a_1 = .3$, the factor for the restoring force is $.2$.

The frequency of the oscillation depends on the constant c; the amplitude depends on the initial values a_0 and a_1. If we set a_0 equal to zero, the displacement a_1 is the initial velocity of the mass point. This velocity results from exciting the mass with an impulse. Therefore we will often consider a_1 as an impulse. In the Csound instrument below we set $a_0 = 0$. The impulse a_1 is given in p4, the constant c in p5. Using the values $p4 = 2000$ and $p5 = .1$, we get a sine wave with a frequency of about 2000 Hz ($sr = 44{,}100$) and an amplitude of about 6000.

```
;8-1-1a
instr 1
a0      init    0               ; initial values:    a0 = 0
a1      init    p4              ;                    a1 = p4
a2 =            (2-p5)*a1-a0    ; calculate the next value, p5 = c
a0 =            a1              ; previous a1 serves as new a0
a1 =            a2              ; previous a2 serves as new a1
        out     a1
endin

i1  0   2   2000    .1
```

To understand the relationship between the constant c and the frequency depending on it (f_s = oscillations per sample period, f = oscillations per second = $f_s \cdot sr$), let us consider a continuous oscillation. The following holds for the required function $y(t)$: the acceleration, that is, the second derivative of the displacement with respect to time, is proportional to the displacement. Hence we must solve this equation:

$$y''(t) = -c \cdot y(t).$$

A particular solution is the sine function:

$$y(t) = \sin(\sqrt{c}\, t).$$

The factor $\sqrt{c}$ corresponds to the angular frequency (oscillations per sample period = f_s). Hence $\sqrt{c} = 2\pi \cdot f_s$. To determine the factor c for a given frequency, we solve the equation for

8.1 Mass-Spring Models

c and get $c = (2\pi \cdot f_s)^2$. By substituting f/sr for f_s and regrouping, we obtain $c = f^2 \cdot (2\pi/sr)^2$ or $f = \sqrt{c}\ sr/2\pi$.

Frequency f when c is given :	$f = \sqrt{c}\ sr/2\pi$
Factor c when f is given :	$c = (f \cdot 2\pi/sr)^2$

When $c = .1$ and $sr = 44{,}100$ Hz, the example above gives a frequency of

$$\sqrt{.1} \cdot 44{,}100/2\pi = 2219.518 \text{ Hz}.$$

But if we actually measure the frequency of the generated tone, we find about 2228 Hz. The discrepancy is due to the sampling process, which can only approximate the continuous function. The discrepancy is small at low frequencies, large at high frequencies. If we calculate a few values, we see that a factor of $c = 3$ gives an oscillation with a period of three sample periods (8.1.1.3). Hence the frequency generated is $sr/3 = 14{,}700$ Hz. The formula above, however, gives 12,156.8 Hz, which is a difference of nearly a minor third. Nonetheless, the approximation is fairly good for a large frequency range, and the discrepancy is barely perceptible even for quite high frequencies (e.g. for the 2220 Hz mentioned above). (→ *MassSpring_1.maxpat*)

We can approximate the amplitude A of the oscillation for the initial values $a_0 = 0$ and $a_1 = a$ as follows. At time $t = 0$ we have $y(0) = \sin(0) = 0$. The first derivative of the function $y = A \cdot \sin(f \cdot 2\pi t)$ is

$$y'(t) = A \cdot (f \cdot 2\pi) \cdot \cos(f \cdot 2\pi t).$$

At $t = 0$ we have

$$y'(0) = A \cdot f \cdot 2\pi \cdot 1.$$

Since the first derivative corresponds to the function's slope, we can approximate the slope at $t = 0$ using the initial value a:

$$y'(0) = \partial y/\partial t = a\ /\ (\text{sample periode}) = a \cdot sr.$$

Equating both expressions leads to

$$A \cdot f \cdot 2\pi = a \cdot sr.$$

From this we derive $A = a \cdot sr/(f \cdot 2\pi)$ and $a = A \cdot (f \cdot 2\pi)/sr$. After substituting the formula for the frequency above, we have $A = a/\sqrt{c}$ and $a = A\sqrt{c}$.

Amplitude A for a given initial value $a_1 = a$:	$A = a/\sqrt{c}$
a for a given amplitude A :	$a = A\sqrt{c}$

In the example above, with $a = 2000$ and $c = .1$, we calculate an amplitude of

$$2000\sqrt{.1} = 6324.4.$$

The measured amplitude is 6405.5.

Now we can give frequency (*p5*) and amplitude (*p4*) in the Csound score and let the instrument calculate the actual parameters it needs for the synthesis ($c = i5, a = i4$).

```
instr 1                 ;8-1-1b
i5 = (p5*2*3.141593/sr)*(p5*2*3.141593/sr)
i4 = p4*p5*2*3.141593/sr
a0          init        0
a1          init        i4
a2=                     (2-i5)*a1-a0
a0=                     a1
a1=                     a2
            out         a1
endin
```

See [6] pp. 1-8, [12] pp. 38-39 and 78-79, [3] pp. 501-507, [5] pp. 194-208, [16] pp. 75-88.

8.1.1.2 Exciting the Oscillation

So far we have only generated a sine wave in a pretty complicated way. But if we excite the system by weak noise rather than by an impulse, an interesting effect results, which also shows the advantages of physical modeling. We first add noise to the signal by adding a random value to the calculated value a_2. In this example the initial values a_0 and a_1 are both zero and the added random value is between $-.05$ and $.05$.

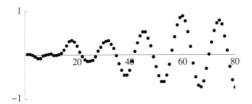

After only a few periods, we see a sine wave with increasing amplitude, and we can use the model for the simple simulation of onset transients. The sound example shows that the amplitude increases neither exponentially nor linearly, but irregularly and more slowly the greater the amplitude. This last effect is due to the decreasing proportion between the excitation and the resulting signal as the amplitude increases.
(→ Sound) (→ Why don't the excitations cancel each other out?)

8.1.1.3 Damped Harmonic Oscillation

There are various ways to damp the oscillation. We will first show two techniques, both of which work but are inadequate for different reasons.

First let us multiply each new value by a factor $d < 1$:

$$a_2 = d((2 - c)a_1 - a_0)$$

Comparing the damped signal with the undamped signal (figures below), we see that the frequency of the damped signal gets higher instead of lower (2.2.1.2). This example shows how cautious one must be when manipulating the results of a simulation. It makes a difference whether one changes the amplitude after the simulation or whether one simulates the

8.1 Mass-Spring Models

cause of the damping. By multiplying the new value by a factor we not only lessened the amplitude but also increased the restoring force.

The figure below shows the damped oscillation $a_2 = .93(1.95a_1 - a_0)$:

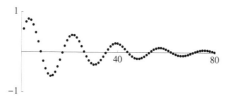

Here is the same oscillation without damping:

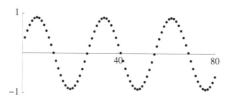

In our second attempt let us set the damping proportional to the velocity. In the following example the oscillation is only damped after 3 seconds. One hears clearly that the damping raises the pitch a bit.

$$a_2 = (2 - c)a_1 - a_0 - d(a_1 - a_0)$$

```
;8-1-1c
...
k6         linseg    0,p7,0,.01,p6,p3-p7,p6
...
a2=                  (2-i5)*a1-a0-k6*(a1-a0)
...

i2 0 5 30000 1000 .0002 3
```

The following example shows the reason for the pitch change and calls attention to another problem that arises at very high frequencies. If we set $c = 3$ in the undamped model, we get an oscillation with the period 3, that is, a frequency of $sr/3$ Hz.

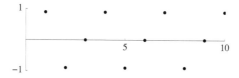

With a damping factor $d = .5$ and an initial value $a_1 = .3$ we obtain the following surprising result.

The frequency rises considerably to $sr/2$ Hz, and instead of being damped the signal is amplified to a certain limiting value.

Here is the derivation of a damping algorithm without the problems described above. We compute the velocity for each new value from the old velocity $a_0 - a_1$ and the acceleration $-c a_1$.

$$(a_1 - a_0) - c a_1 = (1 - c)a_1 - a_0$$

We set the decrease in velocity because of the damping proportional to the velocity and call the factor of proportionality d. Then we obtain for the new displacement

$$a_2 = a_1 + (1 - d) \cdot ((1 - c)a_1 - a_0)$$

In this way we get the required effects, that is, damping at all frequencies and a lowering of the frequency. Compare this with the example above. The undamped oscillation had a period of three sampling periods. With the same value $c = 3$ but with a damping factor $d = .5$ the period is now four sampling periods.

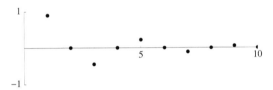

We compute the frequency f_d and the envelope $E(t)$ of the damped oscillation using the formula for continuous oscillation. Once again, we only get an approximation to the real oscillation, and the approximation gets worse as the frequency increases. The velocity corresponds to the first derivative $v = y'(t)$, and the damping is $-d \cdot v = -d \cdot y'(t)$. Hence we have to solve the following differential equation:

$$y''(t) = -c \cdot y(t) - d \cdot y'(t).$$

A particular solution is:

$$y(t) = e^{-(1/2) \, t \cdot d} \sin(\omega_d t).$$

The following holds for the angular frequency $\omega_d = 2\pi f_d$ of the damped oscillation:

$$\omega_d^2 = c - \frac{1}{4} d^2.$$

From this follows for the frequency f_d of the damped oscillation:

$$f_d = \frac{1}{2\pi} \sqrt{c - d^2/4}.$$

The envelope $E(t)$ is described by the factor $e^{-(1/2) \, t \cdot d}$.

Frequency of the damped oscillation :	$f_d = \frac{1}{2\pi} \sqrt{c - d^2/4}$
Envelope :	$E(t) = e^{-(1/2) \, t \cdot d}$

8.1 Mass-Spring Models

Using the values of the example above, we get

$$f_d = \frac{1}{2\pi}\sqrt{3 - (.5)^2/4} = 0.2728.$$

The figure above shows that the damped oscillation has a frequency of .25 Hz/sr. As expected, the approximation is not particularly good because the frequency is very high.

See [3] pp. 508-519, [6] p. 3.

8.1.1.4 Exciting the Damped Oscillation

To investigate the effect of excitation using our damping model, we add a random number to each newly computed value.

$$a_2 = a_1 + (1 - d)((1 - c)a_1 - a_0) + \text{rand}()$$

If we set $c = .06$ and $d = .0004$, we get the oscillation shown below.

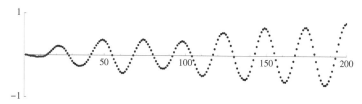

After a short rise time, the oscillation reaches a state where damping and excitation counterbalance each other. If we consider this model a digital system that transforms noise into a nearly sinusoidal signal, we see that it corresponds to a bandpass filter. The filter's center frequency is the frequency we computed above, but it is not straightforward to find an equivalence for the filter's bandwidth. Here we will only demonstrate that the bandwidth depends on the damping coefficient. In the following Csound example, the damping coefficient decreases exponentially. To balance the amplitude, we let the excitation diminish exponentially as well.

```
;8-1-1d
instr 1
i8=                 (p8*2*3.141593/sr)*(p8*2*3.141593/44100)
k6      expon       p6,p3,p7             ; Envelope for the noise signal
k4      expon       p4,p3,p5             ; Envelope for the damping coefficient
k1      linseg      1,p3-.3,1,.3,0       ; Fade out
aexc    rand        k6
a0      init        0
a1      init        0
a2=                 a1+(1-k4)*((1-i8)*a1-a0)+aexc
a0=                 a1
a1=                 a2
        out         k1*a1
endin
```

	Damping coefficient		Excitation		Frequency
;	Beginning	End	Beginning	End	
i1 0 15	.5	.0001	3000	50	2000

The resulting sound begins as broadband noise and becomes a high-pitched whistle with clearly recognizable pitch (Sound Example 8-1-1d). (→ Spectrum)

Now we excite the system with a sine wave. In the first figure below the frequency v of the excitation is close to the system's resonance frequency. In the second figure, the frequency is somewhat higher, in the third somewhat lower.

$$a_2 = a_1 + (1 - d)((1 - c)a_1 - a_0) + \sin(2\pi vt)$$

$v = 1/25$

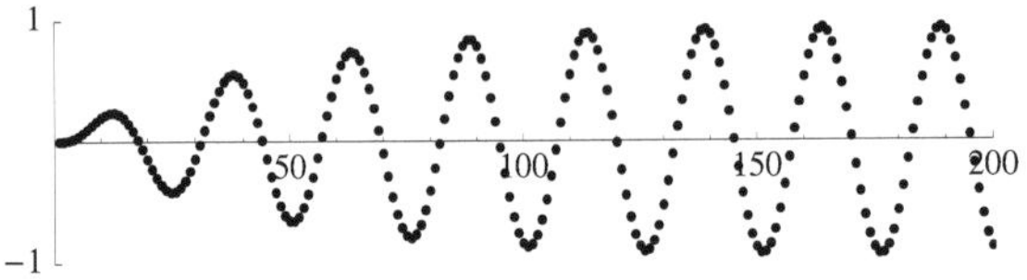

$v = 1/18$

$v = 1/38$

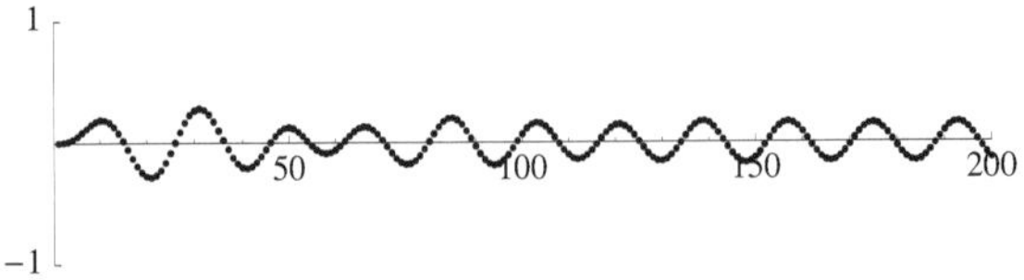

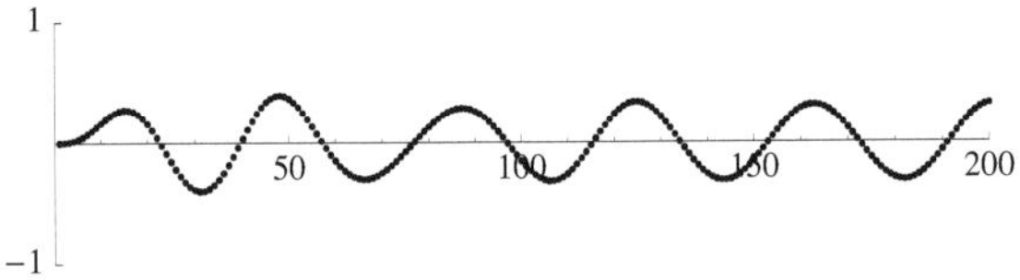

The frequency of the resulting signal always corresponds to that of the excitation, and the signal's amplitude is greater the closer the exciting frequency is to the system's resonant frequency. (→ *MassSpring_1.maxpat*) (→ *8-1-1e.csd*) (→ Excitation by Changing the Parameters c and d)

We can simulate the behavior of a bowed string using a model with only one mass. The simulation is particularly difficult because of the interaction between the bow and the moving string. The essential parameters of bowing are velocity and pressure. We can distinguish two forms of interaction. On the one hand, there are time intervals when the string sticks to the bow hairs so that the bow displaces the string. On the other hand, there are time intervals when the bow glides over the string and has only little effect on it. We need to determine the conditions causing the string to stick to the bow hairs and the conditions causing it to become unstuck. In order for the string to begin to stick to the bow, the instantaneous velocity of the string and that of the bow have to be in the same direction and roughly similar. Whether the string sticks to the bow when direction and velocity are different depends primarily on the bow pressure, which we will call P_1 in what follows. The adhesion ends as soon as the restoring force pulling the string back to the resting position becomes greater than the sticking friction, which is also dependent on the bow pressure and which we will call P_2.

In the animation below we use these values: $c = .3, d = .07, P_1 = .18, P_2 = .7$. The velocity v is initially 0 and increases to .36. (If it were .36 from the beginning, no adhesion could take place.) These conditions are continuously checked, and, depending on the results of the check,

8.1 Mass-Spring Models

the point is either moved on by v or its position is newly computed as for the freely vibrating string. The bowing stops at time $t = 90$, and the string comes to rest. The phases where the string sticks to the bow are indicated by a line showing the direction of the bow movement. (→ Animation)

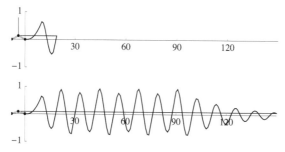

(→ 8-1-1f.csd) (→ 8-1-1g.csd: Simulation of a Bouncing Ball On a Moving Surface)

See [6] pp. 62-93, [3] pp. 520-529, [16] pp. 88-103, [4] pp. 15-19.

8.1.1.5 Oscillation With Nonlinear Acceleration

Even in the simplest systems, the restoring force is not really proportional to the displacement (8.1.5.1). Let us first introduce a term for the calculation of an additional restoring force that is proportional to the cube of the displacement. Then we will show how to use any arbitrary function for the calculation, and finally we will consider how to handle variable functions.

$$a_2 = 2a_1 - ca_1 - c_3 a_1^3 - a_0$$

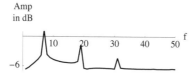

As the illustration shows, we again get a periodic oscillation. Looking closely at the waveform, we see that the waveform rises and falls more steeply and that the curvature at the turning points is narrower than in a sine wave. The spectrum below of the waveform shows a fundamental with odd partials.

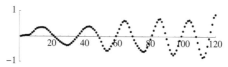

If the oscillation is excited by low-amplitude noise, we see that the frequency of the oscillation is no longer independent of the amplitude but rather increases with it ($c = .01$ and $c_3 = .3$). (→ MassSpring_cub.maxpat) (→ 8-1-1h.csd)

The restoring force can also be an arbitrary function of the displacement. In order to keep the calculations to a minimum and to have a clear representation of the restoring force, we will not compute new values at every step but rather use a table. If the springs in system a) of Chapter 8.1.1.1 are longer than the distance of the masses to the fixed points, then they will be compressed, and in the vicinity of the center position the system will have an acceleration directed away from the center position. The function for the restoring force $F(A)$ in this case can be described approximately as follows. There is an unstable equilibrium in the center position. At the slightest displacement, the force directed away from the center position increases very quickly, then quickly falls to zero at the two rest positions and finally takes on a behavior corresponding to that of a simple spring. We use the function

$$F(A) = -c_1 \text{sign}(A) + c_2 A.$$

If we set $c_1 = -.2$ and $c_2 = 1$, we find the rest positions, that is, those positions where the restoring force is zero, when the displacements are $A_1 = .2$ and $A_2 = -.2$.

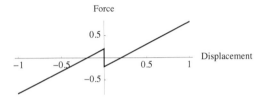

In the following example, we set these initial values: $a_0 = a_1 = .2$, and we excite the oscillation with noise. When the displacement is small, the mass oscillates about one of the rest positions, when the displacement is larger, the mass can swing through the "barrier" at the zero position. (→ 8-1-1i.csd)

See [4], [14] pp. 101-136.

8.1.1.6 Calculations

Let us write $y(k)$ for the displacement at time k. From the equation $a_2 = (2 - c)a_1 - a_0$ (8.1.1.1) we obtain

$$y(k) = (2 - c)y(k - 1) - y(k - 2).$$

We originally used a value great than zero for $y(k - 1)$ (8.1.1.1). In chapter 8.1.1.2 we added noise. Both of these operations can be thought of as adding an input signal $x(k)$. In the first case the input is an impulse, in the second a random sequence. Hence the equation for the undamped oscillation is

$$y(k) = (2 - c)y(k - 1) - y(k - 2) + x(k)$$

and its transfer function is

8.1 Mass-Spring Models

$$H(z) = \frac{1}{1 - (2-c)z^{-1} + z^{-2}}.$$

The poles correspond to the zeros of the denominator. From $1 - (2-c)z^{-1} + z^{-2} = 0$ we get the solutions

$$z_1 = \frac{1}{2}\left(2 - \sqrt{-4+c}\ \sqrt{c} - c\right) \text{ and } z_2 = \frac{1}{2}\left(2 + \sqrt{-4+c}\ \sqrt{c} - c\right).$$

Since the poles are on the unit circle, the amplitude response at the system's resonant frequency is infinite ($\text{Abs}(z_i) = 1$). The resonant frequency (normalized angular frequency Ω) corresponds to the argument of the pole $\Omega = \text{Arg}(z)$. When $c = .1$ we get the value

$$\Omega = \text{Arg}\left(\frac{1}{2}\left(2 + \sqrt{-4+.1}\ \sqrt{.1} - .1\right)\right) = .31756$$

and from this the frequency $f = \Omega sr/2\pi = 2228.87$ (cf. 8.1.1.1).

```
Solve[1 - (2 - c) z⁻¹ + z⁻² == 0, z]
```

$$\left\{\left\{z \to \frac{1}{2}\left(2 - \sqrt{-4+c}\ \sqrt{c} - c\right)\right\}, \left\{z \to \frac{1}{2}\left(2 + \sqrt{-4+c}\ \sqrt{c} - c\right)\right\}\right\}$$

```
c = .1; Arg[1/2 (2 + √(-4 + c) √c - c)]
```
0.31756

.3175604 * 44 100 / (2 * π)   2228.87

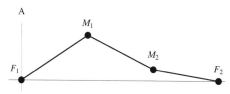

For the derivation of the constant c and for damped oscillation in general, see 3.3.3.5.

8.1.2 Systems With Two Masses

Using a model with two masses, we will show in this chapter how coupled masses affect spectrum, frequency and amplitude of a synthesized signal.

8.1.2.1 The Oscillation of Two Coupled Masses

Let us consider a model having two masses connected by springs to each other and to lateral fixed points. To begin with, let the spring constant be the same for each segment of the spring.

The acceleration affecting the mass M_1 has a component ca_1 from the spring between F_1 and M_1 and a component $c(a_1 - a_2)$ from the spring between M_1 and M_2. We obtain the equation of motion by calculating from the two given values a_{10} and a_{11} the next value a_{12}. We get the equation for the second mass by exchanging indices because of the symmetry of the model.

$$a_{12} = 2a_{11} - c(2a_{11} - a_{21}) - a_{10}$$
$$a_{22} = 2a_{21} - c(2a_{21} - a_{11}) - a_{20}.$$

The next illustration show to the left the instantaneous position of the masses M_1 and M_2 and to the right the displacement of the mass M_2 as a function of time ($t = 0, 4, 8$ and $t = 50$). (→ Animation)

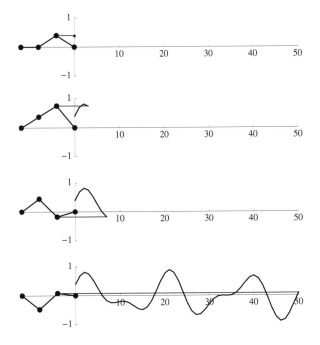

The function is not periodic but consists of two non-harmonic sine waves. In this simple case, it is easy to generate and compute the sine waves, which are the so-called natural oscillations of the function above. If we excite the oscillation by giving both masses the same impulse, they will always oscillate in synchrony, and the period of oscillation will be that of a single mass. The acceleration of M_1 is $-ca_1$, since only the first spring exerts a restoring force. From this we can compute the frequency (8.1.1.1):

$$f_1 = \sqrt{c} \cdot sr/2\pi.$$

8.1 Mass-Spring Models

For the second natural resonance, that is, the second possible symmetrical oscillation, the masses receive the same impulse simultaneously, but in opposite directions. The acceleration of the mass M_1 is $-3a_1$, the frequency

$$f_2 = \sqrt{3c} \cdot sr/2\pi.$$

Because the ratio of the two frequencies $f_2:f_1 = \sqrt{3}:1$ is irrational, the composite oscillation is aperiodic. (→ Correction) (→ 8-1-2a.csd)

8.1.2.2 Excitation and Damping

We saw in Chapter 8.1.2.1 that the natural resonances of the system can be generated individually by appropriate excitations. Let us now calculate the amplitudes of the natural resonances resulting from exciting the mass M_2 with the impulse a. To calculate the effect of this impulse on both resonances, we use a little trick. First we write the impulse a affecting M_2 as $2(a/2)$. Then we imagine that a simultaneous impulse of magnitude $a/2$ affects M_1 in both directions. The partial impulses in the same direction excite the first natural resonance (below left), those in opposite direction the second (below right).

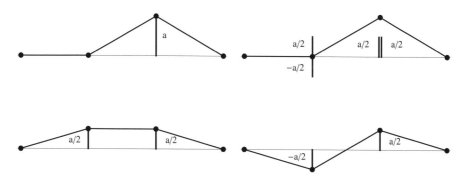

From the formula $A = a/\sqrt{c}$ (8.1.1.1) we get for the first natural resonance

$$A_1 = (a/2)/\sqrt{c_1}$$

and for the second

$$A_2 = (a/2)/\sqrt{c_1 + 2c_2}.$$

Using the parameter values of the example above ($a = .4$, $c_1 = .1$ and $c_2 = .15$), we have $A_1 = .632455$ and $A_2 = .31623$, and the ratio of the two amplitudes is $A_1:A_2 = 2$. If we excite the mass M_1 with the impulse a_1 and the mass M_2 with the impulse a_2, the terms are added together for A_1:

$$A_1 = (a_1/2 + a_2/2)/\sqrt{c_1} = (a_1 + a_2)/(2\sqrt{c_1}).$$

For A_2 the term a_2 is subtracted:

$$A_2 = (a_1/2 - a_2/2)/\sqrt{c_1 + 2c_2} = (a_1 - a_2)/\left(2\sqrt{c_1 + 2c_2}\right).$$

Conversely, from the equations $a_1 + a_2 = 2A_1\sqrt{c_1}$ and $a_1 - a_2 = 2A_2\sqrt{c_1 + 2c_2}$ we can calculate the formulae for a_1 and a_2 when A_1 and A_2 are known:

$$a_1 = A_1\sqrt{c_1} + A_2\sqrt{c_1 + 2c_2}$$
$$a_2 = A_1\sqrt{c_1} - A_2\sqrt{c_1 + 2c_2}.$$

There are two ways to introduce damping into this model. If one damps the oscillation of both masses, the frequencies will be lowered as in real systems, and, since the damping is proportional to the velocity, the lower resonance will be damped more strongly. If one damps the oscillation of only one mass, the frequency will also be lowered, but the lower resonance will be damped more strongly, as we can see in the case of maximal damping. If one fixes the first mass and excites the second with an impulse, only the second produces the second resonance while the first resonance is completely suppressed. The frequency of the first resonance is zero, that of the second only

$$f_2 = \sqrt{c_1 + c_2}\, sr/2\pi \text{ instead of } f_2 = \sqrt{c_1 + 2c_2}\, sr/2\pi.$$

8.1.2.3 Nonlinear Acceleration

If the restoring force is not proportional to the displacement, the same effects appear as with a single oscillating mass. But are both resonances equally affected by the frequency change? In the following Csound example, we generate a function consisting of a linear part and a cubic part (*f1*). This function is used in Instrument 1 to compute the restoring force from the displacements *a11*, *a11* – *a21* for M_1 and *a21*, *a21* – *a11* for M_2. The values of parameters *p5* (c_1) and *p6* (c_2) were chosen so that the frequency ratio of the natural resonances would be 2:1, neglecting the cubic factor. One can clearly hear that this ratio changes with large amplitudes because then the cubic factor becomes more important.

```
;8-1-2b
instr 1
ihtbl=            32768/2              ;Half the length of function f1
k1       linseg   1,p3-.5,1,.5,0
a10      init     0
a11      init     0
a20      init     0
a21      init     0
arnd     rand     p4
at11     table    a11/4,1,0,ihtbl,0
at12     table    (a11-a21)/4,1,0,ihtbl,0
at21     table    a21/4,1,0,ihtbl,0
at22     table    (a21-a11)/4,1,0,ihtbl,0
a12=              2*a11-p5*at11-p6*at12-a10+arnd
```

8.1 Mass-Spring Models

```
       a22=             2*a21-p5*at21-p6*at22-a20
       a10=             a11
       a11=             a12
       a20=             a21
       a21=             a22
              out       a11*k1
    endin

    ;              gen routine     range       x       x^2      x^3
    f1 0 32768     -3              -1   1  0   6000    0        20000
    ;                    Amp. Exc. c1       c2
    i1  0   15           6.7       .1       .15
```

8.1.2.4 Computing the Frequencies of the Natural Resonances

To compute the frequencies of the natural resonances in models with several masses, we need to solve systems of coupled linear differential equations. We will demonstrate the technique in a system with two differential equations and then generalize the principle.

We extract from the equations

$$a_{12} = 2a_{11} - c(2a_{11} - a_{21}) - a_{10}$$
$$a_{22} = 2a_{21} - c(2a_{21} - a_{11}) - a_{20}$$

the terms responsible for the acceleration $-c(2a_{11} - a_{21})$ and $-c(2a_{21} - a_{11})$ and put them into the equations for the required functions $y_1(t)$ and $y_2(t)$:

$$y_1''(t) = -c(2\,y_1(t) - y_2(t))$$
$$y_2''(t) = -c(2y_2(t) - y_1(t)).$$

(→ Direct Solution in Mathematica)

In what follows we demonstrate two methods to solve the system of equations.

The first consists first of removing a function from the equations and then plugging a trial solution into the resulting fourth-order differential equation to convert the differential equation into an algebraic equation. We will simplify $y_1''(t)$ to y_1'' and $y_2''(t)$ to y_2''.

Solving the second equation for y_1 gives

$$y_1 = \frac{1}{c}y_2'' + 2y_2.$$

The derivatives with respect to time are

$$y_1' = \frac{1}{c}y_2''' + 2y_2' \quad \text{and} \quad y_1'' = \frac{1}{c}y_2'''' + 2y_2''.$$

Putting the derivatives into the first equation we obtain

$$\frac{1}{c}y_2'''' + 2y_2'' = -2c(\frac{1}{c}y_2'' + 2y_2) + cy_2.$$

Regrouping and multiplying by c gives

$$y_2'''' + 4cy_2'' + 3c^2 y_2 = 0.$$

As a method of solution we choose an exponential equation $y_2 = e^{\lambda t}$. Its derivatives with respect to time t are

$$y_2' = \lambda e^{\lambda t}, y_2'' = \lambda^2 e^{\lambda t}, y_2''' = \lambda^3 e^{\lambda t} \text{ and } y_2'''' = \lambda^4 e^{\lambda t}.$$

Putting these expressions into the differential equation gives

$$\lambda^4 e^{\lambda t} + 4c\lambda^2 e^{\lambda t} + 3c^2 e^{\lambda t} = 0.$$

After dividing by $e^{\lambda t}$ we get the fourth-power equation $\lambda^4 + 4c\lambda^2 + 3c^2 = 0$ whose left side we can decompose into two factors $(\lambda^2 + 3c)(\lambda^2 + c) = 0$. From this we first derive $\lambda_{1,2}^2 = -3c$ and $\lambda_{3,4}^2 = -c$ and then in a second step

$$\lambda_{1,2} = \pm i\sqrt{3c} \text{ and } \lambda_{3,4} = \pm i\sqrt{c}.$$

Hence the required exponential functions are

$$\exp\left(i\sqrt{3c}\ t\right), \exp\left(i\sqrt{c}\ t\right), \exp\left(-i\sqrt{3c}\ t\right) \text{ and } \exp\left(-i\sqrt{c}\ t\right).$$

Since the general solution of a linear differential equation can be written as a linear combination of the individual solutions, we write:

$$y(t) = C_1 \exp\left(i\sqrt{3c}\ t\right) + C_2 \exp\left(i\sqrt{c}\ t\right) + C_3 \exp\left(-i\sqrt{3c}\ t\right) + C_4 \exp\left(-i\sqrt{c}\ t\right).$$

Alternatively, we can write, as above:

$$y(t) = A_1 \sin\left(\sqrt{c}\ t + \varphi_1\right) + A_2 \sin\left(\sqrt{3c}\ t + \varphi_2\right).$$

Since this method leads to an algebraic equation, and since only algebraic equations of first to fourth order can be solved precisely, systems with more than two differential equations can generally only be solved numerically.

The second method consists in reducing a system with two differential equations of second degree to a system of four differential equations of first degree. In this way we get a system of four equations that can be represented by a matrix of coefficients. Since we are interested in the frequencies of the natural resonances in what follows, we only need to compute the eigenvalues of this matrix. Let us first introduce two new functions $z_1 = y_1'$ and $z_2 = y_2'$. The derivatives of these functions are $z_1' = y_1''$ and $z_2' = y_2''$. The four equations of first order are

$$\begin{aligned} y_1' &= & 1 \cdot z_1 & \\ y_2' &= & & 1 \cdot z_2 \\ z_1' &= -2cy_1' & + cy_2' & \\ z_2' &= cy_1' & - 2cy_2' & \end{aligned}$$

Using matrices we write (A3.4)

$$\begin{pmatrix} y_1' \\ y_2' \\ z_1' \\ z_2' \end{pmatrix} = \begin{pmatrix} 0 & 0 & 1 & 0 \\ 0 & 0 & 0 & 1 \\ -2c & c & 0 & 0 \\ c & -2c & 0 & 0 \end{pmatrix} \begin{pmatrix} y_1 \\ y_2 \\ z_1 \\ z_2 \end{pmatrix}$$

8.1 Mass-Spring Models

or in shortened form

$$y' = A\,y$$

As an approach to the solution we choose $y_1 = K_1 e^{\lambda t}$, $y_2 = K_2 e^{\lambda t}$ etc. The corresponding derivatives are $y_1' = \lambda K_1 e^{\lambda t}$, $y_2' = \lambda K_2 e^{\lambda t}$ etc. Putting these values into the system of differential equations and dividing by $e^{\lambda t}$ gives

$$\begin{aligned}\lambda K_1 &= & 1\cdot K_3 & \\ \lambda K_2 &= & & 1\cdot K_4 \\ \lambda K_3 &= -2cK_1' + cK_2' & & \\ \lambda K_4 &= cK_1' - 2cK_2' & & \end{aligned}$$

To solve the equation system, we have to determine the eigenvalues λ of the coefficient matrix A (see A3.4).

```
       ( 0    0   1 0 )
       ( 0    0   0 1 )
A =    (-2 c  c   0 0 )  ; Eigenvalues[A]
       ( c   -2 c 0 0 )
```

$$\left\{-I\sqrt{c},\; I\sqrt{c},\; -I\sqrt{3}\sqrt{c},\; I\sqrt{3}\sqrt{c}\right\}$$

As a general solution, we once again obtain

$$y(t) = A_1 \sin\left(\sqrt{c}\, t + \varphi_1\right) + A_2 \sin\left(\sqrt{3c}\, t + \varphi_2\right)$$

8.1.2.5 Calculations (→ CD)

8.1.3 The Linear Arrangement of Coupled Masses

If we include further masses in our model, we get a "chain of pearls", that is, a certain number of masses on an elastic cord. We will see that we can simulate the behavior of a musical string using this model. We will examine the spectrum of the oscillations produced by the model, and we will simulate two important techniques of sound production for string instruments, bowing the string and generating harmonics.

8.1.3.1 A Model With Three Masses

To begin with, let us apply the methods of the last chapter to a model with three masses. We treat the masses M_1 and M_3 as we did in the last chapter. If the same force is applied to equal masses, then the acceleration is $-2ca_{11} + ca_{21}$ for M_1 and $-2ca_{31} + ca_{21}$ for M_3. Hence the equations of motion are

$$\begin{aligned}a_{12} &= 2a_{11} - 2ca_{11} + ca_{21} - a_{10} \\ a_{32} &= 2a_{31} - 2ca_{31} + ca_{21} - a_{30}\end{aligned}$$

The accelerations affecting M_2 are $-c(a_{21} - a_{11})$ and $-c(a_{21} - a_{31})$, and the resulting equation of motion for the middle mass is

$$a_{22} = 2a_{21} - 2ca_{21} + ca_{11} + ca_{31} - a_{20}$$

We put these coefficients into a 6×6 matrix and let Mathematica calculate the eigenvalues.

$$A = \begin{pmatrix} 0 & 0 & 0 & 1 & 0 & 0 \\ 0 & 0 & 0 & 0 & 1 & 0 \\ 0 & 0 & 0 & 0 & 0 & 1 \\ -2*c & c & 0 & 0 & 0 & 0 \\ c & -2*c & c & 0 & 0 & 0 \\ 0 & c & -2*c & 0 & 0 & 0 \end{pmatrix};$$

Eigenvalues[A]

$$\left\{ -I\sqrt{2}\sqrt{c},\ I\sqrt{2}\sqrt{c},\ -\sqrt{-2c-\sqrt{2}\ c},\ \sqrt{-2c-\sqrt{2}\ c},\ -\sqrt{-2c+\sqrt{2}\ c},\ \sqrt{-2c+\sqrt{2}\ c} \right\}$$

The function N[%] computes the numerical values of the list above.

N[%]

$$\left\{ -1.414\ I\sqrt{c},\ 1.414\ I\sqrt{c},\ -1.848\sqrt{-c},\ 1.848\sqrt{-c},\ -0.765\sqrt{-c},\ 0.765\sqrt{-c} \right\}$$

The frequency ratios of the natural resonances are approximately 0.765:1.414:1.848 or with the smallest value as one, 1:1.848:2.414. The table below lists the frequency ratios of the natural resonances for models with from one to five masses.

Masses	Resonance 1	2	3	4	5
1	1				
2	1	1.732			
3	1	1.848	2.414		
4	1	1.902	2.618	3.078	
5	1	1.932	2.733	3.346	3.732

As the number of masses increases, the ratio between the lower frequencies approaches the harmonic ratios 1:2:3 ... etc.

8.1.3.2 The String

Let us consider a "chain of pearls" with arbitrarily many masses and see if we can simulate the behavior of a vibrating string with it. Since all the elements of our model are the same, there is a way to calculate the frequencies of the natural resonances without having recourse to the matrix and its eigenvalues. The chain of pearls should have the same form as a vibrating string when it consists of infinitely many points. Then the model's natural resonances will correspond to those of the string. Let us consider the middle point P_0 of a chain with an odd

8.1 Mass-Spring Models

number (n) of masses for the simplest resonance. The neighboring points are P_1 and P_{-1}, and the displacements y_0, y_{-1} and y_1 of the points are functions of time. Let the length of the string be π.

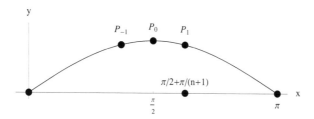

Since at point P_0 it holds that $y_{-1} = y_1$, we have:

$$y_0'' = -c(2y_0 - 2y_1) = -2c(y_0 - y_1).$$

The x-coordinate of P_1 is $\pi/2 + \pi/(n + 1)$, and so at every instant it holds that

$$y_1/y_0 = \sin\left(\frac{\pi}{2} + \frac{\pi}{n+1}\right)/\sin\left(\frac{\pi}{2}\right) = \cos\left(\frac{\pi}{n+1}\right).$$

Hence, $y_1 = y_0 \cos\left(\frac{\pi}{n+1}\right)$. Putting this value into the equation above, we get:

$$y_0'' = -2c\left(1 - \cos\left(\frac{\pi}{n+1}\right)\right)y_0$$

Since $\cos(\alpha) = \cos^2(\alpha/2) - \sin^2(\alpha/2)$, we can also write:

$$y_0'' = -4c\cdot\sin^2\left(\frac{\pi}{2(n+1)}\right)y_0.$$

The solution of this equation is a sine function with the angular frequency

$$\omega = \sqrt{4c\cdot\sin^2\left(\frac{\pi}{2(n+1)}\right)} = 2\sqrt{c}\,\sin\left(\frac{\pi}{2(n+1)}\right).$$

Because $\omega = 2\pi f$, we have for the frequency

$$f = \sqrt{c}\,\sin\left(\frac{\pi}{2(n+1)}\right)/\pi,$$

where f is again the number of oscillations per sampling period. If the sampling rate is sr, then f becomes

$$f_{sr} = sr \cdot f = sr\sqrt{c}\,\sin\left(\frac{\pi}{2(n+1)}\right)/\pi.$$

If we set $n = 2, 3$ and 4, we obtain, even for small and for even values of n, the same results as by calculating the eigenvalues of the corresponding coefficient matrix. To calculate the mth resonance, we imagine that the figure above only shows the mth part of the string and hence that the distance from P_0 to P_1 is m times greater. In that way, we obtain:

$$\omega_m = 2\sqrt{c}\,\sin\left(\frac{\pi \cdot m}{2(n+1)}\right).$$

When the sampling rate is sr, we have for the frequency

$$f_{sr_m} = \frac{sr}{\pi} \sqrt{c} \sin\left(\frac{\pi \cdot m}{2(n+1)}\right).$$

The frequencies of the natural resonances of a pearl chain model with n masses and a sampling rate of sr can be calculated as follows:

Fundamental Frequency:	$f_{sr} = \frac{sr}{\pi} \sqrt{c} \sin\left(\frac{\pi}{2(n+1)}\right)$
Frequency of the mth Partial:	$f_{sr_m} = \frac{sr}{\pi} \sqrt{c} \sin\left(\frac{m\pi}{2(n+1)}\right)$

Conversely, we obtain the constant c for a given frequency from the equation

$$c = \left(\frac{f \cdot \pi}{sr \cdot \sin\left(\frac{m \cdot \pi}{2(n+1)}\right)}\right)^2$$

The relationship between the frequency and the wavenumber is called the *dispersion relation*. The following figure shows the dispersion relation for a chain of 20 masses ($n = 20$) and a constant $c = 1$. The sampling rate is 44,100 Hz.

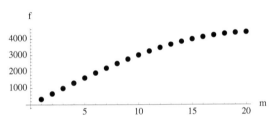

The relation is approximately linear only in its first third. This means that only the frequencies of the lower partials are approximately integer multiples of the fundamental frequency. Above these partials, the increase in the frequencies of the partials diminishes.

The sound in the example below is computed with a model using 19 oscillating masses, the constant $c = .51$ and a damping factor of .001. Initially, the displacement of the masses is random. Because of the non-harmonic spectrum, the example sounds more like a bell than like a plucked string. (→ Sound Example)

See [6] pp. 29-61, [20] pp. 192-212, [13] pp. 65-69.

8.1.3.3 Correcting the Dispersion Relation

Since the computed frequencies are only good approximations when c is small, let us investigate how the dispersion relation looks when c is large. Values for c larger than 1 can lead to an exponential increase in amplitude. The figures below show spectra of simulated string oscillation. The simulation used 8 masses and values for the constant c of .1, 1 and 1.03 respectively. Oscillation was excited by noise.

8.1 Mass-Spring Models

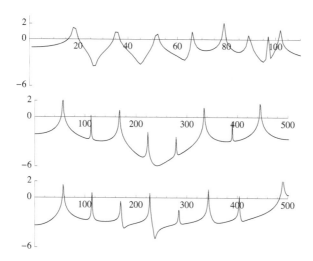

When $c = 1$, the ratio of the frequencies is harmonic, that is, we have strictly periodic motion of the individual points and hence a periodic oscillation of the string. (→ Sound Example)

The following illustration shows images taken from three animations (left $c = 1$, center $c = .3$, right $c = 1.007$).

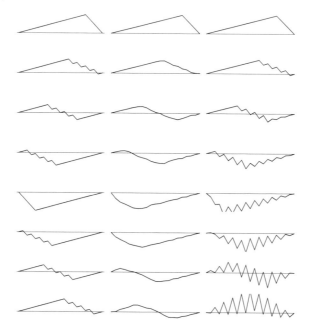

To calculate the frequency μ of the dispersion-free string ($c = 1$), one can count the number of computation steps for one period. If the "string" consists of only one point, the factor for the acceleration is 2 and there are four computation steps. If there are 2 points, one period takes six computation steps, and in general for n points $2(n + 1)$ computation steps are required for one period. Hence the frequency is $sr/(2(n + 1))$.

| Fundamental Frequency of the Dispersion–Free String ($c = 1$) | $\mu = \dfrac{sr}{2(n+1)}$ |

The following table shows the exact frequencies μ and the frequencies f calculated from the approximation formula in Chapter 8.1.3.2. For $c = 1$ the formula is:

$$f = \frac{sr}{\pi}\sqrt{1}\,\sin\!\left(\frac{\pi}{2(n+1)}\right) = \frac{sr}{\pi}\sin\!\left(\frac{\pi}{2(n+1)}\right).$$

n	1	2	3	4	10	20	30
μ	11025	7350.15	5512.5	4410	2004.35	1050.02	711.33
f	9926.9	7018.96	5371.82	4337.68	1997.73	1049.14	710.89

Using the Max patch *MassSpring_String.maxpat*, one can experiment with the parameters of this simulation. User-controlled variables are the number of masses *nmax*, the damping *d*, the position of the sound pickup on the string and the position of the excitation on the string. The code below shows the declaration of the variables and the perform routine used by the mxj-external *cm_massSpring_string*. With the variables $c = 1$, $d = 0$ and $nmax = 80$, we get a sound much like a plucked string. The spectrum is exactly harmonic, but in contrast to a real string the sound remains rich in partials to the very end. For a sampling rate of 44,100 Hz and 78 masses the frequency is $44{,}100 / (2(78 + 1)) = 279.11$ Hz.
($\rightarrow$ *MassSpring_String.maxpat*, preset 2)

```
. . . . .
private int nmax = 100, n = 10, pickup = 1, ex = n-1;
private float c = 0.1f, d = 0.0001f;
private float[] a0 = new float[nmax], a1 = new float[nmax],
                a2 = new    float[nmax];
. . . . .
public void perform(MSPSignal[] ins, MSPSignal[] outs)
{    for(int i = 0; i < in.length;i++)
     {
         a1[ex] += in[i];
         for(int ii = 1; ii < n-1; ii++)
         { a2[ii] = a1[ii] + (a1[ii] - a0[ii] - c*(2*a1[ii]
                   - (a1[ii-1] + a1[ii+1])))*(1 - d);
         }
         for(int ii = 1; ii < n - 1; ii++)
         { a0[ii] = a1[ii]; a1[ii] = a2[ii];
         }
         out0[i] = a2[pickup];
     }
}
```

8.1.3.4 Damping and Nonlinearity

One can damp the high partials more strongly than the low ones by introducing a low-pass filter into the simulation. A simple digital low-pass filter can be made by averaging two successive displacements. We can replace the damping in the example above by substituting

$$a_0[ii] = df \cdot a_1[ii] + (1 - df)a_0[ii] \text{ for } a_0[ii] = a_1[ii] \text{ and}$$
$$a_1[ii] = df \cdot a_2[ii] + (1 - df) \cdot a_1[ii] \text{ for } a_1[ii] = a_2[ii]$$

8.1 Mass-Spring Models

when preparing the variables for the next computation step. Even when c is less than 1, the high non-harmonic partials can be quickly filtered out in this way.

```
private float df = 0.f;
.....
for(int ii = 1; ii < n - 1; ii++)
{   a0[ii]=(1-df)*a1[ii]+df*a0[ii];
    a1[ii]=(1-df)*a2[ii]+df*a1[ii];}
.....
```

The following lines from the mxj-external *cm_massSpring_stringCub* show the implementation of a cubic component for the restoring force (8.1.1.5).
(→ *MassSpring_StringCub.maxpat*).

```
if (getInlet() == 2) cub = f;
.....
  displ = 2*a1[ii] - (a1[ii-1] + a1[ii+1]);
  a2[ii] = a1[ii]+(a1[ii]-a0[ii]-c*displ-cub*displ*displ*displ)*(1-d);
.....
```

8.1.3.5 Picking Up the Sound

Let us find out how the spectra differ when we "record" the oscillation of various masses as sound. The illustration below shows the spectra resulting from picking up the sound at four different masses. The sound is generated by a model using 21 masses, of which the first and the last are fixed points. Of the remaining 19 masses, M_2 is at one end of the string, M_{11} is in the middle, M_6 is at 1/4 of the length and M_5 at 1/5 of the length of the string. The string is excited by random pulses on all the masses, so that a spectrum results containing all partials.
(→ *MassSpring_String.maxpat*)

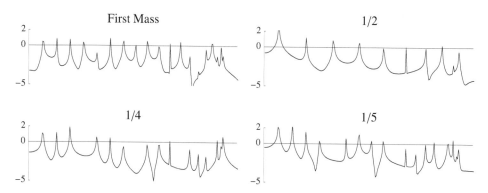

One can see that when we pick up the sound at the end mass, all partials are present. When we take the sound at the middle mass, there are no even partials. When we take the sound at 1/4 of the length, the 4th, 8th and 12th partials are missing, and when we take the sound at 1/5 of the length, the 5th, 10th and 15th partials are missing. In order to understand this behavior, let us consider the movement of the string not as a traveling wave but rather as the superposition of the string's natural resonances. The illustration below shows the nodes of some of the partial waves at the masses M_5, M_6 and M_{11}. One can also see that although the amplitudes of the partials are equal, the ratios of the amplitudes vary from mass to mass. All of the partials are present at the end masses, and their amplitudes increase with the partials' frequencies. In what follows we will always take the sound from the end mass.

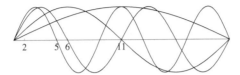

8.1.3.6 Exciting the String

Now let us see how to simulate the usual techniques of exciting the string, taking note of the role of the place of excitation in determining the spectrum of the sound.

It is easy to simulate plucking a string if the string is at rest and the displacement begins slowly. We then have the same situation as in Chapter 8.1.3.2. It is just as easy to simulate striking a string at rest hard at a precise point. The blow can be simulated by a single impulse, and the place where the string is struck or plucked has the same effect on the sound as the place where the sound is picked up. So for instance an impulse in the middle of the string excites only odd partials. The situation is more difficult when the string is already in motion and when the excitation does not occur in a single instant and affects more than a single mass, or when the string does not come to rest during the plucking. In the processing example *mass_spring_string.pde* a "chain of pearls" is excited at one end by a sine wave. One can see how a wave runs the length of the string and is reflected at the other end. If the frequency of the exciting sine wave corresponds to one of the natural resonances of the string, a standing wave arises.

Bowing a string in a model having several masses can be simulated in the same way as with a model having only one mass (8.1.1.4). But it can be considerably more difficult to find appropriate parameters because the instantaneous motion of the point where the string is bowed corresponds to the superposition of several sine waves. In the animation below, a model with eight masses (including the fixed points) is used. The string is bowed at mass M_3. To simulate the friction during the gliding phase of the bowing, we add a small random excitation in the direction of the bow's movement. In the illustration below, the phases when the string sticks to the bow are indicated by a vertical line in the direction of the bow's motion. (→ Animation)

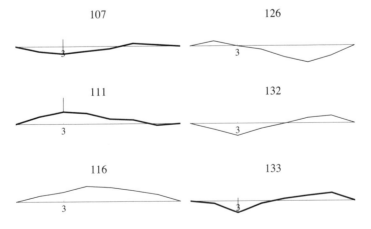

8.1 Mass-Spring Models

The plot of the motion of the first mass shows that a strictly periodic oscillation does not occur, neither at the beginning of the bowing nor later.

Nonetheless, the interaction between string and bow gives rise to a constant period duration that normally corresponds to the period of the fundamental. Hence the sound has a harmonic spectrum, even though the factor *c* is not equal to one.

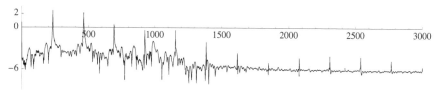

The mxj-external *cm_massSpring_stringBowed* is essentially the same as the mxj-external *cm_MassSpring_1*. In Max-patch *MassSpring_StringBowed.maxpat*, one can additionally indicate position, velocity and bowing pressure. The following lines of code show the inlet function and that part of the perform routine where the effect of the bow on the string is simulated very simply. After computing the new displacement, the velocity of the string at the bowing point is compared to the velocity of the bow. If the difference if smaller than the value *press* for the bowing pressure, the new displacement is corrected by adding to it the velocity of the bow. Since the velocity of the bow and the bowing pressure have to be coordinated with each other, they can be set simultaneously using the object *pictslider*.
(→ *MassSpring_StringBowed.maxpat*)

```
.....
if (getInlet() == 5) bow = (int)f;       // bow position
if (getInlet() == 6) v_bow = f;          // bow velocity
if (getInlet() == 7) press = f;          // bow pressure
.....
public void perform(MSPSignal[] ins, MSPSignal[] outs)
{
....
    for(int ii = 1; ii < n-1; ii++)
    {   a2[ii] = a1[ii] + (a1[ii] - a0[ii] -
                 c*(2*a1[ii] - (a1[ii-1] +  a1[ii+1])))*(1 - d);
    }
    v_str = a2[bow] - a1[bow];
    if( Math.abs(v_bow - v_str) < press )   a2[bow] = a1[bow] + v_bow;
....
}
```

See [20] pp. 217-222, [4] pp. 15-19.

8.1.3.7 Harmonics

We can also simulate harmonics with a model consisting of several masses. To play a natural harmonic on a string instrument sounding a twelfth above the string's fundamental, one touches the string lightly at the point where one would play the fifth above the fundamental.

This point is exactly 1/3 of the distance between the scroll end and the bridge, and touching the string there limits the string's motion at that point. This creates nodes at 1/3 and 2/3 of the string length, which means that the string oscillates at 1/3 its original length and hence sounds at 3 times its original frequency. For the simulation, we choose a model having $n = 3m + 4$ masses ($3m$ for the three oscillating segments and 4 for the two fixed points and the two nodal points). We simulate the limitation of the string's freedom of motion at the mass touched by the finger by multiplying each newly computed displacement by a constant $k < 1$.

The following animation was generated by a model of 37 masses. The two nodes are at the point 13 and 25. To excite the string, we simulate plucking it by displacing a few masses. In the animation, one can see that the second node results from the overlapping of the traveling waves and their reflections. The illustration below shows the state of the string at times $t = 1$, 5, 9, ..., 25 (left column) and $t = 171, 175, ..., 195$ (right column). (→ Animation)

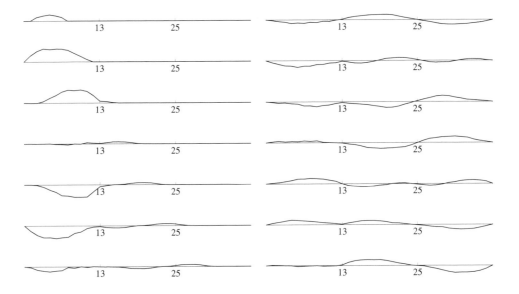

To have a model with which we can generate various partial tones as natural harmonics, we must choose the number of freely oscillating masses so that the equation $n = im_i + (m_i - 1)$ holds for each required partial (m is a natural number and i is the number of the partial). For instance, with $m = 23$, the second, third, fourth and sixth partials can be generated ($23 = 2 \cdot 11 + 1 = 3 \cdot 7 + 2 = 4 \cdot 5 + 3 = 6 \cdot 3 + 5$). In the Max patch *MassSpring_harmonics.maxpat*, we simulate touching the string at a given point by stipulating that at the point the string can only oscillate freely in one direction. The formula $(n + 1)/i$ gives the number i of the point to touch. If the displacement at i is positive, it is reduced by the factor $f\!f < 1$. We see and hear the following behavior: Touching at one of the fixed points allows the entire string to vibrate freely. Touching the string at the points 12, 8, 6 and 4 produces the partials computed above: 2, 3, 4 and 6. Touching the string at a point that does not divide it into an integer ratio produces various effects. Touching at points 1, 2, 3, 7, or 9 causes the sound to be damped quickly and noisily. Touching at points 5 or 10 generates a clearly audible fifth partial. Touching at point 11 generates the second partial, which shows that a partial can sometimes be generated even when the string is not exactly divided. (For example, the points 5 and 10 divided a string with 24 masses at 1/5 and 2/5 of its length respectively.) (→ *MassSpring_harmonics.maxpat*) (→ *string_harmonics.pde*)

8.1.4 Two-Dimensional Arrangements of Coupled Masses

In this chapter, we will investigate the oscillations of flat objects like plates, bars and membranes by modeling them with masses in a plane. The resulting oscillations are more complex than those we have seen so far, but we will see that they can be decomposed into sums of simple natural resonances. How many natural resonances there are will, as before, depend on the number of masses and not on the geometry of their arrangement.

8.1.4.1 An Example With Three Masses

As our first example, let us arrange three masses symmetrically in a plane. The masses are at the same distance from one another and from the fixed points. This symmetry means that the equations of motion will have the same form for all three masses. The acceleration of mass M_1 due to the connection with mass M_2 is $-c(a_{11} - a_{21})$, that due to the connection with M_3 is $-c(a_{11} - a_{31})$ and that due to the connection with the fixed point is $-ca_{11}$. Hence we have for mass M_1

$$a_{12} = 2a_{11} - 3ca_{11} + ca_{21} + ca_{31} - a_{10}.$$

By exchanging indices, we get for M_2 and M_3

$$a_{22} = 2a_{21} - 3ca_{21} + ca_{11} + ca_{31} - a_{21}$$
$$a_{32} = 2a_{31} - 3ca_{31} + ca_{21} + ca_{11} - a_{30}.$$

We enter the indices in a 6×6 matrix and compute the eigenvalues.

$$A = \begin{pmatrix} 0 & 0 & 0 & 1 & 0 & 0 \\ 0 & 0 & 0 & 0 & 1 & 0 \\ 0 & 0 & 0 & 0 & 0 & 1 \\ -3*c & c & c & 0 & 0 & 0 \\ c & -3*c & c & 0 & 0 & 0 \\ c & c & -3*c & 0 & 0 & 0 \end{pmatrix} ;$$

Eigenvalues[A]

$$\left\{ -I\sqrt{c}, I\sqrt{c}, -2I\sqrt{c}, -2I\sqrt{c}, 2I\sqrt{c}, 2I\sqrt{c} \right\}$$

The eigenvalues are in the ratio 1:2:2, that is, they are exactly harmonic but two natural resonances have the same frequency. Since $\sqrt{c}$ is the angular frequency per time unit (here sample period) (8.1.1.1), the frequency can be calculated using the formula $f = \sqrt{c} \, sr/2\pi$.
(→ Animation)

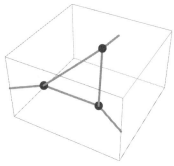

8.1.4.2 Representing the Plane by a Regular Grid

Let us subdivide the plane using a square grid with sides of length n. If we fix the points on the edges, we get the same equation of motion for all the oscillating points within the grid, since each of these points is connected to four neighboring points. We define two-dimensional vector fields for the three successive values $a0$, $a1$ and $a2$, which we can address by two indices (i, k). The neighboring points of the point $P_{i,k}$ are $P_{i,k+1}$, $P_{i,k-1}$, $P_{i+1,k}$, $P_{i-1,k}$, the displacement at the three times is $a0_{i,k}$, $a1_{i,k}$ and $a2_{i,k}$. The acceleration of the mass $M_{i,k}$ due to the connection with the mass $M_{i+1,k}$ is

$$-c(a1_{i,k} - a1_{i+1,k}).$$

The same is true for the other neighboring points, and so we obtain the equation of motion

$$a2_{i,k} = 2a1_{i,k} - a0_{i,k} - c(4a1_{i,k} - a1_{i+1,k} - a1_{i-1,k} - a1_{i,k+1} - a1_{i,k-1}),$$

where i and k go from 2 to $n - 1$.

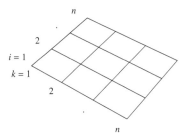

Here is a Mathematica program that shows the oscillations of a square grid. First the constants and variables are declared (lines 1-2). Then we initialize the fields $a1$, $a2$ and $a3$ (lines 3-5). As excitation, we simulate striking the surface at point $P_{7,6}$ by displacing this point and its neighbors (lines 6-7). The outer loop counts the time t (lines 8-17). After the reassignment of variables for the next computation $a0 = a1$ and $a1 = a2$ (lines 12-13), the data for the display is output (lines 14-16). (→ Animation)

```
c = .07; n = 18;                                                      (*1*)
Array[a0, {n, n}]; Array[a1, {n, n}]; Array[a2, {n, n}];
Do[a0[i, k] = 0, {k, 1, n, 1}, {i, 1, n, 1}];
Do[a2[i, k] = 0, {k, 1, n, 1}, {i, 1, n, 1}];
Do[a1[i, k] = 0, {k, 1, n, 1}, {i, 1, n, 1}];                         (*5*)
a1[7, 6] = -1; a1[8, 6] = -.6; a1[6, 6] = -.6; a1[7, 5] = -.6; a1[7, 7] = -.6;
a1[8, 7] = -.3; a1[6, 5] = -.3; a1[8, 5] = -.6; a1[6, 7] = -.6;
Do[
    Do[a2[i, k] = 2*a1[i, k] - c*(4*a1[i, k] - a1[i - 1, k] - a1[i + 1, k]
                    - a1[i, k - 1] - a1[i, k + 1]) - a0[i, k],
                    {k, 2, n - 1, 1}, {i, 2, n - 1, 1}];
    Do[a0[i, k] = a1[i, k], {k, 2, n - 1, 1}, {i, 2, n - 1, 1}];
    Do[a1[i, k] = a2[i, k], {k, 2, n - 1, 1}, {i, 2, n - 1, 1}];
    P[t] = ListPlot3D[Table[a0[i, j], {i, n}, {j, n}],
    Boxed -> False; Axes -> False, BoxRatios -> {5, 5, 1}],           (*15*)
{t, 0, 120, 1}]
```

8.1 Mass-Spring Models

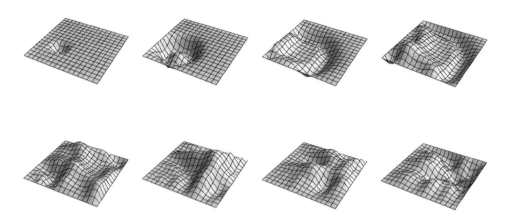

If we store the displacement values in a list, Mathematica can play them as sound. So let us define a new list a of length $T + 1$ where T is the length of the sound in sample periods. Then we replace the graphic commands by the following command, which writes the displacement of a point (here $P_{2,2}$) into the list.

```
a = Table[0, {i, T + 1}];
..........  ...
a[[t + 1]] = a0[2, 2];
..........  ...
, {t, 0, T, 1}]
```

(→ Sound Example)

In more complex cases there is no general solution for the eigenvalue equations, but we can solve them numerically for an arbitrary value, e.g. $c_1 = 1$. The result is a nominal frequency f_1 which we can use in the equation

$$\frac{f}{f_1} = \sqrt{\frac{c}{c_1}} = \sqrt{c}$$

to get a formula with which we can compute the constant c for a given frequency f.

$$\boxed{c = \left(\frac{f}{f_1}\right)^2}$$

(→ Example in C++) (→ *MassSpring_2D.maxpat*)

8.1.4.3 Resonant Frequencies of the Grid

Let us compute the resonant frequencies of a 5×5 grid and show why certain resonances can have the same frequency. Since only the inner nine masses oscillate, let us define an 18×18 matrix for the 2·9 equations to be solved (8.1.2.4) and let Mathematica do the calculations ($c = .1$).

$$M2 = \begin{pmatrix}
0 & 0 & 0 & 0 & 0 & 0 & 0 & 0 & 0 & 1 & 0 & 0 & 0 & 0 & 0 & 0 \\
0 & 0 & 0 & 0 & 0 & 0 & 0 & 0 & 0 & 0 & 1 & 0 & 0 & 0 & 0 & 0 \\
0 & 0 & 0 & 0 & 0 & 0 & 0 & 0 & 0 & 0 & 0 & 1 & 0 & 0 & 0 & 0 \\
0 & 0 & 0 & 0 & 0 & 0 & 0 & 0 & 0 & 0 & 0 & 0 & 1 & 0 & 0 & 0 \\
0 & 0 & 0 & 0 & 0 & 0 & 0 & 0 & 0 & 0 & 0 & 0 & 0 & 1 & 0 & 0 \\
0 & 0 & 0 & 0 & 0 & 0 & 0 & 0 & 0 & 0 & 0 & 0 & 0 & 0 & 1 & 0 \\
0 & 0 & 0 & 0 & 0 & 0 & 0 & 0 & 0 & 0 & 0 & 0 & 0 & 0 & 0 & 1 \\
0 & 0 & 0 & 0 & 0 & 0 & 0 & 0 & 0 & 0 & 0 & 0 & 0 & 0 & 0 & 0 \\
0 & 0 & 0 & 0 & 0 & 0 & 0 & 0 & 0 & 0 & 0 & 0 & 0 & 0 & 0 & 0 \\
-.4 & .1 & 0 & .1 & 0 & 0 & 0 & 0 & 0 & 0 & 0 & 0 & 0 & 0 & 0 & 0 \\
.1 & -.4 & .1 & 0 & .1 & 0 & 0 & 0 & 0 & 0 & 0 & 0 & 0 & 0 & 0 & 0 \\
0 & .1 & -.4 & 0 & 0 & .1 & 0 & 0 & 0 & 0 & 0 & 0 & 0 & 0 & 0 & 0 \\
.1 & 0 & 0 & -.4 & .1 & 0 & .1 & 0 & 0 & 0 & 0 & 0 & 0 & 0 & 0 & 0 \\
0 & .1 & 0 & .1 & -.4 & .1 & 0 & .1 & 0 & 0 & 0 & 0 & 0 & 0 & 0 & 0 \\
0 & 0 & .1 & 0 & .1 & -.4 & 0 & 0 & .1 & 0 & 0 & 0 & 0 & 0 & 0 & 0 \\
0 & 0 & 0 & .1 & 0 & 0 & -.4 & .1 & 0 & 0 & 0 & 0 & 0 & 0 & 0 & 0 \\
0 & 0 & 0 & 0 & .1 & 0 & .1 & -.4 & .1 & 0 & 0 & 0 & 0 & 0 & 0 & 0 \\
0 & 0 & 0 & 0 & 0 & .1 & 0 & .1 & -.4 & 0 & 0 & 0 & 0 & 0 & 0 & 0
\end{pmatrix}$$

N[Abs[Eigenvalues[M2]]]

{0.8263, 0.8263, 0.7358, 0.7358, 0.7358, 0.7358, 0.6325, 0.6325, 0.6325, 0.6325, 0.6325, 0.6325, 0.5085, 0.5085, 0.5085, 0.5085, 0.3423, 0.3423}

Let us call the resonant frequencies belonging to the nine eigenvalues *E1* (0.3422), *E2a* and *E2b* (0.5085), *E3a* through *E3c* (0.6324), *E4a* and *E4b* (0.7358) and *E5* (0.8263). When simulating a string, we find the *n*th resonance by subdividing the string into *n* equal segments. The points of division on the string are the nodes of the oscillation. The segments all oscillate at the same frequency, but adjoining segments oscillate in opposite directions. If we apply this behavior to a surface, we should have lines or branches of lines instead of nodes which divide the surface into symmetrical segments or at least into segments having the same resonant frequency. Individual resonant frequencies can be excited by a sine wave of the corresponding frequency. The following animation shows one of the resonances *E3* being excited by the function $A \cdot \sin(f \cdot t)$ ($f = E3 = .632455$, $A = .06$). (→ Animation)

To produce the different resonances having the same frequency, one can excite different masses with the same signal. The following illustration shows all the natural resonances of a 4×4 surface and their nodal lines. The nodal lines correspond to so-called *Chladni figures* which appear when one sprinkles sand on an oscillating surface. Where the surface is oscillating, the sand is pushed away, it collects where the surface is stationary (i.e. at the nodal lines). The excitation affected $M_{2,3}$ for *E2a*, $M_{2,2}$ for *E2b*, $M_{2,2}$ for *E3a*, $M_{2,3}$ for *E3b*, $M_{3,3}$ for *E3c*, $M_{2,3}$ for *E4a*, and $M_{2,2}$ for *E4b*.

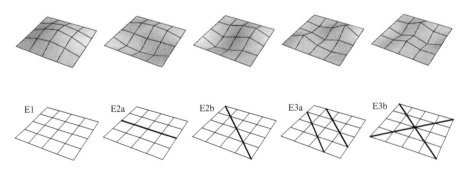

8.1 Mass-Spring Models

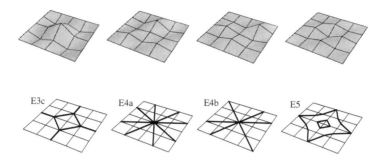

It is easiest to see why various natural resonances can have the same frequency by considering the resonance *E3*. Every oscillating point is surrounded by four points either on the edge or on a nodal line. Hence it is easy to compute the resonant frequency. Since every point is affected by four times the force of the single connection, we have for the angular frequency $f = \sqrt{4c} = \sqrt{.4} = .632455$.

See [20] pp. 177-185.

8.1.4.4 Objects With Curved Edges

The oscillation of circular plates or irregular surfaces can also be simulated using the method above. Only the points lying within the required form are calculated at each step. In the following animation, a circular plate in the middle of the square grid is made to oscillate by striking it in the middle. The running waves are reflected at the edge of the plate and meet each other again exactly in the middle of the plate. It is astonishing that the circular waves produced remain regular after several hundred computation steps, even if the edge is irregular. (→ Animation)

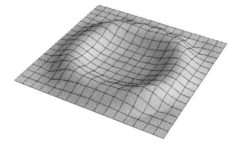

The parameters for the restoring force and for the damping can be modified during the simulation, as can the shape and the size of the object.

8.1.4.5 Objects With Freely Oscillating Edges

For surfaces having freely oscillating edges, we must compute the motion of the masses inside the edges, along the edges and in the corners separately. In the following example we define a rectangular grid with $n_i \times n_k$ points only one of whose edges is fixed, namely the points $P_{i,1}$. The equations of motion for the edge points $P_{1,k}$ and $P_{n_i,k}$ ($k = 2$ to $k = n_k - 1$) are:

$$a2_{1,k} = 2a1_{1,k} - a0_{1,k} - c(3a1_{1,k} - a1_{2,k} - a1_{1,k+1} - a1_{1,k-1})$$
$$a2_{n_i,k} = 2a1_{n_i,k} - a0_{n_i,k} - c(3a1_{n_i,k} - a1_{n_i-1,k} - a1_{n_i,k+1} - a1_{n_i,k-1}).$$

The equation of motion for the edge points P_{i,n_k} ($i = 2, ..., n_i - 1$) are:

$$a2_{i,n_k} = 2a1_{i,n_k} - a0_{i,n_k} - c(3a1_{i,n_k} - a1_{i,n_k-1} - a1_{i+1,n_k} - a1_{i-1,n_k}).$$

The equations of motion for the corners P_{n_i,n_k} and P_{1,n_k} are:

$$a2_{1,n_k} = 2a1_{1,n_k} - a0_{1,n_k} - c(2a1_{1,n_k} - a1_{1,n_k-1} - a1_{2,n_k})$$
$$a2_{n_i,n_k} = 2a1_{n_i,n_k} - a0_{n_i,n_k} - c(2a1_{n_i,n_k} - a1_{n_i,n_k-1} - a1_{n_i-1,n_k}).$$

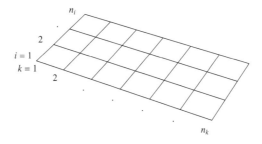

This model can be used to simulate freely oscillating reeds. To excite torsional oscillation, one can strike one of the freely vibrating corners. (→ Animation)

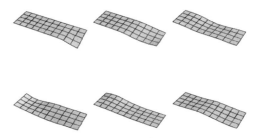

See [20].

8.1.4.6 Rigid Body Motion

When all the masses of a model can oscillate freely, there are usually, besides the oscillations themselves, movements of the entire body like twisting and translational displacement (8.1.5.3). Because these movements are independent of the oscillation and can also occur in undeformed bodies, we speak of rigid body motion. When they occur, the recording of the motion as sound does not oscillate around zero. We solve this problem by subtracting the center of mass of the body, that is, the average distance of the individual masses from their initial position, from the displacement at the point where the sound is picked up.

To show the effect of the freedom of movement of a body on the equation system to be solved, let us look again at the first example from Chapter 8.1.4.1 and imagine that the masses

8.1 Mass-Spring Models

are only connected to one another and are free on the edges of the surface. So that the masses do not collapse into the middle of the system, we imagine that the connections between the masses are springs that have their natural length in the body's initial position, or else we imagine that the masses can only move vertically, like pearls on a string. The matrix for the equations of motion for this system is

$$A = \begin{pmatrix} 0 & 0 & 0 & 1 & 0 & 0 \\ 0 & 0 & 0 & 0 & 1 & 0 \\ 0 & 0 & 0 & 0 & 0 & 1 \\ -2*c & c & c & 0 & 0 & 0 \\ c & -2*c & c & 0 & 0 & 0 \\ c & c & -2*c & 0 & 0 & 0 \end{pmatrix} \quad ; \text{Eigenvalues[A]}$$

$$\left\{ 0, 0, -I\sqrt{3}\sqrt{c}, -I\sqrt{3}\sqrt{c}, I\sqrt{3}\sqrt{c}, I\sqrt{3}\sqrt{c} \right\}$$

The first eigenvalue is zero, that is we get a natural resonance of frequency zero. Since $\cos(0) = 1$, we can interpret this resonance as an arbitrary constant displacement. This corresponds to the fact that the system's equations of motion do not have unique solutions because they are independent of the motion of the entire body along the z-axis. A feature of such so-called singular matrices is that their determinant vanishes ($\text{Det}(A) = 0$).

To avoid rigid body motion, we can use symmetrical excitation, for example noise that on an average gives the masses the same impulse in all directions. Or we can excite the body at symmetrical points with equal but opposite impulses. We will see another way to avoid rigid body motion in the next chapter, namely to excite oscillation by deforming the body. If a body is bent and then let go, no rigid body motion will occur because the body did not receive an impulse.

8.1.4.7 Retroflex Surfaces

The model from the last chapter can easily be modified to simulate oscillation on a retroflex surface. If the surface's curvature is small compared to its thickness and if the surface is not under tension, the waves propagate as on a flat surface. Hence we can simulate the oscillation on a cylinder made by joining together two edges of a surface.

In order to keep the program for the simulation as short as possible, let us proceed as follows. We will overlap two opposing edges by setting the points $P_{i,1}$ equal to the points $P_{i,nk-1}$ and the points $P_{i,2}$ equal to the points $P_{i,nk}$. The edges do not have to be treated specially. We compute the oscillation for $1 < k < nk$ (lines 1-9) and then assign the values in $P_{i,nk-1}$ to $P_{i,1}$ and the values in $P_{i,2}$ to $P_{i,nk}$ (lines 10-11).

```
Do[a2[[1,k]]=a1[[1,k]]+d*(a1[[1,k]]-c*(3*a1[[1,k]]-a1[[2,k]]
        -a1[[1,k-1]]-a1[[1,k+1]])-a0[[1,k]]),
  {k,2,nk-1,1}];
Do[a2[[ni,k]]=a1[[ni,k]]+d*(a1[[ni,k]]-c*(3*a1[[ni,k]]-a1[[ni-1,k]]
        -a1[[ni,k-1]]-a1[[ni,k+1]])-a0[[ni,k]]),           (*5*)
  {k,2,nk-1,1}];
Do[a2[[i,k]]=a1[[i,k]]+d*(a1[[i,k]]-c*(4*a1[[i,k]]-a1[[i-1,k]]
        -a1[[i+1,k]]-a1[[i,k-1]]-a1[[i,k+1]])-a0[[i,k]]),
  {k,2,nk-1,1},{i,2,ni-1,1}];
Do[a2[[i,1]]=a2[[i,nk-1]],{i,1,ni,1}];                     (*10*)
Do[a2[[i,nk]]=a2[[i,2]],{i,1,ni,1}];
```

In the following animation, we set $ni = 5$, $nk = 18$, $c = .18$ and $d = 1$. To avoid rigid body motion, we excite the surface by deforming it. (→ Animation)

(→ Animation With Rigid Body Motion)

8.1.4.8 A Grid With Unequal Distances Between the Masses

If we twist the grid along the plane of its surface and fasten two edges together, we get a disk with a hole in the middle. Let us change that section of the program above dealing with the graphic display. Along a line with constant k a wave results that moves around the ring, meeting itself on the other side along the entire width of the disk. This means that the wave is considerably slower in the middle than at the edge. (→ Animation)

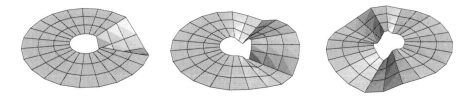

This erratic behavior of the model (namely, that the wave moves simultaneously at different speeds) is due to the fact that the points are not equidistant. Hence, the points represent different masses and the constant $c = cf/m$ (8.1.1.1) is not everywhere the same.

In the model with n linearly arranged masses (8.1.3), we had to distribute the entire mass of the string onto n points. If we subdivide more finely and make a denser distribution by multiplying n by a number p, we obtain for the individual masses $m_{pn} = m_n/p$. But how does the spring force cf of the string change when the subdivision changes?

If we have linearly arranged masses, we can use the formula in Chapter 8.1.3.2 to calculate how c must be changed in order to obtain the same frequency for various subdivisions.

In the formula

$$v_{sr} = \frac{sr}{\pi} \sqrt{c} \sin\left(\frac{\pi}{2(n+1)}\right)$$

we set $c = c_n$ for n masses and $c = c_{pn}$ for $p \cdot n$ masses:

8.1 Mass-Spring Models

$$\frac{sr}{\pi}\sqrt{c_n}\sin\left(\frac{\pi}{2(n+1)}\right) = \frac{sr}{\pi}\sqrt{c_{pn}}\sin\left(\frac{\pi}{2(pn+1)}\right).$$

When n is large,

$$\frac{\pi}{2(n+1)} \approx \frac{\pi}{2n} \text{ and } \frac{\pi}{2(pn+1)} \approx \frac{\pi}{2pn}.$$

When α is small, $\sin(\alpha)$ is approximately equal to α, and so we can simplify the equation above to

$$\frac{\pi}{2n}\sqrt{c_n} \approx \frac{\pi}{2pn}\sqrt{c_{pn}},$$

which implies

$$c_{pn} \approx p^2 c_n.$$

If we set $c = cf/m$ in this equation, we get

$$\frac{cf_{pn}}{m_{pn}} \approx \frac{p^2 \cdot cf_n}{m_n} = \frac{p^2 \cdot cf_n}{p \cdot m_{np}} = \frac{p \cdot cf_n}{m_{np}},$$

from which it follows that

$$cf_{pn} \approx p \cdot cf_n.$$

In other words, the spring constant increases approximately in proportion to the number of masses used, and hence inversely proportionally to the distance between the masses.

Let us add three arrays to our program: one for the masses ($m[i]$), one for the values c of the connections on the ith circle ($ck[i]$) and one ($ci[i]$) for the connections from the points $P_{i,k}$ to the points $P_{i+1,k}$. We set the masses proportional to the surfaces adjoining these points and the radial distances equal to 1. The radius of the hole is R. We set the factors ci and ck inversely proportional to the product mass times distance. We use these constants, whose values depend on their position, in the formulas for computing the new displacements.

```
...
m = Table[π* ((i + R - 1 + .5)² - (i + R - 1 - .5)²)/(nk - 2), {i, 1, ni, 1}];
m[[1]] = π* ((i + R - 1 + .5)² - (i + R - 1)²)/(nk - 2);
m[[ni]] = π* ((i + R - 1)² - (i + R - 1 - .5)²)/(nk - 2);
ci = Table[c/m[[i]], {i, 1, ni, 1}];                    (*5*)
ck = Table[c/ (m[[i]] *2*π (i + R - 1)/(nk - 2)), {i, 1, ni, 1}];
.........
Do[a2[[1, k]] = a1[[1, k]] + d* (a1[[1, k]]
     - ci[[1]] * (a1[[1, k]] - a1[[2, k]])
     - ck[[1]] * (2*a1[[1, k]] - a1[[1, k - 1]]
         - a1[[1, k + 1]]) - a0[[1, k]]), {k, 2, nk - 1, 1}];
Do[a2[[ni, k]] = a1[[ni, k]] + d* (a1[[ni, k]]
     - ci[[ni - 1]] * (a1[[ni, k]] - a1[[ni - 1, k]])
     - ck[[ni]] * (2*a1[[ni, k]] - a1[[ni, k - 1]]
         - a1[[ni, k + 1]]) - a0[[ni, k]]), {k, 2, nk - 1, 1}];
Do[a2[[i, k]] = a1[[i, k]] + d* (a1[[i, k]]
     - ck[[i]] * (2*a1[[i, k]] - a1[[i, k - 1]] - a1[[i, k + 1]])
     - ci[[i - 1]] * (a1[[i, k]] - a1[[i + 1, k]])
```

```
        - ci[[i]] * (a1[[i, k]] - a1[[i - 1, k]]) - a0[[i, k]]),
    {k, 2, nk - 1, 1}, {i, 2, ni - 1, 1}];
```

The animation below shows how waves really behave on a disk. The excitation comes from bending the edge of the disk. The resulting wave moves at the same speed everywhere on the disk forming a flat wavefront, regardless of the distribution of the masses. One can also see how the wave is reflected first by the inner edge and then by the outer edge, and vice versa. (→ Animation)

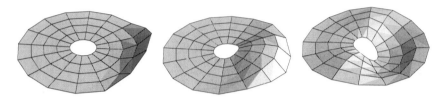

We can simulate waves on other bodies produced by stretching and connecting the edges of regular grids in the same way. Examples are the *torus*, which is a cylinder bent into a ring, and the *Möbius strip*, made by giving a rectangular grid a half twist and then joining opposite edges.

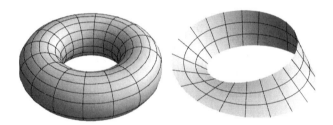

8.1.4.9 Irregular Grids

Certain bodies cannot be completely covered even by distorted grids. For example, if one wanted to transform the ring in the model above into a disk by constricting the inner edge into one point, then the distances between the points on that edge would be zero, and dividing by those distances would give errors. If one only uses rectangular grids, one cannot cover a sphere. Therefore, the next example shows how the surface and the poles of a sphere can be described using *spherical coordinates*, that is circles of latitude and longitude.

As with the cylinder, let us start with a grid of $i \times k$ points. We join two opposing edges by overlapping two rows of points so that the points $P_{i,1}$ are identical to the points $P_{i,nk-1}$ and the points $P_{i,2}$ are identical to the points $P_{i,nk}$. We constrict the upper and lower edges of the cylinder each into a single point so that all the points $P_{1,k}$ fall onto one pole and all the points $P_{ni,k}$ fall onto the other pole.

In the Mathematica program below, we first declare and initialize the constants and the arrays (lines 4-6). Then we calculate the length *di* of a section on a circle of longitude between two circles of latitude (line 7), the radius *r* of the *i*th circle of latitude (line 8) and the length of a segment *dk[i]* on a circle of latitude between two circles of longitude (line 9). We will simplify the calculation of the mass represented by each point by approximating the mass as

8.1 Mass-Spring Models

the area of a trapezoid $dk[i] \cdot di$ with the center line $dk[i]$ and height di. Now we can introduce a new constant cc and, calling the distance to the next mass D, we can write the constant c as $c = c_f/m = cc/(m \cdot D)$. Consequently, we have for c in the latitudinal direction

$$ck[i] = cc/(dk[i] \cdot di \cdot dk[i])$$

and for c in the longitudinal direction

$$ci[i] = cc/(dk[i] \cdot di \cdot di) \quad \text{(lines 11-12)}.$$

The area about the poles is $\pi(di/2)^2$. Hence we have for c at the poles $cp = 4 \cdot cc/(di^3 \pi)$ (line 12). In lines 13-14, we strike the surface from the inside at the point $P_{16,9}$. The outer loop (lines 15-41) counts the time. We first calculate the new displacements at the poles. Because the poles combine all the points of the upper and lower edges, we can choose one point on each edge (e.g. $P_{1,1}$ and $P_{ni,1}$, lines 17 and 22) and assign their displacements to all the other points on the upper and lower edge respectively (lines 19 and 24). We calculate the new displacement at the poles from the sum of the displacements of neighboring points $P_{2,k}$ and $P_{ni-1,k}$ respectively (lines 16-18 and 21-23). The new displacements of all the other points are calculated in the same way (lines 25-29). Because the lateral edges overlap, the displacements of the points $P_{i,nk-1}$ are assigned to the points $P_{i,1}$ and the displacements of the points $P_{i,2}$ are assigned to the points $P_{i,nk}$ (lines 30-31). After the preparations for the next iteration (line 32), the final lines of the program display the results on the surface of the sphere.

```
cc = .007; nk = 16; ni = 19; d = 1; R = 4;                              ; 1
a0 = Table[0, {ni}, {nk}];
a1 = Table[0, {ni}, {nk}];
a2 = Table[0, {ni}, {nk}];
ck = Table[0, {ni}];                                                    ; 5
ci = Table[0, {ni}];
di = R*π/(ni - 1);
r = Table[R*Sin[(π*(i - 1)/(ni - 0.9999))], {i, 1, ni, 1}];
Do[dk[[i]] = 2*Pi*r[[i]]/(nk - 2), {i, 2, ni - 1, 1}];
Do[ck[[i]] = cc/(dk[[i]]^2 *di), {i, 2, ni - 1, 1}];                    ; 10
Do[ci[[i]] = cc/(dk[[i]]*di^2), {i, 2, ni - 1, 1}];
cp = 4*cc/(di^3 *π);
a1[[16, 9]] = .3; a1[[16, 10]] = .14; a1[[16, 8]] = .14;
a1[[17, 9]] = .18; a1[[15, 9]] = .18;
Do[ a = 0;                                                              ; 15
    Do[a += a1[[2, k]], {k, 2, nk - 1, 1}];
    a2[[1, 1]] = a1[[1, 1]] + d*(a1[[1, 1]]
        - cp*((nk - 2)*a1[[1, 1]] - a) - a0[[1, 1]]);
    Do[a2[[1, k]] = a2[[1, 1]], {k, 2, nk, 1}];
    a = 0;                                                              ; 20
    Do[a += a1[[ni - 1, k]], {k, 2, nk - 1, 1}];
    a2[[ni, 1]] = a1[[ni, 1]] + d*(a1[[ni, 1]]
        - cp*((nk - 2)*a1[[ni, 1]] - a) - a0[[ni, 1]]);
Do[a2[[ni, k]] = a2[[ni, 1]], {k, 2, nk, 1}];
Do[a2[[i, k]] = a1[[i, k]] + d*(a1[[i, k]]                              ; 25
    - ck[[i]]*(2*a1[[i, k]] - a1[[i, k - 1]] - a1[[i, k + 1]])
    - ci[[i]]*(2*a1[[i, k]] - a1[[i + 1, k]] - a1[[i - 1, k]])
    - a0[[i, k]]),
```

```
          {k, 2, nk - 1, 1}, {i, 2, ni - 1, 1}];
        Do[a2[[i, 1]] = a2[[i, nk - 1]], {i, 1, ni, 1}];              ; 30
        Do[a2[[i, nk]] = a2[[i, 2]], {i, 1, ni, 1}];
        a0 = a1; a1 = a2;
        p[[t]] = ParametricPlot3D[{
        (4 + a0[[1 + IntegerPart[(u + .5*π) * (ni - 1) / Pi],
              1 + IntegerPart[v * (nk - 2) / (2*π)]]]) * Sin[v] * Cos[u],  ; 35
        (4 + a0[[1 + IntegerPart[(u + .5*π) * (ni - 1) / Pi],
              1 + IntegerPart[v * (nk - 2) / (2*π)]]]) * Cos[v] * Cos[u],
        (4 + a0[[1 + IntegerPart[(u + .5*π) * (ni - 1) / Pi],
              1 + IntegerPart[v * (nk - 2) / (2*π)]]]) * Sin[u]},
        {v, 0, 2*π}, {u, -.5*π, .5*π + .01}]                          ; 40
        , {t, 0, 1, 1}]
```

One can see in the animation that, despite the very different distances between the masses and despite the fact that many circles of longitude meet at the pole, the wave travels circularly over the sphere's surface. (→ Animation)

8.1.4.10 Irregular Density or Elasticity

The last example shows how the oscillation of an irregular plane can be simulated. In the square grid of Chapter 8.1.4.2, the constant c depends on its position on the grid. Reasons for this dependency could be the irregular thickness of the body or the irregular density or elasticity of the material. In our example, the dependency on position consists in the fact that c gets larger as i increases. The waves spread more quickly at the upper edge than at the lower. (→ Animation)

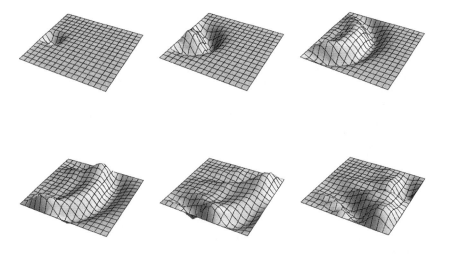

The experiment to simulate waves on a disk from Chapter 8.1.4.8 offers another example of a surface with irregular density.

8.1.5 Three-Dimensional Arrangements of Coupled Masses

As before, the main task of this chapter will be to calculate the acceleration of masses on the basis of their geometrical arrangement and to realize the calculations in a computer program. In the process, we will see to what difficulties incomplete models lead and how to remedy these difficulties. Besides investigating the oscillations produced by given bodies, we will study the bodies' movement in space, the influence of gravity on this movement and elastic impact.

8.1.5.1 Longitudinal and Transversal Oscillations of a Mass

To simulate oscillation in space, we must consider all possible directions of vibration a mass can take. Let us first look at the oscillation of a single mass connected by springs to two lateral fixed points. To describe the mass's spatial motion, we decompose it into motion on the three spatial coordinates. Consider the fixed points to be on the x-axis. When the mass oscillates in the yz-plane, we speak of transversal oscillation, when it oscillates in the x direction, we speak of longitudinal oscillation. Transversal oscillation can be decomposed into y and z components.

Let us first consider the oscillation of a mass along the x-axis. When the springs are attached only to the lateral points, they have the length l_0 (top illustration below). When we connect them to the mass and the mass is in the rest position, they have the length l_r (middle illustration below). If we displace the mass by a along the x-axis, we can express the stretching of the springs as $l_r + a - l_0$ and $l_r - a - l_0$ (lower illustration below). Because the springs' forces act in opposite directions, we take their difference and get a restoring force of $2c_s a$, where c_s is the spring constant. Hence the constant c in our equation of motion is

$$c = \frac{2c_s}{m}.$$

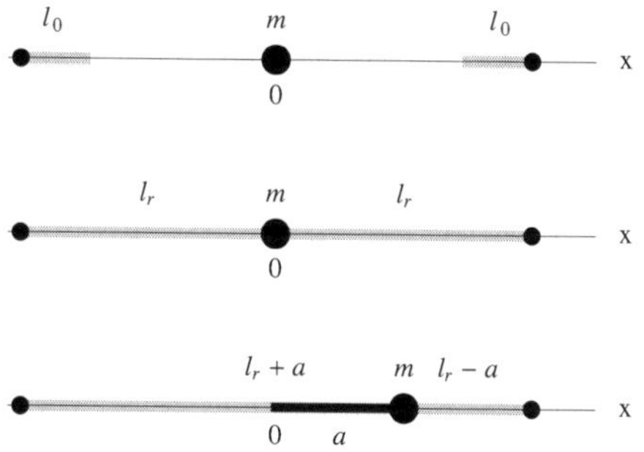

When the oscillation is perpendicular to the x-axis, for example along the y-axis, we have the following situation. The forces act in the direction of the springs, and we call them k_s. They are proportional to the stretching of the springs: $k_s = c_s(l - l_0)$. The y-component of the force is responsible for the acceleration of the mass. Since two springs act on the mass, we have $k = 2k_s \sin(\alpha)$ or, since $\sin(\alpha) = a/l$, $k = 2c_s(l - l_0)/l = 2c_s a(1 - l_0/l)$. For c we get

$$c = \frac{2c_s}{m}\left(1 - \frac{l_0}{l}\right).$$

The factor c is no longer constant, since it depends on the displacement $l = \sqrt{l_r^2 + a^2}$.

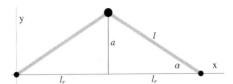

There are two ways to approximate the oscillation using a linear restoring force. The first assumes that the springs are highly stretched and that l_r and l are large compared to l_0. Then the factor $1 - l_0/l$ is practically 1, and we obtain the same equation as for longitudinal oscillation.

If l_0/l is non-negligible, then for small displacements one can set $l = l_r$, giving

$$c = \frac{2c_s}{m}\left(1 - \frac{l_0}{l_r}\right).$$

This formula shows that transversal waves are in general slower than longitudinal ones. (→ Derivation)

The following animation shows the undamped, three-dimensional oscillation of a mass fastened to two fixed points. The thicker lines imposed on the axes correspond to the three spatial components of the instantaneous displacement. In order to allow for the difference between longitudinal and transverse waves described above, let us introduce a second constant c_2, which is a little larger than c. Each individual point has therefore three degrees of freedom and three resonant frequencies. (→ Animation)

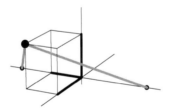

See [6] pp. 4-6.

8.1.5.2 Subdividing Space With a Regular Grid

If we subdivide an arbitrary space using a regular grid, all the points within the space $P_{i,k,l}$ have six neighbors. Each point must be assigned three values for the displacement: $a_{i,k,l,m}$ ($m = 1, 2, 3$). Let us first consider motion in the x direction. Let us call the constants c for the longitudinal and transversal oscillations c_l and c_t. The acceleration of the mass $M_{i,k,l}$ due to the connection with the mass $M_{i+1,k,l}$ is $-c_l(a1_{i,k,l,1} - a1_{i+1,k,l,1})$. The same is true for the connection with the mass $M_{i-1,k,l}$. The acceleration of the mass $M_{i,k,l}$ due to the connections with the other four masses corresponds to the acceleration we calculated for oscillation in a plane. Hence we get the equation of motion for the mass $M_{i,k,l}$ in the x-direction ($m = 1$):

$$\begin{aligned} a2_{i,k,l,1} = & \; 2a1_{i,k,l,1} - a0_{i,k,l,1} \\ & - c_t(4a1_{i,k,l,1} - a1_{i,k+1,l,1} - a1_{i,k-1,l,1} - a1_{i,k,l+1,1} - a1_{i,k,l+1,1}) \\ & - c_l(2a1_{i,k,l,1} - a1_{i+1,k,l,1} - a1_{i-1,k,l,1}) \end{aligned}$$

8.1 Mass-Spring Models

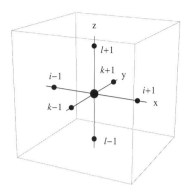

The following excerpt from a Mathematica program shows how the motion in the x-direction is calculated. Damping has been introduced. In addition, the excerpt shows how the motion of the surface points, the edges and the corners can be computed in the same loop as the motion of the inner points. The program can be short and clear, but then many unnecessary calculations are made. A longer but faster program will use a separate program block for each kind of point.

```
a2[[i, k, l, 1]] = a1[[i, k, l, 1]]
    +d*(a1[[i, k, l, 1]]
  -c*(2*a1[[i, k, l, 1]] -
        If[i > 1, a1[[i - 1, k, l, 1]], a1[[i, k, l, 1]]]
         - If[i < ni, a1[[i + 1, k, l, 1]], a1[[i, k, l, 1]]])
  -c2*(4*a1[[i, k, l, 1]]
         - If[k > 1, a1[[i, k - 1, l, 1]], a1[[i, k, l, 1]]]
         - If[k < nk, a1[[i, k + 1, l, 1]], a1[[i, k, l, 1]]]
         - If[l > 1, a1[[i, k, l - 1, 1]], a1[[i, k, l, 1]]]
         - If[l < nl, a1[[i, k, l + 1, 1]], a1[[i, k, l, 1]]])
  - a0[[i, k, l, 1]]);
```

If we excite a rectangular cuboid by striking one corner, a longitudinal wave runs diagonally through the cuboid, and transversal waves appear on the sides. One cannot follow the propagation of the waves for very long, because almost immediately other waves and their reflections overlap. Since the cuboid consists of 4·5·8 points, there are 4·5·8·3 = 480 possible resonant frequencies. (→ Animation)

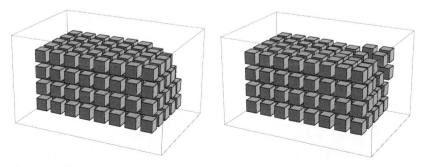

Again, these simulations only give good results when the oscillations are small. The next example shows that not only the relative displacements of neighboring masses have to be small, but also the absolute distances to the rest points. Both a thin bar and a string can easily be twisted. When one lets them go, they perform torsional oscillation. But our model behaves wrong when it is only twisted by 90°. The right end of the bar is twisted back, as it should be, but the bar is first compressed and then pressed apart. If we compare the right and the left end of the bar, we see that the error comes from the too great distance of the points on the right from their rest points. On the left edge, where the points are in the rest position, the bar does not change its diameter. (→ Animation)

8.1.5.3 Bodies in Free Motion

In order to solve the problems we encountered in Chapter 8.1.5.2, let us develop a model in which the motion of the masses is no longer based on constant points of departure. Let us again define three-dimensional arrays for the three successive values $a0$, $a1$ and $a2$. Let the indices be i, k and l, so that we can refer to the mass $M_{i,k,l}$. But now we do not assign displacement values to the arrays $a0$, $a1$ and $a2$ but rather the current coordinates x, y and z. We can compute the force of the springs between two neighboring masses, which in this example can be at any two arbitrary points in space, as follows. If we set the distance d_i between two neighboring masses M_i and M_{i+1} in rest position equal to 1, then the magnitude of the spring force is proportional to $d_i - 1$. Since the masses can be anywhere in space, we decompose the force vector into three components k_x, k_y and k_z. Because $d_{i,x}/d_i = k_x/k$, we get for the acceleration of the mass $M_{i,k,l}$ in the x-direction

$$c(d_i - 1)d_{i,x}/d_i$$

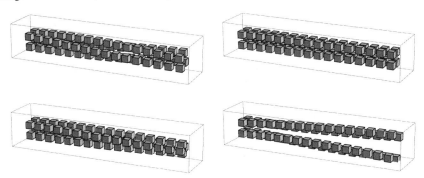

The following excerpt from the Mathematica program used for the next animation shows how to arrange the calculations required in a clear and efficient way. First we calculate the new coordinates for all the masses without acceleration by adding the "velocity" $a1 - a0$ to the old coordinates. (At the start of the calculation, $a2$ corresponds to the old coordinates.) In a next

8.1 Mass-Spring Models

step, we calculate all the position changes due to springs in the i-direction. Hence the counter for i goes only from 1 to $ni - 1$. Program code still has to be written for the other springs. First we compute the distance between the masses $M_{i,k,l}$ and $M_{i+1,k,l}$

$$di = \sqrt{d_{i,x}^2 + d_{i,y}^2 + d_{i,z}^2}.$$

Then we compute the change of position for the three coordinates due to acceleration. The accelerations dv of the two masses have the same magnitude but act in opposite directions.

```
Do[a2[[i, k, l, m]] += d*(a1[[i, k, l, m]] - a0[[i, k, l, m]]);,
    {i, 1, ni}, {k, 1, nk}, {l, 1, nl}, {m, 1, 3}];

Do[ di = √((a1[[i + 1, k, l, 1]] - a1[[i, k, l, 1]])² +
          (a1[[i + 1, k, l, 2]] - a1[[i, k, l, 2]])² +
          (a1[[i + 1, k, l, 3]] - a1[[i, k, l, 3]])²);
    Do[dv = c*(di - 1)*(a1[[i + 1, k, l, m]] - a1[[i, k, l, m]])/di;
       a2[[i, k, l, m]]     += d*dv;
       a2[[i + 1, k, l, m]] -= d*dv;,       {m, 1, 3}];,
    {i, 1, ni - 1}, {k, 1, nk}, {l, 1, nl}];
```

By striking the whole left side of the body, we generate longitudinal waves and cause the body to accelerate to the right. (→ Animation)

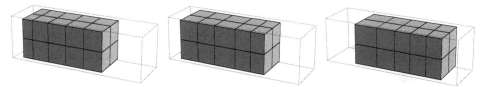

But if we strike the body on the upper edge, we discover a weakness of our model: the whole framework collapses, like a rack built of struts connected by hinges. (→ Animation)

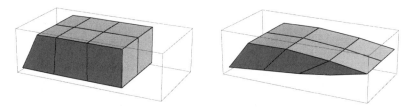

We can give our model greater stability by introducing additional diagonal connections parallel to the faces of the body. This way we get four new connections in each plane. The distance between diagonally connected points in rest position is $D_d = \sqrt{2}$. In order to specify the factor cd more precisely, we would need to know more about the properties of the body simulated by the model and adapt the model's parameters by comparing its behavior with the real object. Let us define $c_d = 1/\sqrt{2}$, assuming that the force are inversely proportional to the distances.

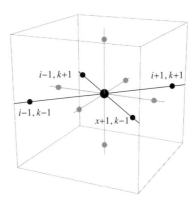

Now we introduce two new computation blocks for each of the three planes, in order to take account of the additional forces. The following program excerpt shows these blocks for connections having constant l. The first block treats connections from $M_{i,k,l}$ to $M_{i+1,k+1,l}$, that is from front left to back right. The second treats connections from $M_{i,k,l}$ to $M_{i+1,k-1,l}$, that is from front right to back left. Hence the indices for i and k in the first block go from 1 to $ni-1$ and 1 to $nk-1$ respectively. In the second block, the indices for k go from 2 to nk.

```
Do[di = √((a1[[i + 1, k + 1, l, 1]] - a1[[i, k, l, 1]])²
        + (a1[[i + 1, k + 1, l, 2]] - a1[[i, k, l, 2]])²
        + (a1[[i + 1, k + 1, l, 3]] - a1[[i, k, l, 3]])²);
    Do[dv = cd * (di - Dd) * (a1[[i + 1, k + 1, l, m]] - a1[[i, k, l, m]]) / di;
       a2[[i, k, l, m]] += d * dv;
       a2[[i + 1, k + 1, l, m]] -= d * dv;,    {m, 1, 3}];,
    {i, 1, ni - 1}, {k, 1, nk - 1}, {l, 1, nl}];
Do[di = √((a1[[i + 1, k - 1, l, 1]] - a1[[i, k, l, 1]])²
        + (a1[[i + 1, k - 1, l, 2]] - a1[[i, k, l, 2]])²
        + (a1[[i + 1, k - 1, l, 3]] - a1[[i, k, l, 3]])²);
    Do[dv = cd * (di - Dd) * (a1[[i + 1, k - 1, l, m]] - a1[[i, k, l, m]]) / di;
       a2[[i, k, l, m]] += d * dv;
       a2[[i + 1, k - 1, l, m]] -= d * dv;,    {m, 1, 3}];,
    {i, 1, ni - 1}, {k, 2, nk}, {l, 1, nl}];
```

We have realized a model that simulates the motion of freely oscillating bodies without taking account of the influence of gravity. If we give the mass $M_{1,1,1}$ an impulse (.4 in the x-direction in the following example), the body oscillates, moving in the x-direction, but also rotating about axes parallel to the y- and z-axes of the coordinate system. (→ Animation)

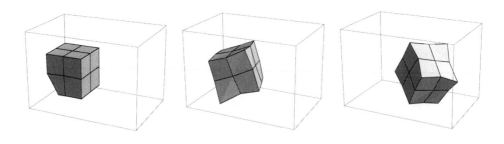

8.1 Mass-Spring Models

We can also simulate torsional oscillation with this model, as the following animation shows. (→ Animation)

8.1.5.4 A Model With Fixed Points

Introducing fixed points is not only the simplest way to avoid rigid body motion; it also brings our model closer to a real situation, where all objects are either attached to or are resting on something. In the C-program that follows we fix the right side of our object. To excite oscillation, we will give all the masses a random impulse. We write the function *dist()* to compute the distance between two points. Because the variables *ni*, *nk*, *nl*, *a1* and *a2* are used in both the main program and the function *dist()*, we will declare them global (lines 1-3). So as to have the same indices as in the Mathematica program above, we declare three arrays *a0*, *a1* and *a2* of size $(ni + 1)(nk + 1)(nl + 1) \cdot 3$ and begin counting the elements of the arrays at 1. We declare the other variables and the function *dist()* in the main program (lines 6-8). We need to give a parameter list when we declare the function. In our example, there will be six whole numbers, namely the coordinates of the two points whose distance the function will calculate. Then we assign the initial coordinates of the points to the arrays *a0*, *a1* and *a2*, adding a random displacement to each (line 11-16). In the outer loop (lines 18-77), the number of computed sample values is counted. In a first inner loop, the velocity is added in (lines 20-24), in the following loops, the acceleration due to the spring forces is added in (lines 26-33), consisting of contributions from the connections in the *i*-direction and contributions from diagonal connections in the *ik*-planes (lines 37-53). Additional computations have to be added for other directions. In order to fix the body's right side, one could change the limits and the ranges of the indices for the computation of the new values. We will choose a clearer and more general approach, however: we calculate new values as though the body could oscillate freely and continuously re-assign the initial values to the fixed points (lines 57-65). When preparing the next iteration by setting *a0* = *a1* and *a1* = *a2*, we interpolate again between *a1* and *a2* to filter out some of the high frequencies. Finally we set *a2* equal to *a1* again, because in this model *a2* has to correspond to *a1* at the beginning of an iteration. After the program block main{} follows the function *dist()*, whose parameter list shows not only the type of the parameter but also the variables used in the following calculations (lines 80-84).

```
//8-1-5a: Cuboid ni*nk*nl, Fixed Points P[ni,k,1],
//       Excitation by a random impulse on all masses
  ....
    const int   ni=9,nk=4,nl=3;                                     ;1
    float a0[ni+1][nk+1][nl+1][3],a1[ni+1][nk+1][nl+1][3],
        a2[ni+1][nk+1][nl+1][3];
void main()
{ ....                                                              ;5
 const int   tt= (int)(44100*10);
 int i,k,l,m,t;
 float dv,di,d=.99997,df=.99993,dd=sqrt(2),c=.2,cd=c/dd;
 float dist(int,int,int,int,int,int);
                                                                    ;10
 for(i=1;i<=ni;i++)                 //Initialization

  { for(k=1; k<=nk; k++)
    { for(l=1;l<=nl;l++)
```

```
    { a0[i][k][l][0]=a1[i][k][l][0]=a2[i][k][l][0]=i+.00001*(rand()-16384);
      a0[i][k][l][1]=a1[i][k][l][1]=a2[i][k][l][1]=k+.00001*(rand()-16384);
      a0[i][k][l][2]=a1[i][k][l][2]=a2[i][k][l][2]=l+.00001*(rand()-16384); }}}

for(t=0; t< tt ;t++)
{
  for(i=1;i<=ni;i++)                    // + Velocity                        ;20
  { for(k=1; k<=nk; k++)
    { for(l=1;l<=nl;l++)
      { for(m=0;m<3;m++)
        {   a2[i][k][l][m]+=d*(a1[i][k][l][m]-a0[i][k][l][m]); }}}}
                                                                             ;25
for(i=1;i<ni;i++)                       // +dv 0
{ for(k=1; k<=nk; k++)
  { for(l=1;l<=nl;l++)
    { di=dist(i,k,l,i+1,k,l);
      for(m=0;m<3;m++)                                                       ;30
      { dv=c*(di-1)*(a1[i+1][k][l][m]-a1[i][k][l][m])/di;
        a2[i][k][l][m]+=d*dv;
        a2[i+1][k][l][m]-=d*dv; }}}}

 ......... similarly for the directions k and l ............                 ;35

for(i=1;i<ni;i++)                       // +dv 0-1
{ for(k=1; k<nk; k++)
  { for(l=1;l<=nl;l++)
    { di=dist(i,k,l,i+1,k+1,l);                                              ;40
      for(m=0;m<3;m++)
      { dv=cd*(di-dd)*(a1[i+1][k+1][l][m]-a1[i][k][l][m])/di;
        a2[i][k][l][m]+=d*dv;
        a2[i+1][k+1][l][m]-=d*dv; }}}}
                                                                             ;45
for(i=1;i<ni;i++)                       // +dv 1-0
{ for(k=2; k<=nk; k++)
  { for(l=1;l<=nl;l++)
    { di=dist(i,k,l,i+1,k-1,l);
      for(m=0;m<3;m++)                                                       ;50
      { dv=cd*(di-dd)*(a1[i+1][k-1][l][m]-a1[i][k][l][m])/di;
        a2[i][k][l][m]+=d*dv;
        a2[i+1][k-1][l][m]-=d*dv; }}}}

 ......... similarly for the planes k-l and l-i ............                 ;55

for(k=1; k<=nk; k++)                    // fixed points 0
{ for(l=1;l<=nl;l++)
  { a2[ni][k][l][0]=ni; }}
for(k=1; k<=nk; k++)                    // fixed points 1                    ;60
{ for(l=1;l<=nl;l++)
  { a2[ni][k][l][1]=k; }}
for(k=1; k<=nk; k++)                    // fixed points 2
{ for(l=1;l<=nl;l++)
  { a2[ni][k][l][2]=l; }}                                                    ;65

for(i=1;i<=ni;i++)
{ for(k=1; k<=nk; k++)
  { for(l=1;l<=nl;l++)
    { for(m=0;m<3;m++)                                                       ;70
      { a0[i][k][l][m]=a1[i][k][l][m];
        a1[i][k][l][m]=df*a2[i][k][l][m]+(1-df)*a1[i][k][l][m];
        a2[i][k][l][m]=a1[i][k][l][m];          }}}}
```

8.1 Mass-Spring Models

```
    spl= 10000*(a1[1][nk][1][2]-1);                                    ;75
    ....
    }
    ....
}
float dist(int x1,int y1,int z1,int x2,int y2,int z2)                  ;80
{ float dx=a1[x2][y2][z2][0]-a1[x1][y1][z1][0];
  float dy=a1[x2][y2][z2][1]-a1[x1][y1][z1][1];
  float dz=a1[x2][y2][z2][2]-a1[x1][y1][z1][2];
    return sqrt(dx*dx+dy*dy+dz*dz);
}
```

8.1.5.5 Variable Fixed Points

Using a three-dimensional model, we can also investigate the oscillations of curvilinear bodies under strain (cf. 8.1.4.7). In this model, the eigenfrequencies will change as the body's shape changes, just as happens with a "Singing Saw". Let us show first that slowly pressing together the edges of a plate in our model causes the entire plate to bend, as is the case for a real plate. In the following animation, the lower left and right edges are slowly pushed together. (→ Animation)

```
Do[a2[[ni, k, 1, 1]] = ni - t*.02;, {k, 1, nk}];
Do[a2[[ni, k, 1, 2]] = k;, {k, 1, nk}];
Do[a2[[ni, k, 1, 3]] = 1;, {k, 1, nk}];
Do[a2[[1, k, 1, 1]] = 1 + t*.02;, {k, 1, nk}];
Do[a2[[1, k, 1, 2]] = k;, {k, 1, nk}];
Do[a2[[1, k, 1, 3]] = 1;, {k, 1, nk}];
```

The following figure shows the deformation at times $t = 10, t = 20$ and $t = 40$.

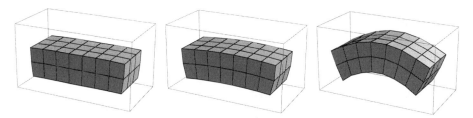

Let us rewrite the section of the C-program 8-1-5a where the fixed points are reset to the initial positions so that the fixed points become variable. In addition, let us write a new section in which we simulate striking the body regularly (C-program 8-1-5b).

```
// 8-1-5b Like 8-1-5a, except that the body is bent
//                    and struck at regular intervals
...
for (t = 0; t < tt ; t++)
{   if (z++ > 44100)                                  // Excitation
    { if (zz++ > 20) {z = 0; zz = 0; cout << .0000227*t << endl;}
      else { a2[2][2][1][2] += .05;          }}
...
for (k = 1; k <= nk; k++) a2[ni][k][1][0] = ni - t*ff; // fixed points
for (k = 1; k <= nk; k++) a2[1][k][1][0] = 1 + t*ff;
for (k = 1; k <= nk; k++) a2[ni][k][1][1] = k;
for (k = 1; k <= nk; k++) a2[1][k][1][1] = k;
```

```
for (k = 1; k <= nk; k++) a2[ni][k][1][2] = 1;
for (k = 1; k <= nk; k++) a2[1][k][1][2] = 1;
...
```

The following figure shows the resulting sound (8-1-5b). (→ Parameters and Commentary)

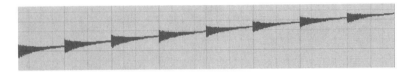

8.1.5.6 The Effect of Gravity

We can simulate the effect of gravity in our model by accelerating every mass downward by a certain amount at every iteration. In the Mathematica program above, we do this by adding this line:

```
Do[a2[[i, k, 1, 3]] -= .004;, {i, 1, ni}, {k, 1, nk}, {l, 1, nl}];
```

We simulate the body's falling onto a surface by limiting its downward movement.

```
Do[a2[[i, k, 1, 3]] = If[a2[[i, k, 1, 3]] < G, G, a2[[i, k, 1, 3]]];,
   {i, 1, ni}, {k, 1, nk}, {l, 1, nl}, {m, 1, 3}];
a0 = a1; a1 = a2;
```

Since our model represents a body that can be twisted, we obtain a simulation of an elastic collision. In the following animation we drop a cube of 3×3×3 points from a height of 0.8 and simultaneously strike it at point $M_{1,1,nl}$. Striking the body causes it to twist (cf. 8.1.5.3). If the body touches the lower surface, it first buckles and then because of its elasticity moves back into its original shape and begins to oscillate. Depending on the body's position, its velocity and rotation can change when it strikes the surface. (→ Animation)

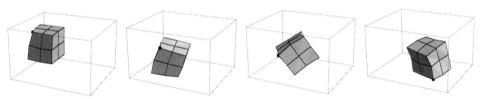

(→ Details and a C-Program)

8.1.5.7 Damping

The damping used thus far affects the displacement and the rotation of the whole body. Damping that affects only the oscillation is proportional to the relative movement of one point to the entire body. Let us introduce two new arrays $m1$ and $m2$ for the coordinates of the midpoint of the body at two successive times. We compute the relative velocity as the difference of the distances $(a2 - m2)$ and $(a1 - m1)$ between the point and the midpoint of the body at the two times. The coordinates of the midpoint can be computed at the start from the coordinates of the body and the height of fall as follows:

8.1 Mass-Spring Models

```
m1[3], m2[3];
..................
m1[0] = 1 + (ni - 1) / 2; m1[1] = 1 + (nk - 1) / 2; m1[2] = 1 + h + (nl - 1) / 2;
```

From the new coordinates we compute the coordinates of the midpoint.

```
for (i = 1; i <= ni; i++)                       // average
{ for (k = 1; k <= nk; k++)
  { for (l = 1; l <= nl; l++)
    { for (m = 0; m < 3; m++)
      {m2[m] += a2[i][k][l][m];}}}}

for (m = 0; m < 3; m++) m2[m] = m2[m] / N;
```

Finally, we subtract from $a2$ a value proportional to the velocity $(a2 - m2) - (a1 - m1)$.

```
a1[i][k][l][m] = a2[i][k][l][m]
                -df * (a2[i][k][l][m] - m2[m] - a1[i][k][l][m] + m1[m]);
m1[m] = m2[m];
```

8.1.6 Arbitrary Configurations and Variations

8.1.6.1 Coupled Strings

If we connect two similar pendula by a weak spring and set one into motion, we observe a remarkable phenomenon. Not only is the motion of the first pendulum transferred to the second until both oscillate equally strongly, the transfer continues until only the second pendulum is oscillating. Then the motion is transferred back to the first pendulum. This process continues until the motion is damped or interrupted by an outside agent. In the following example, we couple two strings with a variable link.
Let the displacements of the points of both strings be a_i (i from 1 to ni) and b_i (i from 1 to mi), let the number of points affected by the coupling be ka and kb respectively, and let the constant for the coupling be ck. Here are the calculations for the interaction between the two strings (8.1.3.2).

```
   ...                     // Calculation for the first string
a2[[ka]] -= ck * (a1[[ka]] - b1[[kb]]);
   ...                     // Calculation for the second string
b2[[kb]] -= ck * (b1[[kb]] - a1[[ka]]);
```

(→ Animation)

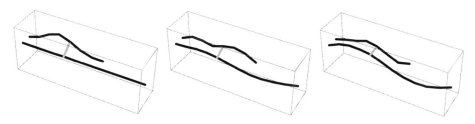

In the Max-Patch *MassSpring_CoupledStrings.maxpat*, we can enter and vary string length and the position and amount of coupling. If we set the number of masses of each string equal to three (one mass at each end of the string and one oscillating mass M_2), we have a system that corresponds to that of the pendula described above. The figure below shows the first 1.5 seconds of the resulting sound. (We see not the individual oscillations but rather the amplitude envelopes of the two strings.)

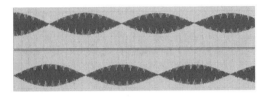

The coupling described above is so straightforward that we could realize it in the real world. But with a virtual string we can simulate a coupling in which the degree of coupling or the position of interaction change during the simulation. We can demonstrate this by defining the position of interaction as a function of time in our Mathematica program. (→ Animation)

```
ka = kb = IntegerPart[2 + (mi - 2) *t/tt];
```

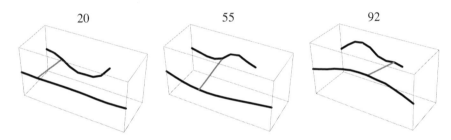

8.1.6.2 Geometrically Impossible Shapes

Below we show two arrangements of masses which are physically impossible, but whose oscillations can be simulated without difficulty because the calculations only require information about the connections between the masses.

In Chapter 8.1.5.5 we simulated bending a bar by pushing the lower edges together. If this movement is continued past the middle of the bar, the bar penetrates itself.

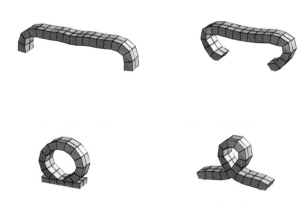

8.1 Mass-Spring Models

Virtual surfaces can also penetrate themselves. The illustration below shows a compressed torus which penetrates itself.

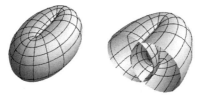

8.1.6.3 Spaces of More Than Three Dimensions

We can generate bodies in more than three dimensions and calculate their oscillations. We can understand how to simulate a four-dimensional cube by considering the extension of our original one-dimensional pearl necklace to the two-dimensional grid and to the three-dimensional grid. In one dimension, a point has two neighbors, in two dimensions two times two neighbors and in three dimensions three times two neighbors. Correspondingly, a point in four dimensions has four times two neighboring points.

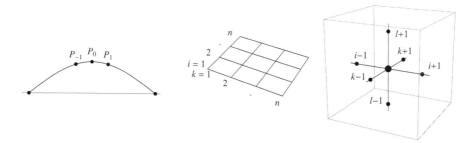

To simulate the oscillations of a four-dimensional cube in the same way as those of a three dimensional cube (8.1.5.2), we first define four-dimensional arrays with four indices (i, k, l, m) for the four spatial directions. Every point in the cube is assigned values for displacement in four directions: $a0_{i,k,l,m,n}$, $a1_{i,k,l,m,n}$, $a2_{i,k,l,m,n}$ ($n = 1$ to $n = 4$). As before, we call the constant factors for the longitudinal and transverse oscillations c_l and c_t. If we consider the movement along the x-axis, the acceleration of the mass $M_{i,k,l,m}$ due to the connection to the neighboring masses in the x-direction $M_{i+1,k,l,m}$ and $M_{i-1,k,l,m}$ is

$$-c_l(2a1_{i,k,l,m,1} - a1_{i+1,k,l,m,1} - a1_{i-1,k,l,m,1}).$$

All other directions are perpendicular to the x-axis and therefore can be calculated in the same way as for the transversal oscillations in the plane. We obtain as equation of motion in the x-direction ($m = 1$):

$$\begin{aligned}a2_{i,k,l,m,1} = {} & 2a1_{i,k,l,m,1} - a0_{i,k,l,m,1} \\ & - c_t(6a1_{i,k,l,m,1} - a1_{i,k+1,l,m,1} - a1_{i,k-1,l,m,1} - a1_{i,k,l+1,m,1} \\ & \qquad - a1_{i,k,l+1,m,1} - a1_{i,k,l,m+1,1} - a1_{i,k,l,m-1,1}) \\ & - c_l(2a1_{i,k,l,1} - a1_{i+1,k,l,1} - a1_{i-1,k,l,1})\end{aligned}$$

The other directions are treated correspondingly.

8.2 Wave Guides

Many natural instruments produce sound by exciting a string or an air column along which a wave propagates that is reflected at both ends. Periodic signals are generated because of the regularly recurring reflections. Wave propagation can be simulated very efficiently using digital delay lines. The first part of this chapter shows how to create delay lines and discusses some of their applications. In the second part, we combine two delay lines into a so-called wave guide in which waves propagate in both directions. In the third part of the chapter, we consider the excitation of the wave guide and the pattern of radiation of the generated sounds.

Introductions to the use of delay lines for sound synthesis can be found in Computer Music *by Charles Dodge and Thomas A. Jerse [2], pp. 277-282, in* Real Sound Synthesis for Interactive Applications *by Perry R. Cook [70] and in* Musical Signal Processing *edited by Curtis Roads [88], pp. 221-264. Primarily, however, articles in the* Computer Music Journal *should be mentioned which trace the development of synthesis using delay lines from the Karplus-Strong algorithm up to the simulation of the most varied instruments. For the Karplus-Strong algorithm see Kevin Karplus and Alex Strong,* Digital Synthesis of Plucked-String and Drum Timbres, CMJ Vol. 7, Nr. 2, 1983 [1-3] *and David A. Jaffe and Julius O. Smith,* Extensions of the Karplus-Strong Plucked-String Algorithm, CMJ Vol. 7, Nr. 2, 1983 [1-4]. *Much of the basic research in wave guide synthesis has been done by Julius O. Smith:* Physical Modeling Using Digital Waveguides, CMJ Vol. 16, Nr. 4, 1992 [1-7], Physical Modeling Synthesis Update, CMJ Vol. 20, Nr. 2, 1996 [1-5]. *In* Digital Sound Synthesis by Physical Modeling Using the Functional Transformation Method, *Lutz Trautman and Rudolf Rabenstein use closed-form expressions for synthesis [82].*

8.2.1 Simple Delay Lines

This chapter shows how to make delay lines in various programs and how to use them to produce periodic signals. It also explains how to determine frequency precisely and demonstrates simple techniques for exciting and damping oscillation.

8.2.1.1 Delay Lines

In the simplest case, if we generate a traveling wave in a string, the wave continues its motion in one direction without changing its form. We can simulate the behavior of an oscillating string by recording the movement of the waves. The animation below was made with a row of points by transferring, at the next computation step, the displacement of each point to the neighboring point to the right. The oscillation was initially excited by moving the leftmost point. The illustration below shows the string at the times $t = 2$ to $t = 5$ and $t = 37$.
(→ Animation)

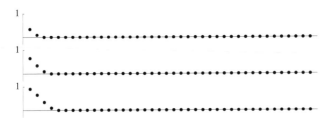

8.2 Wave Guides

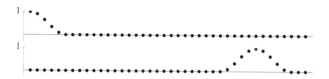

In an undamped string, the wave runs back and forth continuously. If we record the movement of a point on the string as sound, we get a strictly periodic signal. We get the same signal by moving the recording point in the opposite direction from the wave instead of shifting all the displacements. We realize this technically by writing the displacements into computer memory of length N and then reading them out sequentially, jumping back to the beginning when the end of the memory space is reached (5.1.4.2).

In Csound we do this with a delay line. The unit generator *delay* expects as parameters the displacements which have been read in (*a1*) and the time delay (*dt*):

```
a2      delay      a1, dt
```

To repeat the sequence of values, the value *a1* is read back into the delay line. Since this value has to be defined at the beginning of the program, we initialize it. At the beginning of each note, we fill the memory with an arbitrary waveform. To get sounds that are rich in partials, we often use white noise as the initial waveform.

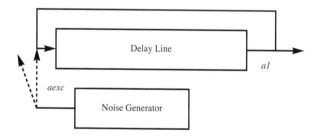

In the Csound instrument below noise of duration $p5$ is generated and read into the delay line as the exciting oscillation. Since the delay cannot be smaller than one control period, we set $ksmps = 1$ in order to be able to generate high frequencies.

```
instr 1
a1      init      0
kexc    linseg    p5,p4,p5,.001,0,p3-p4,0
aexc    rand      kexc
a1      delay     aexc+a1,p4
        out       a1
endin

;       Period  Amplitude
i1 0 5  .01     20000
```

In the program language C we can realize delay with feedback by defining an array with n elements (line 3 below), reading the elements out sequentially (line 6) and jumping back to the beginning when the end of the array is reached (line 7). Since the values in the array are preserved, it is not necessary to feed the values read out back into the delay circuit.

```
const int n = 100;
int i, dur = 44100*2, z = 0;
float a[n];
for(i = 0; i < n; i++) a[i] = rand ()/RAND_MAX - 0.5;
for(i = 0; i < dur; i++)                                          // 5
{    samp = a[z++];
     if(z == n) z = 0;
}
```

In Max the feedback can be realized with the objects *tapin~* and *tapout~*. In Max the delay cannot be shorter than the signal vector size, so we set this to 1 in order to produce high tones (alternatively, one can use the *poly~* object (4.2.2)).

See *[3] pp. 508-519, [6] p. 3.*

8.2.1.2 Simple Damping

If one does the examples above, the result is a fixed, inflexible sound in no way resembling the sound of a string. This changes if we introduce damping that reduces the high frequencies more strongly than the low. This sort of damping corresponds to a low-pass filter. We can make a simple low-pass filter by averaging two successive values *a1* and *a2*. (→ *8.2.1a.csd*)

```
;8-2-1a
ksmps = 1
instr 1
a1        init         0
a2        init         0
kexc      linseg       p5,p4,p5,.001,0,p3-p4,0
aexc      rand         kexc
a1        delay        aexc + .5*(a1 + a2),p4
a2        delay1       a1
          out          a1
endin

i1 0 4 .02 .5
```

Filtering the signal leads to a decrease in amplitude and to continuous smoothing of the waveform. But one can hear and see certain shortcomings of this technique. First, the amplitude decreases quickly at the beginning and then after about a second remains nearly constant. Second, depending on the excitation, the oscillation can contain a DC component.

The damping can be made more flexible by replacing the average of two successive values *a1* and *a2* by the weighted sum $c_1 a1 + c_2 a2$ where $c_1 + c_2 \leq 1$. If we set $c_1 = 1$ and $c_2 = 0$, there is no damping. The closer the values c_1 and c_2 are to 0.5, the stronger the damping will be. When the sum $c_1 + c_2$ is less than 1, there is additional damping independent of frequency, and the damped output signal converges rapidly toward zero, even if the mean value of the signal is not equal to zero. In the Csound program below, c_1 and c_2 have initial values of *p8* and *p10* respectively and change during the duration of the note to *p9* and *p11* respectively. The excitation is a noise signal with a random duration between zero and *p7* appearing at irregular intervals between zero and *p6*. (→ *8.2.1b.csd*)

8.2 Wave Guides

```
;8-2-1b
instr 1
i4=                 1/p4
idd=                p7*sr
idt=                p6*sr
k0      init        0
a1      init        0
a2      init        0
k1      init        idt
kd      init        idd
kc1     expon       p8,p3,p9
kc2     expon       p10,p3,p11
        if          kd   < k0      goto cont1
aexcit  rand        p5
                                   goto cont2
cont1:
aexcit=             0
        if          k1   > k0      goto cont2
k1      rand        idt
kd      rand        idd
cont2:
k1=                 k1-1
kd=                 kd-1
a1      delay       aexcit + kc1*a1 + kc2*a2,i4
a2      delay1      a1
        out         a1
endin

i1 0 5 150 .2 .8 .01 .5 .45 .5 .45
```

The following code excerpt shows the DSP-loop of the mxj~ external *cm_waveGuide_1* taken from Max-Patch *WaveGuide_1.maxpat*. Filtered noise or a pulse train can be used as excitation. The external has inlets for frequency, a feedback factor and the filter coefficients.

```
for(int i = 0; i < in0.length;i++)
{
    x = in0[i];                     // x = input sample
    y = fb_gain*buffer[wr];         // fb_gain = feedback gain
    y = c1*ydel + c2*y;             // low-pass filter
    ydel = y;
    buffer[wr] = y + x;             // wr = read and write position
    out[i] = y;
    wr += 1;
    if(wr >= per) wr = 0;           // per = period }
```

8.2.1.3 Frequency

We can calculate the period P of a tone from the length N of the delay: $P = N \cdot T$ (T = sampling period = $1/sr$) and hence the frequency:

$$f = sr/N$$

Because N is a natural number, not every frequency can be generated with the instrument above. The limitation is striking, even with a sampling rate of 44,100 Hz. The table below shows the frequency for various values of N and for the next larger number $N + 1$. The bottom line shows the frequency ratio between the two frequencies. At frequencies around 2000 Hz we can only tune the tones with an accuracy of about one half-step (the equally tempered half-step corresponds to a frequency ratio of 1.059546).

N	512	256	128	64	32	16	8	4
$f(N)$	86.132	172.26	344.53	689.06	1378.1	2756.2	5512.5	11 025
$f(N + 1)$	85.964	171.59	341.86	678.46	1336.3	2594.1	4900	8820
Interval	1.0019	1.0039	1.0078	1.0156	1.0312	1.0625	1.125	1.25

If we use a value of .000186 for the period *p4* in the Csound example 8.2.1a, we should get a frequency of 6376 Hz. Although this frequency is closer to $f(8)$ than to $f(9)$, Csound apparently chooses $N = 9$, because the frequency of the generated tone is clearly lower than 5000 Hz. If we measure the actual frequency, we get about 4640 Hz, which corresponds roughly to $N = 9.5$. There are two errors here: the first comes from rounding to an integer N, the second from the damping. Damping the signal by averaging the current value *a1* and the previous value *a2* is equivalent to interpolating between $N = 9$ and $N = 10$. The resulting frequency is $sr/(N + .5)$. Conversely, we can adjust the frequency precisely by interpolating. The disadvantage to this technique, however, is that then the damping coefficient is fixed. The differences in the damping of notes of various frequencies are so great as to make interpolation, at least for notes of high frequency, unusable.

To change the frequency during the duration of a tone, we define a delay line length at least equal to the longest period (N_{max}) and vary the tap point. In Csound, the unit generator *delayr* defines a delay line in which values can be read out at an arbitrary position with *deltap* or *deltapi*. The instrument 8.2.1c corresponds to the instrument 8.2.1b except for the code blocks below. The parameter *p4* gives the initial frequency, *p8* is the final frequency. There is no explicit damping in order to make clearer the damping effect of the interpolation. If we use the non-interpolating *deltap*, there is no damping, and the instrument does not produce a glissando. The pitch changes in ever-larger intervals (8.2.1c). (→ *WaveGuide_1.maxpat*)

```
;8-2-1c
instr 1
......
kfr        expon        1/p4,p3,1/p8
......
a0         delayr       .1
a1         deltap       kfr
           delayw       aexc+a1
           out          a1
endin

i1 0 20 1000 6000 .5 .0002 6000
```

If we replace *deltap* by its interpolating form *deltapi* (*8.2.1d.csd*), we get on the one hand a glissando, on the other, however, frequency-dependent damping.

Arbitrary frequencies can be obtained without changing the damping by using an all-pass filter (5.3.1.3), since an all-pass only changes the phases of a signal. Here we will only show how to compute the coefficients of such a filter and refer the reader to Chapter 3.3.3.5 for its derivation. If we call the *n*th value of an input signal $x(n)$ and the *n*th value of an output signal $y(n)$, the equation for a simple all-pass filter is:

8.2 Wave Guides

$$y(n) = a \cdot x(n) + x(n-1) - a \cdot y(n-1) \quad \text{with } 0 < a < 1.$$

The effect of this filter is easy to determine for $a = 0$ and $a = 1$. When $a = 0$, the formula becomes $y(n) = x(n-1)$ which is a delay by one sample period. When $a = 1$, we obtain $y(n) = x(n) + x(n-1) - y(n-1)$. To see the effect of this filter, we put in a few values of a signal $x(1), x(2), ...$ Before the signal begins, all displacements, and particularly $x(0)$, are zero. Hence $y(1) = x(1)$, $y(2) = x(2) + x(1) - x(1) = x(2)$ etc. That is, the filter has no effect. When the coefficient a is between 0 and 1, delays of a fraction of a sample period δ result. We have approximately:

$$\boxed{\delta = \frac{1-a}{1+a} \qquad a = \frac{1-\delta}{1+\delta}}$$

Let us now generate a tone whose frequency is exactly 1000 Hz using the simplest damping. For a sampling rate of 44.1 kHz the exact length of the delay without damping should be 44.1 sampling periods. If we choose $N = 43$, we have an effective length of $L = 43.5$ (see above). In this case we need to lengthen the delay by $\delta = 0.6$ sample periods using the all-pass filter. The formula above gives us the value .25 for the coefficient a.

```
a = .25;
lin = Table[Sin[2*Pi*(i-1)/n], {i, n}]; lallp = Table[0, {i, n}];
lout = Table[0, {i, n}];
Do[lallp[[i]] = .5*(lin[[i]] + lin[[i-1]]);, {i, 2, n-1}];
Do[lout[[i]] = a*lallp[[i]] + lallp[[i-1]] - a*lout[[i-1]];,
   {i, 2, n-1}];
```

The following figure illustrates the phase shift by one half sample period due to the low-pass filter and the phase shift of 0.6 sample periods due to the all-pass filter.

The following code shows the inlet-function and the DSP loop of the mxj~ external cm_waveGuide_2 of the Max-Patches *WaveGuide_2.maxpat*. The period *per*, the fractional part of the period *d* and the coefficient *a* are computed from the frequency input. The all-pass filter described above is implemented in the DSP loop. Since even the least DC-component in the input signal (e.g. for excitation by impulses) can be amplified by feedback, a DC-blocker was included in the code (3.3.3.5).

```
...
public void inlet(float f)
{
    if (getInlet() == 0) {freq = f; per = 44100.0f/freq;
    d = per -   (int)per; a = (1-d)/(1+d);}
    if (getInlet() == 1) fb_gain = f;
    if (getInlet() == 2) { c1 = f; c2 = 1 - f;}
}
...
for(int i = 0; i < in0.length;i++)
{
    x = in0[i];
```

```
        y = buffer[wr];
        y1 = fb_gain*(a*y + ydel - a*y1del);   // feedback and allpass filter
        y1 = c1*y1del + c2*y1;                              // low pass filter
        y2 = 0.5f*(1 + dc_coef)*(dc_coef*y2del + y1 - y1del);// DC blocker
    ...
    }
```

See [2] pp. 304-308, [1-3], [1-4].

8.2.1.4 Nonlinearity

The natural frequencies of strings are slightly high at great amplitude. We can realize this nonlinearity in our synthesis by measuring the amplitude and shortening the delay proportionally to the amplitude. In the Max external *cm_waveGuide_3*, the amplitude is estimated using a low-pass filter for the absolute values of the output signal. The coefficient c_av (inlet 3) determines the reaction time of the filter. When c_av is small, fewer values are averaged, when c_av is large, more values are averaged. The factor f_av (inlet 4) determines the degree of nonlinearity. All the partials are raised proportionally by the same amount so that the spectrum remains harmonic. In real strings the high partials are more affected by nonlinearity than the lower partials and the spectrum becomes inharmonic. (→ *WaveGuide_3.maxpat*)

```
    public void inlet(float f)
    {   ...
        if (getInlet() == 3) c_av = f;   // c_av = coefficient for average filter
        if (getInlet() == 4) f_av = f;   // f_av = strength of the nonlinearity
    }
    ...
    for(int i = 0; i < in0.length;i++)
    {
        x = in0[i];
        y = fb_gain*buffer[wr];            // feedback
        y1 = c2*(c1*y1del + y);            // recursive low-pass filter
        y2 = y1del = y1;
        y3 = 0.5f*(1+dc)*(dc*y3del + y2 - y2del); // normalized dc_blocker
        y2del = y2;
        y3del = y3;
        absy = Math.abs(y3);
        av = (1-c_av)*(c_av*avdel + absy); // recursive lp filter ("average")
        avdel = av;
        buffer[wr] = y3 + x;
        out[i] = y3;
        wr += 1;
        if(wr >= per - f_av*av) wr = 0;
    }
```

One can obtain frequency-dependent nonlinearity using all-pass filters having frequency-dependent phase responses. Because a first-order all-pass filter can produce a phase shift of at most one period, we have to use many first-order all-pass filters in series (or higher-order all-pass filters) to achieve an audible result. (→ Phase response of five first-order all-pass filters in series.)

8.2.1.5 The Excitation

A delay line with feedback is a special case of a comb filter (5.3.1.3) which filters out the frequencies n/dt from the input sound. In the examples above we used uniformly distributed random numbers as the input signal, which gives a spectrum in which all partials are approximately equally strongly present. We get the same spectrum with a random sequence of the numbers +1 and −1 (see the Karplus-Strong algorithm 8.2.1.6), but also with a random sequence that only partially fills the delay line or even with a single impulse. In the illustration below, we see on the left five periods of four waveforms, on the right the corresponding spectra.

Random sequence:

Random switching between +1 and −1:

Impulse:

Random sequence that does not fill the entire delay line:

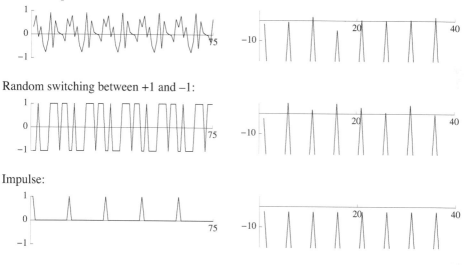

When using random switching between only two numbers, the probability is fairly high that a periodic signal can occur even within the delay period itself. If in the example above the period were three units shorter, there would be two identical waveforms within the period and hence a doubling of the frequency. Because in an ascending glissando the delay period is constantly getting shorter, there can be unexpected leaps of pitch. If one chooses an impulse as input signal to a wave guide (see below), the spectrum can differ greatly depending on where the excitation takes place (8.2.3.1). Again in an ascending glissando, it can happen that the shortening of the delay leads to skipping the impulse. When this occurs, the sound suddenly disappears.

Regardless of the sound input into a delay line, the spectrum of the resulting periodic sound is a line spectrum having the same spectral envelope as the input. The next figure shows white noise consisting of 100 random numbers (1). This signal was filtered to give a signal (2) whose spectrum (3) has a clear peak at 10 Hz. If we now generate a sound of length 10,000 by repeating the input signal 100 times and playing the sound with a sampling rate of 10,000 Hz, we get a sound with a fundamental frequency of 100 Hz and the spectrum (4) illustrated below.

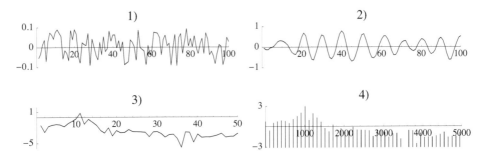

In the first of the following sound examples, one can hear the filtered noise, in the second the sound generated from it. (→ Sound Examples)

8.2.1.6 The Karplus-Strong Algorithm

In 1983 Kevin Karplus and Alex Strong published the article *Digital Synthesis of Plucked-String and Drum Timbres* in the Computer Music Journal (see [1-3]), the first description of the synthesis technique discussed above. Because the algorithm was meant to be used in real time, the authors took care to avoid any unnecessary calculations, in particular multiplications. The Karplus-Strong Algorithm was the first convincing synthesis technique based on a physical model. It is worthwhile looking at the essential points of the article.
If we call the tth value of a signal Y_t, then the synthesis using delays can be represented mathematically as $Y_t = Y_{t-p}$. Here p is the length of the delay and hence the length of the period of the generated sound. A simple low-pass filter gives the equation

$$Y_t = \frac{1}{2}(Y_{t-p} + Y_{t-p-1}).$$

Instead of using white noise for the input signal with uniformly distributed random numbers between $-A$ and $+A$, Karplus and Strong used a random succession of the two value $-A$ and $+A$. If we describe the delay by p samples as z^{-p}, we can draw the following block diagram:

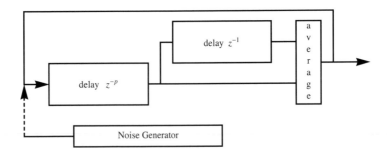

By slightly changing the algorithm in 1979, Kevin Karplus synthesized drum-like sounds. Before feeding the delayed signal back into the delay line, he multiplied some of the values by -1. A factor b indicates the probability that a value will be fed back unchanged. This factor influences the nature of the sound generated.

$$Y_t = \begin{pmatrix} +\frac{1}{2}\left(Y_{t-p} + Y_{t-p-1}\right) & \text{probability } b \\ -\frac{1}{2}\left(Y_{t-p} + Y_{t-p-1}\right) & \text{probability } 1-b \end{pmatrix}$$

When $b = 1$, we have the original algorithm. When $b = 1/2$, we get drum-like sounds. When $b = 0$, all values are multiplied by -1, which gives a strictly periodic signal of double period length. Because of the resulting point symmetry of the period, the sound only contains odd partials.

The Csound example 8-2-1e demonstrates the effect of the parameter b in the algorithm above. At every computation cycle ($kr = sr$) the variable *krnd* takes on a random value between -1 and $+1$. A control variable *kkrnd* goes from -1.3 to $+1.3$ during the duration of the note (line 1). If *krnd* is less than *kkrnd*, the input value to the delay line is multiplied by 1, if *krnd* is greater than *kkrnd*, the input value is multiplied by -1 (line 5). We use filtered noise as the excitation. At the beginning of the note we hear a low string-like sound which gradually changes to a drum-like sound and at the end returns to the string-like sound, this time an octave higher than at the beginning.

```
;8-2-1e
.......
kkrnd      line     -1.3,p3,1.3                                      ;1
.......
aexcflt    reson    aexc,p8,p9,2
krnd       rand     1
a1         delay    aexcflt+(krnd < kkrnd ? 1:-1)*.5*(a1+a2),ifr     ;5
a2         delay1   a1
.......

i1 0 14 150 8000 .5 .01 50 1000
```

Csound provides the unit generator *pluck*, which uses the Karplus-Strong algorithm for synthesis. The frequency parameter *icps* determines the length of the delay line, which usually corresponds closely to the actual frequency of the generated sound *kcps*. This parameter can be varied during the sound, unlike *icps*, which remains constant. The precise frequency is generated by resampling and transposing. At the outset, the delay line can either be filled with a function *ifn* or, if *ifn* = 0, with random numbers. The various methods of damping or changing the timbre (*imeth*) are: 1. Simple average; 2. Slower damping using a stretch factor *iparm1*; 3. Simple drum; 4. Drum with slower damping; 5. Weighted mean with the coefficients *iparm1* and *iparm2*; 6. First-order recursive filter with coefficients 0.5.

```
ar    pluck kamp, kcps, icps, ifn, imeth[, iparm1, iparm2]
```

See [1-3], [1-4], [2] pp. 304-308, [8] pp. 278-290.

8.2.2 Waveguides

In Chapter 8.2.1 we learned about waves which only run in one direction. But in fact sound waves run in both directions along a string or in a tube and are reflected at the ends. Let us first consider an ideal waveguide in which the waves keep their shape and are completely reflected at the ends.

8.2.2.1 The Ideal Waveguide

Let us define two delay lines of the same length, one for each direction of the wave. When a wave reaches the end of one delay line, it is reflected and changes direction. We simulate this process by multiplying the values at the end of one delay line by −1 and feeding them into the other delay line. We pick up the sound by reading out of both delay lines at the same position and adding the signals. Such a structure is known as a *wave guide*.

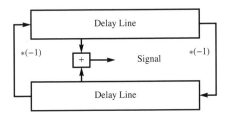

The following animation was made by shifting the values in two delay lines *lr* and *ll* (line 1) one position farther at each computation step (line 3). At the end, the values are multiplied by −1 and read into the other delay line (line 5). To show the state of the entire string, the displacements are not only summed at one position but along the whole string l (line 1). As excitation we use one half sine wave at point 20 (total length of the delay line 45 points).

```
.....
lr = Table[a0, {i, n}]; ll = Table[a0, {i, n}]; l = Table[a0, {i, n}];
.....
Do[lr[[n + 1 - t1]] = lr[[n - t1]]; ll[[t1]] = ll[[t1 + 1]];, {t1, 1, n - 1}];
Do[l[[t1]] = ll[[t1]] + lr[[t1]], {t1, 1, n}];
ll[[n]] = -lr[[n]]; lr[[1]] = -ll[[1]];
```

(→ Animation)

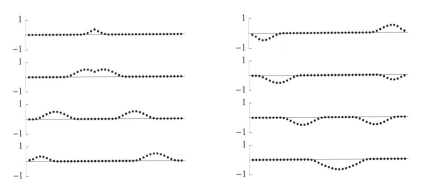

Although we generally use delay lines of fixed lengths, in Csound we will not use the unit generator delay but rather delay lines in which we can read out values at any point. This way we can implement directly the procedure described above and later easily build in extensions to the algorithm.

The delay length for a given frequency *p6* is 1/*p6* (lines 9 and 13 below). The pickup point, expressed as a fraction *p7* of the total length is *p7*/*p6* for one delay line and (1 − *p7*)/*p6* for the other (lines 2-3, 10, 14). So that we can tell which parts of the signal come from the wave moving to the right and which from the wave moving to the left, we use as excitation a short noise burst (line 7) of only positive values with an asymmetrical envelope (line 6).

8.2 Wave Guides

```
;8-2-2a
        instr 1                                             ;1
        iout1=          p7/p6
        iout2=          (1-p7)/p6
a1              init    0
a2              init    0                                   ;5
k1              linseg  0,p5/2,p4,p5,0,p3,0
aexc            rand    k1
a10             delayr  .1
a1              deltap  1/p6
a1out           deltap  iout1                               ;10
                delayw  abs(aexc)-a2
a20             delayr  .1
a2              deltap  1/p6
a2out           deltap  iout2
                delayw  -a1                                 ;15
                out     a1out+a2out
        endin

;               amp exc         dur exc     f       pickup
i1 0 2          15000           .002        200     .3126
```

Unlike the examples in Chapter 8.2.1, where the wave ran only in one direction, the spectrum of this sound contains both odd and even partials. The illustration below shows two periods of the sound.

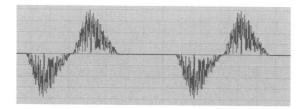

8.2.2.2 Reflection

At the ends of a string or a tube, the waves can be reflected either with the same phase or with opposite phase, depending on how the string or tube ends. Both ends of a taut string are fixed, and the wave is reflected with opposite phase at both ends, as was described above. If the string is hanging from one end and the other end is loose, the wave will not be inverted. The vertical displacement $y(t, x)$ of waves in a tube corresponds to the instantaneous over-pressure $p(t, x)$. The pressure cannot escape at the closed ends of the tube and is reflected back with the same phase. At the open ends of a tube, over-pressure turns into under-pressure and vice versa. The open end of a tube corresponds to the fixed end of a string, the closed end corresponds to the loose end of a string (2.2.1.5).

In waveguides with one open and one closed end, the waves run through the wave guide four times before a period is completed. A wave with positive displacements goes from left to right, is reflected at the right open end as a wave with negative displacements, is then reflected at the left closed end without changing sign. The wave changes sign again at the right end, and the cycle is only complete when the wave is reflected for the second time at the left end. In the following animation, a wave with positive displacements is generated at point 13 in a wave guide of length 20. The wave runs in both directions. (→ Animation)

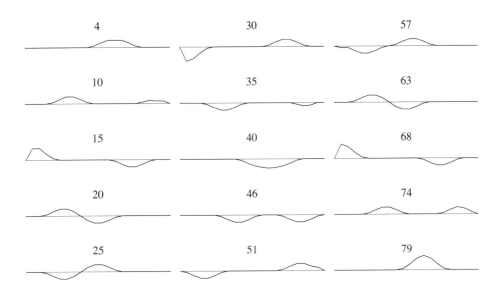

A complete cycle lasts 80 sample periods; the second half of the period is symmetrical to the first. Hence the spectrum of the generated sound only contains odd partials (8.2.1.4). We can observe this phenomenon in gedackt or stopped organ pipes and in clarinets, where the odd partials are considerably stronger than the even ones, at least in certain pitch and dynamic ranges. Trumpets only produce odd partials, too, but the mouthpiece and the bell influence the sound in such a way that it nearly corresponds to a complete partial series (8.2.3.4 and [20] pp. 276-278).

8.2.2.3 The Advancing Wave Front as Solution of the Wave Equation

The following illustration shows a segment of a string, both in a position of rest and in a position of arbitrary displacement. The mass of the segment is $\rho \Delta x$ (ρ = mass density). In the rest position, the same forces act on the segment from both sides through the tension T_0. Because the displaced segment is curved, the angles ϕ_1 and ϕ_2 are not exactly equal.

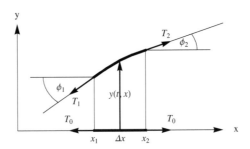

The wave equation results from Newton's second law: force = mass times acceleration. Let $y(x, t)$ be the displacement of a string at point x and at time t. We will show that the force acting on the element is proportional to the second derivative of the displacement with respect to x. The acceleration corresponds to the second derivative of the displacement with respect to time.

8.2 Wave Guides

The force $F_y(t)$ acting on the entire segment in an upward direction is composed of the vertical components of the forces T_1 and T_2 and is equal to $T_2\sin\phi_2 - T_1\sin\phi_1$. For small oscillations, we can consider $T_i\cos\phi_i \cong T_0$. Then, because $\sin\alpha = \cos\alpha\cdot\tan\alpha$, we have

$$F_y(t) = T_2\sin\phi_2 - T_1\sin\phi_1 = T_2\cos\phi_2\cdot\tan\phi_2 - T_1\cos\phi_1\cdot\tan\phi_1 = T_0\tan\phi_2 - T_0\tan\phi_1.$$

Further, since $\tan\phi_2$ corresponds to the derivative $\frac{\partial y(x,t)}{\partial x}$, we can write

$$F_y(t) = T_0\left(\frac{\partial y(x,t)}{\partial x}\right)_2 - T_0\left(\frac{\partial y(x,t)}{\partial x}\right)_1 = T_0\left(\left(\frac{\partial y(x,t)}{\partial x}\right)_2 - \left(\frac{\partial y(x,t)}{\partial x}\right)_1\right).$$

At a first approximation, the difference in parentheses is equal to Δx times the derivative at point x:

$$F_y(t) = T_0\Delta x\frac{\partial}{\partial x}\left(\frac{\partial y(x,t)}{\partial x}\right) = T_0\Delta x\frac{\partial^2 y(x,t)}{\partial x^2}.$$

Putting this into Newton's second law, with $\rho\Delta x$ for the mass of the segment (ρ = linear mass density), yields

$$T_0\Delta x\frac{\partial^2 y(x,t)}{\partial x^2} = \rho\Delta x\frac{\partial^2 y(x,t)}{\partial t^2}.$$

From this follows the wave equation

$$\frac{T_0}{\rho}\frac{\partial^2 y(x,t)}{\partial x^2} = \frac{\partial^2 y(x,t)}{\partial t^2} \quad \text{or} \quad \frac{T_0}{\rho}y''(x,t) = \ddot{y}(x,t).$$

We can describe a wave moving from left to right with velocity c that does not change its shape as a function of $x - ct$. We call the wave $y_r(x - ct)$. Similarly, we call the wave moving from right to left $y_l(x + ct)$. One can also describe the wave as functions of $t - x/c$ and $t + x/c$ respectively. This shortens the shape of the curve in the x-direction by the factor $1/c$. The time shift of one time unit corresponds exactly to one unit on the x-axis. These waves fulfill the wave equation.

To show this, let us define u as $x - ct$ and compute the second derivative of a function $y(u)$ with respect to x and to time t. For the first derivative, the following holds:

$$\frac{\partial y}{\partial t} = \frac{\partial y}{\partial u}\frac{\partial u}{\partial t}$$

and since $\frac{\partial u}{\partial t} = -c$

$$\frac{\partial y}{\partial t} = \frac{\partial y}{\partial u}\frac{\partial u}{\partial t} = -c\frac{\partial y}{\partial u}.$$

Differentiating again yields:

$$\frac{\partial^2 y}{\partial t^2} = -c^2\frac{\partial^2 y}{\partial u^2}.$$

Since $\frac{\partial u}{\partial x} = 1$, the derivative with respect to x gives

$$\frac{\partial^2 y}{\partial x^2} = \frac{\partial^2 y}{\partial u^2}.$$

By substitution in the previous equation, we obtain:

$$\frac{\partial^2 y}{\partial t^2} = -c^2 \frac{\partial^2 y}{\partial x^2}.$$

Representing the advancing wave as a function of $t - x/c$, we get the same result by substituting v for $t - x/c$. This yields $\frac{\partial v}{\partial t} = 1$ and $\frac{\partial v}{\partial x} = -1/c$, and we obtain the same result as above.

By exciting a string, we get two waves moving in opposite directions. Their sum can be written:

$$y(x, t) = y_r(t - \tfrac{x}{c}) + y_l(t + \tfrac{x}{c})$$

We obtain the digital representation of the traveling waves by substituting for continuous time t the expression $t_n = nT$, where T is the sample period $1/sr$. The natural choice for describing a distance in this context is the path X which the wave covers in one sample period: $X = cT$. Thus we can substitute for the continuous segment x the discrete value $x_m = mX$. Putting this into the equation for the traveling waves gives:

$$y_r(t_n, x_m) = y_r(t_n - x_m/c) = y_r(nT - mX/c) = y_r((n - m)T)$$

We can abbreviate $y_r(nT)$ by $y^+(n)$, and we get:

$$y_r(t_n, x_m) = y^+(n - m)$$

The terms $y_r((n - m)T)$ and $y^+(n - m)$ signify a signal $y_r((n)T)$ or $y^+(n)$ delayed by m sample periods (and the term $y_l((n - m)T)$ and $y^-(n - m)$ a signal $y_l((n)T)$ or $y^-(n)$ delayed by m sample periods). Hence we can write for the sum of the waves produced by exciting a string:

$$y(t_n, x_m) = y^+(n - m) + y^-(n - m)$$

See [3] pp. 298-300, [5] pp. 95-104.

8.2.2.4 Other Variables for Representing Waves

Thus far, we have described waves by the behavior of their displacements, because this corresponds to the visible movements of a string. In some contexts, though, it is interesting to consider not the instantaneous displacement, but velocity, acceleration, energy or other variables associated with wave propagation. Since the velocity v of the string segments corresponds to the first derivative of the displacement and the acceleration a corresponds to the second derivative with respect to time, we can convert the variables among themselves as follows:

$$v(t) = \partial y(t)/\partial t \quad a(t) = \partial^2 y(t)/\partial t^2 \quad v(t) = \int a(t)\, dt \quad y(t) = \int v(t)\, dt$$

In digital signals, the velocity corresponds to the difference of successive displacements per time unit, the acceleration to the difference of successive velocities per time unit:

$$v(t) = y(t) - y(t - 1) \qquad a(t) = v(t) - v(t - 1)$$

8.2 Wave Guides

The displacement is the sum of the previous displacement and the instantaneous velocity, the new velocity the sum of the previous velocity and the instantaneous acceleration:

$$y(t) = y(t-1) + v(t) \qquad v(t) = v(t-1) + a(t)$$

Formally these conversions are the equivalent of simple filters. These simple filters only give good results for frequencies considerably smaller than $sr/2$.

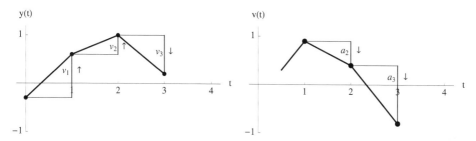

In addition to the derivatives with respect to time, let us consider spatial derivatives, that is derivatives with respect to x. The first derivative is:

$$y'(t,x) = \frac{\partial}{\partial x} y(t,x) = y_r'(t - x/c) + y_l'(t - x/c) = -\frac{1}{c}\dot{y}_r(t - x/c) + \frac{1}{c}\dot{y}_l(t - x/c)$$

or, calling velocity v:

$$y'(t_n, x_m) = \frac{1}{c}(v^+(n+m) - v^+(n+m)).$$

($\rightarrow$ Wave Impedance)

8.2.3 Sound Pickup and Excitation

8.2.3.1 The Positions of Pickup and Excitation

The positions of the excitation and of the sound pickup have the same influence on the spectrum of the generated sound in waveguides as in the spring-mass models (8.1.3.5 and 8.1.3.6). In the simple delay line, excitation by a single impulse gave rise to a spectrum containing all partials, regardless of where in the delay line the impulse occurred. In a waveguide the spectrum of the sound varies depending on the position of the excitation.

The following figures show on the left the waveform at the point of the sound pickup and on the right the waveform's spectrum. In order to record the complete spectrum of the sound, we choose as pickup point the first point of the wave guide, because if the exciting impulse occurs here, all the partials are present. At the same time, taking the sound here means that the partials' amplitudes increase with frequency (8.1.5.3).

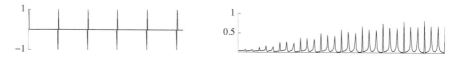

If the exciting impulse occurs in the middle of the waveguide, the spectrum contains only odd partials.

Conversely, let us fix the excitation at the first point of the delay line and vary the pickup position. The figures below show waveform and spectrum when the sound is picked up in the middle of the delay,

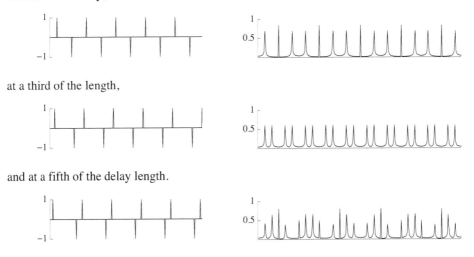

at a third of the length,

and at a fifth of the delay length.

In the following example, the excitation occurs in the middle of the waveguide and the sound pickup at one third of its length. Only odd partials are produced, and the pickup damps every third partial, so that only partials 1, 5, 7, 11, 13, etc. remain.

In the final example, we use a waveguide in which the waves are positively reflected at one end and negatively reflected at the other. This doubles the period of the wave and correspondingly halves its frequency. Because of the wave's symmetry, there are no even partials. The excitation occurs at one fifth of the delay length, which damps the partials 5, 10, 15, etc.

The code example below shows the inlet function and the DPS loop of the mxj~ external *cm_waveGuide_4* of the Max patch *WaveGuide_4.maxpat*. The inlet expects the length of the delay line *per*, the position of the pickup *c_pu* and the position of the excitation *c_exc* in samples. It then computes the pointers for the buffers of the wave traveling to the right (*wr*, *pu*, *exc*) and the wave traveling to the left (*wr*, *pu_l*, *exc_l*). The factor *refl* determines whether the waves are reflected positively or negatively at the ends of the string.

8.2 Wave Guides

```
...
public void inlet(float f)
{
    if (getInlet() == 0) per = (int)f;
    if (getInlet() == 1) fb_gain = f;
    if (getInlet() == 2) { c1 = f; c2 = 1 - f;}
    if (getInlet() == 3) c_pu = (int) f;
    if (getInlet() == 4) c_exc = (int) f;
    if (getInlet() == 5)
    {   if(f >= 0) refl = 1;
        else refl = -1;
    }
}
...
public void perform(MSPSignal[] ins, MSPSignal[] outs)
{   ...
    pu = wr + c_pu; if(pu >= per)  pu = pu % per;
    pu_1 = wr + per - c_pu; if(pu_1 >= per)  pu_1 = pu_1 % per;
    exc = wr + c_exc; if(exc >= per)  exc = exc % per;
    exc_1 = wr + per - c_exc; if(exc_1 >= per)  exc_1 = exc_1 % per;
    for(int i = 0; i < in0.length;i++)
    {
        wr += 1; pu += 1; pu_1 += 1; exc += 1; exc_1 += 1;
        if(wr >= per)  wr = 0;
        if(pu >= per)  pu = 0;
        if(pu_1 >= per)  pu_1 = 0;
        if(exc >= per) exc = 0;
        if(exc_1 >= per) exc_1 = 0;
        x = in0[i];
        y = buf_r[wr];
        y = c1*ydel + c2*y;                           // low pass filter
        y1 = 0.5f*(1 + dc_coef)*(dc_coef*y1del + y - ydel); // DC block
        ydel = y;
        y1del = y1;
        y_1 = buf_l[wr];
        buf_r[wr] = -fb_gain*y_1;
        buf_l[wr] = -refl*fb_gain*y1;
        buf_r[exc] += x;
        buf_l[exc_1] += x;
        out[i] = buf_r[pu] + buf_l[pu_1];
    }
}
```

8.2.3.2 The Duration of the Excitation

Thus far, we have either filled the delay line with the values of an exciting oscillation before the note begins, or we have added the exciting oscillation to the values in the delay line for the duration of the note. If we want to simulate the excitation of a medium already in oscillation, like a bowed string, however, we must proceed differently.

It is easy to simulate excitation when the string comes to rest for a moment as with a plucked string, where the string is first damped then displaced and from this new position let go. It is also easy to simulate excitation that can be reduced to the addition of an impulse in a single moment. But excitation that has an appreciable duration and excitation influenced by the waves themselves poses problems, as the following examples show.

In Chapter 8.2.2.1 we generated a wave by moving a point during the first 13 time units of the oscillation. The result only corresponded to a real action because there were no other waves and the movement returned to the zero position. Any waves already present would be reflected at the string's fixation point. What happens when the movement does not return to the zero position can be seen in the following animation. A sinusoidal motion is generated at point P_m extending only for a quarter period.

$$\text{If}\big[t < 8,\ l[[m]] = ll[[m]] = lr[[m]] = \text{Sin}[t * \text{Pi} / 14]^2,\ 0\big];$$

In a real string, the displaced points would move continuously. In our model however, the waves move in opposite directions, and therefore the points between the waves have a displacement of zero, leading to a discontinuity. (→ Animation)

We address this problem by considering the string's state after the excitation to be a new waveform. We halve this new waveform and put one half in each of the two delay lines. (→ Animation)

$$\text{If}[t == 7,\ \text{Do}[lr[[i]] = ll[[i]] = .5 * l[[i]],\ \{i,\ 1,\ n\}],\ 0];$$

This idea only works for strings initially at rest. If traveling waves are already present, a new error appears. By considering the instantaneous form of the wave as a new initial condition, we effectively cut the traveling waves into two waves moving in opposite directions. We can demonstrate this by repeating the command above at a later time $t = 18$. (→ Animation)

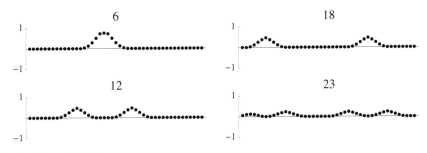

8.2.3.3 Excitation Without Feedback

The feedback between the resonating body of an instrument and that part which excites the oscillation can vary in its strength. In what follows, let us try to simulate the sound of a brass instrument. A good brass player can generate any frequency on his instrument, but normally only those tones sound that are near a natural resonance. In the Csound instrument that

8.2 Wave Guides

follows we simulate the natural resonances of a brass instrument. We construct a waveguide that reflects the waves moving to the left positively and those moving to right negatively. There are two delay lines, one for the waves moving to the left *al* and one for the waves moving to the right *ar* (lines 8 and 9). Since the waves go back and forth twice before completing one period, the delay length is 1/4 period. We introduce a simple damping in both delay lines, which causes an additional delay of 1/2 sample period. We make the excitation using an oscillator (line 7). The amplitude envelope (line 3) turns off the excitation 0.5 second before the end of the sound. The control function for the frequency (line 4) creates a glissando from *p5* to *p6*. We construct a waveguide that reflects the waves moving to the left positively (closed end of the tube at the mouthpiece) and those moving to right negatively (open end of the tube).

```
instr 1
al      init    0                                       ;1
ar      init    0
kamp    linseg  p4,p3-.5,p4,.01,0,.5,0
kfr     expon   p5,p3,p6
ardel   delay1  ar                                      ;5
aldel   delay1  al
aexc    oscil   kamp,kfr,1
ar      delay   .5*(al+aldel)+aexc,.25/p8
al      delay   -.5*p7*(ar+ardel),.25/p8
aldel   delay1  al                                      ;10
        out     ar+al
endin

f1 0 16 2 1 0 0 0 0 0 0 0 0 0 0 0 0 0 0 0
i1 0 10 1500 222 232 .985 230
```

For the first tone (8-2-3a1) we use a square wave impulse as the waveform for the oscillator (*f1*). The initial frequency of the oscillator is 224 Hz. During the duration of the note the frequency rises to 230 Hz. As with a brass instrument, we get the best sound when the frequency of the excitation corresponds to a natural frequency of the resonating body. This is not 230 Hz but about 228 Hz. The discrepancy is due to the lengthening of the delay caused by the damping. (→ Waveform)

Now we will try to simulate overblowing. The figures below show on the left what happens in the tube and on the right the resulting sound wave. Brass players generate the overtone series by producing the corresponding frequencies with the lips. We input to the waveguide running to the right an impulse lasting a fraction of the waveguide's period. As the animation shows, the excitation is amplified, and after only a few periods we have a standing wave. In this example one waveform is 7/4 periods long, and so the tone produced is the seventh partial tone. (→ Animation)

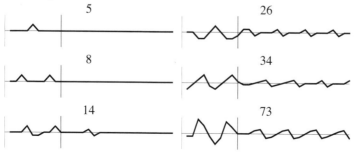

We can produce the overtone series (8-2-3a2) with the Csound instrument above by letting the frequency of the excitation go from 230 to 593 Hz. The waveguide's fundamental is 50 Hz. There are clear resonances at 250, 350, 450 and 550 Hz, that is, at the odd partials.
(→ Waveform)

```
f1 0 1024 7 0 24 1 1000 0
i1 0 40 1500 230 593 .985 50
```

8.2.3.4 Excitation With Feedback

It takes some effort to get a tube to produce the glissando with the odd partials described above. Normally the pitch jumps from one partial to the next. This has to do with the feedback between the pressure waves in the tube and the movements of the lips.

We can simulate this feedback in Csound by making the triggering of the exciting impulse partially dependent on the pressure waves. In the example below, we use the waveguide from the example above (lines 16-19). We first consider the sound production without feedback. The excitation is an impulse (*aexc*) that is triggered when the pressure in the mouth (*kexc*) exceeds a certain value *p4* (line 12). After an impulse is triggered, the pressure falls to zero. We smooth the impulse by letting it decay exponentially (line 20) and passing it through a filter (line 21). We generate the ascending glissando of the exciting oscillation by accelerating the pressure increase (lines 8-9). We simulate the feedback by changing the condition for triggering the impulse: the pressure threshold is no longer constant (*p4*) but varies with the displacement (*al* or *kal*) of the wave entering the tube at the left (line 12). The threshold is inversely proportional to the pressure of the incoming wave. In this way, the triggering of the impulse can be coordinated somewhat with the traveling wave in the tube.

```
        instr 3
        k0      init        0                               ;1
        kexc    init        0
        aexcr   init        0
        al      init        0
        ar      init        0                               ;5
        aout    init        0
        kfr     line        p5,p3,p6
        kd=                 kfr*p4/sr
        kexc=               kexc+kd
        kal     downsamp al                                 ;10

                if          kexc < p4-p9*kal    goto    c1
        aexc=               kexc
        kexc=               0
c1:                                                         ;15
        ardel   delay1      ar
        aldel   delay1      al
        ar      delay       .5*(al+aldel)+aexcr,.25/p8
        al      delay       -.5*p7*(ar+ardel),.25/p8
        aexc=               aexc*p10                        ;20
        aexcr   reson       aexc,40,kfr*p11,2
                out         ar+al
        endin
```

8.2 Wave Guides

```
;          amp    f min  f max   C damp    f       C feedback C Imp   C BW
i3  0  15  2800   90     900     .85       50      .19        .9      6
```

When the excitation frequency moves away from the natural resonance of the instrument in Sound 8-2-3b, the resonance becomes worse and the pitch first remains constant for a certain time and then jumps from one partial to the next, rather than moving in a glissando.
(→ Waveform)

The code below shows part of the DSP-loop of the mxj~ external *cm_waveGuide_5* from the Max patch *WaveGuide_5.maxpat*. The user can indicate the maximal pressure at which the excitation is triggered *p_max*, the pressure increase per sample *dpress* and a factor for the degree of feedback *p_f*.

```
for(int i = 0; i < in0.length;i++)
{
...
    press += dpress;
    if(press > p_max - p_f*y1)
    {   exc = press;
        press = 0;
    }
    buf_r[wr] = -fb_gain*y_1;
    buf_l[wr] = fb_gain*(y1 + exc);
    exc *=0.9f;
...
}
```

In a first attempt to simulate the behavior of a bowed string, we will use the unexact model for feedback described in Chapter 8.2.3.2. In the Mathematica program below, the variables are declared in the first three lines: the length of the delay n; the point of interaction m, the velocity of the bow v; an upper limit for the difference in velocity dv between bow and string at which the bow begins to adhere to the string; the greatest difference ds between the displacement in point P_m, where the adhesion breaks down, and the neighboring points; the duration *tmax*; the damping coefficient d; a control variable f showing whether the bow was sticking to the string at the last computation step; the previous displacement $m0$ at the point of excitation; the previous values at the points ml and mr. The delay lines consist of two lists lr and ll; a further list l contains the sum of the two delay lines. The outer loop counts the time (lines 4-20). In the fifth line, the values in the delay lines are shifted. In the lines 6-10 we simulate the interaction between bow and string. First we test whether the difference of the velocities of the bow (l[[m]] − m0) and the string (v) and the difference of the displacements at point P_m and its neighboring points (l[[m]] − .5(l[[m − 1]] + l[[m + 1]])) is small enough that the bow can stick to the string (line 6). If so, new values for the delay lines are computed from the old values by adding the velocity v (line 7), a vertical line *lin* is generated whose length is proportional to the velocity, and the control variable f is set to one (line 8). If adhesion is not possible and the control variable is not zero, that is, if the bow still stucks at the last step, then the control variable is set to zero and the string is set back to a new initial position by distributing the values in the list l evenly among the two traveling waves. The line showing the adhesion is erased, regardless of whether or not the control variable is set (line 10). Then the values of point P_m are stored, in order to be available at the next computation step (line 11). Next follows the computation of the displacements of the entire string by adding the traveling waves (line 12), simulating the reflection at the ends of the string (line 13) and increasing the bow velocity from zero during the first 10 computation steps (line 14). The graphic representation of the string and the line showing the adhesion is in lines 16-20.

```
n=12;m=8;v=0;dv=.045;ds=.36;                                              //1
tmax=150;d=.993;f=1;m0=mr=ml=0;rm=0;lm=0;
lr=Table[0,{i,n}];ll=Table[0,{i,n}];l=Table[0,{i,n}];
Do[rm=lr[[n]];lm=ll[[1]];
  Do[lr[[n+1-t1]]=d*lr[[n-t1]];ll[[t1]]=d*ll[[t1+1]];,{t1,1,n-1}];        //5
  If[Abs[l[[m]]-m0-v]<dv &&(l[[m]]-.5(l[[m-1]]+l[[m+1]]))<ds,
    lr[[m]]=mr+v/2;ll[[m]]=ml+v/2;
    lin=Line[{{m+1,l[[m]]},{m+1,l[[m]]+6*v}}];f=1,
    If[f==1,f=0;
      Do[ll[[t1]]=lr[[t1]]=.5*l[[t1]],{t1,1,n}];lin={};,lin={}]];         //10
  mr=lr[[m]];ml=ll[[m]];m0=l[[m]];
  Do[l[[t1]]=ll[[t1]]+lr[[t1]],{t1,1,n}];
  ll[[n]]=-lr[[n]];   lr[[1]]=-ll[[1]];
  If[t<10,v+=.01,0];
                                                                          //15
  P1[t]=ListPlot[
  Join[{0},l,{0}],Prolog->lin,PlotJoined->True,Axes->{True,False},
  PlotLabel->t,PlotRange->{{1,n+2},{-1,1}},
  Ticks->{None,None},AspectRatio->.18],
  {t,0,tmax,1}]                                                           //20
```

The pictures below are taken from an animation of the above situation and show that after a certain time the feedback causes a periodic excitation. Each snap shot shows the string at the last moment of adhesion. A period of length 12 establishes itself, and after only a few cycles the oscillations are strictly periodic. The period length is too short, however, 12 instead of 24 as we would expect from the length of the delay. If the parameters are chosen differently, the period can become too long. When $dv = .02$ and $ds = .6$, we get a period of 30. This is a consequence of the inaccuracy of the model and is due to the fact that the string cannot come to rest and that the period length is not determined by the resonance of the string but by the moments of interruption of the adhesion of the bow to the string. (→ Animation)

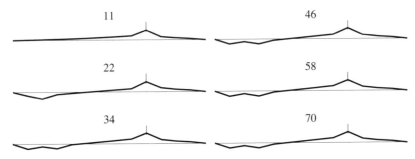

To correct these flaws, we change the program so that the string can oscillate freely by computing separately the additional displacement arising through the bowing, storing it in a new list *ls* and adding it to the traveling wave at the moment when the adhesion is interrupted. The additional displacement is independent of the traveling waves and causes the entire string to become distorted so that it only has two sections, one from the left fixed point to the point of excitation (line 2), the other from the excitation to the right fixed point (line 3). When the adhesion is interrupted, we add half of this additional displacement to each of the traveling waves (line 5) and set the additional displacement to zero (line 6). At this moment the shape of the string consists of the traveling waves and the additional displacement (line 8).

```
.........
If[Abs[l[[m]]-m0-v]<dv &&( l[[m]]-.5(l[[m-1]]+l[[m+1]]))<ds,        ;1
  Do[ls[[i]]+=i*v/m,{i,1,m}];
```

```
Do[ls[[i]]+=(n+1-i)*v/(n+1-m),{i,m+1,n}];
lin=Line[{{m+1,l[[m]]},{m+1,l[[m]]+6*v}}];f=1,
If[f==1,f=0;Do[ll[[t1]]+=.5*ls[[t1]];lr[[t1]]+=.5*ls[[t1]];     ;5
ls[[t1]]=0;,{t1,1,n}];lin={};,lin={}]];
.........
Do[l[[t1]]=ll[[t1]]+lr[[t1]]+ls[[t1]],{t1,1,n}];
.........
If[t<20,v+=.005,0]; If[t==120,v=10,0];                          ;10
.........
```

The diagram below shows the independently computed displacements due to the traveling waves (top) and due to the bow (middle). The lower figure shows their sum at the last moment of adhesion $t = 76$ and at the immediately following moment $t = 77$. (→ Animation)

The following animation shows the result of this simulation. Bowing was interrupted at time $t = 120$, and the moments of maximum positive displacement of the string are shown to prove that the period of the bowed string now corresponds to that of the freely oscillating string. (→ Animation)

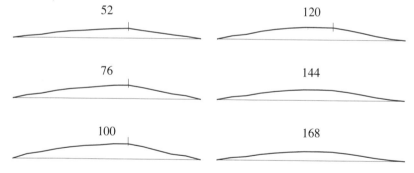

8.2.3.5 Selective Reflection and Sound Radiation

When we listen to instruments, we do not hear the oscillations of a string or of an air column. In string instruments, we essentially hear the sound radiated from the resonating body. In wind instruments, we hear the sound emerging through the various openings of the instrument. The animation below uses a model in which waves are reflected positively at the left end and negatively at the right end. At the right end, indicated by the vertical line, some sound can escape and some is reflected. At the open end of a tube, only those waves are reflected that are significantly longer than the diameter of the opening. Shorter waves can escape without difficulty. To simulate this behavior, we introduce a simple low-pass filter for the reflected waves and at the same time let the high-pass filtered signal escape as sound. The first four figures below show moments during the first period. The last two show the situation two periods later. (→ Animation)

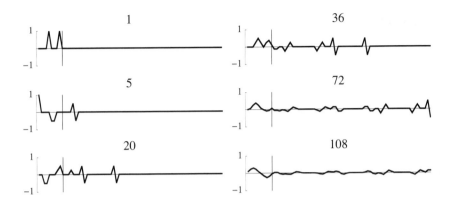

8.2.3.6 Harmonics

Harmonics are produced on string instruments by letting one finger lightly touch a vibrating string at some fraction of its length. The traveling waves are partially reflected at this point. We simulate this behavior by reflecting those parts of waves having positive displacement and letting those parts with negative displacement pass.

```
n = 11; m = 9; fp = 6;                                              ; 1
.......
lr[[m]] = ll[[m]] = 1;
.......
If[lr[[fp]] > 0, ll[[fp]] -= lr[[fp]]; lr[[fp]] = 0, 0];
If[ll[[fp]] > 0, lr[[fp]] -= ll[[fp]]; ll[[fp]] = 0, 0];            ; 5
.......
```

Let us excite a string with an impulse at point m and divide the string in the middle at point fp. The illustrations below show on the left the reflection of the impulse with positive displacement at the point fp, on the right the passing of an impulse with negative displacement. (→ Animation)

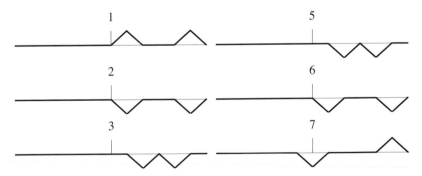

In the example that follows we simulate a plucked string by so deflecting the string that we get two straight segments from the point P_m to the ends of the string (lines 3-4). In addition, we introduce a frequency-dependent damping (lines 6-7).

8.2 Wave Guides

```
n = 13; m = 10;                                                      ; 1
. . . . . . .
Do[lr[[t1]] = ll[[t1]] = -.5*t1/n;, {t1, 1, m}];
Do[lr[[t1]] = ll[[t1]] = -.5*(n - t1)/(n - m);, {t1, m + 1, n}];
. . . . .                                                            ; 5
ll[[n]] = -.5*(rm + lr[[n]]);
lr[[1]] = -.5*(lm + ll[[1]]);
```

The figures below show on the left the initial condition of the string, the moment of touching the string and the effect of touching it. The figures on the right correspond to the beginning, the middle and the end of a period respectively. (→ Animation)

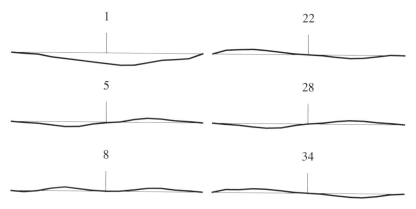

We see that this model is not very good when we divide the string anywhere but in the middle. Because waves with positive displacement are completely reflected at the point of division and every wave with negative displacement is reflected with a change of sign at the ends, after at the most one period there is no longer any interaction between the two parts of the string and they oscillate independently of one another. (→ Animation)

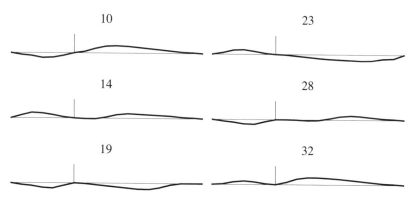

(→ Another animation, with an impulse as excitation, shows that the two parts of the string oscillate independently even when the division is in the middle.)

We obtain a better result when only part of the wave is reflected at the point where the string is touched. To simulate this, we substitute two new lines of code (lines 4-5) for the lines 1-2 so that one half of the wave is reflected at the point of contact and one half passes.

```
If[lr[[fp]] > 0, ll[[fp]] -= lr[[fp]]; lr[[fp]] = 0, 0];
If[ll[[fp]] > 0, lr[[fp]] -= ll[[fp]]; ll[[fp]] = 0, 0];
... ....
If[lr[[fp]] < 0, ll[[fp - 1]] -= d*.5*lr[[fp]]; lr[[fp]] *= .5, 0];
If[ll[[fp]] < 0, lr[[fp + 1]] -= d*.5*ll[[fp]]; ll[[fp]] *= .5, 0];
```

The animation below was generated by putting these new lines into the model of the bowed string from Chapter 8.2.3.4 (last example). In the pictures showing the moments at which the adhesion breaks down, we can see after a short time a regular period and an additional nodal point at 2/3 of the string's length. (→ Animation)

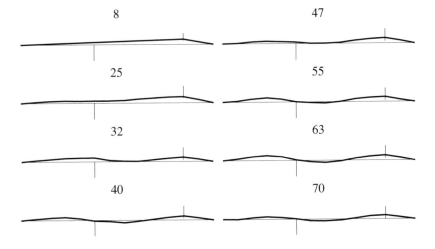

9 Sound and Space

The treatment of space in electroacoustic music is complex. Sounds are produced by loudspeakers, hence spatial perception depends principally on the placement of the loudspeakers. How must the loudspeakers be arranged to create a specific spatial effect? The difficulty here is that in general the sounds appear simultaneously in more than one loudspeaker, and hence different positions in space do not receive the same information. Nonetheless, the possibility of simulating the characteristics of virtual spaces and sound sources has meant that spatial relationships have come to play an important role in much electroacoustic music. Positioning sound events at different places in virtual space makes it both easier to distinguish individual sounds in complex contexts and possible to bring sounds in relation to each other using only one independent parameter.

In order to create a specific spatial effect, one must know what information the auditory perception requires and how this information can be conveyed. The selection and processing of the information are complicated. Besides, experience shows that sometimes even the best simulation does not elicit the desired response. Our idea of the space we are in is principally determined by previous knowledge. We normally know where we are, and this knowledge is continuously substantiated and emended if necessary by our visual and other senses. Acoustic information is usually given less perceptual importance than previous knowledge. For example, one localizes the amplified voice of a speaker at the position of the speaker, even though the perceived sound gives other information. Inversely, a listener hearing a sound coming from a particular direction becomes uncertain about the sound's position if he or she cannot see what produced the sound. Even when a virtual space is very realistically simulated, the characteristics of the actual space in which one is listening, like reflections, reverberation and real noises, cannot be eliminated, and the listener always knows what space he or she is in.

Further Reading: Older books like Dodge and Jerse's Computer Music *[2] and Moore's* Elements of Computer Music *assume computers with very modest computing power and correspondingly limit their discussions to simple models and techniques of spatial simulation. For room acoustics and recording techniques see* Musical Acoustics *by Donald E. Hall [20] and* Handbook for Sound Engineers *by Glen M. Ballou [42]. Introductory texts to spatial simulation can be found in* Current Directions in Computer Music Research *edited by Max Mathews and John Pierce [44] and in various articles of the the* Computer Music Journal, *particularly Volume 19 No. 4. The standard work on spatial hearing is* Spatial Hearing: The Psychophysics of Human Sound Localization *by Jens Blauert [45].*

9.1 Spatial Hearing

Animals use their hearing to locate enemies and prey and need to know as quickly as possible which direction a sound came from and how far away the sound is. Determining the direction is very quick and, at least in the horizontal plane, very accurate. Determining the kind of sound requires more time and often requires visual input. Determining the distance of a sound is normally quite difficult. Because the auditory perception processes a variety of stimuli and continuously checks with other senses and with previous knowledge, it is very difficult to simulate virtual spaces and sound sources with loudspeakers.

When comparing acoustic and visual phenomena, one must be aware of two differences between the respective perceptual systems. First, the retina receives a two-dimensional mapping of three-dimensional reality, but the ear records only a one-dimensional signal from which the brain extracts spatial information. Secondly, sound waves usually come directly from the heard object itself, while the light waves arriving at the eye are normally only reflections of light on the object seen.

Further Reading: Spatial Hearing: The Psychophysics of Human Sound Localization *by Jens Blauert [45].*

9.1.1 Sound Localization

9.1.1.1 Interaural Time Difference ITD

To determine the azimuth of a sound, that is, the angle between the sagittal plane and the direction of the sound, the human auditory system uses the difference in arrival time of the sound at the two ears. This difference is called the *Interaural Time Difference* (ITD). If we indicate the position of the head and that of the sound in Cartesian coordinates (figure left below), then the distance d_l of the sound source at the point (x_0, y_0) to the left ear is

$$\sqrt{(x_0 - h)^2 + y_0^2}\ ,$$

the distance d_r to the right ear is

$$\sqrt{(x_0 + h)^2 + y_0^2}$$

and the difference of the distances is

$$|d_r - d_l| = |\sqrt{(x_0 - h)^2 + y_0^2} - \sqrt{(x_0 + h)^2 + y_0^2}\ |.$$

If the position of the sound source is given in polar coordinates, we can convert them to Cartesian coordinates and apply the formula above, or we can use the following simple approximation. If the sound is far away, its wave front reaches the listener as a (nearly) straight line, and the angle φ between a line connecting the two ears and the wave front is the same as the angle between the sagittal plane and the sound source. The difference of the distances to the two ears is $dh \cdot \sin(\varphi)$, where dh is the diameter of the head. We find the time difference ITD by dividing this difference by the speed of sound.

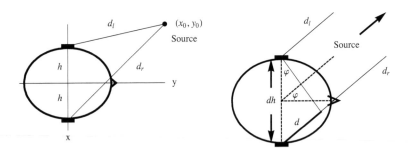

Both of the calculations described above are only approximations, because we need to take account of the fact that the actual path of the sound to the far ear goes around the head. This additional distance obviously depends on the size of the head and varies from listener to listener. Using polar coordinates, we get a better approximation by using the (easier to calculate) factor $dh\cdot\varphi$ (φ in radians) instead of $dh\cdot\sin(\varphi)$. Our auditory system uses time differences from about 20 μs (.00002 s) to about .6 ms to determine direction. When the time differences are larger than a few milliseconds, we perceive separate auditory events. For time differences inbetween, we perceive only singular auditory events, because the sound arriving first determines the direction perceived. This phenomenon, called the *precedence effect* or the "*law of the first wave front*", prevents reflections from disturbing the perception of direction. For work with loudspeakers, the precedence effect means that even when the loudspeakers are at greatly varying distances from the listener, the sound image remains coherent and does not break up into multiple images. On the other hand, it also means that even if the distances to the loudspeakers vary only a bit, the sound seems to come only from the closest loudspeaker.

If we synthesize tones of nearly equal frequency (less than 1000 Hz) and equal amplitude in two channels, we hear the sounds coming alternately from the left and from the right, and we perceive a circular movement. Sometimes it is difficult to determine whether the sounds are circling to the left or to the right (*Sound_Localization.maxpat*). This is because the shifting phase causes the sound to appear earlier first at one and then at the other ear. In the illustration below, the lower waveform first seems to be somewhat "ahead" of the upper one, then a few periods later somewhat "behind" it.

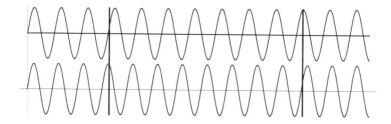

Since the "earlier" sound dominates, one hears the tones alternately, presumably causing the perception of circular movement. The frequency variation is the same as in the Doppler effect that would be produced by the sound's moving toward and then away from the listener. One can test this effect for various frequencies and frequency differences with the program *Sound_Localization.maxpat*. For frequency differences between about 5 and 20 Hz one can hear the tones beating, if one listens with earphones.

9.1.1.2 Interaural Intensity Difference IID

For the reasons mentioned above, it is clear that the difference in arrival time of the sound wave at the ears cannot be used for stereophony with loudspeakers. Accordingly, changing the panorama control on stereo sets or mixing programs only changes the difference in loudness (*Interaural Intensity Difference* IID) between the two channels. The IID is difficult to specify, because it depends not only on the direction and distance of a sound, but also on the sound's frequency (9.1.1.3). Nonetheless, we shall give approximate solutions to the problem of placing a sound virtually between two loudspeakers. Imagine the listener equally far away from two loudspeakers whose directions form a right angle. If we play a sound initially only in the left loudspeaker and then linearly fade the sound out and simultaneously fade in the same sound in the right loudspeaker, the sound seems to move from the left to the right loudspeaker. The movement is not however along a straight line between the speakers, but rather the sound seems first to move away, then after the middle to come closer again.

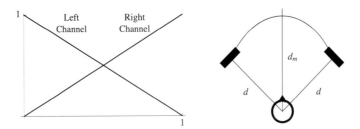

The reason for this is that the intensity of the sum of the two signals is not constant. Because the intensity is proportional to the square of the amplitudes, we have at beginning and end of the experiment an amplitude and intensity of 1, but in the middle the intensity is

$$\sqrt{.5^2 + .5^2} = \sqrt{.5} = \sqrt{2}/2 = .7071...$$

The auditory system interprets the decrease in intensity as an increase in distance. To keep the intensity constant, we use for the fade-in and fade-out the functions $f(x) = \sin(\pi x)$ and $g(x) = \cos(\pi x)$ respectively, the sum of whose squares $f(x)^2 + g(x)^2 = 1$ is constant.

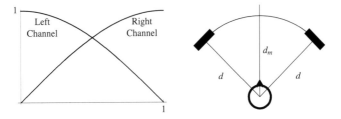

(→ Straight Movement)

To create the illusion that a sound is moving in a straight line, we need to adjust its intensity to correspond to the changing distance. If we add a coordinate system to our set-up of before, we have for the distance D of the virtual sound source at the point $P(x, y) = P(x_t, a)$:

$$D = \sqrt{x_t^2 + a^2}\ .$$

Since the amplitude is inversely proportional to the distance, we divide the functions described above by this term and get the functions illustrated at the left below.

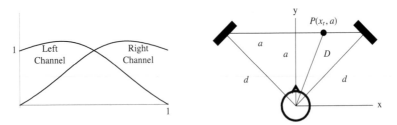

In general, it is difficult to estimate the distance of a sound source. The path taken by a virtual sound source will be very differently perceived by different people, or even by the same

9.1 Spatial Hearing

person in different situations. The speed of the sound usually does not seem constant. If one listens with earphones to room simulations using only intensity differences, the sound clearly moves (for instance) from left to right, but the impression of depth is missing. The reason for this is that the time difference and the frequency dependency of the intensity difference are not taken into account. (→ *Sound_Localization.maxpat*)

9.1.1.3 Head-Related Transfer Function HRTF

The time and intensity differences described above are used by the perception to determine the direction of a sound source in the horizontal plane, but they do not give any information about a sound's vertical position. If the head were completely symmetrical, then the time and intensity differences for a sound source at a given distance would be the same for all points on a circle at that distance whose center is on a line passing through both ears. If the distance is unknown, this is true for all points on a cone extending out from the ears. (→ Calculating the equation for a shape similar to a cone)

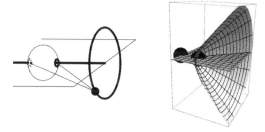

Because the pinna, the head and the torso have asymmetrical shapes, the sound wave arriving at the ears is subject to diffraction and overlapping reflections which have the effect of filtering the sound. The changes in the sound are described by a transfer function (3.3.1.4). One speaks of a *directionally dependent transfer function* DTF or a *head-related transfer function* HRTF. The illustration below shows the sound wave of an impulse arriving at the two ears. The upper figures show the waveform at each ear, the lower figures show the amplitude response at each ear (the figures to the left are for the left ear). The waveforms show not only the time delay and the lesser intensity at the far ear but also individual strong reflections whose time delays indicate the differences in the paths travelled by the sound to each ear. When the sound comes from certain directions, for example, one can see a strong reflection with a delay of about $dt = .06$ ms, corresponding to a difference in path length of about $dt \cdot 340$ m $= 2$ cm. This delay is caused by the outer fold of the pinna. The sum of a signal and its delayed copy corresponds to a comb filter, and so the amplitude response at that ear has clear notches (for $dt \approx .06$ ms at $.5/dt \approx 8333$ Hz).

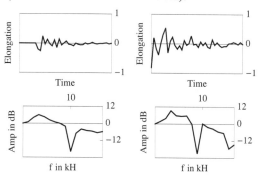

One can make a whole set of the transfer functions by generating a signal with a known spectrum, e.g. white noise, at various points in space, recording the sound at the ear or in the auditory canal and then comparing the spectra of the original and the recorded sound. Since the transfer functions of filters working in series are multiplicative, known or measured distortion in either equipment or in the auditory canal can be divided out of the result, leaving the true transfer function. Hence sounds can be placed in virtual space by filtering them according to such transfer functions.

9.1.2 Distance

9.1.2.1 The Decrease of Sound Intensity With Distance

Sound waves in a homogenous medium propagate uniformly in all directions. In a first approximation, the energy is conserved and can be thought of as being distributed onto the surface of a sphere which increases as the square of its radius, that is, the increasing distance from the sound source. The illustration below shows how the surface is quadrupled when the radius is doubled.

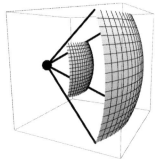

Therefore, the intensity I, defined as energy per unit of surface, decreases inversely as the square of the distance d to the sound source: $I \propto 1/d^2$. Since the intensity is proportional to the square of the amplitude, the amplitude decreases inversely as the distance: $A \propto 1/d$.

There is an additional decrease in sound intensity because part of the energy is absorbed by the medium through which the sound passes. This decrease is exponential. For the intensity as function of the distance from the sound source $I(d)$ we have:

$$I(d) = \frac{1}{d^2} I_0 e^{-ad},$$

For the amplitude we have:

$$A(d) = \frac{1}{d} A_0 e^{-\delta d}.$$

The coefficient a is called the sound *absorption coefficient*, the coefficient δ the *decay coefficient*. From $I \propto A^2$ follows $a = \delta/2$. These coefficients depend on the nature and temperature of the transmission medium. In air, they also depend on the humidity. The following table ([43] p. 614) indicates the sound absorption coefficient a (in m^{-1}) for four frequencies and four humidities.

9.1 Spatial Hearing

	1000 Hz	2000 Hz	3000 Hz	4000 Hz
40 %	.0013	.0037	.0069	.0242
50 %	.0013	.0027	.0060	.0207
60 %	.0013	.0027	.0055	.0169
70 %	.0013	.0027	.0050	.0145

Sound absorption increases with frequency, and so the high components of a spectrum decrease more quickly with distance than do the lower components. Normally, one does not perceive this effect. The auditory perception uses the spectral shift to estimate distance and effectively reconstructs the original timbre. For instance, we do not hear a change in vowel timbre when someone speaking moves away from us. But if we contrast the hiss and crack of a close lightning strike with the rumble we hear when lightning strike far away, the spectral shift is very noticeable. In general, it is difficult to judge loudness differences and hence the distance of a sound source, and it is correspondingly difficult to simulate the effect of distance on a sound source. In the literature we find various functions for the decrease of amplitude (see [45] p. 121). The figures below show the amplitude decrease proportional to $1/d$, proportional to $1/d^{1.5}$ (see [8] p. 370) and proportional to $1/d$ with additional reduction for absorption by the air, for various order of magnitude of distance. The decay coefficient is $\delta = .01$.

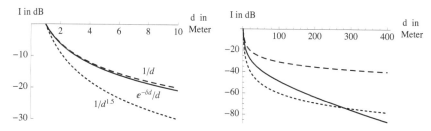

The following illustration shows the amplitude decrease for frequencies of 1000 and 4000 Hz (with $\delta = .0026$ and $\delta = .0414$ respectively).

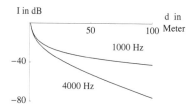

9.1.2.2 Proportion of Indirect Sound

The perception also uses the loudness relationship between reverberation and direct sound to determine the distance of a sound source. The greater the distance, the greater the proportion of reverberation. In a closed room, the loudness of the reverberation, consisting of the sum of all the reflections, is approximately the same everywhere. The loudness of the direct sound, on the other hand, decreases quickly with increasing distance from the sound source. One can see how dominant this factor can be in determining distance in the experience of test subjects who in dark, anechoic rooms always judge a speaker to be closer than he or she is, despite the absorption of the high range of the spectrum and despite the decrease in loudness with distance.

9.1.3 Movement of Sound in Space

9.1.3.1 The Doppler Effect

The Doppler effect was discussed in Chapter 2.2.4.5. Here we will demonstrate its importance for spatial orientation. Unlike the small differences in time and loudness and the often weak reflections from which the auditory perception creates a sense of space, the changes in frequency caused by the Doppler effect are often large and can be consciously perceived. Using the Doppler effect, one can judge the relative velocity between listener and sound source only if the actual pitch of the sound is known or if additional information is available from acoustic reflections, as is often the case with vehicle sirens in a city, for example. If a siren moves straight toward or away from one in front of a reflecting wall, one hears two sounds, for the reflections are transposed in the opposite direction from the direct sound. Assuming a pitch resolution of 5 Cents, one could theoretically determine relative velocities of less than 2 km/h using the Doppler effect.

From the formula $f = f_0/(1 \pm v_s/c)$ we calculate the perceived pitches $f_1 = f_0/(1 + v_s/c)$ and $f_2 = f_0/(1 - v_s/c)$. Then the interval between the pitches is $I = f_1/f_2 = (1 + v_s/c)/(1 - v_s/c)$ and the relative velocity is $v_s = c(I - 1)/(I + 1)$. An interval of 5 Cents ($2^{5/1200} = 1.00289$) is equivalent to a relative velocity of .49 m/s or 1.77 km/h.

To simulate the Doppler effect in sound synthesis, we can compute the frequencies of the sounds using the following formulas. Let the actual frequency of the sound be f_0, the velocity of the sound source v_s and the velocity of the listener v_l. Then the perceived frequency is:

$$f = f_0 \frac{1 \pm \frac{v_l}{c}}{1 \mp \frac{v_s}{c}}$$

and the transposition interval is:

$$I = f/f_0 = \frac{1 \pm \frac{v_l}{c}}{1 \mp \frac{v_s}{c}}$$

If only either the sound source or the listener is in motion, the formula is reduced to

$$I = \frac{1}{1 \mp \frac{v_l}{c}} \quad \text{resp.} \quad I = 1 \pm \frac{v_e}{c}$$

The following diagrams show the interval of transposition as a function of velocity, on the left as a proportion, on the right in octaves.

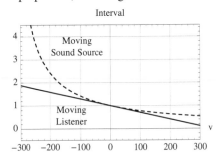
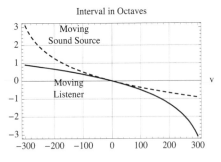

9.1 Spatial Hearing

The tables below give some values for the pitch shift for a stationary listener when a sound approaches (negative velocity) or moves away (positive velocity). The velocities are in meters per second.

v	−340	−100	−10	−1	0	1	10	100	340
I	∞	1.417	1.03	1.00295	1	.997	.9714	.7727	.5

Interval	2 octaves higher	octave higher	maj. third higher	min. second higher	1 Cent higher	min. second lower	fifth lower	octave lower
I	4	2	1.25	1.05946	1.00058	.94387	.6667	.5
v	−255	−170	−68	−19.08	−.197	20.21	169.9	340

In sound synthesis, the frequencies of sounds can be changed corresponding to the Doppler effect. Often, however, the frequencies of the sounds used are unknown. In this case, one obtains the required transposition by simulating the process that causes the transposition. This is done by reading the sound to be transposed into a delay line whose delay corresponds to the time the sound takes to reach the listener. Changing the delay to match the change in distance caused by the sound's movement results in the proper transposition (5.1.4.3). The following illustration shows how the Max patch *Doppler-Effect_2.maxpat* works. Using the mouse, one changes the position of the sound source within a pre-defined region. The distance to the listener and the corresponding delay are computed from the cursor's coordinates. Because of the low resolution of the cursor's position and the mouse's low sampling rate, this information is very discontinuous and must be low-pass filtered before being passed on to the delay line. The stereo placement and the decrease of the sound's loudness as a function of distance are also computed from the coordinates. There is a reflecting wall at the upper edge of the pre-defined area. The reflected sound sounds as though it were behind the wall.
(→ *doppler_effect.pde*)

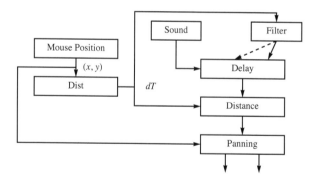

9.1.3.2 Additional Information Through Position Change

Changing one's point of view often gives information about a space and the objects in that space. So for example, if when staring at a snarl of wire one cannot decide which wires are in front, then just a small movement of the head will usually provide enough information to clarify the situation. We study optical illusions from various angles to get an idea of the "real" situation (3.4.3.4). In the same way, the auditory perception constantly acquires information by head movement. We usually do not notice this because we are often in movement anyway, and in situations where we listen consciously, as in a concert, we know the positions of the sound sources in advance. But we can observe how animals constantly change the position of head and ears to localize noises. Synthesized sounds are difficult to localize and so incite the listener to move the head to get more information. Even so, one sometimes gets contradictory signals. If one moves the head while listening with earphones, the entire sound field moves at the same time. This is an obvious effect and is therefore less disturbing than the complicated effects arising when one moves in a virtual sound space created by loudspeakers. There head movements cause factors like time and loudness differences or filter effects to change, often independently of each other and in ways that are mutually contradictory.

9.2 Reflection and Reverberation

9.2.1 Reflections

We have already seen in Chapter 2.3.5.1 that the same laws apply to the propagation and reflection of sound waves as do for the propagation and reflection of light. There are, however, differences between between optical mirroring and acoustic reflection. A beam of light radiating from an object is only reflected from very few surfaces, like mirrors and water, so that the object can be recognized in the reflection. Most surfaces scatter the light beam so much that the light source becomes unrecognizable. Sounds, on the other hand, are usually reflected so that they arrive unchanged at the listener, except that they are softer than the original sound and arrive from a different direction and with some delay. Visual mirror images can be precisely localized and usually do not hide the view of the original. Acoustic reflections, on the other hand, are rarely perceived as such, are difficult to localize and are usually either drowned out by the original sound (cf. the precedence effect 9.1.1.1) or mixed with other reflections to become reverberation. Diffraction is of little importance for visual perception, whereas it causes sound waves to diverge considerably from linear propagation.

9.2.1.1 Geometrical Considerations

If we ever had occasion to hear only a reflected sound, we would in fact hear a mirror image which we would localize elsewhere than at the sound's real source. The position of this virtual source is determined by the fact that the incident sound and the reflected sound form equal angles with the reflecting surface (figure below left). That this is only true for reflection off flat surfaces can be seen in the simple example of a curved surface in the figure to the right below.

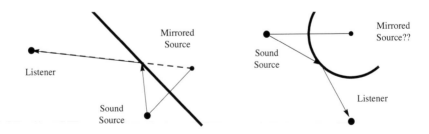

The point of reflection and hence the position of the virtual image obviously depend on the position of the listener. Determining these points for a curved surface means finding the point on the reflecting surface where a line to the sound source and a line to the listener meet the

9.2 Reflection and Reverberation

tangent plane at the same angle. In what follows, we consider only reflections from vertical walls. Seen from above, the walls are shown as curves, the tangent planes as straight lines. If the shape of a curve is given by a sequence of points, we can approximate the direction of the tangent to a point P by substituting the direction of a straight line through the immediately neighboring points (figure below).

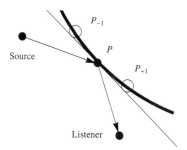

To determine the path of a multiple reflection in a rectangular room, we first mirror the room and the sound source along the plane of each wall and then mirror again each mirror image (see figure below). The dots representing virtual sources get smaller at each reflection, corresponding to the sound's diminishing amplitude. First we choose a position for the listener. To find the path of a reflection from three walls, we draw a line connecting the listener's position with the third virtual source. Where this line crosses the boundary of the room, draw a line to the corresponding second virtual source. Do the same for the first virtual source.

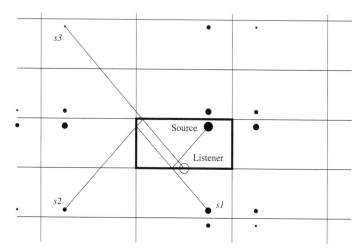

Let us determine the first reflections in an arbitrary space whose (two-dimensional) shape is given by a list of points derived from a known function or from scanning a drawing of the space. In the example that follows, we use the functions

```
rx[r_] := If[r < 1, -1, If[r < 3, r - 2, If[r < 4, 1, Cos[r - 4]]]];
ry[r_] := If[r < 1, -r, If[r < 3, -1, If[r < 4, r - 4, Sin[r - 4]]]];
```

to define the boundary of the space illustrated farther below. The Cartesian coordinates are parameterized by the independent variable r, which is the length of the boundary to (x, y). The following command generates a list (bound) of points along the boundary at a distance $db = .1$

apart and sets the coordinates *sx*, *sy* of the sound source and *lx*, *ly* of the listener (figure left below).

```
db = .1; bound = Table[{rx[r], ry[r]}, {r, 0, 4 + π, dr}];
sx = .6; sy = .5; lx = -.2; ly = -.7;
```

Now we decide for every point on the boundary whether the angles $P_{-1}PS$ and $P_{+1}PS$ are approximately equal by comparing their difference with a given tolerance *dfi*. From these three points we determine the sides of the triangles $P_{-1}PS$ and $P_{+1}PS$, and we calculate the angles at point P using the formula

$$\tan(\alpha/2) = \sqrt{(s-b)(s-c)/(s(s-a))}$$

valid for arbitrary triangles. Here s is half the perimeter of the triangle and α is the angle opposite side a. In order to structure the computations, we define a function to calculate the distance between two points $d[p1\_, p2\_]$, a function to calculate half the perimeter of a triangle $s[a\_, b\_, c\_]$ and a function $fi[a\_, b\_]$ to calculate $\tan^2(\alpha/2)$.

```
d[p1_, p2_] := √((p1[[1]] - p2[[1]])² + (p1[[2]] - p2[[2]])²)
s[a_,b_,c_]:=.5*(a+b+c);
fi[a_,b_]:=(s[a,b,db]-a)*(s[a,b,db]-db)/(s[a,b,db]*(s[a,b,db]-b))
```

Now we generate a list of the first reflections *r1* by checking for every point on the boundary whether the angles $P_{-1}PS$ and $P_{+1}PS$ are approximately equal.

```
r1={};dfi=.05;r=0;
Do[a1=d[bound[[rr]],{lx,ly}];b1=d[bound[[rr-1]],{lx,ly}];
   a2=d[bound[[rr]],{sx,sy}];b2=d[bound[[rr+1]],{sx,sy}];
   If[Abs[fi[a1,b1]-fi[a2,b2]]<dfi,
      r1=Append[r1,Line[{{lx,ly},bound[[rr]],{sx,sy}}]],]
   ,{rr,2,lr-1}]
```

The figure on the right below shows the result of the calculations using a more finely resolved boundary with $db = .03$ and a tolerance of $dfi = .05$.

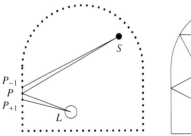
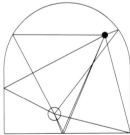

The illustration shows an anomaly of the method, namely that when the tolerance is too small, reflections will be missed and if it is too large, several points next to each other can be listed as reflection points. Actually, the method only works for rooms without corners, since the tangent at a corner is not defined.

From these results one can calculate the direction and the strength (as a function of the distance traveled) of the reflected sound.

It is not possible to determine the reflection points directly for multiple reflections. But one can follow some of the rays from the source to see whether one of them or their reflections

9.2 Reflection and Reverberation

arrive at the listener. We first calculate the points where the rays meet the boundary of the space. Rather than writing equations for rays going in various directions and looking for their intersection with the boundary, we can, as above, go along the boundary, at each point determine the angle to the last ray and, if the angle is sufficiently large, add the point to a list (left illustration below).

```
dfi=2*π/n;r=0;fi0=ArcTan[(ry[r]-sy)/(rx[r]-sx)];i=0;
Do[fi1=ArcTan[(ry[rr]-sy)/(rx[rr]-sx)];
    If[fi0>fi1,fi0-=π];
        If[fi1-fi0>dfi,fi0=fi1;i+=1;
        l[[i]]=Append[l[[i]],{rx[rr],ry[rr]}],]
,{rr,0,4+π,.03}]
```

We can trace the continuation of individual rays in the same way. Again we go along the boundary until angle of incidence and angle of reflection are the same.

```
dfi=.01;rm={{sx,sy},bound[[1]]};rp=2;a1=d[bound[[2]],{sx,sy}];
b1=d[bound[[1]],{sx,sy}];
Do[
    Do[a2=d[bound[[rr]],bound[[rp]]];b2=d[bound[[rr]],bound[[rp+1]]];
    If[Abs[fi[a1,b1]-fi[a2,b2]]<dfi,a1=a2;b1=d[bound[[rr-1]],
        bound[[rp]]];rp=rr;
    rm=Append[rm,bound[[rr]]],]
    ,{rr,3,lr-1}]
,{i,1,5}]
```

The illustration to the right below shows a ray with seven reflections. After the sixth reflection, the ray arrives at the listener. One can see that the approximation here is not quite exact in the fact that the rays are not reflected precisely 180° by two surfaces perpendicular to each other.

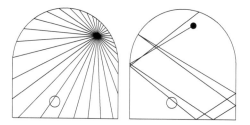

This algorithm is limited to boundaries that are nowhere bent towards the outside. If the boundaries are bent to the outside, several points on the boundary can be on the same ray. If the movement along the boundary is clockwise in the example below, the algorithm chooses the wrong point P_2 as the reflection point for the ray coming from the upper left. If the movement is counter-clockwise, the algorithm finds the right point P_1. In many cases, the algorithm gives better results and runs faster if the direction of the movement is adapted to the direction of the reflection.

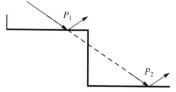

9.2.1.2 Scatter and Absorption

We have assumed in the discussions thus far that the surfaces reflect ideally. If a surface is rough, incident waves are not uniformly reflected but scattered. The critical size for a surface's irregularity is on the order of magnitude of the length of the reflected waves. Since the wave lengths of visible light are between $\lambda = 390$ and $\lambda = 760$ nm, optical mirrors should not have irregularities greater than 1 μm. Sound reflectors, on the other hand, can have irregularities as great as 2 cm and still reflect without distortion, because the wave length of the highest audible sounds is $\lambda = (340$ m/s$)/(20{,}000$ Hz$) = 1.7$ cm. The sound waves of low sounds are reflected even by surfaces with holes greater than a meter in diameter ($\lambda = (340$ m/s$)/(100$ Hz$) = 3.4$ m).

When a sound is reflected, part of its energy is absorbed by the reflecting surface. The absorption coefficient α gives the ratio of absorbed to incident energy and depends on the material of the reflecting surface and on the frequency of the sound (for more extensive tables see [42] p. 110 and [20]).

	125 Hz	250 Hz	500 Hz	1000 Hz	2000 Hz	4000 Hz
Concrete	.01	.01	.02	.02	.02	.03
Glass	.3	.2	.2	.1	.07	.04
Brick	.03	.03	.03	.04	.05	.07
Heavy Carpet	.02	.06	.15	.4	.6	.6
Upholstered Chair	.2	.4	.6	.7	.6	.6
Person Sitting in Upholstered Chair	.4	.6	.8	.9	.9	.9

9.2.2 Reverberation

Reverberation is the sum of individually imperceptible reflections of sound in an enclosed space. Reverberation informs us about the size and properties of the space and about the direction and distance of the sound. The older literature discusses at length simple digital reverberators and the calculation of their characteristics, because in the beginnings of computer music computer memory was limited and performance was modest and so simplicity was called for. We will first study these basic elements and then consider algorithms made possible by greater computing power. Because reverberators correspond to simple filters, their characteristics can be computed with the tools developed in Chapter 3.

Further Reading: The Handbook for Sound Engineers *by Glen M. Ballou [42] contains a good introduction to general acoustics. For more detail see* Musical Acoustics *by Donald E. Hall. More specifically concerning computer music, see* Elements of Computer Music *by F. Richard Moore [8], pp. 380-386,* Computer Music *by Charles Dodge and Thomas A. Jerse [2], pp. 289-307,* About This Reverberation Business *by James A. Moorer in* Foundations of Computer Music, *edited by Curtis Roads and John Strawn [43],* Spatial Reverberation *by Gary S. Kendall et al. in* Current Directions in Computer Music Research *edited by Max Mathews and John Pierce [44] and* The Csound Book *(passim) edited by Richard Boulanger [40].*

9.2.2.1 The Nature of Reverberation

Reverberation consists of direct reflections from walls, ceiling and floor and of indirect reflections whose paths include more than one surface. Particularly in large spaces, the first direct reflections take an appreciable amount of time to arrive at the listener. The time required for the reverberation to reach its maximum, or for constant sounds for it to reach a constant level, is called the *initial reverberation time*. When the sound stops, the reverberation diminishes approximately exponentially, since the arriving reflections travel farther and farther and are reflected more and more often. By definition the *reverberation time* is the time it takes the level of the reverberation to diminish by 60 dB. In a first approximation, the reverberation time can be calculated from the volume of the space V (in cubic meters) and the absorbing surface S by the formula

$$T = .163 V/S \text{ s/m}.$$

The absorbing surface consists of the sum of the individual surfaces S_i multiplied by their absorption coefficients α_i: $S = \alpha_1 S_1 + \alpha_2 S_2 + \alpha_3 S_3 + ...$ (9.2.1.2). The formula

$$T = .163 V/S \cdot (1 + .5\overline{\alpha}) \text{ s/m}$$

gives a better approximation, where $\overline{\alpha}$ stands for the average absorption coefficient $\overline{\alpha} = S/S_t$ over the entire surface S_t. The direct sound predominates close to the sound source, farther away the reverberation becomes more important. The distance from the sound source of the point where direct and reverberated sound are equally strong is called the *reverberation radius*. It can be approximated by the formula

$$R \approx .06 \sqrt{V/T} .$$

Because every enclosed space has resonant frequencies, the actual ambient sound level in the space depends on the listener's position and on the frequency of the sound. Even when many different frequencies are present, the sound level at the walls, where the sound waves all exhibit pressure nodes, can be as much as 3 dB higher than in the room.

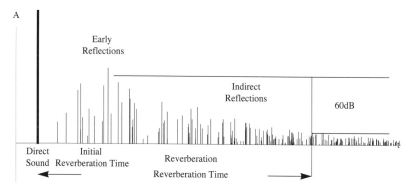

9.2.2.2 Simple Reverberators

Artificial reverberation of a sound is made by generating many copies of the sound, delaying them by varying amounts, reducing their amplitude and adding them together. The simple feedback loop shown at the left below makes copies of an input sound $x(n)$. If we input a pulse to the loop, this simple reverberator generates a sequence of pulses at a constant time interval decreasing in amplitude (below right). The Mathematica program below shows how the "reverberation" *rev* is generated from the sound *snd*. We first define the duration of the reverberation *tt*, a gain factor *g* ($g < 1$) that controls how much weaker each new copy is than the last, and the delay *dd* between consecutive copies. Then we declare lists for the sound *snd*, for the reverberation *rev* and for a ring buffer of length *dd*. The calculations are carried out in a do-loop.

```
tt = 100; g = .8; dd = 10; rev = snd = Table[0, {t, 1, tt}]; snd[[1]] = 1;
buf = Table[0, {d, 1, dd}];
Do[rev[[t]] = buf[[Mod[t, d] + 1]];
    buf[[Mod[t, d] + 1]] = snd[[t]] + g*buf[[Mod[t, d] + 1]],
    {t, 1, tt}]
```

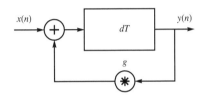
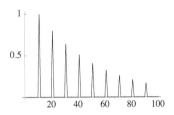

The same feedback loop can be realized in Csound using a delay line of length $p5 = dT$, into which the sound *ain* and the delayed signal *arefl* are read (9-2-2a).

```
....
arefl   init    0
ain     soundin p4
arefl   delay   ain+arefl*p6,p5
        out     arefl
....
```

Since the pulse is multiplied by *g* at each cycle of the loop, the output amplitude decreases exponentially. When the delay is greater than 1/20 s, the reverberation is a flutter echo. When the delay is less than 1/20 s, the output is a periodic waveform with decreasing amplitude. We hear a sound with frequency $1/dT$ and many overtones. This reverberator is in fact a comb filter (3.3.3.5). The filter effect is evident if we input two sine waves, the first with the resonance frequency $1/dT$ (left figure below) and the second with the frequency $1/(2dT)$ but much weaker amplitude.

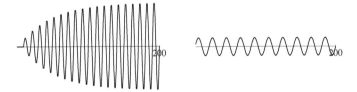

To avoid this filter effect, one can use an all-pass filter in place of the comb filter. In the all-pass filter, the output amplitude is independent of the frequency (3.3.3.5). The illustration below shows the block diagram of an all-pass filter to the left and its impulse response to the right.

```
Do[buf[[Mod[t, dd] + 1]] = snd[[t]] + g*buf[[Mod[t, dd] + 1]];
    rev[[t]] = -snd[[t]] + (1 - g²) *buf[[Mod[t, dd] + 1]];, {t, 1, tt}]
```

9.2 Reflection and Reverberation

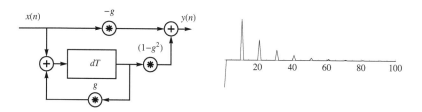

The comparison below of the reverberation of two sine wave with the same frequencies as above shows that after a few cycles both output signals have the same amplitude.

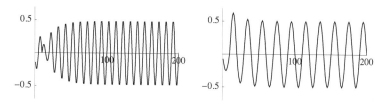

We need to replace the last line of the Csound instrument above by (9-2-2a, instr 2):

```
out      - p6 * ain + (1 - p6 * p6) * arefl
```

The elements we have discussed can be combined in various ways. A good way to avoid the coloring and flutter echo of comb filters is to keep the value of *g* small but to use many elements in parallel. The number of copies of a sound in a parallel system is, however, only proportional to the number of reverberator elements. If we want greater density, we can use serial circuits to reverberate all the preceding copies, taking care that the delays are all different. Comb filters in series produce strong resonances at those frequencies passed by all the filters. For this reason all-pass filters are preferable for circuits in series. If the first all-pass element generates *n* copies of a sound in a given time and the second all-pass *m* copies in the same time, then together they generate *m·n* copies in twice the given time.

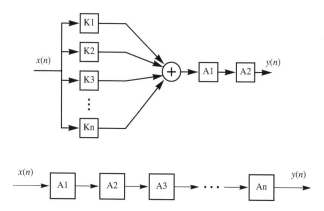

The program *Reverberator1.maxpat* is a kind of workbench for experimenting with reverberation. The default program is a reverberator with four parallel comb filters followed by three all-pass filters. By changing the connections one can test individual elements and other combinations of comb and all-pass filters.

9.2.2.3 Frequency Dependency

Dense and arbitrarily long reverberation can be produced with combinations of several simple elements. The most noticeable difference between artificial and natural reverberation is that in natural reverberation the decrease in amplitude depends on the sound's frequency, since the energy absorption by the air and by the reflecting surfaces (9.2.1.2) affects the high frequencies more strongly than the low frequencies. This effect can be simulated by putting a low-pass filter into the reverberation loop (figure left below). We get a simple low-pass filter by averaging successive samples of a signal (3.3.1.4). The figure to the right below shows the implementation of such a filter using the least computation time possible. The fed-back signal $y(n)$ and the signal $y(n-1)$ delayed by a sample are mixed in the proportion $g1{:}1$ and then scaled by the factor $g2$. The proportion 1:1 gives the strongest filtering, the proportion 0:1 no filtering at all. The factor $g2$ decreases the amplitude of the reverberation and also compensates for the increase in amplitude due to the mixing: $g2 = g/(1 + g1)$.

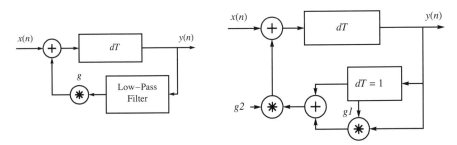

In the following example, we compute the weighted average of $y(n)$ and $y(n-1)$ with the factors df and $(1-df)$, both of which are between 0 and 1. The factor $g2$ is then just g. The code below implements a comb filter and a simple low-pass filter. For the input *snd* we again use a pulse. The figure shows the effect of two different values for df. The greatest filter effect is when $df = (1 - df) = .5$. The factor g is 1 in both examples.

```
g = 1; tt = 200; df = .5;
Do[rev[[t]] = buf[[Mod[t, dd] + 1]];
 buf[[Mod[t, dd] + 1]] = snd[[t]] + df*g*buf[[Mod[t, dd] + 1]] +
 (1 - df)*g*buf[[Mod[t - 1, dd] + 1]];, {t, 1, tt}]
```

We can experiment with various combinations of the factors g and df with the program *Reverberator2.maxpat*. The filter effect is barely audible with low notes and long delay times. But when the frequency $1/dT$ is in the audio range, the feedback large ($g = 1$) and when the input has a high proportion of high-frequency energy like a pulse or a hiss, the filter produces a sound like a plucked and damped string (8.2.1.2).

9.2.2.4 More Complex Filters

Natural reverberation has an appreciable rise time, at least in larger rooms. That means that strength of the reflected sound first increases and that the exponential decay only begins after a certain time. In contrast, in our example above the first reflections were always the loudest. To improve our model, let us first generate a set of unrelated copies of the sound with increasing amplitude and then reverberate them. At the same time we can correct the too low density of the early echoes. In multichannel work, at least these early reflections should be decorrelated, that is, be different for each channel (9.3.2.3) to give the reverberation a natural depth. The figure below shows on the left a few randomly generated early reflections of a pulse and on the right the result after treatment with a comb filter. We can correct the short delay due to the generation of the reverberation by delaying the direct sound by the same amount before mixing the two together.

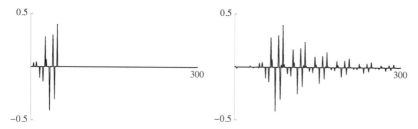

Reverberation in a virtual room can be simulated by calculating early reflections according to the dimensions of the virtual room (9.2.1) and reverberating this signal, correcting for frequency dependency and reverberation time according to classical acoustic approximations.

9.2.2.5 Convolution With an Impulse Response

Many sound programs have so-called room simulators which can treat recorded sounds with the reverberation of real rooms. This is done by using the impulse response of a room, made by recording the reflections of impulse-like sound, e.g. a shot. The sound to be reverberated is multiplied by each of the reflections and the copies are added together. This process is known as convolution and is computationally intensive (3.3.1.3). Positions have to be chosen for the sound source and the recording microphone, and the resulting reverberation is specific for these positions. This means that this technique cannot be used when sound or listeners are meant to move, nor can it be used to simulate several sound sources in different positions in the room.

9.3 Sound Reproduction

9.3.1 Ideal Solutions

In order to create a specific auditory impression by electronic means, all the information used by the perception to create that impression should be recorded and reproduced, at least theoretically. The recording can be made at the sound source, somewhere between the sound source and the listener, at or in the listener's ear.

9.3.1.1 Simulation of the Sound Source

If we replace every natural or virtual sound source in a given space by a loudspeaker, we ought to be able to reproduce recorded or simulated sounds perfectly. If there are few sound sources, this method is theoretically straightforward and feasible. The technical complexity increases however with the number of sources and their movements. The advantage of the method is that the spatial localization is reliable, even if a sound source is in the midst of the audience. If several loudspeakers are used, sound sources between the loudspeakers or moving sound sources can be simulated by interpolation. There are, however, serious problems with the directional characteristics of the loudspeakers. Theoretical discussions and sound synthesis assume sound sources to be point sources from which sound radiates equally in all directions. The situation for natural sound sources is considerably more complicated. The human voice and many instruments have irregular radiation patterns which often can vary with the frequency of the sound emitted. To simulate the acoustics of a specific room, the sound material must be recorded in an anechoic chamber and the reflections of the virtual walls must be generated by additional loudspeakers. This method requires large amounts of data in the form of individual sound tracks either for each loudspeaker or for each sound source. If one prepares a sound track for each source, performances can be realized essentially in any space and with any equipment.

9.3.1.2 Sound Field Reproduction

When reproducing a recorded or computed sound field, one is at the mercy of the sound production, the loudspeaker characteristics and the listener's auditory perception. For example, to reproduce the sound field produced on a stage in the auditorium itself, one needs to record the sound waves passing between sound sources and listeners along a cross-section of the auditorium, putting into practice Huygens' principle (2.2.4.4), which says that a moving sound wave can be considered as the sum of the many secondary waves arising in its wake. Under certain conditions, the information passing through part of the cross-section is sufficient to reconstruct the entire sound field. Sound field reproduction is costly and complicated, because microphones and loudspeakers should not be farther apart than the wave length of the highest tones (about 3.4 cm for 10,000 Hz).
In *Wave Field Synthesis*, WFS, theoretically all the surfaces of the listening room should be covered with loudspeakers, each one with its own signal. In general, however, only flat surfaces or sometimes, since auditory perception is most precise in the horizontal plane, only speakers arranged in a ring around the listener, are used for sound projection.
Ambisonics (9.3.2.2) theoretically reproduces the entire three-dimensional sound field. The information about the sound field is greatly reduced, however, because only a few of the terms of the series expansion of the sound field's analysis are actually used.
Early models of spatial simulation (e.g. [8] p. 373f) already used the idea of a room in which the listener heard the outer sound field through one or more windows. In the simulation, microphones and loudspeakers served as the windows. The sound field could not be reproduced this way; instead the model generated an approximation of the amplitude proportions of the signals at the loudspeakers (9.1.1.2) corresponding to a particular directional perception. Because the windows are treated as points in the model, the information about the sound field is drastically reduced. In addition, the time delays between signals arriving from various directions are increased. Besides the undesirable delays arising when the listener is not exactly in the center of the virtual room, the delays of diagonally arriving waves are greatly exaggerated, as the figure below shows. When the loudspeakers are 10 m or more apart, audible delays arise. Hence the model is useless for recording. When simulating virtual sound sources, the results are better when the time differences are neglected. The model is just as

9.3 Sound Reproduction

useless for moving sound sources because the relative movement is different for each recording position and so the transposition caused by the Doppler effect is different at each position. The direction of the early reflections cannot be determined without exact simulation of the respective delays and without placing the delays properly in the loudspeakers.

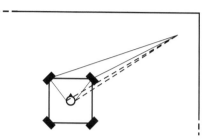

9.3.1.3 The Sound Wave in the Auditory Canal

When sounds are recorded in the auditory canal of a dummy head, the acoustics of the space where the recording was made can be heard clearly. This recording technique makes use of the fact that most of the important sound characteristics used by the perception to judge acoustic spaces, like delays, differences of loudness and filter effects caused by the pinnae, are produced outside the auditory canal. Because the filter effects depend strongly on the shape and size of the pinnae, but also on the torso, recordings made using this technique are often only satisfactory to a limited number of listeners. Theoretically, one could construct a dummy head for every listener, but then there would have to be as many recordings of identical material as there were dummy heads, and the problem would only be shifted to the issue of reproducing an accurate sound field outside the head. The same is true for the generation of sound waves of virtual events. Even if ones knows exactly the filter effects of a listener's anatomy (see HRTF in Chapter 9.1.1.3), one still has to simulate the arriving sound waves. This can be done by measuring the frequency response of the listener's ear for sufficiently many points in the space. The synthesis then modifies the virtual sounds by convolving them with the frequency responses according to the required spatial positions. Of course, this does not take the acoustics of the listening space into account. It requires much more computation if one wants to correct the synthesized sounds for the movements of the head during listening.

9.3.2 Practical Solutions

Until quite recently only two-channel audio signals could be stored on commercially available media, and most audio equipment and computer programs could only work with stereo signals. Developments like Quadraphony found no market, partly for the reasons described in the previous chapters, partly because each set of conditions (space, equipment, etc.) requires its own solution. Even stereo has various norms, each of which has its own advantages and disadvantages. Thanks to the introduction of the DVD and the commercialization of surround techniques, it has become possible to simulate acoustic spaces with simple equipment, even if the number of channels is limited and special arrangements of loudspeakers and certain kinds of listening spaces are required. For computer music, Ambisonics will probably prove to be the best extension of stereophony in the medium term, because it is independent of any particular set-up and and can fall back on the experience of analog sound technology.

9.3.2.1 Stereo

For recording in the so-called *X-Y technique*, two identical directional microphones are placed at the same angle to the recording axis. The characteristics of the recording depend on the angle between the microphones and on their directionality (upper illustration on the left below). The X-Y technique does not attempt to record the sound as it would appear at a listener's ears. Instead, one tries to use the best microphones for a particular recording situation and to find the ideal distance to the sound sources and the best angle between the microphones in order to obtain two different signals. The appropriate distribution of the two signals into the stereo channels depends on the sort of reproduction intended.

In the ORTF technique (Office de Radiodiffusion Télévision Française), the angle between the microphones is fixed at $110°$ (upper illustration on the right below). The distance between the microphones is also fixed, at 17 cm, and so sound coming from one side is recorded by the far microphone with a delay corresponding approximately to the interaural time difference ITD in human binaural audition.

The *M/S stereo technique* uses one microphone pointing towards the sound source (M) and a second microphone perpendicular to the first, pointing to the side (S). The first microphone usually has a cardioid characteristic (solid line, illustration at lower left below), the second must be bi-directional (dotted line, lower left below). The signal from microphone M can be used as a monophonic signal. The signal from microphone S is called the *stereo difference signal*. The relationship between the signals recorded with the M/S technique and the signals for the left and right stereo channels is: $M = 1/2 \cdot (L + R)$ and $S = 1/2 \cdot (L - R)$, or rewritten: $L = M + S$ and $R = M - S$.

The figure on the right below shows the standard listening configuration for stereo.

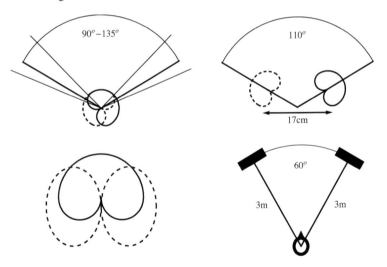

See [42] p. 454.

9.3.2.2 Ambisonics

Ambisonics is a technique of three-dimensional sound projection. The information about the recorded or synthesized sound field is encoded and stored in several channels, taking no account of the arrangement of the loudspeakers for reproduction. The encoding of a signal's spatial information can be more or less precise, depending on the so-called order of the

9.3 Sound Reproduction

algorithm used. Order zero corresponds to the monophonic signal and requires only one channel for storage and transmission. In *first-order Ambisonics*, three further channels are used to encode the portions of the sound field in the three orthogonal directions x, y and z. These four channels constitute the so-called *first-order B-format*. Originally, Ambisonics was a recording technique in which the B-format signal was derived from the signals of four symmetrically placed microphone capsules.

When Ambisonics is used for artificial spatialization of recorded or synthetic sound, the encoding can be of an arbitrarily high order. The higher orders cannot be interpreted as easily as orders zero and one. If one tries to compute the sum of the sound waves produced by a sound source at any point in a space, complicated functions arise. Ambisonics simplifies this situation by assuming that the sound source emits plane waves and that the listener is at the center of the coordinate system of reference. Mathematical problems can often be solved by decomposing complicated functions into sums of simple functions. Ambisonics does the same thing by describing the sound waves in space using spherical coordinates and encoding them as sums of functions called spherical harmonics. In a two-dimensional analogy to Ambisonics (called *Ambisonics2D* in what follows), sound waves in the horizontal plane are described by polar coordinates and encoded using sums of sine and cosine functions.

The loudspeaker feeds are obtained by decoding the B-format signal. The resulting panning is amplitude panning, and only the direction to the sound source is taken into account. In order to simulate distance, movement and any directional characteristics of the sound source, the signal has to be treated before being encoded.

The illustration below shows the principle of Ambisonics. First a sound is generated and its position determined. The amplitude and spectrum are adjusted to match the distance, the latter using a low-pass filter. Then the Ambisonic encoding is computed using the sound's coordinates. Encoding mth order B-format requires $2m + 1$ channels in two dimensions and $(m + 1)^2$ channels in three dimensions. By decoding the B-format signal one can obtain the signals for any number of loudspeakers in any arrangement. (→ *Max/ICST/Ambisonics*)

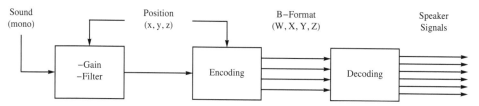

We will first explain the encoding process in Ambisonics2D. The position of a sound source in the horizontal plane is given by two coordinates. Cartesian coordinates (x, y) are most frequently used in Ambisonic theory, where the listener (or the microphone) is at the origin of the coordinate system $(0, 0)$, and the x-coordinate points to the front, the y-coordinate to the left. The position of the sound source can also be given in polar coordinates by the angle φ between the line of vision of the listener (front) and the direction to the sound source, and by their distance r. Cartesian coordinates can be converted to polar coordinates by the formulas

$$r = \sqrt{x^2 + y^2} \text{ and } \phi = \arctan(x, y),$$

polar to Cartesian coordinates by

$$x = r \cdot \cos(\phi) \text{ and } y = r \cdot \sin(\phi).$$

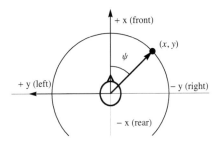

It can be shown that at the origin of this coordinate system the sound field of a plane wave produced by signal S coming from the direction ψ can be described as the product of the signal with a function $f(\varphi)$ which is equal to 1 when $\varphi = \psi$ and equal to zero otherwise. The independent variable φ is the azimuth angle and is 2π-periodic. The illustration to the left below shows the unit circle in the horizontal plane and the function $f(\varphi)$ with $-\pi < \varphi \leq \pi$. The product of this function with a sound is the sound itself when $\varphi = \psi$ and is zero otherwise. The illustration on the right shows the same function in a two-dimensional representation.

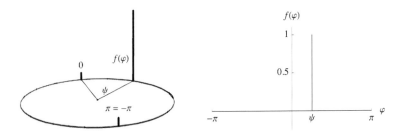

There are two ways to represent this function discretely, as there are with sounds: either as a sequence of samples or as a sum of sine and cosine functions, that is as the function's spectrum. If the function is represented by n samples (that is, sampling from n discrete directions), then impulses from directions between two samples can be assigned to the neighboring samples. Multiplying a signal by such a function gives us a function for pair-wise panning over n channels.

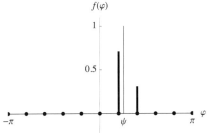

If we describe the function $f(\varphi)$ using Fourier Analysis as a sum of sine and cosine functions, we get Ambisonic2D. By using a sum of cosine functions

$$\frac{1}{m+1}\sum_{k=0}^{m}\cos(k\cdot\varphi) = \frac{1}{m+1}(\cos(0\cdot\varphi) + \cos(1\cdot\varphi) + \cos(2\cdot\varphi) + ...)$$

we can generate an arbitrarily narrow pulse at $\varphi = 0$. The number of terms in the sum is the order of the approximation. Because the development of the series usually ends after a few

9.3 Sound Reproduction

terms, the function has other local minima and maxima besides the pulse at $\varphi = 0$. We will see later that these artefacts can seriously distort the sound field.

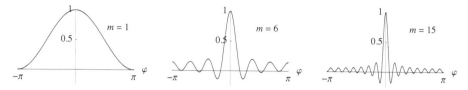

Let us generate a pulse at $\varphi = \psi$ with $\frac{1}{m+1}\sum_{k=0}^{m}\cos(k(\varphi - \psi))$

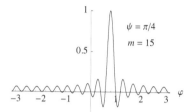

Then we can write the sound field (with signal S) of a plane wave with an incident angle of ψ at the origin of the coordinate system as

$$S\frac{1}{m+1}\sum_{k=0}^{m}\cos(k(\varphi - \psi)).$$

To derive the formulas for Ambisonic encoding and decoding, let us consider the following situation. N loudspeakers in the same plane are arranged in a circle. We want to synthesize the sound waves from a virtual source arriving at the center of the circle from an arbitrary direction ψ as well as possible with the n loudspeakers.

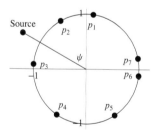

The sum of the sound waves from the n loudspeakers should recreate the sound wave at the center of the circle that would be emitted from the virtual source. N equations can be established for the unknown amplitudes p_i of the n loudspeakers. Since the sum of the amplitudes p_i must give the amplitude S of the simulated sound source, we have as first equation

$$S = \sum_{i=1}^{n} p_i.$$

For the remaining $n - 1$ equations we must consider the directions to the virtual source and to the loudspeakers. Let us call the direction from the center to the simulated source ψ and the directions to the loudspeakers φ_i. As before, we write for the n loudspeaker signals p_i

$$p_i\frac{1}{m+1}\sum_{k=0}^{m}\cos(k(\varphi - \varphi_i)).$$

If we set the signal of the virtual source equal to the sum of the n loudspeaker signals, we get

$$S \frac{1}{m+1} \sum_{k=0}^{m} \cos(k(\varphi - \psi)) = \sum_{i=1}^{n} p_i \frac{1}{m+1} \sum_{k=0}^{m} \cos(k(\varphi - \varphi_i)).$$

Using the formula $\cos(n(\alpha - \beta)) = \cos(n\alpha)\cos(n\beta) + \sin(n\alpha)\sin(n\beta)$ we rewrite the equation above as

$$S \frac{1}{m+1} \sum_{k=0}^{m} \cos(k\varphi)\cos(k\psi) + \sin(k\varphi)\sin(k\psi) =$$
$$\sum_{i=1}^{n} p_i \frac{1}{m+1} \sum_{k=0}^{m} \cos(k\varphi)\cos(k\varphi_i) + \sin(k\varphi)\sin(k\varphi_i).$$

Because the functions $\cos(\alpha)$, $\cos(2\alpha)$, $\cos(3\alpha)$, ... $\sin(\alpha)$, $\sin(2\alpha)$, $\sin(3\alpha)$, ... are linearly independent, the terms in sine and cosine on both sides of the equation must correspond for every individual i. So we have for the $2m$ equations

$$S \cdot \cos(k\psi) = \sum_{i=1}^{n} p_i \cos(k\varphi_i)$$
$$S \cdot \sin(k\psi) = \sum_{i=1}^{n} p_i \sin(k\varphi_i).$$

The $2m + 1$ equations have unique solutions when the number of loudspeaker signals p_i is equal to the number of equations, that is, when $n = 2m + 1$ loudspeakers are used. The left sides of the equations above describe the encoding of the signal S. In Ambisonics, the encoded signals are usually denoted by W, X, Y, Z, U, V, ... The order of the encoding is m. For sound sources on the unit circle we can set $x = \cos(\psi)$ and $y = \sin(\psi)$.

$$W = S$$
$$X = S \cdot \cos(\psi) = S \cdot x$$
$$Y = S \cdot \sin(\psi) = S \cdot y$$
$$U = S \cdot \cos(2\psi) = S(\cos^2(\psi) - \sin^2(\psi)) = S(x^2 - y^2)$$
$$V = S \cdot \sin(2\psi) = S \cdot 2\cos(\psi)\sin(\psi) = S \cdot xy$$

(→ Cartesian Coordinates for Higher Orders)

We must decode the $2m + 1$ Ambisonic signals to get the amplitudes of the loudspeaker signals p_i. If the $n = 2m + 1$ loudspeakers are set up symmetrically on a circle, the decoding formula is:

$$p_i = \frac{1}{n}(W + 2X\cos(\varphi_i) + 2Y\sin(\varphi_i) + 2U\cos(\varphi_i) + 2V\sin(\varphi_i) + ...)$$

If we define channel $W = \frac{1}{\sqrt{2}} S$, we get

$$p_i = \frac{1}{n}(\frac{1}{\sqrt{2}} W + X\cos(\varphi_i) + Y\sin(\varphi_i) + U\cos(\varphi_i) + V\sin(\varphi_i) + ...)$$

If the virtual sound source is precisely in the direction of a loudspeaker ($\psi = \varphi_i$), only loudspeaker i gets a signal. If the loudspeakers are arranged as in the left figure below, and if $\psi = 0$ (figure center), then we have for the amplitudes the values (1, 0, 0, 0, 0, 0, 0). If the virtual sound source is between two loudspeakers, these loudspeakers receive the strongest signals, all the other loudspeakers have weaker signals, some with negative amplitude (that is, reversed phase).

9.3 Sound Reproduction

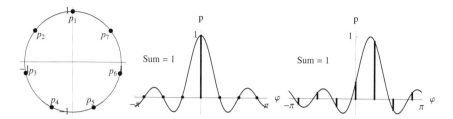

The decoding formulas can also be used for symmetrical arrangements of more than $2m + 1$ speakers (see the left illustration below). If fewer than $2m + 1$ speakers are used, the sum of the speaker amplitudes is still 1 until about $m + 2$ loudspeakers, but the quality of the panning diminishes continually. (→ Manipulate)

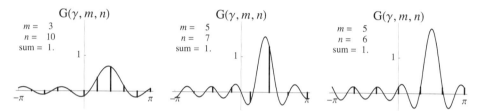

If one solves the $2m + 1$ equations for an asymmetrical arrangement of $n = 2m + 1$ loudspeakers, the sum of the amplitudes is 1, but amplitudes of individual speakers can be greater than 1, and amplitudes of speakers that are not in the direction of the virtual sound source can be great. Sometimes the results are better when one decodes using the formula for the symmetric arrangement. In this case, the sum of the amplitudes is generally not 1. (→ Example 5.1)

To avoid having loudspeaker sounds that are far away from the virtual sound source and to ensure that negative amplitudes (inverted phase) do not arise, the B-format channels can be weighted before being decoded. The weighting factors depend on the highest order used (M), the order of the particular channel being decoded (m) and the number of loudspeakers. For Ambisonics2D the factors for the first four orders are:

$$g_m = g_0(m)\frac{(M!)^2}{(M+m)!\cdot(M-m)!}, \text{ with } g_0(1) = 1.5, g_0(2) = 1.944, g_0(3) = 2.310, g_0(4) = 2.627.$$

The illustration below shows a third-order B-format signal decoded to 13 loudspeakers first uncorrected (so-called *basic decoding*, left), then corrected by weighting (so-called *in-phase decoding*, right).

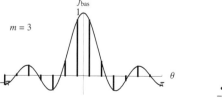 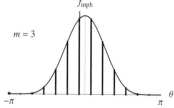

The formulas for encoding and decoding signals in three-dimensional Ambisonics are essentially the same as those derived above for two-dimensional Ambisonics, except that spherical harmonic functions take the place of the trigonometric functions in two-dimensional

Ambisonics. Complicated functions can often be represented as the sum of simple functions and can be treated according the rules of these simple functions. For example, both trigonometric functions and exponential functions can be expanded into polynomial series. Just as every vector in three-dimensional space can be decomposed into three orthogonal components, certain functions can be decomposed into what are called orthogonal functions, which can be thought of as infinite dimensional vectors. Orthogonal vectors or functions are mutually independent, that is, no component can be represented as a sum of the others. Hence, every vector in three-dimensional space can be uniquely represented as the sum of three basic vectors, and every function can be represented as the infinite sum of basic functions. Two vectors are orthogonal when their scalar product (the sum of the products of their components) is zero. Two functions $f(x)$ are orthogonal when the integral of the product of the functions is zero:

$$\int f(x)g(x)dx = 0.$$

A function $f(x)$ is called normal when $\int f(x)^2 dx = 1$.

All functions of period $2\pi/\omega$ can be represented by the independent functions $\sin(n\omega t)$ and $\cos(n\omega t)$. This is the basis of the theorem of Fourier which states that every periodic function can be decomposed into a sum of harmonics. In the derivation of the encoding and decoding formulas for Ambisonics2D above, we used the functions $\sin(n\varphi)$ and $\cos(n\varphi)$ as functions on the unit circle and not as periodic functions of time.

In contrast to the circle, which can be subdivided into arbitrarily many symmetrical segments, a sphere can only be subdivided into similar regular polygons in five ways. Therefor the circle can be approximated by regular polygons of arbitrarily many angles, but the sphere can only be approximated symmetrically by the five Platonic solids. It is impossible to construct a completely symmetrical coordinate system for the sphere. The more complex geometry of the sphere means that its orthogonal functions are more complex than those of the circle.

A circular disk can be subdivided into arbitrarily many equal segments. Like the circle, it has resonant frequencies where the disk is subdivided into segments, but unlike the circle, it also has resonant frequencies where nodal lines (lines which do not move during oscillation) form concentric circles. Although the sphere is a completely symmetrical solid, it has resonances similar to those of the disk. If the nodal lines are longitudinal, that is run from pole to pole, we speak of *sectoral oscillation* (left figure below), if they are latitudinal we speak of *zonal oscillation* (right figure). If both kinds of nodal lines appear, we speak of *tesseral oscillation* (middle figure).

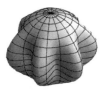 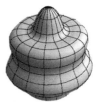

Functions on the surface of the sphere can be represented as sums of so-called *spherical harmonics*. Spherical harmonics form a complete orthogonal system of functions and are the product of sectoral and zonal functions. The sectoral functions are the harmonic functions

9.3 Sound Reproduction

sin($n\varphi$) and cos($n\varphi$), the zonal functions are the so-called *associated Legendre polynomials*. The figure to the left below shows a sectoral harmonic function in cross section through the equator, the figure to the right shows a Legendre polynomial in longitudinal section.

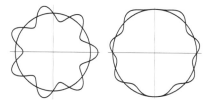

The position of a point in space can be given by its Cartesian coordinates x, y and z or by its spherical coordinates. In mathematics and the natural sciences, a point is described in spherical coordinates by its radial distance r from the origin of the coordinate system, the *inclination angle* θ and the *azimuth angle* φ (in the figure below these terms are indicated in parentheses). The inclination angle θ is measured from the positive z-axis (the zenith direction) to r and lies between 0 and π. The azimuth angle φ is measured from the positive x-axis counterclockwise to the projection of r onto the x-y plane. In the Ambisonics literature on the other hand, the inclination θ is replaced by the elevation δ which lies between $-\pi$ and π. The azimuth angle is usually called θ.

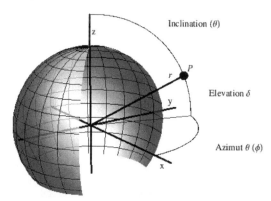

The formulas for transforming coordinates in Ambisonics are as follows:

$$x = r \cdot \cos(\delta)\cos(\theta) \qquad y = r \cdot \cos(\delta)\sin(\theta) \qquad z = r \cdot \sin(\delta)$$

$$r = \sqrt{x^2 + y^2 + z^2} \qquad \theta = \arctan\left(\frac{y}{x}\right) \qquad \delta = \operatorname{arccot}\left(\frac{\sqrt{x^2+y^2}}{z}\right)$$

The spherical harmonics $Y_{m,n}^{\sigma}(\theta, \delta)$ with azimuth angle θ and elevation δ are defined as the product of the associated Legendre polynomials for sin(δ), the spherical harmonics for θ and normalization factors.

The spherical harmonics can be visualized in two ways, either as functions on the unit sphere or as *directional characteristics*, that is, as the distance from the origin of the coordinate system in the direction of the angles θ and δ. Visualization on the unit sphere is the clearer of the two because it corresponds to the natural resonances of the sphere. In addition, it shows in which directions the function values are positive and in which they are negative. The following illustrations show both representations for fifth-order spherical harmonics.

$1 + .5\,\mathrm{Re}[Y_5^0]$	$1 + .5\,\mathrm{Re}[Y_5^1]$	$1 + .5\,\mathrm{Re}[Y_5^2]$
$\mathrm{Abs}[\mathrm{Re}[Y_5^0]]$	$\mathrm{Abs}[\mathrm{Re}[Y_5^1]]$	$\mathrm{Abs}[\mathrm{Re}[Y_5^2]]$
$1 + .5\,\mathrm{Re}[Y_5^3]$	$1 + .5\,\mathrm{Re}[Y_5^4]$	$1 + .5\,\mathrm{Re}[Y_5^5]$
$\mathrm{Abs}[\mathrm{Re}[Y_5^3]]$	$\mathrm{Abs}[\mathrm{Re}[Y_5^4]]$	$\mathrm{Abs}[\mathrm{Re}[Y_5^5]]$

9.3 Sound Reproduction

The following illustration shows how one direction can be emphasized by adding a few spherical harmonics.

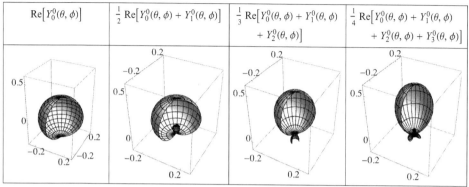

| $\text{Re}[Y_0^0(\theta,\phi)]$ | $\frac{1}{2}\text{Re}[Y_0^0(\theta,\phi) + Y_1^0(\theta,\phi)]$ | $\frac{1}{3}\text{Re}[Y_0^0(\theta,\phi) + Y_1^0(\theta,\phi) + Y_2^0(\theta,\phi)]$ | $\frac{1}{4}\text{Re}[Y_0^0(\theta,\phi) + Y_1^0(\theta,\phi) + Y_2^0(\theta,\phi) + Y_3^0(\theta,\phi)]$ |

The channels of the Ambisonic B-format are computed as the product of the sounds themselves and the spherical harmonics representing the direction to the virtual sound sources. The spherical harmonics can be normalized in various ways. *Semi-normalized* spherical harmonics are often used in Ambisonics. The following table shows the encoding functions up to and including the third order.

m	B	n	$Y_{mn}(\theta, \delta)$	$Y_{mn}(x, y, z)$
0	W	0	1	1
1	Z	0	$\sin[\delta]$	z
	X	1	$\cos[\delta]\cos[\theta]$	x
	Y	-1	$\cos[\delta]\sin[\theta]$	y
2	R	0	$\frac{1}{2}(-1 + 3\sin[\delta]^2)$	$\frac{1}{2}(-1 + 3z^2)$
	S	1	$\frac{1}{2}\sqrt{3}\cos[\theta]\sin[2\delta]$	$\sqrt{3}\,xz$
	T	-1	$\frac{1}{2}\sqrt{3}\sin[2\delta]\sin[\theta]$	$\sqrt{3}\,yz$
	U	2	$\frac{1}{2}\sqrt{3}\cos[\delta]^2\cos[2\theta]$	$\frac{1}{2}\sqrt{3}(x^2 - y^2)$
	V	-2	$\sqrt{3}\cos[\delta]^2\cos[\theta]\sin[\theta]$	$\sqrt{3}\,xy$
3	K	0	$\frac{1}{8}(3\sin[\delta] - 5\sin[3\delta])$	$\frac{1}{2}z(-3 + 5z^2)$
	L	1	$\frac{1}{8}\sqrt{3/2}\,(\cos[\delta] - 5\cos[3\delta])\cos[\theta]$	$\frac{1}{4}\sqrt{6}(-x + 5xz^2)$
	M	-1	$\frac{1}{8}\sqrt{3/2}\,(\cos[\delta] - 5\cos[3\delta])\sin[\theta]$	$\frac{1}{4}\sqrt{6}(-y + 5yz^2)$
	N	2	$\frac{1}{2}\sqrt{15}\cos[\delta]^2\cos[2\theta]\sin[\delta]$	$\frac{1}{2}\sqrt{15}(z - 2y^2z - z^3)$
	O	-2	$\sqrt{15}\cos[\delta]^2\cos[\theta]\sin[\delta]\sin[\theta]$	$\sqrt{15}\,xyz$
	P	3	$\frac{1}{2}\sqrt{5/2}\cos[\delta]^3\cos[3\theta]$	$\frac{1}{4}\sqrt{10}(x^3 - 3xy^2)$
	Q	-3	$\frac{1}{2}\sqrt{5/2}\cos[\delta]^3\sin[3\theta]$	$\frac{1}{4}\sqrt{10}(3x^2y - y^3)$

(→ Computation of the Encoding Formulas up to $m = 7$)

To compute the decoding formulas we require the sum of the sound waves p_i of the N loudspeakers located at (θ_n, δ_n) to reproduce the original sound wave S emitted at (θ, δ):

$$S \cdot Y^{\sigma}_{m,l}(\theta, \delta) = \sum_{n=1}^{N} p_n Y^{\sigma}_{m,l}(\theta_n, \delta_n), \quad m = 1, 2, \ldots, M$$

(N = number of loudspeakers, n = index of the loudspeakers, m = order of the spherical harmonics, $l = 1, 2, \ldots, M$, $\sigma = \pm 1$, M = highest order)

Since the functions $Y^{\sigma}_{m,l}(\theta, \delta)$ are linearly independent, we have the system of equations

$$\boldsymbol{B} = \boldsymbol{C} \cdot \boldsymbol{p} \quad \text{with}$$

$$\boldsymbol{p} = [p_1, p_2, \ldots, p_N]^T$$

$$\boldsymbol{B} = \left[Y^1_{0,0}(\theta, \delta), Y^1_{1,0}(\theta, \delta), \ldots, Y^{-1}_{M,M}(\theta, \delta) \right]^T \cdot S$$

$$\boldsymbol{C} = \begin{pmatrix} Y^1_{0,0}(\theta_1, \delta_1) & Y^1_{0,0}(\theta_2, \delta_2) & \cdots & Y^1_{0,0}(\theta_N, \delta_N) \\ Y^1_{1,0}(\theta_1, \delta_1) & Y^1_{1,0}(\theta_2, \delta_2) & \cdots & Y^1_{1,0}(\theta_N, \delta_N) \\ \cdot & \cdot & & \cdot \\ \cdot & \cdot & & \cdot \\ \cdot & \cdot & & \cdot \\ Y^{-1}_{M,M}(\theta_1, \delta_1) & Y^{-1}_{M,M}(\theta_2, \delta_2) & \cdots & Y^{-1}_{M,M}(\theta_N, \delta_N) \end{pmatrix}$$

If the number of Ambisonics channels L ($2M + 1$ for two dimensions, $(M + 1)^2$ for three) is the same as the number of loudspeakers N, then the solution of the system of equations is

$$\boldsymbol{p} = \boldsymbol{C}^{-1} \cdot \boldsymbol{B}$$

If the loudspeakers are arranged in a regular polyhedron, this relationship holds:

$$\frac{1}{N} \boldsymbol{C} \cdot \boldsymbol{C}^T = \boldsymbol{I} \quad \text{(See [92] p. 176)}$$

If $N > L$, the matrix $\boldsymbol{C}$ is not square, and hence the system of equations has no unique solution. One can obtain a least squares solution by multiplying $\boldsymbol{C}$ by its so-called *pseudo-inverse*, called here $\boldsymbol{D}$ ($\boldsymbol{D} = \text{pinv}(\boldsymbol{C})$ or $\boldsymbol{D} = \boldsymbol{C}+$ or $\boldsymbol{D} = \boldsymbol{C}^{(-1)}$).

$$\boldsymbol{D} = \text{pinv}(\boldsymbol{C}) = \boldsymbol{C}^T \cdot \left(\boldsymbol{C} \cdot \boldsymbol{C}^T \right)^{-1}$$

If the loudspeaker arrangement is regular, $\boldsymbol{D}$ is simplified to

$$\boldsymbol{D} = \text{pinv}(\boldsymbol{C}) = \boldsymbol{C}^T \cdot \left(\boldsymbol{C} \cdot \boldsymbol{C}^T \right)^{-1} = \boldsymbol{C}^T \cdot (N \boldsymbol{I})^{-1} = \frac{1}{N} \boldsymbol{C}^T.$$

We obtain the loudspeaker signals by multiplying the matrix with the B-format signals by the transposed matrix $\boldsymbol{C}^T$ of the speaker positions and dividing by N.

$$\boldsymbol{p} = \frac{1}{N} \boldsymbol{C}^T \cdot \boldsymbol{B}$$

9.3 Sound Reproduction

For in-phase decoding, all the channels of the same order m are multiplied by the weighting factor g_m:

$$p = \frac{1}{N} C^T \text{Diag}[...g_m...] \cdot B$$

The weighting factors for in-phase decoding of Ambisonics3D are:

$$g_k = g_0 \frac{m!\,(m+1)!}{(m+k+1)!\,(m-k)!} \quad \text{with} \quad g_0 = \frac{\sqrt{n(2m+1)}}{m+1}.$$

If we combine the encoding and the in-phase decoding, we obtain the following panning functions for sound sources on the unit circle ([92] p. 183)

$$G_{2D}(\gamma, m) = \frac{1}{n}\left(g_0 + 2\sum_{k=1}^{m} g_k \cos(k\gamma)\right) \qquad \text{for 2D,}$$

$$G_{3D}(\gamma, m) = \frac{1}{n}\sum_{k=0}^{m}(2k+1)\,g_k\, P_k(\cos\gamma) \qquad \text{for 3D.}$$

Using the functions $G(\gamma, m)$, the loudspeaker amplitudes can be computed directly from the angle between loudspeaker and virtual sound source γ. The P_k are the Lengendre polynomials of the kth order.

The panning functions above, together with the in-phase factors g_k, are equivalent to the simple function

$$P_{\text{inph}}(\gamma, m) = \left(\frac{1}{2} + \frac{1}{2}\cos\gamma\right)^m = \left(\cos\frac{\gamma}{2}\right)^{2m}$$

for integral orders m (Ambisonics Equivalent Panning AEP) ([3-6], [3-7]).

The order indicated in the function does not have to be an integer, however. This means the order can be continuously varied during decoding. The figures below show examples from a program for Ambisonics2D in which the order (here p), the number of loudspeakers and the angle θ can be varied. The maximum amplitude of the loudspeakers is 1, and the sum of the amplitudes of all the speakers is constant for $n > p + 1$. ($\rightarrow$ Manipulate)

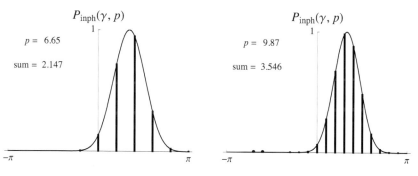

The function $P_{\text{inph}}(\gamma, p)$ can be used in both Ambisonics2D and Ambisonics3D. The factor $(p+1)/n$ normalizes the sum of the loudspeaker amplitudes in Ambisonics3D:

$$\frac{p+1}{n}\sum_{i=1}^{n} P_{\text{inph}}(\gamma_i, p) = 1$$

The following figures show the arrangement of 12 loudspeakers in the form of an icosahedron. The virtual sound source is on a circle of latitude of the corresponding sphere, the amplitude of the loudspeakers is indicated by the size of the dots at the vertices of the icosahedron. The figures show the amplitudes of the loudspeaker signals for $p = 2.6$ (left) and $p = 6.5$ (right). (→ Manipulate)

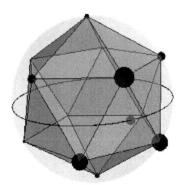
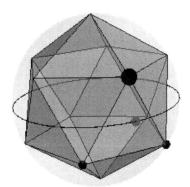

The implementation of the panning functions $P_{inph}(\gamma, p)$ is straightforward. In order to produce the signal for a certain speaker at position $X_s = (x_s, y_s, z_s)$ a sound at position $X = (x, y, z)$ is multiplied by $P_{inph}(\gamma, p)$ where γ denotes the angle between the sound source and the speaker. If the speakers are positioned on a unit sphere the cosine of the angle γ is calculated as the scalar product $(x, y, z).(x_s, y_s, z_s)$.

$$P_{inph}(\gamma, p) = P_{in}(X_s, X, p) = \left(\frac{1}{2} + \frac{1}{2}\cos\gamma\right)^p = \left(\frac{1+X_s.X}{2}\right)^p$$

$$= \left(\frac{1 + x x_s + y y_s + z z_s}{2}\right)^p = \frac{1}{2^p}(1 + x x_s + y y_s + z z_s)^p$$

For a sound source at distance r we get

$$P_{inph}(\gamma, p) = \left(\frac{1}{2} + \frac{X_s.X}{2r}\right)^p = \left(\frac{r + x x_s + y y_s + z z_s}{2r}\right)^p$$

In spherical coordinates we get for a speaker at position $X_s = (\theta_s, \delta_s, r_s)$ and a sound at position $X = (\theta, \delta, r)$

$$P_{inph}(\gamma, p) = \left(\frac{1}{2}(1 + \cos(\theta - \theta_s)\cos(\delta)\cos(\delta_s) + \sin(\delta)\sin(\delta_s))\right)^p$$

$$= \frac{1}{2^p}(1 + \cos(\theta - \theta_s)\cos(\delta)\cos(\delta_s) + \sin(\delta)\sin(\delta_s))^p$$

This system of panning is called Ambisonics Equivalent Panning. It has the disadvantage of not producing a B-format representation, but its implementation is straightforward and the computation time is short and independent of the Ambisonics order simulated. Hence it is particularly useful for real-time applications, for panning in connection with sequencer programs and for experimentation with high and non-integral Ambisonic orders.

9.3.2.3 Decorrelation

In electroacoustic music, signals coming from several loudspeakers, and hence from several directions, can be identical. Usually, however, the signals on different tracks are different and are changed even more by reflection and reverberation before they reach the listener. Two identical signals have a maximum degree of correlation, two very different signals a high degree of *decorrelation*. As the examples above show, a certain decorrelation occurs naturally in sound projection, recording and sound treatment as well as in the simulation of resonant spaces. Highly correlated signals can cause these problems:

- When using headphones, the sound source is often localized in the head.
- When mixing only partially decorrelated signals (for example, when generating a monophonic signal from two signals with different delays) the sound is comb filtered and coloration results.
- The superposition of signals from loudspeakers at different distances and directions from the listener also creates coloration. The superposition of the two signals from control monitors in a highly absorbent control room is a special case of so-called crosstalk (crosstalk arises when a signal in one channel or circuit produces an undesired effect on a signal in another channel or circuit).
- Because of the precedence effect (9.1.1.1), the source of identical signals is localized at the nearest loudspeaker.

These problems can be solved either by adding natural or artificial reverberation (different for each channel) or by decorrelating the signals. Decorrelation corresponds to adding to the sound short, barely perceptible reverberation different for each channel, that is, the sum of several slightly delayed and variously loud copies of the original signal.

To avoid coloration of the sound, the amplitude response of the system defined by the decorrelation must be constant. Therefore we need all-pass filters with various phase responses. Before we show how to construct such filters, let us see how the correlation of two signals is measured. We are looking for a quantity q that is 1 when a signal is compared with itself and 0 when two independent signals are compared. The sum of the products of the simultaneous values of two signals $y_1(k)$ and $y_2(k)$

$$S = \sum y_1(k)\, y_2(k)$$

increases the more the signals agree. But the sum also increases with the length of the two signals. Therefore we divide the sum by

$$\sqrt{\sum y_1(k)^2 \cdot \sum y_2(k)^2}$$

and obtain the required quantity

$$q = \frac{\sum y_1(k)\, y_2(k)}{\sqrt{\sum y_1(k)^2 \cdot \sum y_2(k)^2}}.$$

The measure q is only valid for signals with a mean value of zero. If we replace $y_i(k)$ in the definition above by $y_i(k) - m_i$, then the formula is generally valid and we obtain the correlation coefficient r, which in statistics describes the mutual dependence of two samples. By setting $y_1 = y_2 = x$, we can maximize the value of r:

$$r = \sum x(k)x(k) \Big/ \sqrt{\sum x(k)^2 \cdot \sum x(k)^2} = 1.$$

The following examples show how to calculate the correlation of a signal y_1 with itself

```
Cor[l1_, l2_] := Total[l1*l2] / √(Total[l1²] * Total[l2²])
y1 = {1, 2, 3, 1}; Cor[y1, y1]                                              1
```

and the correlation of two independent signals y_1 and y_2

```
y1 = {1, 1, 1, 1}; y2 = {-1, 1, 0, 0}; Cor[y1, y2]                          0
```

as well as the correlation of two symmetrical signals y and $-y$

```
y1 = {1, 2, 3, 1}; y2 = {-1, -2, -3, -1}; Cor[y1, y2]                      -1
```

The correlation of two signals y_1 and y_2 with reference to a delay of n time units is defined as

$$r(n) = \sum y_1(k)\, y_2(k+n).$$

Using the same normalization as above, we get the correlation coefficient

$$c(n) = \frac{\sum y_1(k)\, y_2(k+n)}{\sqrt{\sum y_1(k)^2 \cdot \sum y_2(k)^2}}.$$

The following illustration shows left a signal and right the correlation coefficient $c(n)$ of the signal with itself. The coefficient has its greatest value ($c(n) = 1$) when there is no delay ($n = 0$). If $|n|$ is greater than the length of the signal, $c(n) = 0$.

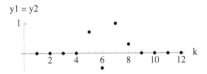
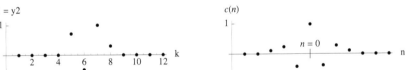

If we calculate the correlation coefficient of two random signals y_1 and y_2, we have the situation illustrated below. The correlation coefficient does not reach its maximum possible value 1 anywhere. The actual maximum is in general not at the delay $n = 0$, but for a signal of finite duration it will be near $n = 0$. As $|n|$ increases, $c(n)$ tends to decrease and becomes zero when $|n|$ is greater than the length of the signal.

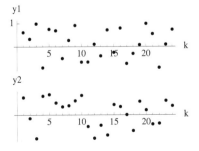
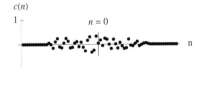

9.3 Sound Reproduction

In order to generate two decorrelated signals from a single signal $x(k)$, we treat the signal with two different all-pass filters so as not to change the signal's spectrum. If we keep the delays short, and hence the impulse responses of the filters h_1 and h_2 short too, the signal will sound the same despite the decorrelation. We then convolve the signal $x(k)$ with h_1 and h_2 and obtain two decorrelated signals y_1 and y_2 whose decorrelation is approximately that of the signals h_1 and h_2.

In order to avoid coloration, we generate a transfer function H having a constant amplitude response and an arbitrary phase response. Using the Fourier transform we calculate from H an impulse response $h(k)$. Note that we cannot realize exact all-pass filters with FIR filters but only approximations.

Let a function Hw with constant amplitude response $Abs(Hw)$ and arbitrary phase response $Arg(Hw)$ be given. From it we generate the sequence $h(k)$ (3.3.2.4). We obtain transfer functions by choosing an arbitrary function f and setting $Hw(\Omega) = \cos(f(\Omega)) + i\cdot\sin(f(\Omega))$. The following illustration shows the impulse response $h(k)$ (left) and the amplitude response (right) for $f(\Omega) = 3\Omega^2 - 3\Omega$.

$$\mathtt{Hw[\Omega\_] = N[Cos[3*\Omega^2 - 3*\Omega] + \mathtt{i}*Sin[3*\Omega^2 - 3*\Omega]];}$$

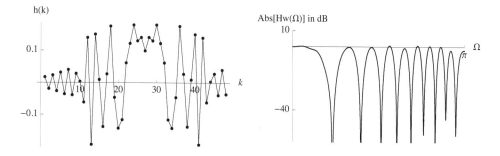

We can use a discrete spectrum in place of the continuous function Hw and calculate the sequence $h(k)$ from it. The illustration below shows amplitude and phase response for a random sequence of 32 values $h(k)$.

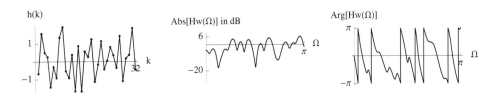

Let us assume a complex spectrum in which the amplitudes are equal to 1 and the phases random. We obtain a real sequence after taking the inverse Fourier transform of the spectrum:

```
φ = RandomReal[];
InverseFourier[{1, Sin[φ] + i*Cos[φ], 1, Sin[φ] - i*Cos[φ]}]

{1.46404, 0.885817, 0.535965, -0.885817}
```

If we generate 32 values $h(k)$ this way, we obtain a much smoother amplitude response than we had above.

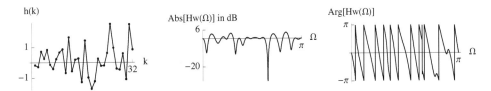

IIR all-pass filters can be realized by generating pairs of poles and zeroes in the z-plane which lie on a straight line and whose distances to the unit circle are inversely proportional to each other (3.3.3.5). The filter coefficients are straightforward to compute and the amplitude response is constant. To generate signals whose correlation is close to zero, the filter must be of a high order (see [1-12] p. 77).

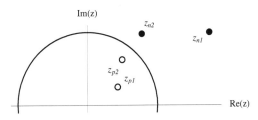

10 Computers and Composition

There are many ways to use the computer in music. In this chapter, we will only discuss the fundamentals of probability theory (Chapter 10.1), basic principles of random processes (Chapter 10.2) and certain special mathematical models and algorithms.

10.1 Chance and Probability

Further Reading: Quite brief and not very systematic introductions to combinational analysis and probability theory along with applications to computer music can be found in Computer Music *by Charles Dodge and Thomas A. Jerse [2], pp. 341-361 and pp. 374-382 and in* Elements of Computer Music *by F. Richard Moore [8], pp. 408-429. In* Fractals in Music *[29] Charles Madden shows examples of using statistics and probability theory for the analysis of music in various styles, pp. 97-117 and pp. 139-159.*

10.1.1 Fundamentals of Combinatorics

Combinatorics is concerned with combinations, permutations and arrangements in sets and with counting possible configurations. Combinatorics makes it possible to compute the number of possible orderings of discrete entities and thus provides an essential basis for probability theory.

10.1.1.1 Introductory Examples

To begin with, let us consider three examples to which we shall frequently make reference. The first, the so-called urn model, is the standard example in all introductions to combinatorics and probability. The second and third examples show applications of combinatorics to music.

For the urn model, we imagine a container holding a certain number of balls. The balls can be similar or dissimilar from each other, for example they can have the same or different colors or bear the same or different numbers. The main task of combinatorics is to determine all the possible orderings of part or all of the balls. One does this by successively drawing balls from the container, either putting each ball back or not before drawing the next ball.

We can construct a twelve-tone row by choosing a pitch from the 12 chromatic tones of the octave, noting it and striking it from the list of available pitches. This is like drawing 12 balls from the container of our urn model without putting any back.

Constructing a musical motif or theme, where the same pitch can appear more than once, is like drawing a ball from the container and putting it back before drawing the next one. From the set of pitches $T_1 = \{c\#, d, e, c, g, a\}$ we can form the theme of Bach's Art of the Fugue, if we are allowed to use pitches more than once: d, a, f, d, c#, d, e, f, g, f, e, d. We can also form the theme without using any pitch more than once from the extended list of pitches $T_2 = \{c\#, d, d, d, d, e, e, f, f, f, g, a\}$.

10.1.1.2 Permutations

The various orderings of discrete elements are called *permutations*. Let us determine how many orderings of n elements there are. For two balls, one black, one white, there are two possibilities: $(b; w)$ and $(w; b)$. For three balls, 1, 2 and 3, there are six possibilities: $(1; 2; 3)$, $(1; 3; 2)$, $(2; 1; 3)$, $(2; 3; 1)$, $(3; 1; 2)$ and $(3; 2; 1)$. To determine the number of permutations of n elements, we consider drawing n balls. For the first ball we choose, there are n possibilities, for the second $n - 1$ possibilities, for the third $n - 2$ possibilities, and so forth, until we reach the last ball left over. If one has a variety of possible choices in succession, the total number of choices is equal to the product of the individual choices, as can be seen in the following diagram showing the simple cases for two and three possibilities.

Hence, for n elements, the total number of permutations is $P(n) = n(n-1)(n-2) \cdot \ldots \cdot 1$, written as $n!$ (n factorial).

$$P(n) = n! = n(n-1)(n-2) \cdot \ldots \cdot 1$$

Thus, the number of possible 12-tone rows is 12!. But since in the classical 12-tone tradition, groups of 12 rows can be considered transpositions of the same row, we can divide this number by 12, leaving $11! = 39,916,800$ different rows.

If not all the elements are different from one another, fewer permutations are possible. For two white balls and one black ball there are six permutations, pairs of which are identical. So the number of unique permutations is $6/2 = 3$. In general, if $n_1, n_2, \ldots$ elements cannot be distinguished from each other, then $n_1!, n_2!, \ldots$ permutations cannot be distinguished from each other, and the total number of permutations has to be divided by this number. The number of permutations of n elements of which $n_1, n_2, \ldots, n_k$ elements cannot be distinguished from each other is

$$P(n; n_1, n_2, \ldots, n_k) = \frac{n!}{n_1! n_2! \ldots n_k!} := \binom{n}{n_1, n_2, \ldots, n_k} \quad (10.1.1.6)$$

To determine the number of permutations of $T_2 = \{c\#, d, d, d, d, e, e, f, f, f, g, a\}$, we set $n = 12, n_1 = 4$ (the d's), $n_2 = 2$ (the e's) and $n_3 = 3$ (the f's). We then have

$$\frac{12!}{4! \cdot 2! \cdot 3!} = 1,663,200.$$

10.1.1.3 Combinations

If from a container with n different balls one draws k balls ($k \leq n$) without considering their order, one speaks of a *non-ordered sample* or a *combination of order k without repetition*. To calculate the number of possible combinations $C(n; k)$, we first consider drawing k balls in a specific order. There are n possibilities for the first ball, $n - 1$ for the second, and so forth to the kth ball, for which there are $n - k + 1$ possibilities. Hence the number of possibilities of drawing the balls in a specific order is:

10.1 Chance and Probability

$$n(n-1)(n-2)\ldots(n-k+1) = \frac{n!}{(n-k)!}.$$

This is the number of *arrangements* of k elements (see 10.1.1.4 below). But since in our example the order is not important, we can divide this result by the number of permutations of k balls. For the number of possible combinations of k ($k \le n$) elements without repetition we then have

$$C(n; k) = \frac{n!}{k!\,(n-k)!} = \binom{n}{k}$$

A triad formed from any three tones of the chromatic scale fulfills the conditions for a non-ordered sample of $k = 3$ elements from a set of $n = 12$ distinct elements. Therefore, there are $12! / (3! \cdot 9!) = 220$ possible triads. The number of pairs of voices that need to be checked for parallel voice-leading in a ten-part counterpoint exercise is

$$\binom{10}{2} = 45.$$

If in the example above each ball is put back in the container after being drawn and so can be drawn again, one speaks of *combination* of order k *with repetition*. The number of possible combinations of k ($k \le n$) elements with repetition is

$$C_r(n; k) = \frac{(n+k-1)!}{k!\,(n-1)!} = \binom{n+k-1}{k}$$

10.1.1.4 Arrangements

If one considers in what order k balls are drawn from a container of n balls, one speaks of an ordered *sample of order k* or of an *arrangement*. We have already derived the formula for the number of arrangements above (10.1.1.3). The number of arrangements of k elements ($k \le n$) without repetition is

$$A(n; k) = \frac{n!}{(n-k)!}$$

Hence the number of motifs that can be formed from any three tones ($k = 3$) of the chromatic scale ($n = 12$) without repetition is $12!/(12-3)! = 1320$.

If we allow the balls to be drawn more than once, there are always n possibilities for each draw. Then the number of possible *arrangements* of k elements *with repetition* is

$$A_w(n; k) = n^k$$

For a theme of 12 notes with any repetitions from the set of pitches $T_1 = \{c\#, d, e, f, g, a\}$ there are $6^{12} = 2{,}176{,}782{,}336$ possibilities.

10.1.1.5 Ordering Permutations

Most of the sets we consider in connection with permutations have a specific principle of ordering (a set of numbers by size, letters in alphabetical order, pitches according to a scale). When the order is lexicographical, permutations are ordered with the first element first, the second next, etc. The lexicographical order of the permutations of the letters a, b and c is the list {abc, acb, bac, bca, cab, cba}, that of the permutations of the numbers 0, 1 and 2 {012, 021, 102, 120, 201, 210}.

10.1.1.6 Binomial and Polynomial Coefficients

The nth power of the binomial $(a + b)$, i.e. $(a + b)^n$, gives an algebraic sum of $n + 1$ terms. Each of these terms is the product of a so-called *binomial coefficient* and of a and b raised to the same sum of exponents. Arranging the coefficients as below, one gets what is called Pascal's triangle, in which each number is the sum of the two numbers above it.

$$(a+b)^0 = 1$$
$$(a+b)^1 = 1a + 1b$$
$$(a+b)^2 = 1a^2 + 2ab + 1b^2$$
$$(a+b)^3 = 1a^3 + 3a^2b + 3ab^2 + 1b^3$$

$$\begin{array}{ccccccc} & & & 1 & & & \\ & & 1 & & 1 & & \\ & 1 & & 2 & & 1 & \\ 1 & & 3 & & 3 & & 1 \end{array}$$

In general, the coefficients of $(a + b)^n$ are

$$1, \frac{n}{1!}, \frac{n(n-1)}{2!}, \frac{n(n-1)(n-2)}{3!}, \ldots \quad \text{or}$$

$$1, \frac{n!}{1!(n-1)!}, \frac{n!}{2!(n-2)!}, \ldots$$

If we define $\binom{n}{k} = \frac{n!}{k!(n-k)!}$, then we can write for the binomial coefficients

$$\binom{n}{0}, \binom{n}{1}, \binom{n}{2}, \ldots$$

(→ Polynomial coefficients)

10.1.2 Fundamentals of Probability Calculus

10.1.2.1 Standard Examples and Definitions

Let us add two further examples to those of Chapter 10.1.1, namely throwing one die and throwing two dice.

Example 1. When throwing one die, six *outcomes* ω_1 to ω_6 are possible. The totality of the outcomes is the *sample space* $\Omega = \{\omega \mid \omega \text{ is a possible outcome of the random experiment}\} = \{\omega_1, \ldots, \omega_6\} = \{1, 2, 3, 4, 5, 6\}$. Subsets of a sample space are called *events*. The event of throwing an odd number with one die corresponds to the set $\{1, 3, 5\}$, the certain event of

10.1 Chance and Probability

throwing a number between 1 and 6 is Ω, the impossible event that no number is thrown corresponds to the empty set $\{\} = \emptyset$. The events $\{\omega_1\}, ..., \{\omega_6\}$ are called *elementary events*.

Example 2. When two dice are thrown simultaneously, 36 outcomes are possible. We can represent an elementary event as an ordered number pair, e.g. (2, 5). The sample space Ω contains 36 such number pairs. $\Omega = \{(a, b) \mid 1 \le a \le 6 \text{ and } 1 \le b \le 6\}$. The event of throwing the same number twice is $A = \{(1, 1), (2, 2), (3, 3), (4, 4), (5, 5), (6, 6)\}$, throwing the sum of the two dice equal to 4 is $B = \{(1, 3), (2, 2), (3, 1)\}$. The event of throwing a sum equal to 12 is $C = \{(6, 6)\}$, that of a sum of 2 is $D = \{(1, 1)\}$ and the event of a sum greater than 9 is $E = \{(4, 6), (5, 5), (5, 6), (6, 4), (6, 5), (6, 6)\}$.

A random experiment has these properties: 1) the experiment can be repeated arbitrarily often under the same conditions; 2) several mutually exclusive outcomes are possible; 3) the outcome of an instance of the experiment cannot be predicted and depends on chance.

10.1.2.2 Combining Events

Combining events logically corresponds to the basic set operations. The table below shows the possible operations together with an example each based on the second standard example above (10.1.2.1).

Operation	Name	Meaning	Example
$M_1 \cap M_2$	Intersection	Set of events belonging to M_1 and M_2	The event of throwing the same number twice and a sum of 4 = $A \cap B = \{(2, 2)\}$
$M_1 \cup M_2$	Union	The set of events belonging to M_1 or M_2 or to both	The event of throwing a sum of 12 or a sum of 2 = $C \cup D = \{(6, 6), (1, 1)\}$
$M_1 \setminus M_2$	Set Difference	The set of events belonging to M_1 but not to M_2	The event of throwing the same number twice but a sum < 10 = $A \setminus E = \{(1, 1), (2, 2), (3, 3), (4, 4)\}$
$\overline{M}$	Complement	The set of events not belonging to M	The event of throwing a sum > 2 = $\overline{D}$
$M_1 \triangle M_2$	Symmetric Difference	The set of events belonging to M_1 or M_2 but not to both	The event of throwing either a sum of 4 or the same number twice (but not both) = $A \triangle B = \{(1, 1), (3, 3), (4, 4), (5, 5), (6, 6), (1, 3), (3, 1)\}$

10.1.2.3 The Terminology and Axioms of Probability Theory

In a random experiment with finitely many equally likely outcomes ω_i, the probability of any event can be defined as its relative frequency of occurrence. When throwing a fair die for a few times only, the frequency of 5's is indeterminate, but if one makes very many throws, the relative frequency = (number of 5's thrown) / (total number of throws) will tend toward 1/6. Since this is so for all outcomes, we can represent their probability $p(\omega_i)$ as shown in the diagram below.

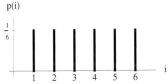

The probability of any one of n equally probable elementary events occurring is $p(\omega_i) = 1/n$. The sum of the probabilities is $n \cdot 1/n = 1$.

$$p(\omega_i) = 1/n$$

For equally likely outcomes the probability of an event A occurring is equal to the number of favorable outcomes $|A|$ divided by the number of possible outcomes n, the size of the sample space Ω.

$$P(A) = \frac{|A|}{|\Omega|}$$

To compute the probability of the events of throwing the numbers 2, 3, ... as the sums of two dice, we determine the number of favorable cases for each sum and divide by the total number of cases (36). The sum 2 corresponds only to the elementary event (1, 1), the sum 3 to the elementary events (1, 2) and (2, 1), etc. This gives the diagram below.

The following axiomatic formulation of the concept of probability is by A. N. Kolmogorov (1903 – 1987):

Every event A of a random experiment having the sample space Ω corresponds to a real number $P(A)$, called the probability of event A, so that the following axioms are satisfied:

Axiom 1: $P(A)$ is a non-negative real number.
Axiom 2: $P(\Omega) = 1$.
Axiom 3: Any countable sequence of pairwise disjoint events $A_1, A_2, ...$ satisfies
$P(A_1 \cup A_2 \cup ....) = P(A_1) + P(A_2) + ...$

From this further properties can be derived:

1) The impossible event $\emptyset$ has $P(\emptyset) = 0$.
2) The probability of the complementary event $\overline{A}$ to event A is $P(\overline{A}) = 1 - P(A)$.
3) For two arbitrary events A and B: $P(A \cup B) = P(A) + P(B) - P(A \cap B)$.

10.1.2.4 Conditional Probability and Stochastic Independence

In general, an event's probability changes if it is known that another event has occurred. For example, if we know that an even number was thrown with a die, then the conditional probability of this number's being a 6 is no longer 1/6 but 1/3. The conditional probability of B, given A, is written $P(B|A)$, and we have

$$P(B \mid A) = \frac{P(A \cap B)}{P(A)} \quad \text{with} \quad P(A) \neq 0$$

Let us consider again our second standard example, throwing two dice simultaneously. The conditional probability $P(A|E)$ of throwing the same number twice (A), under the condition that the sum of the numbers is greater than 9 (E), is 1/2. From $A = \{(1,1), (2,2), (3,3), (4,4), (5,5), (6,6)\}$ and $E = \{(4,6), (5,5), (5,6), (6,4), (6,5), (6,6)\}$ we have $P(A \cap E) = P(\{(5,5), (6,6)\}) = 1/18$ and $P(E) = 1/6$, from which we get

$$P(A|E) = \frac{1/18}{1/6} = \frac{1}{3}.$$

We can derive the multiplication theorem of probability calculus from the equation above. The probability of the simultaneous occurrence of two event A and B is

$$P(A \cap B) = P(A) P(B \mid A) = P(B) P(A \mid B)$$

Two events are said to be *stochastically independent* of each other when the occurrence of one event does not influence the occurrence of the other. There are several ways to formulate stochastic independence:

Formulation 1: $P(A \cap B) = P(A)P(B)$
Formulation 2: $P(B|A) = P(B)$ and $P(A|B) = P(A)$
Formulation 3: $P(B|A) = P(B|\overline{A})$ and $P(A|B) = P(A|\overline{B})$

10.1.2.5 Examples

To calculate the probability that the random choice with repetition of 12 pitches will yield a 12-tone row, we divide the number of favorable cases $N = 12!$ by the number of possible cases $N_p = 12^{12}$ and get $12!/12^{12} = 0.0000537232$.

The probability of getting the theme of J. S. Bach's Art of the Fugue (10.1.1.1) by choosing at random from the set $\{c\#, d, e, f, g, a\}$ is $4! \cdot 3! \cdot 2! / 6^{12} = 0.000000132305$.

10.1.3 Probability, Density and Distribution Functions of Random Variables

If one notes the probability with which certain values x occur, probability or density functions can be generated that show the expected frequency with which the values occur. In some cases, for instance to generate random numbers with a particular distribution, we need a further function whose value at x corresponds to the probability that a number smaller than x will occur. The two functions each completely describe a random distribution.

10.1.3.1 Random Variables and the Probability Function

By *random variable* we mean a function that assigns exactly one real number $X(\omega)$ to each outcome ω in the result set Ω of a random experiment. The *probability function* assigns each value of the random variable its probability. A random variable is called discrete when it can assume a finite (or countably infinite) number of values, continuous when it can assume any value within a given range. We represent random variables with capital letters, their values with small letters.

The second example in Chapter 10.1 (10.1.2.1, 10.1.2.4) can be described using these terms as follows. An elementary event is a number pair generated by throwing two dice, e.g. (2, 5). The result set Ω consists of the 36 possible outcomes $\Omega = \{(a, b) \mid 1 \leq a \leq 6 \text{ and } 1 \leq b \leq 6\}$. The function $X(\omega)$ assigns each elementary event ω a number, namely the sum of the pips of the dice thrown (e.g. $X((2, 5)) = 7$). Since only 11 values (2 to 12) can be thrown, X is a discrete random variable. The probability function $p(x)$ assigns each whole number x between 2 and 12 its probability and all other values zero.

ω	(1, 1)	(1, 2) (2, 1)	.	.	.	.	.	.	.	(5, 6) (6, 5)	(6, 6)
$X(\omega)$	2	3	4	5	6	7	8	9	10	11	12

10.1.3.2 Continuous Random Variables and Their Density Function

A *wheel of fortune* is a large circular wheel divided in n sections and having a revolving pointer. The probability that the pointer will come to rest in one of the sections is $1/n$. If the wheel is divided into ever smaller sections, the probability of hitting a particular section becomes smaller and smaller. The probability that the pointer will come to rest exactly at a given number x is zero, but one can indicate the probability that it will come to rest within a particular interval. If we assume that the circumference of the circle is 1, then the probability that the pointer will come to rest between .5 and .6 is .1 and the probability that it will come to rest somewhere is 1. The probability function is continuous and takes the same value for all x ($0 \leq x \leq 1$), even though this value is always zero. We therefore define a new function, called the *probability density function* (or simply *density function*) $f(x)$, where the area under the curve between two x-values a and b corresponds to the probability that the random variable X will have a value between a and b. In our example the area under the curve $f(x)$ between $x = 0$ and $x = 1$ is equal to 1, and we have the situation shown below:

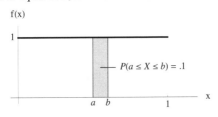

10.1 Chance and Probability

Density functions have the following properties:

$$f(x) \geq 0$$

$f(x)$ is normalized, that is the area under the curve is equal to : $\int_{-\infty}^{\infty} f(x)\,dx = 1$

When the probability function is discrete, $p(x)$ is always less than or equal to 1. The following figure shows that for continuous density functions $f(x)$ can be greater than 1.

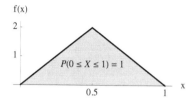

(→ *Probability_Density_Function.maxpat*)

10.1.3.3 Distribution Functions

The *cummulative distribution function*, or simply the *distribution function*, $F(x)$ of a random variable X describes the probability of X having a value smaller than or equal to x.

$$F(x) = P(X \leq x)$$

The distribution of the first example in Chapter 10.1.3.2 can be described as follows. If x is less than 0, $F(x) = 0$. If x is greater than 1, $F(x) = 1$. Between these values, x increases linearly.

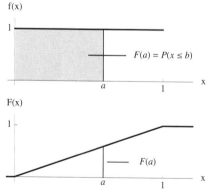

Distribution functions have these properties:

$F(x)$ is a monotone increasing function with $0 \leq F(x) \leq 1$

$\lim_{x \to -\infty} F(x) = 0$ (impossible event)

$\lim_{x \to -\infty} F(x) = 1$ (certain event)

Since the value of the distribution function $F(x)$ corresponds to the area under the density function $f(x)$ up to the point x, we have

$$F(x) = P(X \leq x) = \int_{-\infty}^{x} f(u)\,du$$

Distribution functions of discrete random variables are step functions. So the bar graph of Chapter 10.1.2.3 (left below) gives the distribution function shown to the right below.

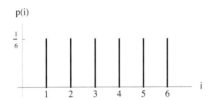
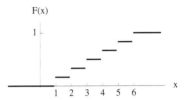

10.1.3.4 Continuous Density Functions and Their Distributions

With the help of some examples, let us now derive the appropriate distribution functions from various density functions $f(x)$.

First we construct a density function $f(x)$ from horizontal line segments. The area under the function is $.2 \cdot 2 + .6 \cdot 1 = 1$. The appropriate distribution function $F(x)$ is zero when $x < 0$, increases constantly to $.4$ when $x = 2$ (corresponding to the area under $f(x)$ until $x = 2$), and from there increases constantly to 1 when $x = 3$.

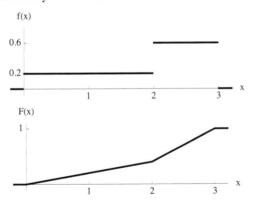

In the following linear density function $f(x) = .125x$ ($0 \leq x \leq 4$), we can no longer straightforwardly estimate the distribution function and have to have recourse to the formula

$$F(x) = \int_{-\infty}^{x} f(u)\,du.$$

The antiderivative of the function $f(x) = ax$ is $ax^2/2$. Since $f(x) = 0$ for $x \leq 0$, we can set the lower limit of integration at 0, and we get

$$F(x) = \int_{-\infty}^{x} f(u)\,du = .0625x^2 \Big]_0^x = .0625x^2.$$

10.1 Chance and Probability

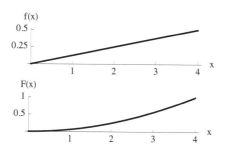

(→ Second example)

10.1.3.5 Functions of Random Variables

Using a function equation $Z = g(X)$, we can assign new random variables Z to random variables X. So, for instance, if X is a uniformly distributed random variable in the interval $[0, 1]$, then $Z = 10X + 5$ is a uniformly distributed random variable in the interval $[5, 15]$. The random variable $Y = X^2$ will have a distribution with greater density in the smaller numbers (10.1.4.2).

Using the linear function $Z = a + (b - a)X$, we can generate from a random variable in the interval $[0, 1]$ a random variable in the interval $[a, b]$ whose density function has the same shape as that of the random variable X. For example, if we multiply a random variable with a triangular density function by 10 and add 10, we get a random variable with a triangular density function in the interval $[10, 20]$.

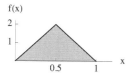 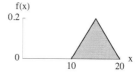

10.1.3.6 Measuring Probability Distribution

The *expected value* $E(X)$ or the *mean value* μ of a random variable indicates the on the average expected value of a random experiment. The mean value of the discrete and equally probable random values occurring when throwing dice is the arithmetic mean of the possible values $(1 + 2 + 3 + 4 + 5 + 6)/6$. The expected value of a random variable whose values do not all have the same probability is called a *weighted mean*. The expected value of a random variable with the discrete probabilities $p(2) = .2, p(3) = .2, p(4) = .6$ and $p(x) = 0$ for all other x is $.2 \cdot 2 + .2 \cdot 3 + .6 \cdot 4 = 3.2$.

The expected value E or the mean value μ of a discrete random variable X having the probability function $f(x)$ is

$$E(X) = \mu = \sum x_i f(x_i)$$

The expected value E or the mean value μ of a continuous random variable X having the density function $f(x)$ is

$$E(X) = \mu = \int_{-\infty}^{\infty} x f(x) \, dx$$

An example of a continuous random variable is the life expectancy T of an electronic component. It can be approximated as a random variable with the exponential density function $f(t) = \lambda e^{-\lambda t}$.

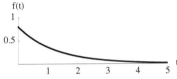

The mean lifetime expectancy corresponds to the mean value

$$E(T) = \int_{-\infty}^{\infty} t \cdot f(t) \, dt = \int_{0}^{\infty} t \lambda e^{-\lambda t} \, dt$$

$$= \lambda \int_{0}^{\infty} t e^{-\lambda t} \, dt = \lambda \left(\frac{-\lambda t - 1}{\lambda^2} e^{-\lambda t} \right)_{0}^{\infty} = \frac{1}{\lambda}$$

To get an appropriate measure of the spread of the values x_i of a random variable about its mean value μ, we take the sum of the squares of all deviations from the mean value $(x_i - \mu)$ multiplied by the frequency of the individual values. (By summing the squares of the deviations, we weight the large deviations more strongly than the small ones.) The corresponding function of the random variable X is called the variance and is denoted by σ^2.

The variance σ^2 for discrete random variables is defined as

$$\sigma^2 = E\left[(X - \mu)^2\right] = \sum_i (x_i - \mu)^2 f(x_i)$$

The variance σ^2 for continuous random variables is defined as

$$\sigma^2 = \text{Var}(X) = \int_{-\infty}^{\infty} (x - \mu)^2 f(x) \, dx$$

It is usually more convenient to compute the variance for both discrete and continuous random variables using this formula:

$$\sigma^2 = E(X^2) - \mu^2$$

The standard deviation describes the mean deviation of the random variable from its mean value and is defined as $\sqrt{\text{Var}(X)} = \sigma$.

The mean value of the example in Chapter 10.1.3.1 is 7 because the probability function is symmetrical. The variance is

$$\sigma^2 = \sum_i (x_i - \mu)^2 f(x_i) = (2-7)^2 \cdot \frac{1}{36} + (3-7)^2 \cdot \frac{2}{36} + (4-7)^2 \cdot \frac{3}{36} + \ldots = 5.83$$

10.1 Chance and Probability

The standard deviation is

$$\sigma = \sqrt{\text{Var}(X)} = \sqrt{5.83} = 2.42$$

10.1.3.7 Parametric Control of Functions

The density functions of some distributions can be represented in such a way that the position and the shape of the curve can be controlled by parameters, that is, by certain constants in the function equation. The most important function or curve parameters are the location parameters, which control the curve's position, the scale parameters, which control the scaling of the curve and hence the variance, and the form parameters, which influence the shape of the curve.

The density function of the *normal distribution*, also known as the *Gauss distribution* (10.1.5.5), has two parameters, viz. as typical location parameter the mean value μ and as scale parameter the standard deviation σ. The equation for this density function $f(x)$ is:

$$f[\mu\_, \sigma\_] := \frac{1}{\sigma \sqrt{2*\pi}} \text{Exp}\left[-\frac{(x-\mu)^2}{2*\sigma^2}\right]$$

$\mu = 2, \sigma = .4$ $\mu = 3, \sigma = 1$

The density function of the *Weibull distribution* has three parameters, viz. the location parameter a, the scale parameter b and the form parameter c (10.1.5.8).

10.1.4 Generating Random Numbers With a Given Density or Distribution

One rarely finds instructions for generating random values with a specific distribution in the literature. Here we will present various methods in considerable detail, including showing how to generate random numbers when one has a visual impression of their density but cannot specify an equation for the corresponding function.

10.1.4.1 Pseudorandom Numbers

A computer cannot generate truly random numbers. There are, however, various methods for generating so-called pseudorandom numbers, that is, numbers that are apparently randomly distributed (5.1.2.3).

Mathematica provides the functions RandomReal[], RandomInteger[] and RandomComplex[].

RandomReal[]

0.154674

This command generates a table with six random numbers between 0 and 1:

```
RandomReal[{0, 1}, 6]
```

{0.954634, 0.576965, 0.402883, 0.660969, 0.218734, 0.0434067}

This command generates a list of 10 random integers between –6 and 6:

```
RandomInteger[{-6, 6}, 10]
```

{-6, -2, -3, -2, -1, 2, -2, 2, 2, -4}

The random function rand() in the C programming language generates random integer numbers between 0 and RAND_MAX = $2^{15} - 1 = 32767$, or 0 and RAND_MAX = $2^{31} - 1 = 2147483647$, depending on the compiler (4.4.1.4). The following program in C++ generates three pseudorandom numbers (namely {16807, 282475249, 1622650073}) and displays them on the monitor.

```
int main ()
{    int i;
     for (i = 0; i < 3; i++) cout << rand () << endl;
     return 0;    }
```

Each time the program is run, it produces the same three numbers. To avoid this, we can give the random algorithm a different starting value using the function srand(value). By using the current system time, we have a value that is always different.

```
void main ()
{    int i;
     srand ( (unsigned) time ( NULL ) );
     for (i = 0; i < 3; i++) cout << rand () << endl;     }
```

By dividing the pseudorandom numbers by the largest possible value, we should get numbers between 0 and 1. But when we write rand()/*max*, we always get 0 because the program performs integer division. To fix this, we declare *max* as a floating point number: float *max* = 2147483647.0.

To distribute the random numbers in an arbitrary interval [*a*, *b*], we multiply the numbers from the interval [0, 1] by the size of the required interval (*b* – *a*) and add the lower limit of the range *a*. This command generates four random numbers between 70 and 80.

```
cout << 70 + 10*(rand()/max) <<endl;
```

Integers in an arbitrary range [*a*, *b*] can be generated from the real random numbers from the range [0, 1] as follows:

```
floor (a + (b - a + 1) * (rand () / max))        // or
a + (int) ((b - a + 1) * (rand () / max))
```

In Csound, the unit generator rand requires an amplitude *kamp*, that is, a maximum value, for the random numbers generated. The random numbers are then distributed between –*kamp* and *kamp*.

```
arnd          rand      kamp
```

10.1 Chance and Probability

To get values between 0 and *kamp*, we take the absolute value of the numbers generated. To generate numbers in the range $[a, b]$, we can proceed as above, or we can take the midpoint of the interval $m = .5(a + b)$ and add to it random numbers whose maximum amplitude is half as great as the range $b - a$.

```
arnd        rand        .5 * (ib - ia)
arndab =                .5 * (ia + ib) + arnd
```

10.1.4.2 Direct Methods for Generating a Given Distribution

Direct methods for generating random numbers having a distribution are based on the simulation of random processes. To generate random numbers corresponding to the probability function of Chapter 10.1.2.3 (simultaneously tossing two dice), we add two random numbers between 1 and 6. The following diagrams show the results of two trials of 50 throws each (top row) and the results of two trials of 1000 throws each (bottom row).

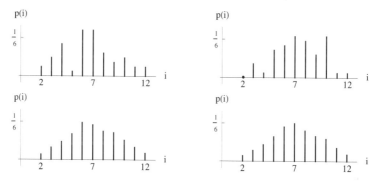

Similarly, we generate a continuous triangular density function in the interval [0, 1] by adding k times two real random numbers from the range [0, .5]. We store the results in a list l displayed below (left). Each result is a point whose x-coordinate is the sum of the two real numbers and whose y-coordinate is the number k of the trial. To display the density, we divide the display into arbitrarily many (vertical) sections and count the number of results $n(x)$ in each section. The figure to the right shows the relation $n(x)/k$ as a bar graph.

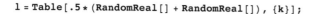

```
l = Table[.5 * (RandomReal[] + RandomReal[]), {k}];
```

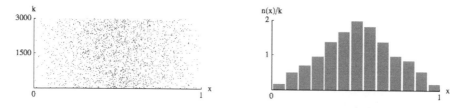

Many different distributions can be obtained by arithmetic means. In the first example below, we first distribute random numbers from the range [0, 1] into the range [0, 3] and square them. This gives numbers in the interval [0, 9] with a predominance of smaller numbers. Finally, we add 2 two each number.

```
l = Table[2 + (3 * RandomReal[])², {k}];
```

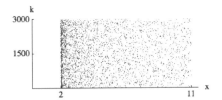 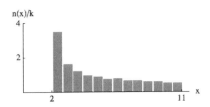

If of two random choices one always takes the smaller number, one has the following distribution:

```
l = Table[If[(a = RandomReal[]) > (b = RandomReal[]), b, a], {k}];
```

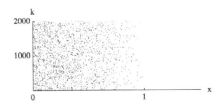 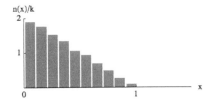

10.1.4.3 Inverting the Distribution Function

The density function $f(x) = 2x$ ($0 \leq x \leq 1$) has the distribution function $F(x) = x^2$. If we subdivide the y-axis equally and take the corresponding x-values of the distribution function, we see that they are distributed according to the density function. If we now exchange the x- and the y-coordinates, we obtain a function, $F^{-1}(x) = x^{1/2}$, which transforms the uniformly distributed random numbers from the range [0, 1] into random numbers with the density $f(x)$.

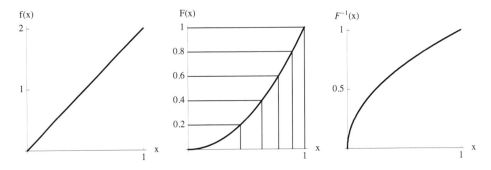

If we generate 1000 uniformly distributed random numbers in the unit range [0, 1] and transform them by taking their square root, we get the following distribution:

```
l = Table[√RandomReal[], {k}];
```

10.1 Chance and Probability

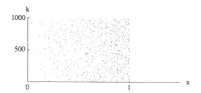
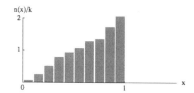

Exchanging the roles of the *x*- and *y*-axes corresponds to transforming a function into its inverse. The general procedure can be described as follows:

> To generate a random distribution with the density $f(x)$, compute the distribution function $F(x)$ and apply its inverse function to uniformly distributed random numbers from the unit range.

The procedure can be explained as follows. If we define a set Z of uniformly distributed random numbers z_i from the unit range $[0, 1]$, then the random variable $X = F^{-1}(Z)$ has the required distribution function because $P(X \leq x) = P[F^{-1}(Z) \leq x] = P[Z \leq F(x)] = F(x)$.

In the following example we generate a function of exponentially decreasing density $g(x) = e^{-x}$ over the interval $[0, 2]$. This density function is not normalized. When we integrate it, we get

$$\int e^{-x} dx = -e^{-x}$$

which is 0.864665.... for $x = 2$. We obtain the normalized density function by dividing $g(x)$ by 0.8647. The distribution function is then

$$F(x) = \int_0^x (e^{-x}/0.8646) dx = -1.1566(e^{-x} - 1)$$

and the inverse function is $F^{-1}(x) = -1 \cdot \ln[0.8646 (1.1566 - x)]$.

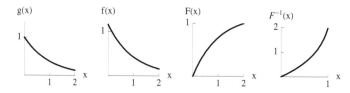

An experiment with 1000 random numbers shows the required distribution.

```
l = Table[ -Log[.8646 (1.1566 - RandomReal[])], {k}];
```

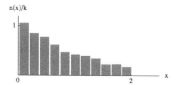

(→ *Rand_Inv_Func.maxpat*)

10.1.4.4 Rejection Sampling

The idea of *rejection sampling* is first to generate uniformly distributed random numbers in the range required and then to accept only those values corresponding to a given density function, rejecting the others. Let us demonstrate the method using a density function $f(x)$ with the form $g(x) = \sin(x)^2$ (left below) over the interval $[0, 2\pi]$. To obtain a normalized density function $f(x)$, we calculate the area

$$A = \int_0^{2\pi} \sin(x)^2\, dx = \pi$$

under the function $g(x)$ and set $f(x) = g(x)/A = (\sin(x))^2/\pi$.

We first determined a value h that is greater than $f(x)$ for all x, for example $h = .4$ (center figure). Now we generate uniformly distributed random numbers *rnd* over the interval $[0, 2\pi]$. Of these we accept a proportion corresponding to $f(rnd)/h$. For example, the value 2.01 will only be accepted in the proportion of $f(2.01)/.4 = .26/.4$. We do this by generating a new random number *rh* between 0 and h and accept *rnd* only if $rh < f(rnd)$ and reject it otherwise.

```
k = 4; l = {}; i = 0;
While[i < k , rnd = 2 * Pi * RandomReal[];
    If[rh = RandomReal[] * h ≤ f[rnd], l = Append[l, rnd]; i += 1,]]
```

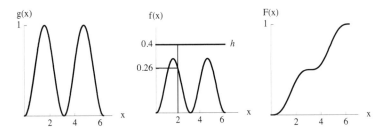

A test with a list of 4000 random numbers and a subdivision into 25 (vertical) sections led to the results below.

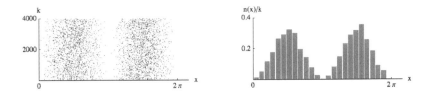

Since with this method we only care about the relationship between the values of a density function and a control value, the function doesn't need to be normalized. In the example above, we could just as well have used the function $g(x)$ and a control value of 1.
(→ Rejection_Sampling.maxpat)

The Csound instrument below shows one way to produce random numbers according to a given density function. We generate a sine wave which takes a new frequency value between 440 and 880 Hz five time each second (lines 2 and 10) as follows. We define the density by a function composed of exponential line segments (*f2*), generate a random number *krnd* from the range [0, 1] (lines 6-7) and use it as an index into *f2* to read out the number's probability

10.1 Chance and Probability

kdens (line 8). We compare this with another random number *kv* from the unit range. We repeat this procedure until *kv* < *kdens* (lines 4-9).

```
.....                                                    ;1
kr=5
.....
instr 1
c1:
kv       rand     1                                      ;5
k2       rand     1
krnd=             abs(k2)
kdens    table    krnd,2,1,0,0
         if       kv > kdens  goto c1
a1       oscil    30000,440*(1+krnd),1                   ;10
         out      a1
endin


f1 0 32768 10 1
f2 0 2048 5 1 1800 10 248 10
i1 0 10
```

10.1.4.5 Tables of Elements With Specified Frequency of Occurrence

By constructing a table containing 10 ones, 3 twos and 1 three and choosing items from the table at random we can generate integers between 1 and 3 with a relative frequency of occurrence $p(1):p(2):p(3) = 10:3:1$.

```
l = {1, 1, 1, 1, 1, 1, 1, 1, 1, 1, 2, 2, 2, 3};
r = Table[l[[RandomInteger[{1, 14}]]], {40}]

{1, 2, 1, 1, 1, 2, 1, 1, 2, 1, 1, 1, 1, 1, 1, 1, 2, 1,
 2, 2, 1, 1, 1, 1, 1, 1, 1, 1, 1, 1, 1, 1, 3, 2, 1, 1, 2}
```

We can generate real random numbers with the density shown below by creating a list of integers corresponding to their relative frequency of occurrence $ll = \{1, 2, 3, 3, 3\}$, choosing an integer randomly from this list and adding a real random number from the range $[n-1, n]$.

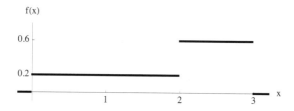

```
ll = {1, 2, 3, 3, 3};
l = RandomReal[{rnd = ll[[rnd = RandomInteger[{1, 5}]]]; rnd - 1, rnd}, 20]

{2.41771, 2.366, 0.749389, 0.773394, 2.74869, 2.29225,
 1.86919, 0.32268, 1.93633, 2.26436, 1.57882, 1.35381, 0.24358,
 2.09821, 2.55508, 0.631928, 2.58889, 1.14126, 2.24069, 2.71628}
```

A test with 3000 numbers gives the following result:

 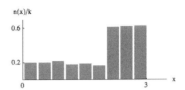

We could get the same result by making a long list of 2000 real numbers between 0 and 2 and 3000 real numbers between 2 and 3 and then choosing directly at random from this list.

```
l1 = Table[i*.001, {i, 2000}];
l2 = Table[2 + i*.000333, {i, 3000}];
l1 = Join[l1, l2];
l = Table[l1[[RandomInteger[{1, 5000}]]], {16}]

{2.60939, 1.928, 2.47486, 0.901, 2.71029, 2.87313, 1.706, 2.77556,
 2.27572, 2.08758, 2.68165, 0.733, 0.779, 2.18648, 2.59141, 2.50882}
```

We have constructed a table using this method (left figure below), which corresponds to the inverse function of the distribution function of the example above (right figure below).

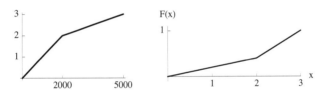

In this way, we generate tables of inverses of distribution functions using a Csound instrument. In Csound we will limit ourselves to random numbers Z in the unit range [0, 1], since we can transform them into random numbers in any interval $[a, b]$ (10.1.4.1). (In the figure above left, the unit range (x-axis) is divided into 5000 values.)

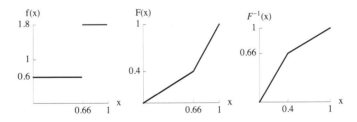

Generating the inverse F^{-1} of a function corresponds to exchanging the axes of its graph. This means that the slope sl of a segment of the slope of the function F^{-1} is reciprocal to the slope of the corresponding segment of the function F. We can generate functions with a given slope in Csound by reading a table of the function $y = x$ with the frequency sl. Since the slope of F at point x corresponds to the value of the density function $f(x)$ at point x, the inverse function has the slope $1/f(x)$. We generate a table of the function F^{-1} having 32768 points by synthesizing a sound of length $32768/44100 = .743039$ seconds that we can read as a table thanks to the Csound function GEN 1. We need two tables, the first ($f1$) containing the required density function, the second ($f2$) for the function $y = x$. The instrument shown below uses an oscillator

10.1 Chance and Probability

to read the value $k1 = f(x)$ from the first table (line 6) to compute the oscillator's frequency. The same frequency is used by a second oscillator to read from the second table (line 7).

```
;10-1-4a.orc
sr = 44100                                              ;1
kr = 44100
....
instr 1
k1        init    1                                     ;5
k1        oscil   1,1/(k1*p3),1
a1        oscil   1,1/(k1*p3),2
          out     a1
endin

f1 0 32768 -7 .6 21845 .6 1 1.8   10922   1.8
f2 0 32768 -7 0 32768 32768
i1 0 .743039
```

In the example above, we assumed that the density function is normalized. Often, however, we only have an idea of the "shape" of a density function and would like to generate it, either by using a GEN function in Csound or an external drawing program, or by deriving it from a sound or a control function. Therefore we need to show how a function can be normalized. To be used as a density function, an arbitrary function has to be everywhere positive. We can determine the area under the curve of a GEN function of length 32768 by adding all the function values and dividing the sum by 32768. If we display the summation, we can get an idea of the distribution function and its greatest value, which corresponds to the sum.

```
;10-1-4b.orc
sr = 44100
kr = 44100
.....
instr 1
k1        init    0
k2        oscil   1,1/p3,1
k1=               k1+k2
          display k1/32768,.743
endin

f1 0 32768 -7 .6 21845 .6 1 1.8   10922   1.8
i1 0 .743039
```

When we apply this to our example, we get a curve corresponding to the function $F(x)$ above with a maximal value of $im = .998$. When we apply it to the arbitrary curve below, we get $im = 5.231$.

```
f1 0 32768 -7 1 9000 2 5000 1 15000 15 3768 1
```

In order to normalize the density function, we can either divide the entire function by this value, or, more simply, since we only care about the function F^{-1}, replace $k1$ by $k1/im$ in the instrument generating this function.

10.1.5 Particular Distributions

10.1.5.1 Continuous Uniform Distribution

A random variable X that can assume any value between a and $a + b$ with equal probability is said to have *continuous uniform distribution* in the interval $[a, a + b]$. This is often written as $X \sim U(a, b)$. The density is given by

$$f(x) = \begin{cases} 1/b & \text{for } a \leq x \leq a + b \\ 0 & \text{otherwise} \end{cases}$$

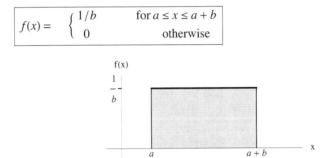

Two parameters are:

Mean	$\mu = a + b/2$
Variance	$\sigma^2 = b^2/12$

The equation of the distribution function is:

$$F(x) = \begin{cases} 0 & \text{for } x < a \\ (x-a)/b & \text{for } a \leq x \leq a + b \\ 1 & \text{for } x > a + b \end{cases}$$

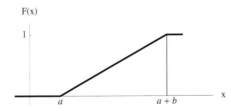

Every programming language provides functions for generating uniformly distributed random numbers (10.1.4.1). Uniformly distributed random numbers Z over the interval $[0, 1]$ can be transformed into random numbers over the interval $[a, a + b]$ using $X = a + bZ$.

10.1.5.2 Trapezoid Distribution

The density function $f(x)$ of the trapezoid distribution $Tr(a, b, c, d)$ rises linearly from 0 at $x = a$ to a maximum at $x = a + cb$, remains constant until $x = a + db$ and then falls linearly to 0 at $x = a + b$ where $a \in \mathbb{R}, b > a, 0 < c < d < 1$.

10.1 Chance and Probability

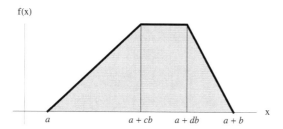

We note the following special cases:

Uniform Distribution	$c = 0; d = 1$
Symmetrical Triangular Distribution	$c = .5; d = .5$
Asymmetrical Triangular Distribution	$c = d \ne .5$

Random numbers having this distribution can be generated as sums of two uniformly distributed random numbers $X_1 = U(a_1, b_1)$ and $X_2 = U(a_2, b_2)$:

$$X = X_1 + X_2 \sim Tr(a_1 + a_2; b_1 + b_2; \frac{b_1}{b_1 + b_2}; \frac{b_2}{b_1 + b_2})$$

10.1.5.3 Binomial Distribution

Repeating a random experiment under equal conditions can be viewed as a compound random experiment. If the outcomes of the compound experiment are defined as the number of occurrences of outcom ω of the individual trials baring the constant probability P, the binomial distribution is obtained. Let $q = 1 - p$ be the probability that ω does not occur in the individual trial.

When tossing a coin, the probability of throwing "heads" is equal to that of throwing "tails": $p = q = .5$. Let us consider the probability $f(x)$ that in n tosses of a coin A occurs x times. By using the urn model of Chapter 10.1.1.1 and putting the result thrown back into the urn each time, we can derive the discrete probability function $f(x)$ of the so-called binomial distribution (10.1.1.6).

$$f(x) = \binom{n}{x} p^x q^{n-x}$$
Mean $\mu = np$
Variance $\sigma^2 = npq$

The distribution is called binomial because the probabilities correspond to the development of $(q + p)^n$.

$$(q + p)^n = q^n + \binom{n}{1} q^{n-1} p + \binom{n}{2} q^{n-2} p^2 + \binom{n}{3} q^{n-3} p^3 + \dots + p^n$$

If we toss a coin 10 times, we have this probability function:

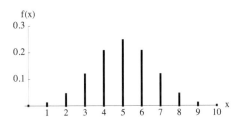

For a random test with $p = .8$ and $q = .2$ we get this probability function:

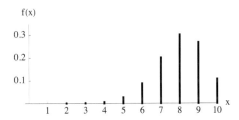

10.1.5.4 Poisson Distribution

When there are many trials n and a small probability p, the binomial distribution can be replaced by the Poisson distribution. (The probability function can be derived from the probability function of the binomial distribution by letting $n \to \infty$ and $p \to 0$.)

$f(x) =$	$\dfrac{\mu^x}{x!} e^{-\mu}$
Mean	μ
Variance	$\sigma^2 = \mu$

In a random experiment of 5000 trials with two possible results A and $\overline{A}$ and a probability for the result A of $p = .001$, the mean of the binomial distribution is $\mu = 5$. Putting that into the formula for the Poisson distribution gives us:

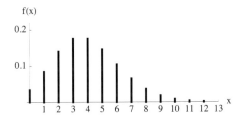

If Z_i are equally distributed random numbers from the unit interval, then

$$X = \max \{j: \ln Z_0 + \ln Z_1 + \ldots + \ln Z_j \geq -\lambda\}$$

are random numbers with the Poisson distribution $P(\lambda)$. In other words, one generates random numbers and sums their logarithms, testing whether the sum is less than $-\lambda$. The number j of random numbers Z_i generated until the test fails gives the random variable X.

10.1.5.5 Normal Distribution

The *normal* or *Gaussian distribution* plays an important role in probability theory and in statistics, because for certain parameter values many distributions converge to it, and because other distributions are based on it.

$$f(x) = \frac{1}{\sigma\sqrt{2\pi}} \exp\left(-\frac{(x-\mu)^2}{2\sigma^2}\right)$$

Mean μ
Variance σ^2

The graph of the function is known as the *Gaussian function* or the *bell curve*. The peak of the curve is at $x = \mu$, the points of inflection at $x = \mu + \sigma$ and $x = \mu - \sigma$. It is written $N(\mu, \sigma^2)$.

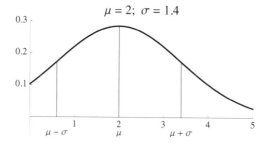

The distribution function cannot be represented with elementary functions but can only be given as an integral corresponding to its definition.

$$F(x) = \frac{1}{\sigma\sqrt{2\pi}} \int_{-\infty}^{x} \exp\left(-\frac{(t-u)^2}{2\sigma^2}\right) dt$$

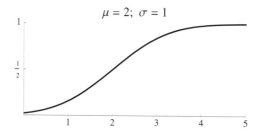

If μ is set to 0 and σ to 1, one speaks of the standard normal distribution $N(0, 1)$. The density function simplifies to:

$$f(x) = \frac{1}{\sqrt{2\pi}} \exp\left(-\frac{x^2}{2}\right)$$

If Z_1 and Z_2 are equally distributed random numbers from the unit interval, then

$$X = \sqrt{-2\ln(Z_1)}\,\cos(2\pi Z_2) \qquad \text{(and similarly } X = \sqrt{-2\ln(Z_1)}\,\sin(2\pi Z_2)\text{)}$$

creates random numbers with the normal distribution $N(0, 1)$.

The following formula gives random numbers with standard normal distribution as n goes to infinity. One already obtains a good approximation of the distribution when n is 12:

$$X_n = \frac{\sum_{i=1}^{n} Z_i - n/2}{\sqrt{n/12}} = N(0; 1); \qquad X_{12} = \sum_{i=1}^{12} Z_i - 6 \approx N(0; 1)$$

where Z_i are equally distributed random numbers. By using the following transformation, one can generate approximately normal distributed random numbers Y:

$$Y = \mu + \sigma X_{12} \sim N(\mu; \sigma^2).$$

10.1.5.6 Exponential Distribution

The exponential distribution has the following density function:

$f(x) =$	$\begin{cases} \lambda e^{-\lambda x} & \text{for } x > 0 \\ 0 & \text{for } x < 0 \end{cases}$
Mean	$\mu = 1/\lambda$
Variance	$\sigma^2 = 1/\lambda^2$

The figure below shows the development of the density function for three parameter values $\lambda = .6, \lambda = 2$ and $\lambda = 9$.

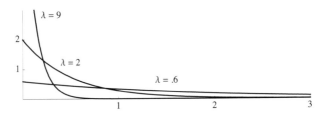

We then have for the distribution function:

$F(x) =$	$\begin{cases} 0 & \text{for } x < 0 \\ 1 - e^{-\lambda x} & \text{for } x \geq 0 \end{cases}$

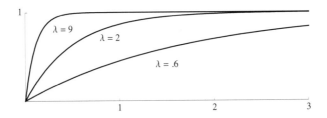

The inverse function of the density function is:

$$F^{-1}(P) = -\frac{1}{\lambda} \ln(1 - P) \quad ; 0 \leq P \leq 1.$$

10.1 Chance and Probability

Random numbers X having exponential distribution can be generated using uniformly distributed random numbers $Z \sim U(0,1)$ by the formula:

$$X = -\frac{\ln(Z)}{\lambda}.$$

10.1.5.7 Gamma Distribution

The *gamma distribution* has the density function:

$$f(x) = \begin{cases} \dfrac{\lambda (\lambda x)^{c-1} e^{-\lambda x}}{\Gamma(c)} & \text{for } x > 0 \\ 0 & \text{for } x < 0 \end{cases}$$

Mean $\quad \mu = c/\lambda$
Variance $\quad \sigma^2 = c/\lambda^2$

The figure below shows the development of the density function for the parameter values $\lambda = 3, c = .5, 1.5$ and 3.

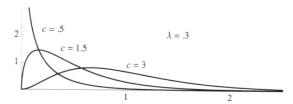

The distribution function can only be computed numerically. The figure below shows its development for the parameter values above.

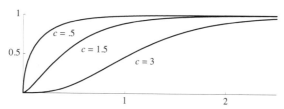

10.1.5.8 Weibull Distribution

The density function of the *Weibull distribution* is given by the equation

$$f(x) = \begin{cases} \dfrac{c}{b}\left(\dfrac{x-a}{b}\right)^{c-1} \exp\left(-\left(\dfrac{x-a}{b}\right)^c\right) & \text{for } x \geq a \\ 0 & \text{for } x < a \end{cases}$$

Mean $\quad \mu = a + b\,\Gamma\!\left(1 + \dfrac{1}{c}\right)$
Variance $\quad \sigma^2 = b^2\,\Gamma\!\left((1 + 2/c) - \Gamma^2(1 + 1/c)\right)$

The figures below illustrate various values of the individual parameters.

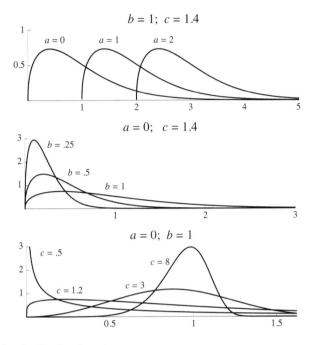

We have for the distribution function:

$$F(x) = \begin{cases} 0 & \text{for } x < a \\ 1 - \exp\left(-\left(\frac{x-a}{b}\right)^c\right) & \text{for } x \geq a \end{cases}$$

The inverse function of the distribution function can be written explicitly:

$$F^{-1}(P) = a + b\left(\ln\left(\frac{1}{1-P}\right)\right)^{1/c}$$

By using the equation for the inverse function of the distribution function, one can generate random numbers X having the Weibull distribution directly from the uniform distribution $U(0, 1) = Z$:

$$X = F^{-1}(Z) = a + b\left(\ln\left(\frac{1}{1-Z}\right)\right)^{1/c} \quad \text{or, since } 1 - Z = Z: \quad X = a + b\left(\ln\left(\frac{1}{Z}\right)\right)^{1/c}.$$

10.1.6 Applications

10.1.6.1 Sound Examples From Random Number Sequences

A sequence of mutually independent random numbers generates white noise. Their distribution does not matter, only that they be mutually independent of one another. The following sounds all have the same spectrum.

1. Sequence of uniformly distributed random numbers:

```
t = RandomReal[{-1, 1}, 200];
```

2. Sequence of random numbers with a triangular distribution:

```
t = Table[0.5` (RandomReal[{-1, 1}] + RandomReal[{-1, 1}]), {200}];
```

3. Sequence of exponentially distributed random numbers (the factor Sign[Random[Real,{–1,1}]] causes the positive random numbers –log(Random())/λ to be multiplied by 1 or –1 randomly):

$$\lambda = 8; t = \text{Table}\left[-\frac{\text{Sign}[\text{RandomReal}[\{-1, 1\}]] \, \text{Log}[\text{RandomReal}[]]}{\lambda}, \{200\}\right];$$

4. Sequence of random numbers from the set {–1, 0 1}:

```
λ = 20; t = RandomInteger[{-1, 1}, 200];
```

White noise is almost always generated using uniformly distributed random numbers. Generating noise from the set {–1, 0, 1) has the advantage of being very fast. The differences between the various noises become evident when the noise is processed (8.2.1.5). When we measure over a sufficiently long duration, we obtain a flat spectrum even from distributions that differ markedly from uniform distribution. The noise can, however, sound rougher when single large values are heard as pulses, as is the case in the exponential distribution.

10.1.6.2 The Application of Random Numbers to Musical Parameters

In order to experiment with random numbers in the realm of pitch, we must first distribute the frequencies exponentially, because pitch is perceived logarithmically. To distribute pitches randomly in the three-octave range between A2 and A5, we generate uniformly distributed random numbers $X = U(0, 3)$ and get the uniformly distributed pitches by using $110 \cdot 2^X$.

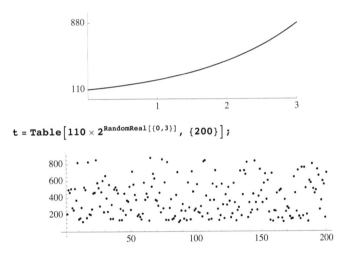

$$t = \text{Table}\left[110 \times 2^{\text{RandomReal}[\{0,3\}]}, \{200\}\right];$$

Pitch in Csound can be given in the so-called "octave point pitch class" format pch, where the number to the left of the decimal point indicates the octave and the number to the right the pitch classes from 0 to 11. To realize the example above, we start from the lowest pitch A2, written in pch-format 6.09 and add random numbers from the range [0, 36], corresponding to the 36 half-tones in three octaves. With the value conversion function cps = cpspch(pch) we can convert the pch-values (between 6.09 and 6.45) to the corresponding frequencies.

It is simpler to realize a given temporal distribution of the attack times of a set of randomly distributed notes. For example, over a total duration of 10 seconds we can first get an increase in the number of attacks per time unit and then after 5 seconds a decrease in the number of attacks over 5 more seconds by computing the attack times directly using a triangular distribution:

$$t = \text{Table}[\text{Point}[\{\text{RandomReal}[\{0, 5\}] + \text{RandomReal}[\{0, 5\}], 0\}], \{40\}];$$

We perceive the duration of individual tones and relationships among tones in very different ways. For instance, one can hardly determine the distribution of the durations of tones between 8 and 10 seconds in length because the differences in length are too small and the durations too long. In general, in complexes of several tones it is unclear whether we perceive individual durations of tones or rather the tones' relationships to each other (see the discussion in connection with techniques of serial composition, e.g. P. Boulez [30] p. 83f). Sometimes a tone's beginning and end are perceived, rather than its duration. For example, if we generate a 12-note chord of simultaneously beginning tones with uniformly distributed random durations of between 8 and 10 seconds, we hear for two seconds a uniform distribution of the ends of the tones.

10.1.6.3 Applications Using Variable Distributions

If we always use the same random distribution to determine musical parameters, the music soon sounds monotonous. For this reason, we usually change random distributions frequently. If we have a direct method to generate a given distribution, we can vary the distribution's parameters while using it.

The position and range of a uniform distribution $U(a, a + b)$ can be varied continuously using the parameters a and b. In the example below, the minimal value $a = .02t$ and the size of the range $b = 1 + .02t$ are linear functions of time t.

```
l = Table[0.02 t + (1 + 0.02 t) RandomReal[], {t, 200}];
```

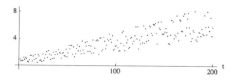

In the next example, we use the algorithm from Chapter 10.1.5.5 to generate a normal distribution with mean $\mu = 10 \cdot \sin(t)$ and standard deviation $\sigma = \sqrt{.01\, t}$.

```
l = Table[z1 = RandomReal[]; z2 = RandomReal[];
    10 Sin[0.003 t] + √0.01 t √-2 Log[z1]  Cos[2 π z2], {t, 2000}];
```

We can create transitions between various types of distribution by "cross-fading" from one to another. Whether we perceive this as a variation within one and the same distribution or whether the change of distribution is audible as such, depends on how strongly the types and ranges of the distributions differ and on how recognizable the distributions are in themselves. (Similar problems arise as with the cross-fading of spectra.) The transition from a uniform to a normal distribution with the same mean value will probably be heard as a variation of the same distribution (Example 1 below), whereas the transition between two distributions having very different ranges will more likely be heard as a cross-fade between distributions (Example 2).

Example 1: Transition from the uniform distribution $U(-1, 1)$ to the normal distribution $N(0, 1)$:

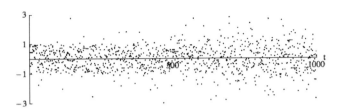

Example 2: Transition from the uniform distribution $U(0, 10)$ to the normal distribution $N(15, 5)$. Whether or not we can identify a distribution and tell the difference between distributions depends principally on the number of random numbers generated.

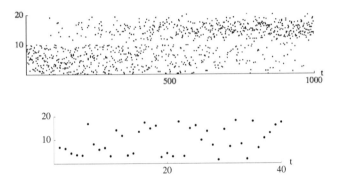

We can also vary a density function over time and generate random values as described in Chapter 10.1.4. The density function can be given either explicitly as an equation or by a table which is frequently changed. In the example which follows we use the function

$$g(x, t) = e^{-\text{abs}(x)} \cos(\pi x c t)$$

to determine the density of the random numbers in the range $[-\pi, \pi]$. As the figures below show, not all values of the function are positive and the function is not normalized, and so it does not actually fulfill the requirements for a density function. But if we generate the random numbers by rejection sampling (10.1.4.4), it suffices that the function be limited and that a part always be positive. No random numbers are generated in those areas where $g(x, t)$ is less than zero, but in the areas of positive function values $g(x, t)$, the relative frequency of occurrence of the random numbers generated corresponds to these function values. (→ Animation)

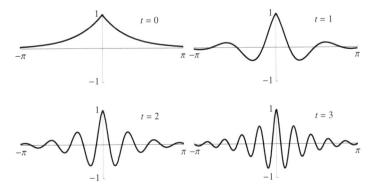

10.1.6.4 Choosing Among Random Values

Up to now, we have always assumed that the random values are determined only by whatever distribution is used. But very often, we want the values to fulfill other conditions as well. For example, if to a given sequence of notes we want to generate a second voice that should be random but nonetheless should satisfy certain rules of counterpoint, we usually cannot integrate the rules in the random process. Instead we generate random values and then test whether they satisfy the rules. If not, we repeat the process. Since "correct" values are, so to speak, being sifted out of the random values, the algorithms used are called *random sieves*.

In random sequences of tones or numbers, especially when only a few elements are used, there are often repetitions.

```
l = RandomInteger[{0, 4}, 20]
```
{1, 4, 2, 4, 3, 2, 3, 3, 3, 3, 2, 4, 2, 0, 1, 2, 1, 2, 2, 2}

To avoid repetitions, the last value generated can be stored ($xx = x$) and compared with the new value generated.

```
x = 5; tmax = 20; l = Table[0, {t, tmax}];
Do[While[xx = x; l[[t]] = x = RandomInteger[{0, 4}]; x == xx, Null],
   {t, 1, tmax}]; l
```
{2, 4, 0, 2, 1, 2, 3, 1, 4, 0, 4, 0, 4, 2, 3, 1, 3, 4, 2, 4}

10.2 Stochastic Processes

10.2.1 Introductory Examples and Concepts

10.2.1.1 Games of Chance

Games of chance provide good examples of stochastic processes. Let us consider a dice game with the following rules. One's capital (equal to 1 at the beginning) is doubled (d) each time one tosses a 6, it remains the same (s) if one tosses 4 or 5, and it is cut in half (h) if one tosses 1, 2 or 3. In each round the probability of winning is 1/6, that of losing 3/6 and that of staying put 2/6. In n rounds there are 3^n possible different games ω. In the illustration below, one can see the $3^2 = 9$ different games leading to the five outcomes possible at $t = 2$. We will number them according to the possible throws of the die: $\omega_{s,s} = 1$, $\omega_{s,d} = 2$, etc.

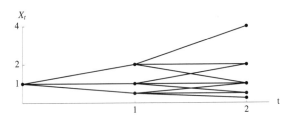

Every game ω has a certain value X_t at every moment t. For a fixed ω^* at a fixed time t^* we get a number $X(\omega^*, t^*) = x^*$. In our example we have $X(1, 1) = .5$, $X(1, 2) = .25$, etc. For a fixed ω^* and a variable time t, we get a sequence of numbers representing one possible game. This sequence is called a path or trajectory of the process (it is also called a time series). If in our example one wins all the time, we get the time series 1, 2, 4, 8, etc. For a fixed t^* and variable ω, we get a random variable $X(\omega, t^*) = X_{t^*}$ with the appropriate distribution dependent on t^*. For example, when $t^* = 2$ we have the discrete distribution of the random variable $X(\omega, 2) = X_2$ with the probability function shown below. We can compute the individual probabilities $P(x)$ from the sum of the probabilities of all games ω that lead to the result $X = x$ after two rounds. Only one game leads to the result $X = .25$, namely $\omega_{h,h}$ with the probability $3/6 \cdot 3/6 = .25$. Three games lead to the result $X = 1$, namely $\omega_{s,s}$, $\omega_{h,d}$ and $\omega_{d,h}$, with the probability $2/6 \cdot 2/6 + 3/6 \cdot 1/6 + 1/6 \cdot 3/6 = 10/36$. The sum of all the probabilities of obtaining one of the possible results must be 1.

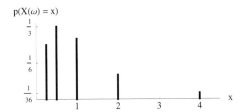

When both ω and t are variable, we have the entire stochastic process, which can be understood as a collection of time series (X_t) one of which is realized in a game, or else as a succession of random variables $X_1, X_2, ..., X_t, ...$ in which a random variable is attributed to every point in time. The following diagram show a possible randomly generated game.

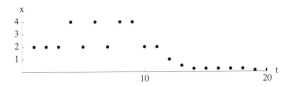

10.2.1.2 White Noise

In Chapter 10.1 we ordered random numbers in time to produce noise or to generate time-variable distributions. Let us describe the Examples 1) and 4) from Chapter 10.1.6.1 as realizations of stochastic processes using the concepts introduced above. In Example 1) we generated noise by randomly choosing real numbers between −1 and 1 and placing them in sequence. In Example 4), we used randomly chosen members of $\{-1, 0, 1\}$. Every sequence of this kind is a time series that can be understood as a realization, path or trajectory of a stochastic process called a *white noise process* or simply *white noise*. The diagrams below

10.2 Stochastic Processes

show the random variable X_t for $t = 0$ to 3 for Example 4) (left) and Example 1) (right). A realization of the process corresponds to a path through one point for each of the possible values of X_t.

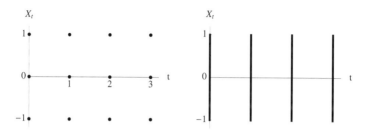

Since the distribution of the random variables $X(\omega, t)$ is not dependent on time, we can visualize for all possible times the probability function of Example 4) (figure below left) and the density function for Example 1) (figure right below). As described in Chapter 10.1.6.1, any distribution can be used to generate white noise.

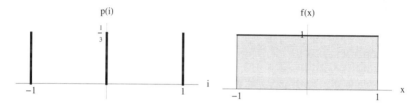

10.2.1.3 General Formulation of the Concepts

A function X that maps $\Omega \times T$ to $\mathbb{R}$ is called a (one-dimensional real) *stochastic process*. Here Ω is the sample space, that is, the set of possible outcomes, of a random experiment (10.1.2.2), $t \in T$ is a parameter, for example time, and $\Omega \times T$ is the Cartesian product of the two sets, in other words the set of ordered pairs (ω, t) of elements from Ω and T. We speak of a *point process* when the state space, that is, the set of $X(\omega, t)$, is finite or countably infinite. We speak of a *real-valued* or *diffusion process* when the state space is uncountable. Depending on whether discrete time or continuous time is used as a parameter, one speaks of *discrete* or *continuous processes*.

Various mean values are defined. The *temporal mean* $\mu_\omega := E_T\{X(\omega, t)\}$ is the mean or expected value of the random variable of the trajectory belonging to ω. The *ensemble mean* $\mu_t := E_\Omega\{X(\omega, t)\}$ is the mean or expected value of the random variable X at time t. The *total mean* $\mu := E_\Omega E_T\{X(\omega, t)\} = E_T\{\mu_t\} = E_\Omega\{\mu_\omega\}$ is the mean or expected value of all realizations of a stochastic process.

The *variance function* $\sigma_t^2 := E_\Omega\{(X(\omega, t) - \mu_t)^2\} = E_\Omega\{X^2(\omega, t)\} - \mu_t^2$ shows the scatter of the random variable X_t with reference to the mean at this point for every point in time t.

To measure the degree of independence of two random variables at times t_1 and t_2, we define the *autocovariance function* (usually called the *covariance function*) $\gamma(t_1, t_2)$ and the *autocorrelation function* (usually *correlation function*) $\rho(t_1, t_2)$ as follows:

$$\gamma(t_1, t_2) := E_\Omega \{(X(\omega, t_1) - \mu(t_1)) \cdot (X(\omega, t_2) - \mu(t_2))\}$$
$$= E_\Omega \{X(\omega, t_1) X(\omega, t_2)\} - \mu(t_1)\mu(t_2)$$

$$\rho(t_1, t_2) \equiv \frac{\gamma(t_1, t_2)}{\sigma(t_1)\sigma(t_2)}$$

When $t_1 = t_2 = t$, we obtain, for example, the covariance of a random variable with itself $\gamma(t, t) = \sigma_t^2$ and the autocorrelation of a random variable with itself $\rho(t, t) = 1$. For processes whose random variables are pairwise independent, so for example for white noise, we have $\gamma(t_1, t_2) = \rho(t_1, t_2) = 0$ for $t_1 \neq t_2$.

10.2.2 Markov Chains

The *Markov chain* is a type of stochastic process often used in computer music. Markov chains are Markov processes with the particularity that both time and state space are discrete.

10.2.2.1 Introductory Examples

Example 1. Let us consider a system that at any moment can be in one of n states $(A_1, A_2, \ldots A_n)$. One can often indicate the probability with which the system will change state after a certain time, either on the basis of given rules or by statistical analysis. If we reduce the harmonic movement of a tonal piece of music to the succession of the three harmonic functions subdominant (A_1), tonic (A_2) and dominant (A_3), we might find these probabilities for the transition (or not) from one state to another:

T → S	T → T	T → D	S → S	S → T	S → D	D → S	D → T	D → D
.2	.5	.3	.3	.4	.3	.1	.4	.5

We can show the transition probabilities either as a table (left below) or as a matrix (right). The sum of the probabilities in any one column is always 1.

Previous function

Next function

	S	T	D
S	.3	.2	.1
T	.4	.5	.4
D	.3	.3	.5

$$\begin{pmatrix} .3 & .2 & .1 \\ .4 & .5 & .4 \\ .3 & .3 & .5 \end{pmatrix}$$

The following program shows a realization of this process. We store the numbers for the harmonic functions (1, 2, 3) in the list f and generate the random numbers with the required probabilities using the method of Chapter 10.1.4.5. In addition, for each harmonic function we need a list (s, t and d) containing the number of the next function to choose in the frequency of occurrence given by the transition probabilities. We call the previous (actually the current) function x. We declare a list l to be filled with 20 harmonic functions. Before determining each next function, the routine *Switch* examines the previous function. If it was subdominant (1), for example, the number of the next function is chosen randomly from the list s. Finally, we use our list of 20 numbers l to read out the appropriate harmonic functions from f.

10.2 Stochastic Processes

```
f = {S, T, D}; s = {1, 1, 1, 2, 2, 2, 2, 3, 3, 3};
t = {1, 1, 2, 2, 2, 2, 2, 3, 3, 3}; d = {1, 2, 2, 2, 2, 3, 3, 3, 3, 3}; x = 2;
l = Table[Switch[x, 1, x = s[[RandomInteger[{1, 10}]]], 2,
    x = t[[RandomInteger[{1, 10}]]], 3, x = d[[RandomInteger[{1, 10}]]]],
    {i, 20}]; Table[f[[l[[i]]]], {i, 20}]

{T, T, T, D, D, D, D, T, T, S, T, S, S, D, T, T, S, T, S, D}
```

From the table of transition probabilities, using the method shown in Chapter 10.2.1.1, we can compute the probabilities after two or more steps in the process by summing the probabilities of all possible paths from the first harmonic function to each of the three functions. If we start with the tonic, then the probability of reaching the tonic again after two steps is:

$$P[T \to X \to T] = P[T \to T \to T] + P[T \to S \to T] + P[T \to D \to T] = .5 \cdot .5 + .2 \cdot .4 + .3 \cdot .4 = .45.$$

Example 2. One can use various methods to generate a random melody in a particular style. Using distributions of pitches and rhythms derived from analysis gives bad results. The method of random sieves (10.1.6.4) gives "arbitrarily good" results if the melodic structure can be described by rules and if all the rules are known. This method is, however, time-consuming, and the rules are not straightforward to program. In what follows, we do not generate individual random pitches according to a given distribution. Instead, we indicate for each pitch the probability of its occurring after each possible previous pitch. The probability data below comes from Gregorian chants in the first mode found in the Kyrial of the Graduale Triplex and show how often each pitch occurs in the corpus. The pitch range corresponds to the diatonic scale from c3 to d4. Since b-natural is usually lowered to b-flat, and even where b-natural is written usually b-flat is sung, we shall use only b-flat for our example. We shall also disregard e4 and f4, since they each only occur once.

	c	d	e	f	g	a	b	c	d
c	2	50	3	0	0	0	0	0	0
d	32	38	58	32	2	11	0	0	0
e	12	31	11	59	2	1	0	0	0
f	5	30	25	6	63	4	0	0	0
g	0	4	20	35	11	79	7	5	2
a	0	13	0	7	63	46	24	6	4
b	0	0	0	0	6	20	0	13	0
c	0	0	0	0	1	16	12	5	17
d	0	0	0	0	2	4	0	17	2

$$m = \begin{pmatrix} 2 & 50 & 3 & 0 & 0 & 0 & 0 & 0 & 0 \\ 32 & 38 & 58 & 32 & 2 & 11 & 0 & 0 & 0 \\ 12 & 31 & 11 & 59 & 2 & 1 & 0 & 0 & 0 \\ 5 & 30 & 25 & 6 & 63 & 4 & 0 & 0 & 0 \\ 0 & 4 & 20 & 35 & 11 & 79 & 7 & 5 & 2 \\ 0 & 13 & 0 & 7 & 63 & 46 & 24 & 6 & 4 \\ 0 & 0 & 0 & 0 & 6 & 20 & 0 & 13 & 0 \\ 0 & 0 & 0 & 0 & 1 & 16 & 12 & 5 & 17 \\ 0 & 0 & 0 & 0 & 2 & 4 & 0 & 17 & 2 \end{pmatrix}$$

To convert this table into a matrix of transition probabilities, we need to divide all the elements of each column by the sum of the elements in that column. We first compute the sums *sm* of the columns

```
sm = Table[0, {9}]; Do[sm += m[[i]], {i, 9}]; sm
```

{51, 166, 117, 139, 150, 181, 43, 46, 25}

and after dividing each entry by the sum of the entries in its column, we obtain the matrix *mp* of transition probabilities.

```
mp = TraditionalForm[Transpose[Transpose[m] / sm]]
```

$$mp = \begin{pmatrix} \frac{2}{51} & \frac{25}{83} & \frac{1}{39} & 0 & 0 & 0 & 0 & 0 & 0 \\ \frac{32}{51} & \frac{19}{83} & \frac{58}{117} & \frac{32}{139} & \frac{1}{75} & \frac{11}{181} & 0 & 0 & 0 \\ \frac{4}{17} & \frac{31}{166} & \frac{11}{117} & \frac{59}{139} & \frac{1}{75} & \frac{1}{181} & 0 & 0 & 0 \\ \frac{5}{51} & \frac{15}{83} & \frac{25}{117} & \frac{6}{139} & \frac{21}{50} & \frac{4}{181} & 0 & 0 & 0 \\ 0 & \frac{2}{83} & \frac{20}{117} & \frac{35}{139} & \frac{11}{150} & \frac{79}{181} & \frac{7}{43} & \frac{5}{46} & \frac{2}{25} \\ 0 & \frac{13}{166} & 0 & \frac{7}{139} & \frac{21}{50} & \frac{46}{181} & \frac{24}{43} & \frac{3}{23} & \frac{4}{25} \\ 0 & 0 & 0 & 0 & \frac{1}{25} & \frac{20}{181} & 0 & \frac{13}{46} & 0 \\ 0 & 0 & 0 & 0 & \frac{1}{150} & \frac{16}{181} & \frac{12}{43} & \frac{5}{46} & \frac{17}{25} \\ 0 & 0 & 0 & 0 & \frac{1}{75} & \frac{4}{181} & 0 & \frac{17}{46} & \frac{2}{25} \end{pmatrix}$$

We use the rejection method (10.1.4.4) to generate the random pitches:

```
sc = {c, d, e, f, g, a, h, c, d}; x = 1;
l = Table[While[r = RandomInteger[{1, 9}]; RandomReal[] > mp[[r, x]]];
  x = r, {i, 20}]; Table[sc[[l[[i]]]], {i, 20}]
```

(→ Further examples)

10.2.2.2 Definition

Stochastic processes in which the distribution of $X(t_n) = x_n$ only depends on the previous state x_{n-1} are called *Markov processes* (after the Russian mathematician Andrei A. Markov, 1856-1922).

A Markov process with discrete time and a finite number of possible states produces a *Markov chain*. The elementary events A_i of a complete event system $\{A_1, A_2, ...\}$ (10.1.2.1) constitute the possible states of that system. The system is monitored at the times t_i. We write the *transition probability*, that is, the probability that at time $t + 1$ the system will be in state A_k, as

$$P[A_k(t+1) \mid A_i(t)].$$

If the transition probability does not depend on time, the Markov chain is said to be *time-homogeneous* or *stationary*. We then have

$$P[A_k(t+1) \mid A_i(t)] = P[A_k(1) \mid A_i(0)]$$

for all t, and we write the transition probability as

$$P_{i,k} := P[A_k(1) \mid A_i(0)].$$

10.2 Stochastic Processes

The probability $P_{i,k}(n)$ of the transition from state A_i to state A_k after n steps, that is

$$P_{i,k}(n) := P[A_k(n) \mid A_i(0)],$$

is called an *n-step transition probability*.

A state is said to be *recurrent* or *persistent* if the system can take on the state arbitrarily often. State A_i is said to be *transient* if it can only be reached a finite number of times. This means that another state A_k must exist from which a return to A_i is impossible and which, in turn, cannot be reached from A_i. A state is called *absorbing* if, once reached, it can never be left ($P_{i,i} = 1$). A Markov chain is called *ergodic* if it is possible to go from any state to any other state, it is called *irreducible* if it is possible to get to any state from any other state in a single time step.

10.2.2.3 Transition Matrix and State Vectors

In a finite homogenous Markov chain having N states, the fixed transitional probabilities $P_{i,k}$ can be represented as an $N \times N$-Matrix, the so-called *transition matrix* $\mathbf{P}$. In the same way, we define the *n-step transition matrix* $\mathbf{P}(n)$.

$$\mathbf{P} = \begin{pmatrix} P_{11} & P_{12} & \cdots & P_{1N} \\ P_{21} & P_{22} & \cdots & P_{2N} \\ \vdots & \vdots & & \vdots \\ P_{N1} & P_{N2} & \cdots & P_{NN} \end{pmatrix} \qquad \mathbf{P}(n) = \begin{pmatrix} P_{11}(n) & P_{12}(n) & \cdots & P_{1N}(n) \\ P_{21}(n) & P_{22}(n) & \cdots & P_{2N}(n) \\ \vdots & \vdots & & \vdots \\ P_{N1}(n) & P_{N2}(n) & \cdots & P_{NN}(n) \end{pmatrix}$$

The two-step transition matrix $\mathbf{P}(2)$ can be computed from $\mathbf{P}$ as shown in Example 1) above as a matrix product:

$$P_{i,k}(2) = \sum_{l=1}^{N} P_{i,l} P_{l,k} = \mathbf{P}^2.$$

For the n-step transition matrix $\mathbf{P}(n)$ we continue the calculations and obtain $P_{i,k}(n) = \mathbf{P}^n$.

The state of a system at time t, or more precisely, the probability distribution of its states at time t, is described by the state vector $\mathbf{p}(t)$. The initial state is given by the initial vector $\mathbf{p}(0)$, indicating the probability that the system begins in one or another state, or alternatively by a unit vector specifying the initial state.

$$\mathbf{p}(t) = \begin{pmatrix} P_1(t) \\ P_2(t) \\ \vdots \\ P_N(t) \end{pmatrix} \quad \text{where } P_i(t) := P[A_i(t)] \qquad \mathbf{p}(0) = \begin{pmatrix} P_1(0) \\ P_2(0) \\ \vdots \\ P_N(0) \end{pmatrix} \quad \text{where } P_i(0) := P[A_i(0)]$$

The state vectors can be computed from the initial vector and the transition matrix.

$$\mathbf{p}(1) = \mathbf{P}\mathbf{p}(0)$$
$$\mathbf{p}(t) = \mathbf{P}\mathbf{p}(t-1) = \mathbf{P}(t)\mathbf{p}(0) = \mathbf{P}^t \mathbf{p}(0)$$

A Markov chain is said to be *stationary* when a so-called *eigenvector* or *steady state vector* $\mathbf{p}$ exists such that $\mathbf{p} = \mathbf{P}\mathbf{p}$. If the sequence $\mathbf{p}(t)$ beginning with $\mathbf{p}(0)$ converges to a limit vector $\mathbf{p}(\infty)$, this vector is called the *limit distribution vector*.

10.2.2.4 Applications

Let us apply the concepts and calculations to our first example above (10.2.2.1). We have already defined the transition matrix:

$$P = \begin{pmatrix} .3 & .2 & .1 \\ .4 & .5 & .4 \\ .3 & .3 & .5 \end{pmatrix}$$

We can obtain the state vector at time $t = 1$, that is, the probabilities of the three possible states at time $t = 1$, called $p(1)$, by multiplying the transition matrix P by the initial vector $(0, 1, 0)$. This choice of the initial vector indicates that we begin the harmonic sequence with the tonic function.

$$\text{MatrixForm}\left[P \cdot \begin{pmatrix} 0 \\ 1 \\ 0 \end{pmatrix}\right] \qquad \begin{pmatrix} 0.2 \\ 0.5 \\ 0.3 \end{pmatrix}$$

If we take as initial vector $\{1/3, 1/3, 1/3\}$, that is, if it is equally probable that any of the three functions could begin the sequence, then we obtain the state vector $p(1)$:

$$\text{MatrixForm}\left[P \cdot \begin{pmatrix} 1/3 \\ 1/3 \\ 1/3 \end{pmatrix}\right] \qquad \begin{pmatrix} 0.2 \\ 0.433 \\ 0.366 \end{pmatrix}$$

Let us compute the n-step transition matrix $P(n)$ for $n = 2, 3, 4$ and 10. We find the probability we computed above (10.2.2.1) $P[T \to X \to T] = .45$ as element P_{22} in the matrix.

$$\text{MatrixForm}[\text{MatrixPower}[P, 2]]$$

$$\begin{pmatrix} 0.2 & 0.19 & 0.16 \\ 0.44 & 0.45 & 0.44 \\ 0.36 & 0.36 & 0.4 \end{pmatrix}$$

$$P^3 = \begin{pmatrix} 0.184 & 0.183 & 0.176 \\ 0.444 & 0.445 & 0.444 \\ 0.372 & 0.372 & 0.38 \end{pmatrix} \qquad P^4 = \begin{pmatrix} 0.1812 & 0.1811 & 0.1796 \\ 0.4444 & 0.4445 & 0.4444 \\ 0.3744 & 0.3744 & 0.376 \end{pmatrix}$$

$$P^{10} = \begin{pmatrix} 0.180556 & 0.180556 & 0.180555 \\ 0.444444 & 0.444444 & 0.444444 \\ 0.375 & 0.375 & 0.375 \end{pmatrix}$$

The coefficients of the n-step transition matrix $P(n) = P^n$ converge as n increases. This reflects the fact that after many steps the initial vector no longer affects the state vector. We refer to $\lim_{n \to \infty} P^n$ as P_∞.

$$P\infty = \begin{pmatrix} 0.180556 & 0.180556 & 0.180556 \\ 0.444444 & 0.444444 & 0.444444 \\ 0.3745 & 0.3745 & 0.3745 \end{pmatrix}$$

Multiplying P_∞ by an arbitrary initial vector gives the limit distribution vector $p(\infty)$.

$$P\infty \cdot \begin{pmatrix} 1/3 \\ 1/3 \\ 1/3 \end{pmatrix} \qquad \{\{0.180556\}, \{0.444444\}, \{0.3745\}\}$$

10.2 Stochastic Processes

$$P\infty . \begin{pmatrix} 1 \\ 0 \\ 0 \end{pmatrix} \qquad \{\{0.180556\}, \{0.44444\}, \{0.3745\}\}$$

This gives us the relative frequency of each state over many steps n. The lists below show the relative frequencies of the three states for 20, 100 and 1000 steps.

$$\{0.15,\ 0.75,\ 0.1\} \qquad \{0.23,\ 0.4,\ 0.37\} \qquad \{0.183,\ 0.452,\ 0.365\}$$

The power sequence of a transition matrix does not always converge, as the following example demonstrates.

$$P2 = \begin{pmatrix} .1 & 0 & 0 \\ 0 & 0 & 1 \\ .9 & 1 & 0 \end{pmatrix}$$

MatrixForm[MatrixPower[P2, 20]]

$$\begin{pmatrix} 1. \times 10^{-20} & 0. & 0. \\ 0.909091 & 1. & 0. \\ 0.0909091 & 0. & 1. \end{pmatrix}$$

MatrixForm[MatrixPower[P2, 21]]

$$\begin{pmatrix} 1. \times 10^{-21} & 0. & 0. \\ 0.0909091 & 0. & 1. \\ 0.909091 & 1. & 0. \end{pmatrix}$$

To find the steady state vector of the first example, we have to solve the equation $p = Pp$, that is, we need an eigenvector with the eigenvalue 1. The Mathematica function Eigensystem[*P*] generates a list of eigenvalues and the associated eigenvectors.

Eigensystem[P]

$$\{\{1., 0.2, 0.1\},$$
$$\{\{0.296529, 0.729917, 0.615867\}, \{0.707107, 2.87324 \times 10^{-16}, -0.707107\},$$
$$\{-0.707107, 0.707107, 1.58521 \times 10^{-16}\}\}\}$$

The first eigenvector is the one we are looking for, as a check confirms.

$$P . \begin{pmatrix} 0.296529 \\ 0.729917 \\ 0.615867 \end{pmatrix}$$

$$\{\{0.296529\}, \{0.729917\}, \{0.615867\}\}$$

The components of this vector do not add up to 1. We get the limit distribution vector $p(\infty)$ by dividing the steady state vector by the sum of its components.

$$\begin{pmatrix} 0.296529 \\ 0.729917 \\ 0.615867 \end{pmatrix} \Big/ (0.296529 + 0.729917 + 0.6158673)$$

$$\{\{0.180556\}, \{0.444444\}, \{0.375\}\}$$

In the same way, we can compute the limit distribution vector $p2(\infty)$ of the non-convergent matrix $P2$.

Eigensystem[P2]

{{-1., 1., 0.1}, {{0., 0.707107, -0.707107},
{0., 0.707107, 0.707107}, {0.738272, -0.671156, -0.0671156}}}

Here the second vector is the eigenvector we are looking for.

p2 (∞) = {0., 0.7071, 0.7071}

$$p\infty = \begin{pmatrix} 0 \\ 0.707107 \\ 0.707107 \end{pmatrix} \Big/ (0 + 0.707107 + 0.707107)$$

{{0}, {0.5}, {0.5}}

Let us now modify our 3×3 matrix for the states 1, 2 and 3 so that the states have different characteristics (10.2.2.2). If we put a zero in the main diagonal ($P_{i,i} = 0$), the state i is left immediately. With the following transition matrix, only the state 3 can be repeated.

$$\begin{pmatrix} 0 & .5 & .1 \\ .4 & 0 & .4 \\ .6 & .5 & .5 \end{pmatrix}$$

{3, 1, 3, 2, 3, 1, 2, 3, 3, 3, 2, 3, 3, 2, 1, 2, 3, 3, 3, 3}

There can only be a one in the transition matrix when all the other elements in the same column are zero. If the one is in the main diagonal ($P_{i,i} = 1$), the state i is absorbing. With the following transition matrix, the Markov chain will sooner or later get stuck in the third state.

$$\begin{pmatrix} .5 & .8 & 0 \\ .4 & .1 & 0 \\ .1 & .1 & 1 \end{pmatrix}$$

{1, 1, 1, 2, 1, 2, 1, 1, 2, 1, 3, 3, 3, 3, 3, 3, 3, 3, 3, 3}

If all the elements of the same line of the matrix are zero except that of the main diagonal ($P_{i,i}$), the state i is transient. In the following chain, the initial state occurs several times at the beginning, because the probability for its repetition is great (.9), but once it is left, it cannot be reached again.

$$\begin{pmatrix} .9 & 0 & 0 \\ 0 & .3 & .6 \\ .1 & .7 & .4 \end{pmatrix}$$

{1, 1, 1, 1, 1, 1, 1, 1, 3, 2, 3, 2, 3, 3, 2, 3, 3, 2, 3, 2, 3, 2, 3}

10.2.2.5 Markov Chains With Variable Transition Probabilities

Until now, we have considered only Markov chains whose transition probabilities are not time-dependent. Such chains are said to be *time-homogenous*. Without going into the theoretical details, let us now show how to generate Markov chains for which individual probabilities, or even the entire transition matrix, vary in time.

First we represent the transition probabilities as functions of time. Let us assume four states whose probabilities for a transition to state A_j are given by these functions f_i:

```
f[1][x_] := 1 + Sin[π * (x + t / tmax)]
f[2][x_] := 1 + Cos[2 * x * t / tmax]
f[3][x_] := 2 - (2 * (x - 1 - 3 * t / tmax))^2
f[4][x_] := 2 - (2 * (x - 2.5) * t / tmax)^2
```

The illustration below shows the four functions at time $t = 1$, *tmax*/2 and *tmax*. Their values do not add up to 1, hence they are not genuine probability functions. But because the function values are never greater than 2, we can use the rejection method to generate distributions corresponding to the relationships among the function values.

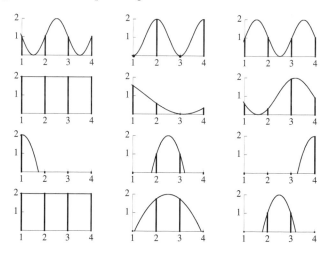

To realize the Markov chain, we use the program of the second introductory example (10.2.2.1), substituting RandomReal[{0,2}]>f[x][r] for Random[]>mp[[r,x]].

```
x = 2;
l = Table[While[r = RandomInteger[{1, 4}];
    RandomReal[{0, 2}] > f[x][r], Null]; x = r, {t, 120}]
{4, 4, 4, 4, 2, 3, 1, 2, 4, 4, 4, 1, 4, 4, 2, 4, 3, 2, 2, 3, 2, 1, 4, 2, 1,
 4, 2, 3, 2, 4, 4, 1, 1, 4, 1, 4, 2, 1, 4, 3, 2, 2, 2, 1, 4, 3, 2, 1, 4,
 3, 2, 1, 4, 3, 2, 1, 2, 2, 1, 4, 2, 4, 2, 1, 2, 1, 2, 1, 2, 4, 2, 2, 1,
 4, 3, 3, 3, 3, 3, 3, 3, 3, 3, 3, 3, 3, 3, 3, 3, 3, 3, 3, 4, 3, 3, 4,
 2, 4, 2, 3, 4, 2, 4, 2, 3, 3, 4, 3, 4, 2, 3, 4, 2, 4, 2, 3, 4, 3, 4}
```

It is easiest to predict the behavior of the chain after the state A_3: Initially A_3 can only be followed by A_1, then A_2 (line 2), A_3 and finally A_4 (4). Initially A_4 can be followed by any state (1), after the half duration however only by A_2 or A_3.

The transition matrix $mw(t)$ can vary as often as one wishes. In the following example, there are five states and the initial transition matrix $mw(1)$. After every step of the chain two randomly chosen elements of a randomly chosen column of the matrix are exchanged.

$$mw(1) = \begin{pmatrix} 0.5 & 0 & 0 & 0 & 0.5 \\ 0.5 & 0.5 & 0 & 0 & 0 \\ 0 & 0.5 & 0.5 & 0 & 0 \\ 0 & 0 & 0.5 & 0.5 & 0 \\ 0 & 0 & 0 & 0.5 & 0.5 \end{pmatrix}$$

We generate a realization of the chain with the following program which writes 100 numbers into a table (lines 2 and 8). To exchange two elements of the matrix, we choose randomly an index i for the column and two indices a, b for the row numbers of the elements to be exchanged (lines 3 and 4). After the elements are exchanged, the new matrix is stored as $mr(t)$ (lines 4-5). The sequence of numbers is generated using the same method as in the previous example (lines 6-7).

```
x=3;                                                                    (*1*)
l=Table[
    i=Random[Integer,{1,5}];a=Random[Integer,{1,5}];
    b=Random[Integer,{1,5}];mem=mw[[a,i]];mw[[a,i]]=mw[[b,i]];
    mw[[b,i]]=mem;mr[t]=mw;                                             (*5*)
    While[r=Random[Integer,{1,5}];
        Random[Real,{0,.5}]>mw[[r,x]],];x=r,
{t,100}];l
```

{4, 5, 1, 2, 2, 2, 2, 3, 4, 4, 4, 5, 5, 5, 1, 1, 1, 2, 4, 5, 4, 5, 5, 4, 4, 5, 5, 5, 4, 5, 4, 5, 2, 3, 2, 4, 5, 2, 1, 3, 2, 4, 5, 4, 3, 4, 3, 4, 4, 4, 4, 4, 4, 2, 4, 1, 1, 1, 5, 2, 4, 4, 2, 4, 4, 4, 1, 1, 1, 1, 5, 4, 1, 1, 5, 2, 4, 1, 1, 4, 4, 4, 1, 1, 1, 1, 4, 4, 4, 4, 3, 5, 2, 4, 2, 4, 3, 3, 1, 4}

Initially, the transition matrix shows that at each next step the previous number can either repeat or be augmented by 1 (except 5, which can go back to 1). In the course of time, new characteristics appear. The long period between $t = 19$ and $t = 32$ where only 4 and 5 appear stems from the fact that at this time the four elements at the lower right of the matrix were all .5 (see $mr(20)$ below). For some time, the third line of the matrix contained only zeros (see $mr(55)$), which meant that no threes appeared in the chain. The almost complete lack of five and the domination of four (in line 4 of the result) came from a later constellation of the matrix (see $mr(80)$) where the fourth line of the matrix contained only .5 and the fifth line only 0.

$$mr(1) = \begin{pmatrix} 0.5 & 0 & 0 & 0 & 0.5 \\ 0.5 & 0.5 & 0 & 0 & 0 \\ 0 & 0.5 & 0.5 & 0 & 0 \\ 0 & 0 & 0.5 & 0.5 & 0 \\ 0 & 0 & 0 & 0.5 & 0.5 \end{pmatrix} \quad mr(5) = \begin{pmatrix} 0.5 & 0 & 0 & 0 & 0.5 \\ 0.5 & 0.5 & 0 & 0 & 0 \\ 0 & 0.5 & 0.5 & 0 & 0 \\ 0 & 0 & 0.5 & 0.5 & 0 \\ 0 & 0 & 0 & 0.5 & 0.5 \end{pmatrix} \quad mr(10) = \begin{pmatrix} 0.5 & 0 & 0 & 0 & 0.5 \\ 0.5 & 0 & 0.5 & 0 & 0 \\ 0 & 0.5 & 0 & 0 & 0 \\ 0 & 0.5 & 0.5 & 0.5 & 0 \\ 0 & 0 & 0 & 0.5 & 0.5 \end{pmatrix}$$

$$mr(20) = \begin{pmatrix} 0 & 0 & 0.5 & 0 & 0 \\ 0.5 & 0.5 & 0.5 & 0 & 0 \\ 0.5 & 0 & 0 & 0 & 0 \\ 0 & 0.5 & 0 & 0.5 & 0.5 \\ 0 & 0 & 0 & 0.5 & 0.5 \end{pmatrix} \quad mr(55) = \begin{pmatrix} 0.5 & 0.5 & 0 & 0 & 0 \\ 0 & 0 & 0.5 & 0.5 & 0.5 \\ 0 & 0 & 0 & 0 & 0 \\ 0 & 0.5 & 0.5 & 0.5 & 0 \\ 0.5 & 0 & 0 & 0 & 0.5 \end{pmatrix} \quad mr(80) = \begin{pmatrix} 0.5 & 0 & 0.5 & 0 \\ 0 & 0 & 0.5 & 0 & 0.5 \\ 0 & 0.5 & 0 & 0 & 0 \\ 0.5 & 0.5 & 0.5 & 0.5 & 0.5 \\ 0 & 0 & 0 & 0 & 0 \end{pmatrix}$$

10.2 Stochastic Processes

10.2.2.6 Markov Chains With Variable States

Let us generate a sequence of states A_i, using the Markov chain from the first example of Chapter 10.2.2.1. This time, however, we will interpret the states not as harmonic functions but as time-dependent intervals of real random numbers. We generate a realization of the Markov chain, store it as a list l and in a second step add to the numbers 1, 2 and 3 real random numbers between $-.7i/tmax$ and $.7i/tmax$.

```
tmax = 200; s = {1, 1, 1, 2, 2, 2, 2, 3, 3, 3};
t = {1, 1, 2, 2, 2, 2, 2, 3, 3, 3}; d = {1, 2, 2, 2, 2, 3, 3, 3, 3, 3};
x = 2; l = Table[Switch[x, 1, x = s[[RandomInteger[{1, 10}]]],
    2, x = t[[RandomInteger[{1, 10}]]], 3,
    x = d[[RandomInteger[{1, 10}]]]], {i, tmax}];
ll = Table[l[[i]] + RandomReal[{-.7*i/tmax, .7*i/tmax}], {i, tmax}];
```

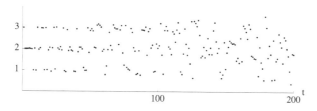

The definition of a Markov chain requires that the events A_i constitute a complete event system. Among other things, this means that the events $A_1, A_2, ...$ are pairwise disjoint, that is they have no elementary events in common. After a certain time in our example, the ranges of the three possible events overlap. We can still continue the chain, but if we only have the resulting real numbers, it is no longer always possible to determine which event generated which real number.

10.2.2.7 Other Examples

Let us generate a Markov chain in which, with low probability, switches occur between two different sorts of states, here between numbers from {1, 2, 3} and from {12, 13, 14, 15}. The transition table, shown below, leads to these properties for realizations of the process: 1) no number repeats because the main diagonal contains only zeros; 2) each number is most often followed by a neighboring number; 3) the switch from big numbers to small numbers can occur only after the maximum, the switch from small to big numbers only after the minimum occurs.

	1	2	3	12	13	14	15
1	0	.5	.3	0	0	0	.2
2	.6	0	.7	0	0	0	0
3	.2	.5	0	0	0	0	0
12	0	0	0	0	.4	.2	.1
13	0	0	0	.6	0	.4	.3
14	0	0	0	.3	.4	0	.4
15	.2	0	0	.1	.2	.4	0

{15, 12, 14, 15, 14, 13, 12, 13, 14, 15, 1, 2, 3, 2, 3, 1, 2,
 1, 2, 3, 2, 1, 15, 12, 13, 14, 15, 13, 12, 13, 12, 13, 14, 13,
 15, 13, 14, 13, 12, 13, 15, 13, 14, 15, 1, 2, 1, 15, 14, 13}

We can determine the relative frequency of occurrence of the individual states by computing P_∞ and multiplying it by an arbitrary initial vector. To get a good approximation of P_∞, we have to choose a large n in P^n.

$$P2 = \begin{pmatrix} 0 & .5 & .3 & 0 & 0 & 0 & .2 \\ .6 & 0 & .7 & 0 & 0 & 0 & 0 \\ .2 & .5 & 0 & 0 & 0 & 0 & 0 \\ 0 & 0 & 0 & 0 & .4 & .2 & .1 \\ 0 & 0 & 0 & .6 & 0 & .4 & .3 \\ 0 & 0 & 0 & .3 & .4 & 0 & .4 \\ .2 & 0 & 0 & .1 & .2 & .4 & 0 \end{pmatrix}$$

P∞ = MatrixPower[P2, 100]

TraditionalForm[P∞]

$$\begin{pmatrix} 0.139502 & 0.139505 & 0.139505 & 0.13948 & 0.13948 & 0.139481 & 0.139484 \\ 0.158821 & 0.158825 & 0.158826 & 0.158789 & 0.15879 & 0.158791 & 0.158795 \\ 0.107312 & 0.107315 & 0.107315 & 0.10729 & 0.10729 & 0.107291 & 0.107294 \\ 0.116861 & 0.116859 & 0.116859 & 0.116879 & 0.116878 & 0.116878 & 0.116875 \\ 0.176549 & 0.176546 & 0.176545 & 0.176574 & 0.176574 & 0.176573 & 0.176569 \\ 0.161471 & 0.161468 & 0.161467 & 0.161493 & 0.161492 & 0.161492 & 0.161488 \\ 0.139484 & 0.139482 & 0.139482 & 0.139496 & 0.139496 & 0.139496 & 0.139494 \end{pmatrix}$$

$$P\infty \; . \; \begin{pmatrix} 0 \\ 0 \\ 0 \\ 1 \\ 0 \\ 0 \\ 0 \end{pmatrix}$$

{{0.13948}, {0.158789}, {0.10729},
 {0.116879}, {0.176574}, {0.161493}, {0.139496}}

Some realizations of this process may deviate considerably from the distribution calculated above, even after many steps, as these relative frequencies for three chains of different length n show:

$n = 30$:   {0.333333, 0.366667, 0.3, 0, 0, 0, 0}
$n = 100$: {0.2, 0.3, 0.17, 0.07, 0.08, 0.09, 0.09}
$n = 1000$: {0.157, 0.193, 0.136, 0.101, 0.151, 0.135, 0.127}

The probability of a strong deviation from the calculated distribution only becomes small for very long chains (e.g. $n = 10{,}000$).

{0.1387, 0.1627, 0.1096, 0.1148, 0.1737, 0.1639, 0.1366}

As a second example, let us generate a Markov chain in which sequences of numbers occur like motifs. The sequences can be identical or can vary from one another slightly. We define a matrix in which every column contains a 1 and one or two numbers between 0 and .2.

10.2 Stochastic Processes

```
tmax = 200; mw = Table[0, {i, 10}, {j, 10}];
Do[mw[[RandomInteger[{1, 10}], j]] = N[RandomReal[{0, 0.2`}], 2],
  {i, 2}, {j, 10}]
Do[mw[[RandomInteger[{1, 10}], j]] = 1, {j, 10}]
mw
```

$$\begin{pmatrix} 1 & 0 & 0 & 0.049 & 0.13 & 0.18 & 0.15 & 0 & 0 & 0 \\ 0 & 0 & 0 & 0 & 1 & 1 & 0 & 0 & 0.0032 & 0 \\ 0.045 & 0 & 0 & 0 & 0 & 0 & 1 & 0 & 0 & 0 \\ 0 & 1 & 0 & 0 & 0 & 0 & 0 & 0 & 0 & 0 \\ 0 & 0.12 & 0 & 0 & 0.16 & 0 & 0.073 & 0 & 0 & 0 \\ 0 & 0 & 0.069 & 0 & 0 & 0 & 0 & 0.083 & 0 & 0 \\ 0 & 0 & 0 & 0 & 0 & 0 & 0 & 0 & 1 & 0 \\ 0 & 0 & 0 & 1 & 0 & 0 & 0 & 1 & 0 & 0 \\ 0 & 0 & 1 & 0 & 0 & 0.19 & 0 & 0 & 0 & 0.017 \\ 0 & 0 & 0 & 0 & 0 & 0 & 0 & 0.068 & 0.0093 & 1 \end{pmatrix}$$

{1, 1, 1, 1, 1, 1, 1, 1, 1, 1, 1, 1, 1, 1, 1, 3, 9, 7, 3, 9, 7, 3, 9, 7, 3, 9, 7, 3, 9, 7, 3, 9, 7, 3, 9, 7, 5, 2, 4, 8, 10, 9, 7, 3, 9, 7, 3, 9, 7, 3, 9, 7, 3, 6, 2, 4, 8, 8, 8, 8, 8, 8, 8, 8, 8, 10, 10, 10, 10, 10, 10, 10, 10, 10, 10, 10, 10, 10, 10, 10, 10, 10, 10, 10, 10, 9, 7, 3, 9, 7, 1, 1, 1, 1, 1, 1, 1, 1, 1, 1, 1, 3, 9, 7, 3, 9, 7, 3, 9, 7, 3, 9, 7, 3, 9, 7, 1, 1, 1, 1, 1, 1, 3, 9, 7, 3, 9, 7, 3, 9, 7, 3, 9, 7, 1, 1, 1, 1, 1, 1, 1, 1, 1, 1, 1, 1, 1, 1, 1, 1, 1, 1, 1, 1, 1, 1, 1, 1, 1, 1, 1, 1, 1, 1, 1, 1, 1, 1, 1, 1, 3, 9, 7, 3, 9, 7, 3, 9, 7, 3, 9, 7, 3, 9, 7, 3, 9, 7, 3}

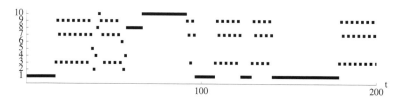

There are many repetitions of the numbers 1, 8 and 10, since the diagonal elements corresponding to these numbers are equal to one. At the same time, we can see a repeated pattern of three numbers: 3-9-7-3-9 ... The illustration below shows realizations generated by the same program but using two different matrices.

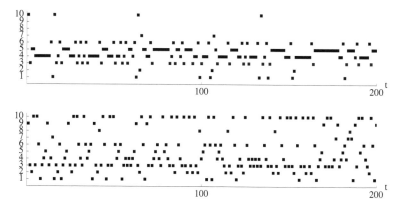

10.2.3 More Stochastic Processes

10.2.3.1 Processes With Independent Increments

Processes with independent increments are special Markov processes. They are usually described by the distribution of the increments rather than by that of the random variables. As before, we shall consider only discrete processes here.

Let us consider first a process with real numbers in which the increment $U_t = X_t - X_{t-1}$ decreases exponentially over time, for example $U_t = \text{Random}[0, e^{-.1\,t}]$. If we call the initial value x_0, we have $X_1 = x_0 + U_1, X_2 = x_0 + U_1 + U_2$, etc.

```
x = 0; tmax = 30; l = Table[x += RandomReal[{0, π^-0.1`t}], {t, tmax}]
```

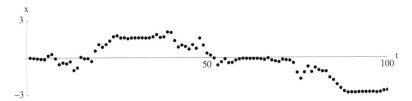

As an example for an oscillating increment we have $U_t = \text{Random}[-\sin^2(.1t), \sin^2(.1t)]$.

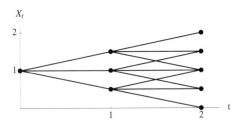

Most of the Markov chains we saw in Chapter 10.2.2 are not processes with independent increments because the random variables can almost always only take on values within a finite range. This means that for a positive increment at time t the probability for a "negative increment" at time $t + 1$ increases. Nor are the increments of the introductory example in Chapter 10.2.1.1 independent. The probabilities for winning and losing are always the same, but the magnitude of the increment increases with each win and decreases with each loss. We could easily change the game so that the increment does not depend on winning or losing by adding .5 to the capital rather than doubling it for each win and subtracting .5 from the capital rather than cutting it in half for each loss.

(→ Example)

10.2 Stochastic Processes

10.2.3.2 Random Walk

Random Walk is the name for a discrete process with stationary independent increments, that is, with increments that do not change over time. In Chapter 10.2.3.1 only the last example is a random walk. We can write $X_1 = x_0 + U_1, X_2 = x_0 + U_1 + U_2$, etc., where the U_t are independent both of each other and of time. Hence, U_t is white noise (10.2.1.2) with mean μ_U and variance σ_U^2. We distinguish between a *random walk with drift* when $\mu_U \neq 0$ and the *random walk without drift* when $\mu_U = 0$. If we define the so-called centered random variable ϵ_t as the difference between U and μ ($\epsilon_t = U_t - \mu_U$), then we can also write the sequence X_t as

$$X_1 = x_0 + \mu_U + \epsilon_1, X_2 = x_0 + \mu_U + \epsilon_1 + \mu_U + \epsilon_2, \ldots\ldots X_t = x_0 + t\mu_U + \sum_{i=1}^{t} \epsilon_i.$$

The expected value of the sequence X_t is $E(X_t) = x_0 + t\mu_U$, for a random walk without drift it is $E(X_t) = x_0$. In both cases, the variance always increases $V(X_t) = t\sigma_U^2$.

In a first example we generate a sequence of numbers by adding a real random number between –2 and 3 to the previous number.

```
x = 0; tmax = 40;
l = Table[x += RandomReal[{-2, 3}], {t, tmax}];
```

This is a random walk with drift. The mean value μ_U of the increment is .5 which gives an expected value of $E(X_t) = x_0 + t\mu_U = 0 + t.5$. So, for example, at time $t = 40$ the expected value is a number around 20. We can generate a similar sequence by using the centered random variable $\epsilon_t = U_t - \mu_U = U_t - .5$.

```
x = 0; tmax = 40;
l = Table[x += (.5 + RandomReal[{-2 - .5, 3 - .5}]), {t, tmax}];
```

The individual increments can differ extremely from one another, but they must be independent of one another. The following example illustrates this. Here the increments follow the bilateral exponential distribution (10.1.6.1, Example 3).

```
x = 0; tmax = 100;
l = Table[x += Sign[RandomReal[{-1, 1}]] *Log[Random[]], {t, tmax}];
```

In the following example, we use a random walk to generate movement on a plane. Given is a rectangle of length 20 and width 10 in which a point moves randomly. We create a time series each for x and y by adding to the previous values a uniformly distributed random number, for x from the range [–2, 2], for y from the [–1, 1]. If the next coordinate value would be outside the given rectangle, the random value is subtracted from the previous value.

```
If[x + (r = Random[Real, {-2, 2}]) > 10 || x + r < -10, x -= r, x += r],
If[y + (r = Random[Real, {-1, 1}]) > 5 || y + r < -5, y -= r, y += r]
```

The figure to the left shows a pattern in which the coordinates were calculated at random at each step. The figure to the right shows a pattern corresponding to the calculation above. One can see that the figure to the right corresponds better to random movement, but that there are changes of direction at every step, causing the movement to stay pretty much in the same place. This is how Brownian motion is modeled, except that there the increments are normally distributed.

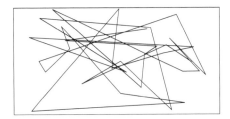

We can get a less confused movement by adding the random numbers not to the coordinates but to the velocities v_x and v_y (below left).

```
x = 3; y = 3; vx = -.5; vy = 0;
If[x + (vx += (r = Random[Real, {-.4, .4}])) > 10 || x + vx < -10,
    x += (vx *= -1), x += vx],
If[y + (vy += (r = Random[Real, {-.4, .4}])) > 5 || y + vy < -5,
    y += (vy *= -1), y += vy]}
```

The movement becomes even smoother and clearer when we add the random values to the accelerations a_x and a_y (below right).

```
x = 3; y = 3; vx = -.3; vy = -.3; ax = 0; ay = 0;
If[x + (vx += (ax += (r = Random[Real, {-.03, .03}]))) > 10 || x + vx < -10,
    x += (vx *= -1), x += vx],
If[y + (vy += ay += (r = Random[Real, {-.03, .03}]))) > 5 || y + vy < -5,
    y += (vy *= -1), y += vy]}
```

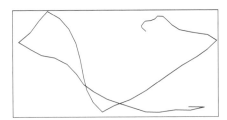
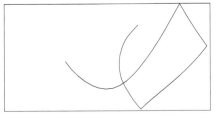

(→ *Random_Walk.maxpat*)

10.2.3.3 Describing a Process By Its Spectrum

Let us first consider a Markov chain defined by the transition matrix below. The three states are the numbers 1, 2 and 3.

$$m = \begin{pmatrix} .1 & 0 & .9 \\ .9 & .1 & 0 \\ 0 & .9 & .1 \end{pmatrix}$$

Since 1 is usually followed by 2, 2 by 3, and 3 by 1, we get a cyclic movement that is only slightly disturbed by occasional repetitions of single numbers. These disturbances are due to the probabilities .1 in the main diagonal.

Now we generate a Markov chain of 8192 points. When played as sound at a sampling rate of 8192 Hz, this gives one second of colored noise with the spectrum below.
(→ Sound Example)

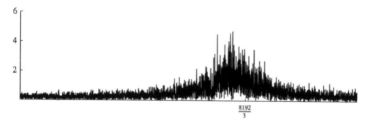

The signal's mean period is 3.1, hence the spectrum has a maximum at 8192/3.1 Hz. Because the signal is not centered around 0 Hz, the spectrum shows a strong component at that point. The mean deviation from the zero position in sound signals is called the *Direct Current* (DC) *bias* or *component*. It corresponds to the mean μ of the distribution and of the process.

As we showed in Chapter 10.1.6.1 and 10.2.1.2, many different distributions of random numbers yield white noise. The differences are usually not audible when the random numbers are used to generate the signal of white noise itself. They become significant, however, when the random numbers are used to generate musical parameter values.

The spectrum of the random walk without drift shows a decrease of amplitude as the frequency increases according to the formula Amp $\simeq 1/f^2$. (→ Sound Example)

```
x = 0; tmax = 8192;
l = Table[x += RandomReal[{-1, 1}], {t, tmax}];
```

Using filters we can generate random sequences having specific spectra.

10.2.3.4 Processes Involving Previous Events

Let us generate a process in which the distribution of the random variables depends not only on the last value but also on earlier values. We do so by constructing a table showing the transition probabilities for all possible realizations of the last n steps. For possible states 1, 2 and 3 there are nine possible sequences for the last two steps.

	1 – 1	1 – 2	1 – 3	2 – 1	2 – 2	2 – 3	3 – 1	3 – 2	3 – 3
1	0	0	0	.5	0	0	0	.5	1
2	1	.5	1	.5	0	.5	1	.5	0
3	0	.5	0	0	1	.5	0	0	0

$$P = \begin{pmatrix} 0 & 0 & 0 & .5 & 0 & 0 & 0 & .5 & 1 \\ 1 & .5 & 1 & .5 & 0 & .5 & 1 & .5 & 0 \\ 0 & .5 & 0 & 0 & 1 & .5 & 0 & 0 & 0 \end{pmatrix};$$

In the following program we call the new random number r, the previous random number x and the random number before that xx. We declare two initial values (line 1). At each next step of the calculation xx is replaced by x, x is replaced by r and r is re-computed. The index for the column of the transition table is calculated using the formula $(xx - 1)\cdot 3 + x$. We compute the next random number just as in the previous examples.

```
x = 1; r = 1;
l = Table[xx = x; x = r;
        While[r = Random[Integer, {1, 3}];
              Random[] > P[[r, (xx - 1) * 3 + x]],];
        r, {i, 25}];
{2, 2, 3, 3, 1, 2, 2, 3, 3, 1, 2, 3, 3, 1, 2, 2, 3, 3, 1, 2, 3, 3, 1, 2, 2}
```

But we can also construct a transition matrix **Pm** for a given number of previous values m and calculate the next random number by multiplying the last m state vectors with the corresponding matrices and summing the probabilities with a certain weighting. The following matrices indicate the transition probabilities from state $n - 2$ to state n (**P2**) and from state $n - 1$ to state n (**P1**).

$$P1 = \begin{pmatrix} .1 & 0 & .9 \\ .9 & .1 & 0 \\ 0 & .9 & .1 \end{pmatrix}; P2 = \begin{pmatrix} .9 & 0 & .1 \\ .1 & .9 & 0 \\ 0 & .1 & .9 \end{pmatrix};$$

10.2 Stochastic Processes

If we want the next random number to be influenced four times more strongly by the previous value than by the value before that, we weight the summands by the factors .8 and .2 respectively. We get for the transition probabilities after two successive states

$$\text{MatrixForm}\left[.8 * P1 \cdot \begin{pmatrix} 0 \\ 1 \\ 0 \end{pmatrix} + .2 * P2 \cdot \begin{pmatrix} 0 \\ 1 \\ 0 \end{pmatrix}\right] \qquad \begin{pmatrix} 0. \\ 0.26 \\ 0.74 \end{pmatrix}$$

In the following realization, the frequent recurrence of the sequence 1, 2, 3, 1, ... can be traced back to the dominance of the previous state and to the probabilities .9 in the matrix **P1**. Descending sequences like 3, 2 cannot be traced back to the matrix **P1** and are due to the influence of next to the last state.

```
x = 1; r = 1; l = Table[xx = x; x = r; While[r = RandomInteger[{1, 3}];
   RandomReal[] > 0.8` P1〚r, x〛 + 0.2` P2〚r, xx〛, Null]; r, {i, 25}]

{2, 2, 3, 1, 2, 1, 2, 3, 3, 1, 2, 1, 2, 3, 1, 1, 2, 2, 3, 1, 2, 3, 2, 1, 2}
```

10.2.3.5 Processes With Sieved Random Variables

When we use stochastic processes, we can change or reject the random numbers generated according to specific requirements, just as with other random processes. In Chapter 10.2.3.2 we limited the random walk by requiring it to take place within a rectangle of a certain size. In the example below, the realization of the process provides the requirements for the new values, creating a process with a "memory". We generate a random walk within a rectangle (lines 5-9 below), using the bilateral exponential distribution (lines 5 and 7) (10.2.3.2, Example 2). The limiting condition is that the newly generated points must not lie too near the points already visited on the walk. We fulfill the condition by rejecting those points whose x- or y-coordinates are within one unit of any point already generated (lines 9-10).

```
x = 0; y = 0; tmax = 30; lin = {{x, y}};
Do[
  lin = Append[lin,
    While[fl = 0;        If[
       x + (rx = Sign[Random[Real, {-1, 1}]] * .5 * Log[Random[]]) > 10
        || x + rx < -10, rx *= -1,];
      If[y + (ry = Sign[Random[Real, {-1, 1}]] * .5 * Log[Random[]]) > 5
        ||      y + ry < -5, ry *= -1,];
      Do[If[Abs[x + rx - lin[[i, 1]]] > 1 ||
          Abs[y + ry - lin[[i, 2]]] > 1, fl += 0, fl += 1], {i, 1, t}];
      fl > 0];
    x += rx; y += ry;
    {x, y}];
  ls = {Line[{{-10, -5}, {10, -5}, {10, 5}, {-10, 5}, {-10, -5}}],
    Line[lin]};
  g4[t] = Show[Graphics[ls], PlotRange -> {{-10, 10}, {-5, 5}}],
  {t, tmax}]
```

10.3 Other Techniques Used for Composition

10.3.1 Cellular Automata

Cellular automata are feedback machines. They consist of individual cells that can take on a finite number of states. Changes take place in single steps and are determined by rules and by the current state of each cell. The theory of cellular automata was developed in connection with the first computers in the 1940's. Cellular automata are used in various fields to model and simulate artificial and natural processes.

Further Reading: In A New Kind of Science *[58], Stephen Wolfram shows how mathematical models, and in particular cellular automata, have changed science. There is a short general introduction to the subject in* Fractals for the Classroom: Introduction to Fractals and Chaos *by Heinz-Otto Peitgen, Hartmut Jürgens and Dietmar Saupe [22]. For applications to computer music see* Digital Synthesis of Self-Modifying Waveforms by Means of Linear Automata *by Jacques Chareyon in the* Computer Music Journal *[1-15] and* Composing Music with Computers *by Eduardo Reck Miranda, pp. 121-135 [60].*

10.3.1.1 One-Dimensional Automata With Two States

Let us consider a chain of cells, each in one of two possible states. Let us call the states 0 and 1. We can represent them graphically by coloring the cells representing state 0 white and coloring those cells that represent state 1 black. In what follows, we consider a chain having two cells in state 1 and all the others in state 0.

```
im = 20; t1 = Table[0, {i, im}]; t1[[2]] = 1; t1[[9]] = 1; t1
```

{0, 1, 0, 0, 0, 0, 0, 0, 1, 0, 0, 0, 0, 0, 0, 0, 0, 0, 0, 0}

Let us calculate the next generation of the chain using the following rule: A cell takes on state 1, if it or one, but not both, neighbor cells had the state 1 in the previous generation. Otherwise, it takes on the state 0.

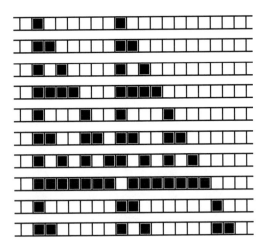

10.3 Other Techniques Used for Composition

The Mathematica command CellularAutomaton[*rule*, *init*, *t*] generates lists of the following *t* states, starting from state *init* according to *rule*. One derives the rule, the so-called *Wolfram code*, by listing all the combinations of states of the active neighbors of a cell, assigning to each combination the state in which it leaves the cell and then interpreting the resulting series of states as a binary number. For the example above, we get the code 00111100 = 60.

All states	111	110	101	100	011	010	001	000
Rule	0	0	1	1	1	1	0	0

```
CellularAutomaton[60, {0, 0, 0, 1, 0, 0, 0, 0, 0, 0, 1, 0, 0, 0, 0, 0}, 2]

{{0, 0, 0, 1, 0, 0, 0, 0, 0, 0, 1, 0, 0, 0, 0, 0},
 {0, 0, 0, 1, 1, 0, 0, 0, 0, 0, 1, 1, 0, 0, 0, 0},
 {0, 0, 0, 1, 0, 1, 0, 0, 0, 0, 1, 0, 1, 0, 0, 0}}

ArrayPlot[CellularAutomaton[60,
    {0, 0, 0, 1, 0, 0, 0, 0, 0, 1, 0, 0, 0, 0, 0}, 10]]
```

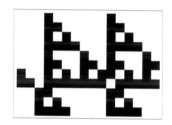

Stephan Wolfram defined four classes into which cellular automata can be divided depending on their behavior.

Class 1: Nearly all initial patterns evolve quickly into a stable, homogeneous state. Any randomness in the initial pattern disappears.

Class 2: Nearly all initial patterns evolve quickly into stable or oscillating structures. Some of the randomness in the initial pattern may filter out, but some remains. Local changes to the initial pattern tend to remain local.

Class 3: Nearly all initial patterns evolve in a pseudo-random or chaotic manner. Any stable structures that appear are quickly destroyed by the surrounding noise. Local changes to the initial pattern tend to spread indefinitely.

Class 4: Nearly all initial patterns evolve into structures that interact in complex and interesting ways. Class 2 type stable or oscillating structures may be the eventual outcome, but the number of steps required to reach this state may be very large, even when the initial pattern is relatively simple. Local changes to the initial pattern may spread indefinitely. Wolfram has conjectured that many, if not all class 4 cellular automata are capable of universal computation.

The illustration below shows a typical example for the behavior of class 3.

```
ArrayPlot[CellularAutomaton[30, {{1}, 0}, 300]]
```

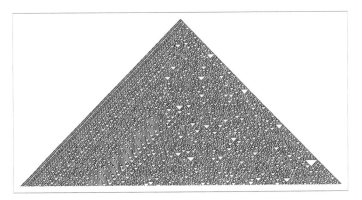

The illustration below shows a typical example for the behavior of class 4.

```
ArrayPlot[CellularAutomaton[110, {{1}, 0}, 500]]
```

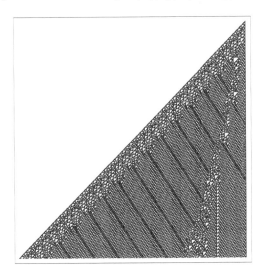

10.3.1.2 One-Dimensional Automata With Many States

In the next example, let us allow arbitrarily many different states and also close the chain of cells. For the initial states A_i let us choose random numbers from [0, 1] (line 1 below). The rules for computing the next generation are: $A_i = .7A_i + .3A_{i-1}$ for $i = 2$ to 16 (line 3 below) and $A_1 = .3A_1 + .7A_{16}$ (line 4 below). If we represent the numbers by shade of grey, the contrast between cells decreases continuously.

```
im=16;t1=Table[Random[],{i,im}];                            ;1
.......
Do[ t1n[[i]]=.3*t1[[i-1]]+.7*t1[[i]],{i,2,im}];
t1n[[1]]=.3*t1[[1]]+.7*t1[[im]];t1=t1n;rec={};
Do[rec=Append[rec,                                          ;5
    {GrayLevel[t1[[i]]],Rectangle[{i,0},{i+1,1}]}]
.......
```

10.3 Other Techniques Used for Composition

Cellular automata can be thought of as sequences of numbers that are filtered again and again. If we plot the values of the grey shades as a function, we obtain a curve that becomes smoother and smoother and moves one position to the right with each successive generation. This process can be interpreted as the repeated filtering of sounds. Time plays a role in two senses: first, the sequence of numbers represents a time series, and secondly, the processing of the series takes place in successive steps. We can also think of the process as a continuous transformation of a table, which could be used, for example, as the density function of a statistical distribution.

In this example, we can avoid too great a decrease in the curve's "amplitude" by finding the greatest value after each calculation step and dividing all the values in the series by this maximum. In this way, the curve always maintains a maximum value of one. (→ Animation)

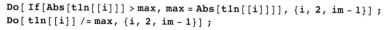

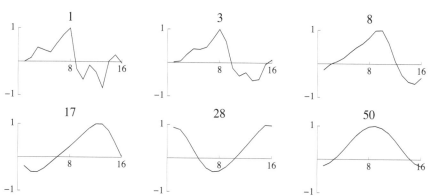

In the example that follows, we assign all the cells of a chain random numbers between −1 and +1, except the ends, which are zero. The rule for computing each new generation is: $A_i = .3A_{i-1} + .4A_i + .3A_{i+1}$ for $i = 2$ to 15. This avoids the curve's "wandering" over time. The end cells remain empty, but all the other cells eventually have either positive or negative values, depending on the initial conditions. (→ Animation)

```
Do[ t1n[[i]] = .3*t1[[i - 1]] + .4*t1[[i]] + .3*t1[[i + 1]],
 {i, 2, im - 1}]; t1 = t1n;
```

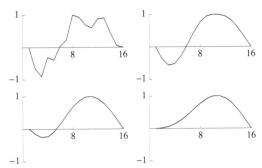

The simulations of oscillating strings discussed in Chapter 8.1.3 are special cases of cellular automata.

In the Mathematica command CellularAutomaton[*rule*, *init*, *i*] there are many different ways to define the rule. The example below shows a definition given by code as above, the number of possible states and the number of active neighbors (3/3 means two neighbors left and two neighbors right).

```
CellularAutomaton[{366 142 353 346 368, 4, 3/3}, {{1}, 0}, 12]

{{0, 0, 0, 1, 0, 0, 0, 0, 0, 0, 0, 0, 0, 0, 0, 0},
 {0, 0, 0, 3, 1, 0, 0, 0, 0, 0, 0, 0, 0, 0, 0, 0},
 {0, 0, 1, 2, 0, 1, 0, 0, 0, 0, 0, 0, 0, 0, 0, 0},
 {0, 0, 0, 1, 0, 3, 1, 0, 0, 0, 0, 0, 0, 0, 0, 0},
 {0, 0, 0, 3, 0, 2, 0, 1, 0, 0, 0, 0, 0, 0, 0, 0},
 {0, 0, 1, 0, 0, 0, 0, 3, 1, 0, 0, 0, 0, 0, 0, 0},
 {0, 0, 3, 1, 0, 0, 1, 2, 0, 1, 0, 0, 0, 0, 0, 0},
 {0, 1, 2, 0, 1, 0, 0, 1, 0, 3, 1, 0, 0, 0, 0, 0},
 {0, 0, 1, 0, 3, 1, 0, 3, 0, 2, 0, 1, 0, 0, 0, 0},
 {0, 0, 3, 0, 2, 0, 0, 0, 0, 0, 3, 1, 0, 0, 0, 0},
 {0, 1, 0, 0, 0, 0, 0, 0, 0, 1, 2, 0, 1, 0, 0, 0},
 {0, 3, 1, 0, 0, 0, 0, 0, 0, 1, 0, 3, 1, 0, 0, 0},
 {1, 2, 0, 1, 0, 0, 0, 0, 0, 0, 3, 0, 2, 0, 1}}

ArrayPlot[CellularAutomaton[{366142353346368,4,3/3},{{1},0},142]]
```

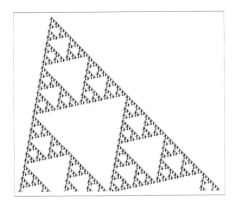

10.3.1.3 Two-Dimensional Automata With Two States

Cellular automata were made popular by the "Game of Life", invented by the British mathematician John Horton Conway. The cells of this two-dimensional automaton can take on two states, living (1) or dead (0). A cell remains alive in the next generation if it has two or three live neighbors in the current generation, otherwise it dies. A dead cell becomes a live cell if it has exactly three neighbors.

To play the Game of Life, we begin by storing the initial states $t[[i, k]]$ in a two-dimensional array t with im lines and km columns (line 1 below). To determine the states in the next moment $tn[[i, k]]$ of the game, we generate a list u containing the states of the neighbor cells $t[[i \pm 1, k \pm 1]]$ (lines 3-4 below). The cell's new state is determined by the conditional statements in lines 5-8 which inspect the states of the neighbors.

```
t = Table[0, {i, im}, {k, km}];                                          ; 1
... ...
Do[u = {t[[i - 1, k - 1]], t[[i - 1, k]], t[[i - 1, k + 1]], t[[i, k - 1]],
       t[[i, k + 1]], t[[i + 1, k - 1]], t[[i + 1, k]], t[[i + 1, k + 1]]};
   If[Count[u, 1] == 2 && t[[i, k]] == 1, tn[[i, k]] = 1,                 ; 5
   If[Count[u, 1] == 3, tn[[i, k]] = 1,
   If[Count[u, 1] < 2, tn[[i, k]] = 0,
   If[Count[u, 1] > 3, tn[[i, k]] = 0,]]]];,
 {i, 2, im - 1}, {k, 2, km - 1}] ; t = tn;
```

The illustrations below are snapshots from an animation. The three-cell bar in the first pictures turns 90 degrees farther at each step. The figure at the upper right, a so-called *glider*, takes on its original form again after four steps but has moved down one cell. The third figure dissolves. When the figures touch, they influence each other. After 32 steps, only the bar remains. (→ Animation)

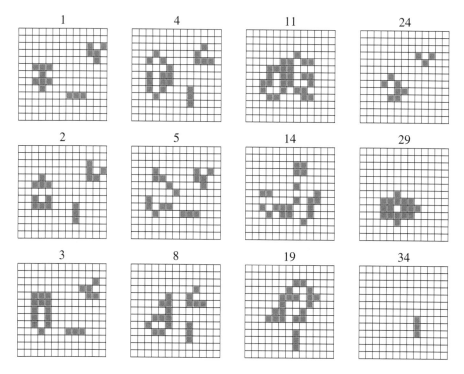

The rule for the Game of Life using the Mathematica command CellularAutomaton[ ] is:

`GameOfLife = {224, {2, {{2, 2, 2}, {2, 1, 2}, {2, 2, 2}}}, {1, 1}};`

With this rule, the movement of the glider below can be represented by giving this initial state: $\{\{\{1,0,1\},\{1,1,0\},\{0,1,0\}\},0\}$.

`CellularAutomaton[GameOfLife, {{{1, 0, 1}, {1, 1, 0}, {0, 1, 0}}, 0}, 6]`

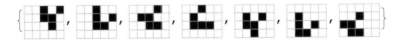

The next example shows the evolution of random initial conditions:

`CellularAutomaton[GameOfLife, RandomInteger[1, {16, 16}], 8]`

The simulations of oscillating surfaces using the spring-mass model shown in Chapter 8.1.4 are two-dimensional automata in which the new state of each cell is computed from the states of the four neighboring cells.

10.3.2 The Golden Ratio

The *golden ratio* was originally a special proportion known in geometry and used in architecture and painting. The golden ratio appears in processes and algorithms as the limiting value of the *Fibonacci sequence*. The golden ratio has long been observed in natural processes and forms, but only recent research has shown some reasons for its prevalence. The significance of the golden ratio for classical music is disputed, but the processes discussed in Chapter 10.3.2.5 suggest some ways to use it for synthetic sound synthesis.

Further Reading: Two easy general introductions are The Golden Section *by Hans Walser [50] and* The Golden Ratio: The Story of Phi, the World's Most Astonishing Number *by Mario Livio [76]. For the occurrence of the Golden Section in biology see the interesting article* Der Goldene Schnitt in der Natur, Harmonische Proportionen und die Evolution *by Peter H. Richter and Hans-Joachim Scholz in* Ordnung aus dem Chaos *edited by Bernd Olaf Küppers [51].* Fractals in Music *by Charles Madden [29] has a brief introduction to the subject, a discussion of applications in music and an exhaustive bibliography.*

10.3 Other Techniques Used for Composition

10.3.2.1 Definition and Classical Construction

A line segment is divided in the golden ratio when the length of the shorter segment is to the length of the longer segment as the longer segment is to the length of the whole. In the illustration below, the unit segment AC is divided by the point B in the golden ratio. If we call the segment $AB\ x$, we have $(1-x)/x = x/1$ from which we can derive the quadratic equation

$$x^2 + x - 1 = 0$$

with the solutions

$$x_1 = \frac{-1+\sqrt{5}}{2} = .61803... \text{ and } x_2 = \frac{-1-\sqrt{5}}{2} = -1.61803...$$

The first solution corresponds to the required segment. The second solution gives a kind of external divider for which the proportion $CD:DA = DA:1$ holds. The value $|x_2| = 1.61803...$ is the reciprocal of x_1. In what follows, we will call the golden ratio g, its reciprocal $1/g$.

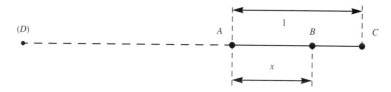

From the formula for the golden ratio

$$g = \left(-1 + \sqrt{5}\right)/2 = \sqrt{5/4} - 1/2,$$

we can derive its classical geometrical construction. Let there be a right triangle ABC with the sides $AB = 1$ and $BC = .5$ and a circle of radius $.5$ around point C. Then the line segments AD and AE have the lengths g and $1/g$ respectively.

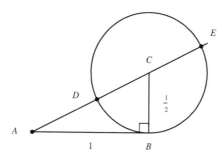

The illustration below shows geometric figures in which the golden ratio appears. The isosceles triangle ABC at the top left has a base angle of $\alpha = 72°$. If one extends the bisecting line of the angle B to the opposite side at D, one obtains the triangle ABD that is similar to the large triangle. From the similarity, we can derive the equation $c/1 = (1-c)/c$, from which follows that c corresponds to the golden ratio. Since $72°$ is one fifth and $36°$ one tenth of $360°$, we can expect to find corresponding triangles in regular polygons of five or ten sides.

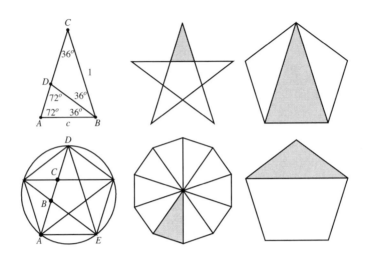

10.3.2.2 Fibonacci Numbers

The numbers in the sequence 1, 1, 2, 3, 5, 8, 13, 21, ... are called *Fibonacci numbers* (B.4). The Fibonacci sequence is defined by the recursion formula $a_n = a_{n-2} + a_{n-1}$ with initial values $a_1 = a_2 = 0$. The ratio a_{n-1}/a_n of two successive Fibonacci numbers tends toward the golden ratio.

a_n	1	1	2	3	5	8	13	21	34
a_{n-1}/a_n		1.	0.5	0.666667	0.6	0.625	0.615385	0.619048	0.617647

55	89	144	233	377	610	987	...
0.618182	0.617978	0.618056	0.618026	0.618037	0.618033	0.618034	...

Taking arbitrary initial values a_1 and a_2 in the above formula leads to the same limit of ratios a_{n-1}/a_n.

a_n	1	11	12	23	35	58	93
a_{n-1}/a_n		0.0909091	0.916667	0.521739	0.657143	0.603448	0.623656

151	244	395	639	1034	1673	2707	...
0.615894	0.618852	0.617722	0.618153	0.617988	0.618051	0.618027	...

10.3.2.3 Continued Fractions, Surds and Golden Ratio

The decomposition of the golden ratio into a continued fraction shows that in a certain sense the golden ratio is the opposite of a simple harmonic ratio. A number is called irrational if it cannot be written as the quotient of two integers p/q. The decimal representation of an irrational number never terminates and never repeats. (An example of a number that never terminates but repeats, and hence is rational, is 3/11 = .2727272727...) The decimal representation does not give the best approximation of an irrational number. For example, the irrational number π can be more precisely written as 22/7 than in decimal form with two places of

10.3 Other Techniques Used for Composition

accuracy and as 355/113 more precisely than with seven places of accuracy. The so-called continued fraction gives the quickest approximation of an irrational number. The expansion of π is obtained as follows. Let a number have an integer part and a remainder between 0 and 1: $\pi = 3 +$ remainder. The reciprocal of the remainder is greater than 1 and so can be written as an integer plus a remainder: $\pi = 3 + 1/(7 +$ remainder), etc.

$$\pi = 3 + \cfrac{1}{7 + \cfrac{1}{15 + \cfrac{1}{1 + \cfrac{1}{\cdots}}}}$$

We obtain the best approximations of π using simple fractions by interrupting the expansion at some point. We have: $3 + 1/7 = 22/7$, $3 + 1/(7 + 1/15) = 333/106$ and $3 + 1/(7 + 1/(15+1)) = 355/113$. The smaller the remainder the larger its reciprocal will be (for π the numbers 7, 15, 1, 292 ...). For the golden ratio, we have the equation $g = 1/(1 + g)$. If we substitute $1/(1 + g)$ for g in the denominator of a continued fraction we obtain:

$$g = \cfrac{1}{1+g} = \cfrac{1}{1+\cfrac{1}{1+g}} = \cfrac{1}{1+\cfrac{1}{1+\cfrac{1}{1+g}}} = \cfrac{1}{1+\cfrac{1}{1+\cfrac{1}{1+\cfrac{1}{1+g}}}} = \ldots = \cfrac{1}{1+\cfrac{1}{1+\cfrac{1}{1+\cfrac{1}{1+\cfrac{1}{1+\cdots}}}}}.$$

If we interrupt the expansion after n steps, we get the same approximation as with the Fibonacci numbers: 1/2, 2/3, 3/5, ...

In the expansion of the golden ratio by a continued fraction, the integer part of the remainders is always equal to one. That means that g is the number that can least well be approximated by harmonic proportions. If we interrupt the expansion after 5 steps, we get the value .625 which is only correct to one decimal place. The "continued square root" below also leads to the golden ratio:

$$\sqrt{1 + \sqrt{1 + \sqrt{1 + \sqrt{1 + \sqrt{1 + \ldots}}}}} = 1/g = 1.618\ldots$$

The "golden rectangle" (figure b below), for most people the ideal rectangle, is often cited as an example of the aesthetic effect of the golden ratio. It is very different from the square, and because of the irrational proportions of its sides it does not break down visually into smaller squares.

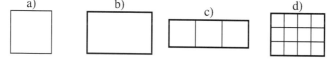

If one subtracts a square from a golden rectangle, a smaller golden rectangle remains. This procedure can be repeated infinitely.

10.3.2.4 Fractals

A fractal can be generated very easily using the following rule. At the end of a given line segment one draws two shorter segments each at an angle of 120 degrees to the original segment (figure left below). If this process is continuously repeated, the fractal shown in the middle below emerges. (→ *fractal.pde*)

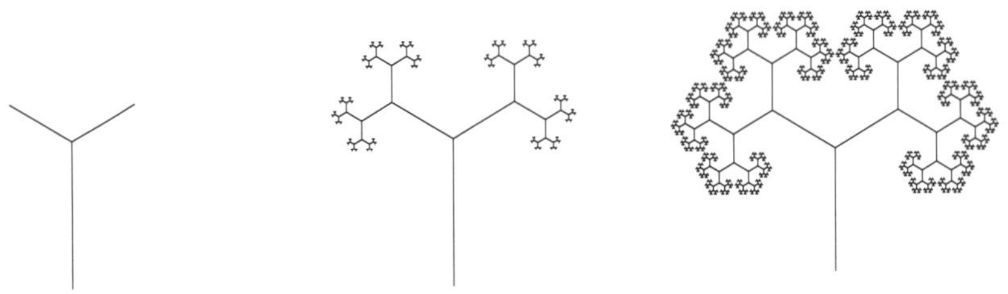

For the fractal in the middle above, each new segment is factor $f = 0.5$ as long as the preceding segment. If we gradually increase this factor, there is a value at which the branches of the tree will eventually touch (right figure above).

This value corresponds to the golden ratio, as the figure below illustrates. The branches touch when the heavy lines moving left ($f^3 + f^4 + $ ...) reach the middle. Since all the line segments (except the vertical segment f^2) are at an angle of 30 degrees to the horizontal, we have $f \cdot \cos(30°) = f^3 \cos(30°) + f^4 \cos(30°) + $ ... or $f = f^3 + f^4 + f^5 + $... $= f^3/(1 - f)$. This leads directly to the equation for the golden ratio $1 - f = f^2$.

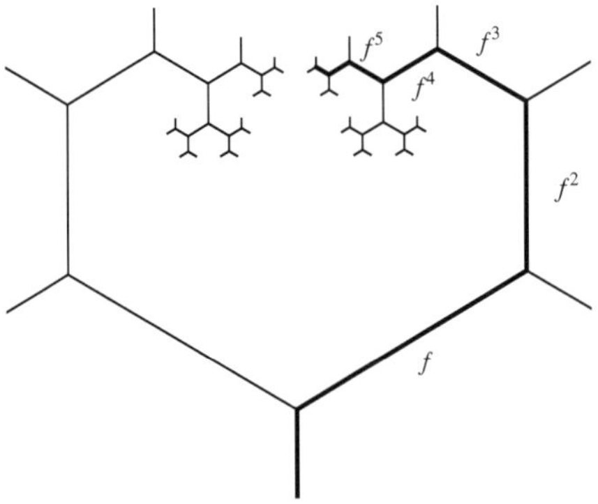

10.3.2.5 A Process of Natural Growth

This example can serve as a model for natural events like the growth process of branches or roots that fill the available space most effectively when the relationship of successive segments is the golden ratio. Of course, this model is both too simple and too abstract to describe natural growth. Surprisingly, however, sometimes a simple model can create patterns that we can find in nature. The middle figure below shows the arrangement of the seeds of a sunflower. Here the space is ideally used. The left-handed and right-handed spirals are so entwined in each other that one sees both simultaneously. This entwining also reduces the likelihood of fissures and thus stabilizes the structure. (→ Animation)

10.3 Other Techniques Used for Composition

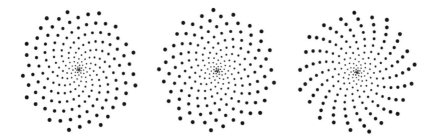

The left figure below shows how the arrangement arises. The angle of rotation always divides the circle in the golden ratio and is equal to $g \cdot 2\pi$ or $g \cdot 360° \approx 222.492°$. Stability and ideal utilization of available space are considered evolutionary advantages. The right figure shows an arrangement found in leaves or in the scales of pine cones. Every new leaf grows into the largest gap between two other leaves. The angles used for the left and right figures above deviate from the golden ratio by only 0.1%. (→ Animation)

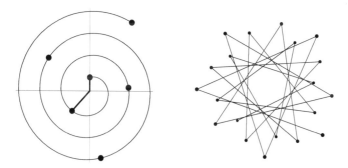

Cellular automata can easily simulate the growth process. Let a ring of "cells" be given. After a certain time, one of the cells produces a "leaf", and immediately an inhibition to produce another leaf appears at this position. In nature, this can be due to chemical inhibitors or to lack of nutrients. In our model, every cell stores the magnitude of the inhibitor. The inhibitor is passed on to the neighboring cells and is gradually reduced over time ("metabolized", as it were). After a certain time, another leaf is produced at the spot where the concentration of the inhibitor is the weakest, directly opposite the first leaf. The next leaf is produced between the first two, but closer to the first, etc. The arrangement described above can be seen already after a few steps. (→ Animations)

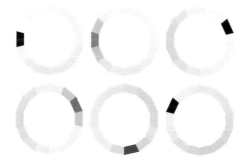

When parameter values are used simulating inhibitors that wear off quickly, the model produces simple, objective arrangements (left figure below). Within a fairly large range of parameter values, spiral arrangements are produced similar to those above. Since the above simulation has 70 cells and the successive points are either 27 or 43 cells apart, the result comes as close to the golden ratio as is possible with this simple model (43/70 = .614...). Two parameters were used for the simulation. Factor $a < 1$ is multiplied with the concentration of inhibitor in the cells at each iteration and determines how fast the inhibitor "wears off". Factor b determines the amount of inhibitor that is passed to the neighboring cells. The figure left below was generated using the values $a = .4$ and $b = .2$. For the the middle figure, the values are $a = .48$ and $b = .24$, for the figure at the right $a = .7$ and $b = .1$. The same spiral arrangement occurs despite the very different parameter values. In addition to the 13 obvious spiral arms in the middle and right figures, one can also see five spirals turning in the opposite direction (Fibonacci numbers).

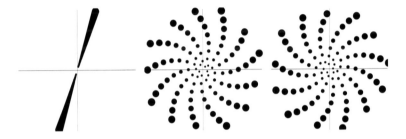

10.3.2.6 Applications

Because the golden ratio is the opposite of a harmonic proportion, one can find it where simple proportions, resonances and the like are to be avoided. For example, if one damps an oscillating string at a point that divides the string in a simple proportion, some related harmonics sound, that is, not all the partials are damped. On the other hand, if one damps the string at a point that divides it in an irrational proportion, it will be strongly damped. As we will see below, damping at the golden ratio gives good but not optimal results, and only damping at the simplest harmonic proportions gives clearly inferior results compared with irrational proportions.

Here is why. Let the damping at position x be proportional to the displacement of the string $a(x)$ at that point. We set the amplitude of the nth partial equal to A/n because in many natural sounds the partials diminish in amplitude with frequency.

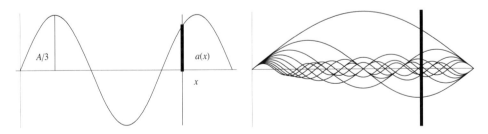

The illustration below shows the sum of the amplitudes of the first 100 damped partials of a string as a function of the point of damping x.

10.3 Other Techniques Used for Composition

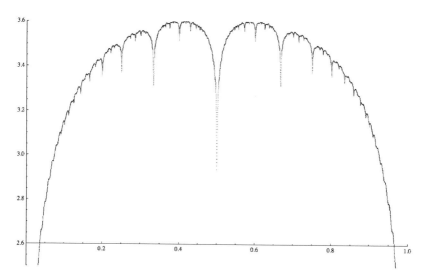

(→ Excerpt .578 < x < .63)

At the point $x = g = .61803...$ we get the value $\Sigma(g) = 3.59953$. The value is somewhat higher at $x = 586/1000$, a rational number. The results are similar for other amplitudes, damping points and precision of calculation.

The golden ratio damps well in this example, but it is certainly not exceptional among the non-harmonic proportions. There is a clear hierarchy of harmonic proportions, and they are usually easy to hear and to produce.

Traditional tone systems are based on an equal or at least regular division of the octave. The differences between pure and equally tempered intervals are so small that the ear generally tolerates them. In what follows, we will consider only the regular division of the octave. The tones, and the relationships between them, are the same in every octave. By adding the same interval to itself, one eventually lands an octave higher than at the outset. (Using the tritone only gives two tones before the octave is reached, thirds give three or four tones, and fifths, fourths and half-tones give all twelve tones of the chromatic scale.) In traditional systems, the octave is divided into twelve equal intervals (figure left below). If we want to generate new tones having different frequencies, we can use a random number generator. The figure on the right shows a randomly generated sequence of tones where the size of the point corresponds to the order of its generation. One sees clearly how irregular the random distribution is. Our perception will always detect a certain arrangement or grouping in a random distribution. In order to generate as "regularly irregular" a distribution as possible, we can use the interval that divides the octave into the golden ratio, namely 2^g. The result is similar to the arrangements in Chapter 10.3.2.5 (middle figure below).

To demonstrate what is special about this distribution of tones, let us compare it with systems of generation using other irrational intervals. Three arbitrarily chosen intervals are shown below. Patterns can be seen in all of them.

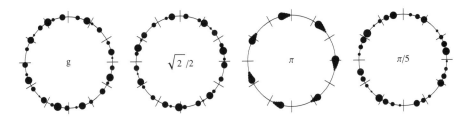

The following Csound example makes audible the particular nature of the golden ratio. The tones have the frequency *p5*. The amplitude starts at zero, increases linearly to *p4* and then decreases linearly back to zero.

```
;10-3-2a.sc
instr 1
kamp        linseg      0,p3/4,p4,3*p3/4,0
a1          oscil       kamp,p5,1
a2          tone        a1,1000
            out         a2
endin
```

We make the score using a Mathematica program (4.3.1). We generate $nn = 30$ tones in intervals of 0.5 seconds. Each tone's frequency is computed by multiplying the last frequency by 2^g. The fundamentals of the tones lie between 20 and 40 Hz. Frequencies over 40 Hz are halved.

```
g = (√5 - 1)/2.0; nn = 30; ig = 2ᵍ; fr = 40;
score = OpenWrite["csc1"]; Do[fr = fr*ig; If[fr > 40, fr /= 2];
  WriteString[score, "i1 ", n/2.0, "\t", 8, "\t", 5000, "\t", fr, "\n"],
  {n, 0, nn}]; Close["csc1"];
```

The first notes of the score are:

```
i1 0      8    5000    30.6956
i1 0.5    8    5000    23.5556
i1 1.     8    5000    36.1527
i1 1.5    8    5000    27.7432
i1 2.     8    5000    21.2899
...
```

We need to add to the score the indications for generating the waveform to be used by the oscillator. We define a waveform *f1* with a fundamental and nine higher partials, each at the interval of an octave from the last. This will generate the tone swarms. The low pass filter in the instrument makes the high tones softer and distributes the energy of the entire swarm more evenly over the spectrum.

f1 0 65536 9 1 1 0 2 1 0 4 1 0 8 1 0 16 1 0 32 1 0 64 1 0 128 1 0 256 1 0 512 1 0

The synthesis gives a dense cloud of sounds that quickly fills the frequency space and then remains very stable (Sound example 10-3-2a3). If we replace the factor $ig = 2^g$ in our program with

10.3 Other Techniques Used for Composition

$$ig = 2^{\pi/5} \text{ or } ig = 2^{\sqrt{2}/2},$$

we get the Sound examples 10-3-2a1, a downwards spiral, and 10-3-2a2, an upwards spiral. If we generate the new frequencies using 1 + Random[0, 1] (→ Score), we get a sound that constantly changes (Sound example 10-3-2a4).

The example can easily be modified as follows. We distribute sine waves over the entire frequency space from 40 to 10,240 Hz, that is, over eight octaves. Now these eight octaves correspond to the circle of tones above. We continuously multiply the frequency by $2^{8 \cdot g}$ and divide by 2^8 when the result is greater than 10,240. Because of the large frequency range and the limitation through the use of sine waves, the sound changes constantly, but it densely fills the entire sound space without moving up or down (Sound example 10-3-2a6). (→ Program and score excerpt)

If we replace the golden ratio in the factor $2^{8 \cdot g}$ with $\pi/5 = .628319...$, we also get an upwards spiral and the sound is less dense (Sound example 10-3-2a7).

In the example that follows, we apply the principle of "regular irregularity" discussed above to rhythm and loudness. We use the circle shown above and interpret the distances as time intervals and the size of the points as loudness.

The individual tones are sine tones with exponentially decreasing amplitude. The low pass filter for the amplitude envelope smooths the attacks of the tones.

```
;10-3-2b.sc
instr 1
aamp1    expon    p4,p3,2
aamp     tone     aamp1,50
a1       oscil    aamp,p5,1
         out a1
endin
```

We generate the score again with Mathematica. Because the notes' starting times and their frequencies should be computed independently of one another using the golden ratio, we first make a list *tbl* containing starting time, duration and amplitude for 300 tones. Then we sort this list by starting time (*stbl*).

```
g = (√5 - 1)/2.0; nn = 300; t = .17;
tbl = Table[t += g; If[t > 1, t -= 1]; {N[10*t, 4],
    N[e^-.008*n, 4], 15 000 *e^-.008*n}, {n, 0, nn}]; stbl = Sort[tbl];
```

Now we generate the frequencies as we did in the previous example and write the score (Sound example 10-3-2b).

```
g = (√5 - 1)/2.0; nn = 300; ig = 2^g; fr = 400;
score = OpenWrite["csc4"]; Do[fr *= ig; If[fr > 800, fr /= 2];
  WriteString["csc4", "i1 ", stbl[[n, 1]], "\t", stbl[[n, 2]],
  "\t", stbl[[n, 3]], "\t", fr, "\n"], {n, 1, nn}]; Close["csc4"];
```

The figure below shows the onset and loudness of the tones in Sound example 10-3-2b1.

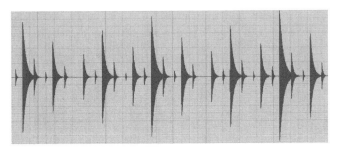

The upper figure below shows the same schematically. The lower figure below shows the distribution obtained using a factor that differs from the golden ratio by .00005. (→ Manipulate)

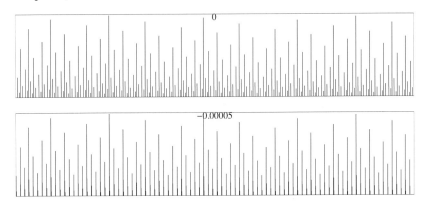

In the final example we generate a regular sequence of sounds. The fundamental of the nth sound is derived from the fundamental of the previous sound as $f_n = g f_{n-1}$. As with the acoustic illusions of R. Shepard and J. C. Risset (5.2.1.5), every fundamental generates a swarm of octaves so that the partials are always distributed throughout the entire spectrum from 20 to 20,000 Hz. The envelopes for the amplitude of the sounds (at the left below) and for the amplitudes of the partials as a function of their frequencies (shown logarithmically at the right below) are computed using the formula $(1 - e^{-at})e^{-bt}$.

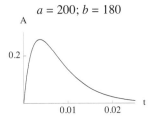
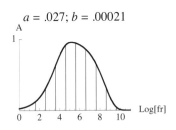

10.3 Other Techniques Used for Composition

At first one only hears the sounds leaping about in seemingly arbitrary fashion, but after a few moments, one can hear ascending or descending sequences moving more slowly, depending how one focusses one's attention.

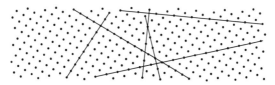

The lines above correspond to the right-handed and left-handed spirals of varying speed in the arrangement described above. In the figures below, every nth point is marked.

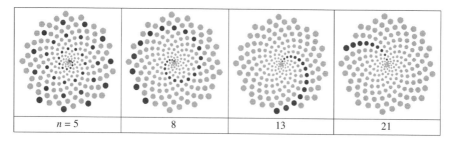

| $n = 5$ | 8 | 13 | 21 |

In the Study 18.1 (→ CD), the circular movement of the sound sequence accelerates, constantly causing other more slowly moving frequency patterns to emerge. We hear ever new lines and rhythms (18.1.mp3 can be played with any conventional mp3 player or, if one wants to hear the movement in three dimensions, with the Fraunhofer mp3 Surround Player, available from the Internet). (→ *Phyllotaxis.maxpat*)

10.3.3 Chaos Theory

Numerous publications dealing with so-called *chaos theory* have found enthusiastic readers among both scientists and non-scientists. The fascination for the subject has mostly to do with the fact that simple formulas and algorithms can produce unexpectedly complex patterns. These patterns can easily find expression in colorful, decorative pictures, but their transformation into sound or compositional structures presents great problems.

Further Reading: Out of the substantial literature on Chaos Theory, Fractals for the Classroom: Introduction to Fractals and Chaos *by Heinz-Otto Peitgen, Hartmut Jürgens and Dietmar Saupe [22] can be recommended.* Chaos in dynamischen Systemen *by Willi H. Steeb und Albrecht Kunick [14] and* Mathematik der Selbstorganisation *by Gottfried Jetschke [4] presuppose considerable mathematical knowledge.* Fractals in Music *by Charles Madden [29] and the article* Chaos, Self-Similarity, Musical Phrase and Form *by Gerald Bennett [3-1] have direct connections with music.*

10.3.3.1 Concepts

The primary meaning of the word chaos is disorder, unpredictability, arbitrariness. In a mythological context, chaos is the origin, the *prima materia*, of all development. In chaos theory, processes are simultaneously unpredictable and determined, in fact, one speaks of deterministic chaos. The more chaotic the results of a process are, the more information is required to describe them completely. Chaotic, random oscillation produces white noise. To describe white noise completely means giving a sample value for every instant, an enormous number of values, depending on the duration of the noise. A sine wave is far more organized and can be described fully with only a few indications.

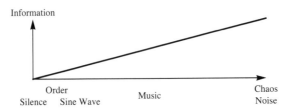

The information necessary to record a sound is not the same as the information we perceive in a sound. Our perception hears no information whatsoever in uniform noise. It might seem easy to define what is meant by "musical information" in traditional music, namely what has to be written down into order that the music can be played, that is, the score. But when music is played, it usually contains more information than is written in the score. The musician contributes additional knowledge in order to interpret the score. On the other hand, the score can sometimes be redundant, for example when repeated accompaniment figures are written out.

Complexity is defined differently in various disciplines: as one possible converse of simplicity; as those details that so differ from other details of a situation that there is no simplifying abstraction; as a measure of the effort required for the algorithmic treatment of certain problems, etc. Formerly, one assumed that complexity was based on intricately linked processes, but chaos theory shows that even very simple processes can behave in complex ways. It is difficult to define complexity in relation to art, but complexity belongs to the most significant aesthetic attributes of many historical musical styles. Complex musical works often contain relatively little information. For instance, certain contrapuntal pieces by Johann Sebastian Bach can be reduced to simple arrangements of a few motifs and can described using fewer signs than musical notation requires.

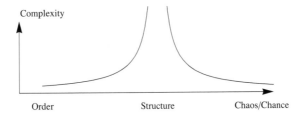

10.3.3.2 The Theory of Dynamic Systems

The term *chaos theory* includes theories and models that were worked out at various times and in various fields of endeavor. The similarities and relationships between these theories and models only emerged over time. Many of the processes these theories spoke of could not be calculated analytically but could be simulated on the computer. Computer simulation was essential for the development of chaos theory.

Nonlinearity and discontinuity play an important role in all areas of chaos theory. The American meteorologist, Edward N. Lorenz, realized in the 1950's that minimal variation in nonlinear atmospheric phenomena could cause large deviations in linear weather models (3.4.3.4). In 1972, he gave a talk at a meeting of the American Association for the Advancement of Science with the title: "Predictability: Does the Flap of a Butterfly's Wings in Brazil set off a Tornado in Texas?" Since then one speaks of the the "butterfly effect". Another basic element of chaos theory is feedback, that is, the influence of the results of a process on the process itself. When random unpredictable behavior occurs in systems in the absence of any external random influence, one speaks of *deterministic chaos*. The reverse phenomenon, the spontaneous generation of order in a chaotic system, is the subject of *synergetics*, the theory of self-organization of patterns and structures. Feedback plays a central role in both deterministic chaos and synergetics.

The *logistic equation* (also called the Verhulst model or the logistic growth curve) calculates each new value based on the previous value with the formula $x_n = rx_{n-1}(1 - x_{n-1})$ (3.4.3.3). Despite its simplicity, the equation shows complex behavior and illustrates typical characteristics of many chaotic systems. The equation is explained in every book on chaos theory, and in some computer programs, it is used to generate music. Many composers are attracted by the economy of the equation: a simple formula driven by a single parameter.

10.3.3.3 Self-Similarity and Fractals

The fascinating images mentioned at the outset, the so-called *fractals*, may seem to have nothing to do with chaos, since at least the simpler examples clearly reflect order. One way to make fractal curves is to start with an arbitrary graphic model and then add ever smaller structures by continuously reducing and reproducing the model. The best-known example of such a construction is the *Koch curve*, named after the Swedish mathematician Helge von Koch, who published instructions for its construction in 1904. The illustration below shows the first steps in the construction of the Koch curve. (→ *koch_curve.pde*)

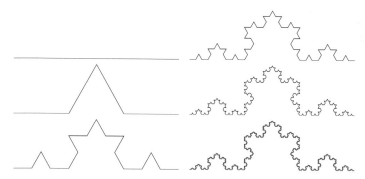

At each step, the curve's length increases by the factor 4/3. If this process is continued, the result is an infinitely long curve. After sufficiently many iterations, the curve no longer has straight segments and has a form somewhere between a line and an area. It is possible to assign a dimension to such curves that is between that of a line and that of a surface, that is, between one and two. The Koch curve has the dimension 1.2691... (see [22] p. 249). A figure having a non-integral dimension is called a *fractal*. Many fractals are *self-similar*, which means that a part of the curve is similar to the whole curve.

Certain *attractors* that arise in connection with deterministic chaos, for example, the attractor of the logistic equation (3.4.3.3), are fractals and show self-similarity. The term self-similarity is also used in musical contexts, for example, in the description of contrapuntal techniques, of formal organization or of harmonic sequences at various levels, without, however, implying any connection to chaos theory. On the contrary, often a familiar and straightforward musical event is simply given a fashionable name. Applying extra-musical concepts to musical processes is often problematic, because these concepts seem to link phenomena that are at best outwardly similar to each other, without taking historical or stylistic characteristics into consideration. For example, if one speaks of self-similarity in regard to the form of dance movements in classical music (two groups of eight measures, divided into phrases of four measures, each of which consists of motifs of two measures, etc., all the way down to the even subdivision of the note values), one mixes together aspects of the piece such as melodic structure, historical style and personal style, that have not much to do with each other.

The best-known fractal is the so-called *Mandelbrot set*, named after the French mathematician Benoît B. Mandelbrot. The illustration on the left below shows the entire set, the illustration on the right shows an enlargement of part of the figure on the left. An animation that zooms deeper and deeper into the set always shows new forms. The delicate structures remind one of organic branchings and ramifications that never repeat but are always similar. (→ *Mandelbrot1.pde*, *Mandelbrot2.pde*)

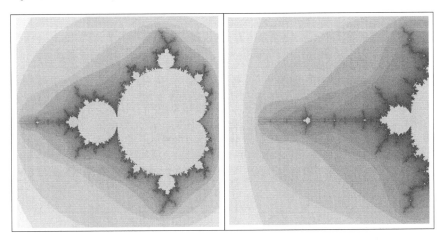

10.3.3.4 Applications to Music

It is hardly surprising that people often try to realize the structures described in Chapter 10.3.3.3 in sound. Fractals are two- or three-dimensional geometric figures while music may seem two-dimensional in the sense that it has traditionally been written showing the parameters time and pitch. However, to attempt to translate geometric figures into music by letting one spatial dimension represent time, the other pitch, is doomed to failure because the two musical dimensions differ fundamentally and profoundly both from each other and from geometric dimensions (10.3.5).

There are, however, many ways to represent concepts and models taken from chaos theory meaningfully in music. To begin with, let us mention those composers who feel inspired by the images and ideas of chaos theory without making use of its technical details. The most prominent of these composers was György Ligeti, to whom (the German translation of) the book "Fractals for the Classroom" is dedicated. Even before the chaos theory boom, Ligeti's music of the 1960's "dealt with complex structures. (...) However, this tendency did not lead to superficial translations of scientific insights into compositional models." Most of the "composition programs" found on the Internet derive their parameters, usually limited to pitch and rhythm, from algorithms taken from chaos theory. The procedures resemble serial music, but stylistically the music is simpler and more traditional, coming as it does from the environment of popular and minimal music, rather from the world of the serial or electronic music

10.3 Other Techniques Used for Composition

from the nineteen fifties and sixties. The programs are usually limited to traditional rhythmic patterns and temperament systems and are always based on the individual note as material. The treatment of noise and sounds and the structuring of compound events is usually not possible.

Self-similar sounds have the same spectrum in different enlargements, that is, in different transpositions. White noise remains the same at every transposition. Spectra that are periodic in logarithmic representation can merge into one another at certain transpositions. The endless glissandi (5.2.1.5) correspond to the animation of a fractal that is constantly being enlarged. One can see the self-similarity of the Koch curve in a single picture (see the figures below). With sound, the repetitions only become clear after a certain time. If one uses sounds with simple frequency ratios between the partials, one only hears the self-similarity after one repetition because the partials fuse. If one uses sounds with irrational ratios between the partials, one only hears the self-similarity after one repetition because the sameness of the intervals is hard to perceive. (→ koch_curve.pde)

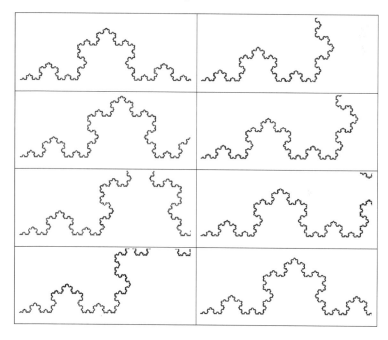

In the Shepard tones as in Risset's glissandi (5.2.1.5), the self-similarity can only be heard and seen in the waveform after one repetition, because the partials are faded in and out and the oscillation is only enlarged in time and not in amplitude. We obtain a sound that is produced by the enlargement of the waveform in time and in displacement by using an exponential series not just for the frequencies but also for the amplitudes. The illustration below shows the spectrum of a sound whose partials are in octaves. The amplitude of each partial is a factor of two less than that of the partial before. Both frequency and amplitude are shown logarithmically. Theoretically, it is not necessary to fade the sound out and in as with Risset's glissandi, because the high partials are very weak and frequencies under 20 Hz cannot be heard. Enlarging the waveform in time and displacement results in shifting the spectrum in the direction shown by the arrow.

$$\sum_{n=0}^{\infty} \frac{\cos(2^n \cdot 2\pi t)}{2^n} = \cos(2\pi t) + \frac{\cos(2 \cdot 2\pi t)}{2} + \frac{\cos(2^2 \cdot 2\pi t)}{2^2} + \ldots$$

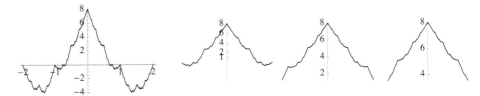

The illustration below shows the waveform $\sum_{n=-2}^{12} \frac{\cos(2^n \, 2\pi t)}{2^n}$ for $-2 < t < 2$. On the right, one can see three enlargements of the region around $t = 0$. (→ Animation) (→ Sounds and Spectra)

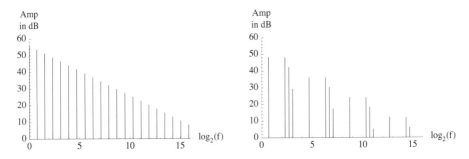

This effect can be modified in various ways. When the intervals between the partials are irrational, octaves do not arise, and the individual partials become audible (left figure below). If the spectrum does not repeat until after several partials, the sound does not repeat until after a correspondingly large transposition (right figure below).

Another self-similar structure was discussed in the chapter about the golden ratio (10.3.2.6). Enlargements (that is, transpositions) of certain sections of the tone sequence sound like the entire sequence.

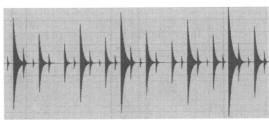

10.3 Other Techniques Used for Composition

The theory of dynamic systems frequently finds application in sound synthesis. It is difficult to generate sounds that are as lively and rich as natural sounds, primarily because the pitch, timbre and loudness of natural sounds are constantly varying. If we synthesize sounds with arbitrary variations of these parameters, the variations themselves appear unnatural. A more natural effect can be obtained by simulating physical processes, for instance the nonlinear feedback between bow and string in a bowed instrument (8.1.1.4, 8.1.3.6). Of course, one can approximate the oscillation of a string using linear equations, but the interaction of string and bow is nonlinear, and the switching between the phases where the bow hair is on the string and the phases where the bow glides are discontinuities. If one bows with too much pressure, even if the pressure remains constant, chaotic oscillation of the string results (8.1.1.4).

A nonlinear system can generate sounds whose spectra depend on the system's initial values (6.2.3). When used as a filter, a nonlinear system can generate noise whose spectrum depends on the amplitude of an input signal (8.1.1.5), or it can generate noise with audible irregularities (6.2.3 and 6.2.4).

Results from chaos theory are not only used to synthesize sound. Many composers employ them to generate scores or to control parameters of interactive music. Chaotic oscillation, random generators, complex deterministic processes and nonlinear interaction between computer programs, between musicians and between the computer and musicians, have all been part of computer music since its beginnings. Hence, in computer music, chaos theory did not lead to any essential aesthetic transformation, but rather provided new, rational and appropriate tools for analyzing and realizing sounds and compositions.

10.3.4 Simulating Swarm Behavior

Swarm simulators are *multi-agent systems*, in which several subsystems interact with each other. These subsystems, called the *agents*, can be very simple, but the entire system can nevertheless display complex emergent behavior. *Emergence* is the spontaneous development of new features or structures at the macro-level of a system as a consequence of the interplay of its elements. Examples where emergence occurs are insect populations, swarms, herds, crowds of people in certain situations like road traffic, associations of biological cells (in particular the brain), communicating computer programs and robots.

Various aspects of swarm simulation can be of interest in a musical context. Simple elements can bring forth higher structures through self-organization. In contrast to more abstract agent systems, the movement of the agents in swarms is concrete and clear and can be interpreted as spatial movement of sounds or virtual objects or as movement through a parameter space (10.3.5). All stages of systemic behavior from that of a completely *autonomous system* (algorithmic composition) to that of an entirely *reactive system* (improvisation and live electronics) can be implemented by defining the interaction between the agents and their environment. In contrast to other models, for example cellular automata, we have a great deal of everyday experience with "swarms", both from the inside, so to speak, as agents (crowds, traffic, public events) and from the outside, as observers or influencers of human or animal groups.

10.3.4.1 The Classical Boids Algorithm

In 1986, Craig Reynolds simulated the coordinated movement of animals using a simple computer model, later called the *Boids algorithm*. The algorithm uses three simple rules to determine the movement of agents, called *boids* (from "bird" and "-oid", meaning "like", but also "birds" pronounced with a New York accent). The agents are programmed to move toward the average position of the swarm (*cohesion*), to avoid colliding with their neighbors (*evasion*), and to move towards the average heading of their neighbors (*alignment*). Especially in large swarms or herds, animals react only to their immediate neighbors. For this reason, the algorithm defines an angle of view and a maximal distance at which another agent can be seen.

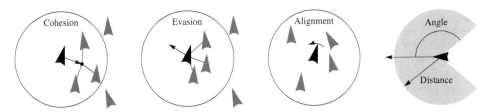

The illustration below shows a program in which the parameters for the three behaviors, the angle of view, the range of random acceleration and a damping factor for the boids' movement can be manipulated. The angle of view for one boid is shown by the semicircle.
(→ Manipulate)

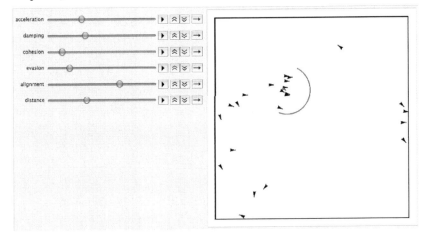

The figures below illustrate the effect of each of the three behavior principles. The left figure shows strong alignment, the middle figure shows strong evasion and the right figure shows strong cohesion.

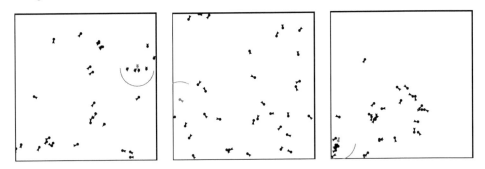

The simulation described above can provide considerable information that can be used in musical implementations: the position parameters of the boids, their velocities and accelerations, the number of visible neighbors, the size of the group, etc.

10.3.4.2 Extensions of the Swarm Algorithm

The simple swarm algorithm can be extended in various ways. One can define other behaviors than the three originally proposed by Reynolds. The temporal behavior of the agents can be extended by taking into consideration laziness and inactivity or the accumulation of agents' needs (hunger, greater independence). The agents' spatial behavior can be changed by specifying non-Euclidian or non-three-dimensional spaces. In addition, all behaviors can evolve over time. As is usually the case in nature, the behavior of single agents or of an entire swarm can be limited by external circumstances or influenced by boundaries, topographical features, force fields, etc. There can be several kinds of agents at the same time, influencing or antagonizing each other. Attributes and behaviors can be optimized by submitting the agents to "natural selection".

The following examples are taken from the research project "Interactive Swarm Orchestra ISO" at the Institute for Computer Music and Sound Technology ICST [I-6][I-7]. The project has developed C++ libraries for swarm simulation and visualization as well as for sound synthesis and video tracking. The libraries and these examples and illustrations are the work of Daniel Bisig [3-10, 11, 14]. The three-dimensional sound projection was realized with Ambisonics (9.3.2.2) using 20 loudspeakers arranged in a dodecahedron. An OSC interface allows the exportation of swarm data to any sound synthesis program.

The following three examples are taken from "flowspace", an installation realized in the framework of the ISS project. The first example demonstrates the influence of force fields on the agents. Force fields can emulate natural fields like gravity or they can be defined by arbitrary differential equations. The left illustration below shows a swarm following the *Rössler attractor* (3.4.3.4). Alternatively, lines or planes can be defined that attract or repel agents. The right illustration below shows the interaction of various kinds of agents. The smaller agents are attracted to the larger agents like moths to a lamp.

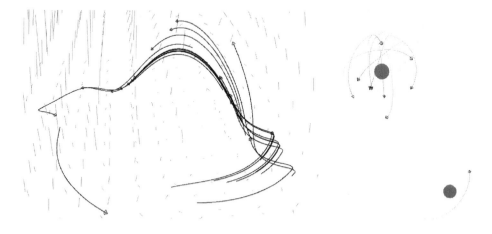

10.3.4.3 Implementations

The simplest way to generate sounds with the data from swarm simulations is to map the data onto parameters of software instruments. If each agent is assigned an oscillator, that agent's coordinates, speed, etc. can, after appropriate scaling, be interpreted as the oscillator's frequency, amplitude, etc. During the mapping, the data can not only be scaled but also transformed in any way one desires. Requirements for the synthesis parameters, like the discreteness of pitches or the limitation of the position of sound sources to the horizontal plane can be implemented in the swarm itself or in the mapping of data to the synthesis process.

In the example shown below, the application "sample triggering" from the ISO project, sample playback units were each placed in a specific zone of three-dimensional space. When an agent was in a given zone, that zone's playback unit played a stored sound sample with amplitude inversely proportional to the agent's distance from the center of the zone. The position and size of each zone could be varied, as could the choice of sampled sounds.

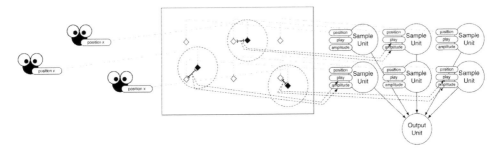

In the application "Strings", each agent was considered the mass of a *mass-spring system* (8.1). The agents were dynamically connected to their neighbors so that groups of agents formed a network, generating sounds resembling those of strings (8.1.3) or surfaces (8.1.4). The agents' oscillations were continuously excited by noise whose amplitude corresponded to the agents' average movement. In addition, there were individual strong excitations every time a connection was made or broken. The visualization showed not only the agents and their tracks, but also the current groupings. The installation's visitors could influence the constantly changing network by moving a second kind of agent, shown as a circle, that repelled the oscillating agents.

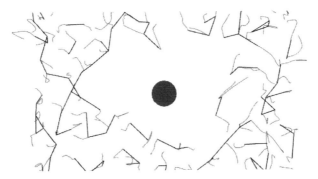

It is particularly interesting to apply features of swarm behavior to structures of sound synthesis. In the application "generative patch", the grouping of swarms influences the signal flow in a network of sine wave oscillators. The illustration below shows schematically one implementation. The constantly changing grouping of swarms in the simulation defines the signal flow in the synthesis patch, whose complexity, like the complexity of the generated sounds, correlates directly with the current state of the swarm.

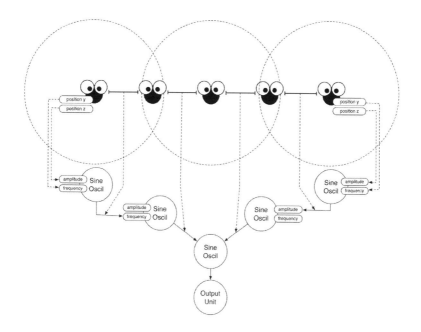

10.3.5 Towards a Topology of Sounds

This chapter deals with the characteristics of musical parameters, their outward representation and their inner representation to the imagination. It reflects my experiences in electroacoustic music, in using mathematical models for sound generation and particularly in using algorithms for composing music. The term "sound" has both a broader and a more specific meaning. In what follows, "sound" generally means all kinds of sounds from noises to strictly organized music. But sounds in a narrower sense, namely sounds of recognizable pitch and characteristic timbre, are also of importance here.

To begin with, we explain some general terms and concepts and discuss their importance for the representation and imagination of sounds and illustrate these ideas using the parameter pitch. Then we introduce the parameters time, space and timbre and discuss mappings between these parameters. This discussion should not be construed as criticism or as evaluation of techniques employed in electroacoustic music, nor should it be understood as a tutorial for working with sounds. The observations made here are meant to be a contribution to the understanding of the fundamental parameters of sound, of their structure and perception. As such, they seek to awaken a sensibility for the complexity of sonic events and of the perception of multimedial art [75].

10.3.5.1 Mental Representation

The mental representation of space is primarily geometric. Accordingly, parameters, functions and structures are imagined and represented spatially. In this process, certain aspects are simplified, left out or added. Spatial representations, whether in print or on a screen, are two-dimensional or virtually three-dimensional. Seemingly linear, homogenous parameters such as time and pitch are much more complicated in structure than geometrical dimensions. Accordingly, circles, tori, grids etc. are sometimes used for their representation. Although in topology (a subdiscipline of mathematics) there are models for spaces with an arbitrary number of dimensions, with various concepts of distance, with the differentiation between discrete and compact sets, and so on, all these structures are so complex and so lacking in any demonstrative quality that they are of no help neither for the artist in the production of his works nor for the recipient of those works. Even if such models can be used to demonstrate characteristics of sound material, they fail completely in the description of higher structures such as motifs and textures in traditional classical music or of sound spectra, sound transformation and the semantic quality of sounds in electroacoustic music. Relations of meaning and reference can at best be visualized as networks and diagrams. However, such methods replace straightforward representation with a symbolic notation.

Space and time are the classical categories of perception. In classical physics they are the constant independent variables of states and changes. Physical space, in order to distinguish it from more abstract spaces, is called *position space*. It is three-dimensional, Euclidian (i.e. not curved), homogenous, continuous and isotropic (i.e. its geometry is valid regardless of direction). Accordingly, the position space in mathematics is described by Cartesian coordinates, i.e. by three linear and continuous axes perpendicular to each other with uniform units of length. Just as the perception of space is primarily visual, imagination, too, is primarily visual. Graphic representations of perceived or imagined things are two-dimensional visualizations. Visualizations of objects and situations can be so realistic that the three-dimensional space seems to appear while we look at them, or they can be so abstract that only experts will be able to understand the reality being visualized. Remembering and anticipating motion are part of imagination. Movements are much more difficult to represent visually than static situations. Realistic representation of motion became possible only with the advent of film, its abstract representation only with computer-generated animations.

We always have a preset image of the world in us. This image is only modified by sensory sensations. Conversely, our knowledge or prejudices influence how we interpret sensory data. All the senses contribute to the modification of our image of the world; their individual functions however are very different from each other. Visual perception most directly complements the geometric properties of what we imagine. Hearing as a warning system quickly provides information on direction, distance and the nature of sounding objects or events and directs our attention and hence our visual perception to important events. But we must not forget the perception of the atmosphere and the general characteristics of a space, like its size and its materials, through acoustic events such as reverberation and reflections, as well as the perception of the immediate environment around us through background noises. According to the functions of our individual senses, the mental representation of their stimuli and the recognition of their objects vary. For the visual registration of spatial conditions, imagination and recognition are equal capabilities. However, with highly differentiated objects like faces the imagination often fails, while recognition works well. Corresponding to its warning function, recognition dominates in audition. Even skilled musicians have to acknowledge that the much talked-about mystical ability to imagine sounds and even entire musical compositions can be practiced only to a very limited extent, whereas familiar sounds are immediately recognizable, even if they have been acoustically or electronically altered. Only a small number of people claim to be able to imagine sounds, colors and odors as intensely and precisely as they can forms and movements.

It is difficult to imagine position space other than Euclidian. Since graphical representations are two-dimensional or virtually three-dimensional, they can only represent functions of one or two variables. We have become so used to two-dimensional reduction that we hardly realize how drastic the simplifications involved are. Most natural and artificial systems process several input values and produce several output values. Graphical representations are based on mathematical functions in which the relation between precisely one output value (function value) and one (in two-dimensional representations) or two (in three-dimensional representations) input values is visualized. If choosing one visualized output value and

thereby suppressing all the other values present in a system is already tendentious and possibly misleading, the reduction to one or two input values mostly leads to senseless diagrams if a causal relation between input and output values is suggested. Furthermore, in complex systems, the input values are themselves often dependent on the output values.
Graphical representations are often more abstract than the trained eye realizes. At first glance, the Cartesian coordinate system seems adequate for the representation of music: time and the frequencies of the oscillation which we perceive as pitches can be considered and recorded as linear, homogenous and independent parameters. While pitches, scales, intervals and harmonies can be unambiguously described by frequencies and frequency relations, they all have much more complex musical properties than is evident in their physical representation (10.3.5.2).

10.3.5.2 Complex Topologies to Represent Pitch Space

Since it is hard to imagine position space in other than Euclidian terms, we represent three-dimensional parameter spaces that way, too. Two-dimensional surfaces, show the multitude of forms and properties inherent in non-Euclidian spaces. Surfaces can be curved, curved back on themselves (such as spheres or tori), limited in one direction and unlimited in the other (like a strip), have only one side (like the Möbius strip), can have holes, consist of one piece or of several pieces joined together etc. Certain musical parameters display complex topological structures and can simultaneously show various aspects having different structures. The octave identity of tones can easily be represented by a circular coordinate axis. At the same time, however, absolute pitch also has musical significance. Absolute pitch and octave identity can be represented together in the so-called *pitch helix* (left figure below). Intervals, on the other hand, are different. Acoustics and history have given them properties that cannot be visualized in a meaningful way, either on a linear or on a circular axis. They have their own individual harmonicity and harmonic function and are defined both by their use in the course of history and by individual listener experience. It is impossible to represent simultaneously all the properties of pitches and intervals within an adequate parameter space. For some aspects of music theory, adequate topologies have been devised, such as the circle of fifths, a Möbius strip as a "harmonic band" for triad relationships [78] or the torus for relationships of fifths and thirds (right figure below).

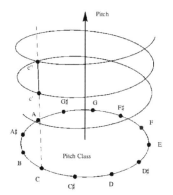
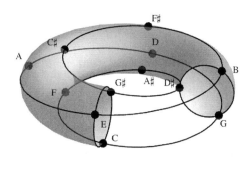

By comparing pitch and visual color, we can demonstrate how differently the perception processes data and how differently this data is represented. Pitch corresponds to the frequency of mechanical oscillation between about 20 and 20,000 Hz. Color corresponds to the frequency of electromagnetic oscillation between 400 and 800 trillion Hz (THz). In both cases, we perceive only part of a scale of oscillation open at both ends. But perception treats the extremes very differently. We generally do not recognize perceptual limits because we cannot imagine what is beyond these limits (we cannot imagine a sound softer than the softest sound we can hear or a lower tone than the lowest tone we can hear). If the frequency of a tone rises while the tone's energy remains constant, we hear the tone becoming softer until it disappears. In addition, the perception distorts the representation of frequency as pitch: with very high and very low tones we perceive intervals between tones as smaller than they are (the more extreme the pitch, the greater the interval distortion). To represent the perceived pitch scale properly would require not a straight line, but rather one either compressed or curved back in space at either end of the scale. We perceive colors very differently, not as linear parameters but rather as forming the chromatic circle. The extremes, that is, the colors at either end of the visible spectrum, join to form a circle. In addition, our perception constructs characteristics of the color scale having no physical analogy, namely the complementary colors and the neutral "color" white. It can be shown that this particular topology makes possible color constancy, a feature of the perception allowing the perceived color of an object to remain constant under changing light conditions. In contrast to tones, which are produced by sounding objects themselves, the sensation of the color of an object has to be reconstructed by the perception from the spectrum of the particular lighting and the reflection of the light off the object.

Some physical parameters serve to determine more than one mental parameter, depending on the range of their physical values. The perception of non-synchronous events is only possible when the time difference between the events is greater than several milliseconds. At the same time, shorter arrival time differences of sound waves from reflections are used to characterize rooms and other acoustic spaces, and time differences on the order of fractions of milliseconds in the arrival times of a wave front at the two ears serve to determine the direction of a sound source. Normally, only large changes of timbre are perceived as such, but the smallest changes of timbre are used to determine the position of a sound source.

The perception and the representation of certain parameters can influence or perturb each other (2.3.4). Very high pitches seem to get higher as they get louder, very low pitches lower (2.3.4.4). The distortion of pitch perception at the extremes of the range of hearing depends on loudness. The fading-out of high and low frequencies in the perception means that in soft sounds the low partials seem weaker than those in the middle range. That is why loud sounds seem more strongly colored than soft sounds with the same spectrum. On the other hand, the same parameters of different sounds can influence each other. Loud tones can mask soft tones sounding simultaneously or shortly before or after (2.3.4.2). A tone whose frequency is a multiple of a second, simultaneously sounding tone blends with the second tone and loses its independence (5.2.1.3). We never hear the individual partials of natural sounds in a natural environment because they are produced at the same time and place. But we can often hear individual partials of synthetic sounds in virtual spaces, and if there are several simultaneous sounds, it is not always clear to which sound a partial belongs.

But also higher level properties of sounds can be perceived differently depending on context. A certain piano note can be heard either as g-sharp or as a-flat, and even as correspondingly higher or lower, depending on the harmonic context. For many perceptive tasks, the ear evaluates several features of a sound. It is often difficult to determine the effect of individual features. The more complex the perceptive or cognitive task, the more acoustic stimuli are evaluated, the greater the influence of the other senses becomes and previous knowledge takes on greater significance. Many studies show how visual stimuli can influence auditive judgment, and vice-versa. Here one speaks of *cross-modal effects*.

10.3.5.3 Time

In classical physics, time is the evenly flowing independent function variable. In our everyday perception, but especially in the aesthetic perception, time is extremely complexly structured. The fundamental difference to the spatial coordinates – that time passes and cannot be stopped or reversed – cannot be represented spatially. Time as experience passes irregularly. This is already so in the present, but especially time remembered is rather a succession of moments than a continuum. The temporal sequence of independent events is of little importance in our recollection and can often only be reconstructed with the help of abstractions like dates. Although the future is undetermined, its mental anticipation, planning and hopes for it, premonitions about it, etc. all play a vital role in our perception. In light of the complex and different structures of the time and pitch dimensions described above, it becomes clear why the translation of two-dimensional geometric forms into musical scores so rarely works. Fascinating fractal structures, like the Mandelbrot Set (10.3.3.3), often give rise to such experiments. Geometric structures are essentially based on the premise that the two-dimensional coordinates are similar and equivalent. Complex self-similar structures are only recognizable when all elements of an image can be compared at all times, which is not possible in the time dimension. We described above how much more complex the structure of pitch perception is than that of a geometric dimension.

A sequence of images that succeed each other at intervals shorter than about 20 ms are perceived as continuous movement. Similarly, rapid sequences of (nearly similar) acoustic events are perceived as continuous sound. The difference between the perception of separate events and that of sounds is so fundamental that the projection of the proportions between overtones (at the micro level) onto rhythms or tempo changes (as described by Karlheinz Stockhausen in his text "Wie die Zeit vergeht" and practiced for example in his composition "Zeitmaße" are just as abstract and speculative as mapping between categorically different parameters. The shift from continuous experiencing to the perception of individual events can be precisely described and measured for different modes of perception. Much more difficult to measure is the difference between what is perceived as the present or as a moment in the past or future and what is segmented into a series of events. The so-called *specious present*, that is, the interval during which one's perceptions are judged to be in the present, can last up to 3 seconds. The "unit in regard to content" of the present is the word, the gesture, the action, the affective reaction, etc. To split the perception into different modes for time spans of varied length is apparently useful in nature and was adopted in language and music. Traditional music shows similarities to language not only because it developed with language, but also because it has the same preconditions and because playing instruments takes place in temporal dimensions similar to those of language: beats, bow strokes and the movements of beating time are short, simple actions. To this day, articulation and phrasing in singing and speaking serve as the model for instrumental phrasing. This is where the rhetorical element in contemporary instrumental music, which often feels somewhat baroque, stems from. In electroacoustic music, however, the boundaries of perceptual modes can be made perceptible and can even be transcended. Transient phenomena can be enlarged into gestures, gestures can be accelerated to just below the level of individual perception. Figures can be enlarged at will, because they are not generated through actions. Electronic music can "express" movements and states that have nothing to do with gestures and language and that unfold within completely different time spans. Electroacoustic music can thus avoid the need to be interpreted, that is, to mean something.

It would seem obvious to use new media to replace traditional musical notation by animations in which musical time would run in real time and in which two or three spatial dimensions would represent additional parameters. That such a course of action would seem very alien to the musician points to further aspects of the inner representation of music. In many aspects, musical notation resembles a choreography or a screenplay. These all have in common that they do not record or depict every detail of the aesthetic object, but rather they represent that object in the clearest and most straightforward manner possible using symbols. In music practice as well as in analysis, passages that are temporally remote from one another are constantly being compared, one skims through scores forwards and backwards and forwards again. To understand music means to understand the significance of a moment in the context of a whole piece.

10.3.5.4 Timbre

Timbre is a particularly important parameter in electroacoustic music. In the narrower sense, the term *timbre* refers to the character of sounds with distinguishable pitch. Physically, timbre can be described with the sound spectrum. However, correlation between spectrum and perceived sound quality are difficult to determine. The structure of the physical spectrum, theoretically made up of an infinite number of partial tones, creates a sound space of infinitely many dimensions. However, if one takes into account that only partials not falling into the same critical band (2.3.4.1) can be distinguished, the number of dimensions is reduced to about 24. Several attempts have been made to describe sound space. For the Austrian composer and theorist Josef Mathias Hauer (1883–1959), the interval between the fundamental note and the strongest overtone was decisive for timbre [77]. With octave reduction, this leads to a one-dimensional, cyclic representation of timbres. By taking into account the phases of the partials, other authors derived doubly cyclical timbral structures that can be visualized on a torus [78]. But all these approaches reveal little about the structure of the perceived sound quality. Timbre as described above is not very interesting, neither for ordinary nor for aesthetic listening. It turns out that it is the changes in sounds that bear information. The physical properties of variable spectra can be visualized three-dimensionally with amplitude as a function of time and frequency (2.3.2.3). With some practice, much can be read from such visualizations with respect to the effects of the represented sounds. On the other hand, nobody will be able to imagine sounds in that space. More plausible results can be obtained by multidimensional scaling, a statistical technique where similarity judgments of stimulus pairs allow one to calculate the ideal number of dimensions for representing the differences perceived in the stimuli. For timbre, this method generally yields dimension numbers smaller than five. The psychologist John Grey has done experiments to measure the timbre of musical instruments in terms of aesthetic quality. He has found three relevant dimensions which he names *spectral centroid, flux* and *attack time* [3-15].

These considerations, however, have little to do with the "real" perception of timbre. In a natural listening situation, we deduce from timbre what object is producing the sound. Only when in doubt do we hear sounds as sounds. For example, we do not say "I hear a humming sound", but "I hear a car". Here, too, we see that sounds do not form a homogenous space but primarily point to objects or actions. Only in an aesthetic context do they have a life of their own. Whether properties of sounds can be imagined depends on whether we can produce them ourselves and whether our actions directly influence these properties. For many people, imagining music is limited to the silent inner singing of melodies or to beating rhythms. The difficulties of classifying sounds are evidenced not only by the fact that they cannot just be placed in a sound space, but also by the fact that there are no genuine adjectives for them in everyday language. Adjectives from other perceptual domains are used, like hard/soft, light/dark, raw/fine, etc. Neither in music theory nor in disciplines involving sound processing is there a developed and generally recognized sound typology. In sound synthesis, the problem becomes apparent in that the adjustable parameters often have no intuitively obvious correlate in the sound. In additive synthesis, for example, the amplitudes of the partial tones have to be defined individually. At the other extreme, a simple increase in the so-called *index* in nonlinear techniques like frequency and amplitude modulation leads to an altogether unspecific intensifying of timbre. The outlined difficulties show that there can be no straightforward mapping of structured data onto timbre.

10.3.5.5 Position Space

The space in which music is played has always influenced how that music is performed. Instrumentation, tempo and dynamics are, unconsciously or systematically, adapted to the acoustics of the performance space. As a parameter in composition or performance, space has been used in responsorial chants, in introits and recessionals, in the use of multiple choirs and especially in the opera since 1600. Only in electroacoustic music, however, has it become possible to create virtual spaces. There are fundamental differences between virtual visual and virtual acoustic spaces. They are partly technical and partly conditioned by the differences between visual and acoustical perception. Our hearing first perceives what is sounding where. Although visually we perceive everything from one point and always look in one direction, everything we see is interpreted as being in Euclidian space. Things we only hear and cannot associate with known objects in space are normally imagined in spherical coordinates: "the sound is coming from above", "the sound is coming from far away". We look in one direction and can close our eyes at any time, whereas hearing is immersive: we hear sounds from all directions and cannot close our ears. We also hear sounds whose sources are optically hidden, since sound waves are inflected and reflected (we do not see light, but rather objects that reflect light, but when sound is reflected we hear its source and not the objects reflecting it). We get a visual impression of the environment instantly, but aural perception always takes some time. In our modern world, movements are so omnipresent that their visual perception only subliminally causes stress. Moving sounds however, real or virtual, are always perceived in the foreground. There is only one position space. Objects that penetrate each other can be graphically represented, but they are always in the same space. Different virtual acoustic spaces, on the other hand, can coexist. We can simultaneously perceive music with the acoustics of a cathedral and speech with the acoustics of a living room.

Whether the simulation of a sound space works depends mainly on the plausibility of the simulation and on the ratios of the various acoustic properties. For example, we are so used to adjusting the volume of a device like a radio that a fading voice is interpreted as a technical manipulation rather than as a speaker who is moving away. Since loudness, timbre and reverberation are not only used to determine the position of a sound but can also be independent aesthetic parameters, they can be used to create interesting ambivalences, transitions from perceived musical properties to subliminally perceived sound properties, as well as surreal scenarios. For example, if one plays exactly the same signal on both channels of a headphone or on speakers placed symmetrically around the listener, the sound is not heard to be coming from the outside space, but is localized in the head. Even so, the impression of distance can be simulated. If a sound source is simulated to sound distant by adding reverberation and darkening the timbre, the loudness can be treated separately so that for example unnaturally loud sounds in an otherwise seemingly realistic sound scenery appear to be electronically amplified. Since the distance cues (volume, absorption and reverberation) are independent of one another, they can be represented in a three-dimensional parameter space. Combinations of parameters that create a consistent spatial feeling form a subspace within the parameter space. If combinations of parameter values not belonging to this space are used, the spatial impression can collapse or effects like the ones described above can occur.

The acoustic space is immersive. It is always there and surrounds us. Therefore it is an important aspect of the atmosphere of both natural and virtual environments. The reflected parts of the sound, perceived partly consciously as echo or reverberation and partly on a subliminal level, provide us with information about the size and quality of the space around us. The sounds themselves provide a general impression of actions and things present in the space. Although fully determined by the space, the atmosphere is marked mainly by sound properties that cannot be localized. It cannot be described with spatial parameters or other measurable properties. Therefore, an atmosphere cannot be imagined; at best, a situation evoking a certain atmosphere can be imagined. The power of atmosphere created through sound is well known to the producers of shopping mall music, so-called *Muzak*, and is used to manipulate consumers and clients. The purposeful creation of an acoustic atmosphere can, however, also be an important or even exclusive aim in sound installations. The natural immersion of the acoustic space can be suspended artificially. For example, if sounds that in a natural environment could not be coming from the same direction or reverberation that normally comes from all directions are played over a single speaker, the sense of immersion breaks down.

10.3.5.6 Mapping

Data sets occuring in natural sciences, economics, etc., can be visualized using suitable graphic tools. This visualization is called mapping. Mappings use both continuous parameters like spatial coordinates, colors or levels of gray and discrete or symbolic objects like arrows, geometric symbols, etc. The great advantage of spatial coordinates is that three parameters can be represented in a unified space, thus making the relations between the parameters evident. The easiest way to expand the mapping space is to use gray scales. Like spatial coordinates, gray scales are continuous and linear. In a sonogram, for example, a sound spectrum recorded with two spatial coordinates for time and frequency and with a gray scale (or color intensity) for the amplitude. The use of color as a further parameter, however, is problematic because of the circular topology of color described above. The mapping of time to a spatial coordinate is familiar from functions and diagrams. We only realize how abstract this representation is in a diagram where time does not run horizontally from left to right. Conversely, it is possible to map a spatial coordinate to time. To show the complex structure of an organ of the body, for example, an animation can show cross sections of the organ in two dimensions, mapping the third dimension to time. One experiences such visualizations as a journey through the body.

Although there are as many parameters available for the sonification of data as for its visualization (spatial coordinates time, volume, timbre), the sonic representation of data is hardly ever used. On the one hand, this has to do with the general dominance of the visual in our culture, but it is also a consequence of the complexity of sound parameters discussed above and particularly of the fact that sounds require time. For several years there have been attempts to supplement data visualizations with sonfications, since the quantities and structures of the data to be represented can overtax our visual perception and sounds sometimes fit the data more precisely. In other cases, the ear as a warning system might trigger a quicker reaction than the eye (for example in cars). We use the term *audification* to denote the playing of data as sound. In this case, the data is subjected only to basic transformation such as transposition or filtering. *Auditory display*, on the other hand, also includes the use of *auditory icons*, acoustic symbols that are mainly used as feedback in technical equipment from Geiger counters to computers. The term *earcon* was coined in order to denominate extensions of auditory icons that also provide information on sizes and states.

There are more or less convincing sonifications. Physical and psychoacoustic phenomena, as well as habits play an important role. In terms of psychoacoustics, high pitches are generally localized higher within a space, and all music notation systems indicate pitch on a vertical scale. That higher pitches are associated with smaller objects has to do with the fact that lower pitches can only be effectively amplified and diffused by large resonating bodies.

The difficulty of making data, or more precisely, the information contained in the data and their structure audible through simple mapping has been touched upon above. However, as the following example illustrates, structures that cannot be heard in natural sounds in their natural environment can be mapped in such a way that they become audible. In additive sound synthesis sounds are synthesized by adding sine waves. Normally, the sine waves blend into one sound. In principle, the sound spectrum has infinitely many dimensions, however, timbre is described mostly by less than five parameters (10.3.5.4). If the individual partial tones are assigned virtual sound sources in a space, the blending effect disappears. Although there are only three dimensions in space, there can be any number of independent spatial movements forming a parameter space with infinitely many dimensions to which spectra can be mapped.

Besides one-to-one mapping, where the transformation of data parameters into sound parameters is pre-defined, many extensions are possible. Instead of using raw data like the position of a listener, higher order data can be generated, like gestures or the intensity of movements, and this higher order data in turn can be mapped, not to basic sound parameters, but to more complex structures of sound generation like the intensity of sound excitation. Another possibility is to map the data directly but to give the listener the opportunity to learn how to interpret the data through interaction with it.

10.3.5.7 Systems

The properties of sound discussed thus far are parameters that take on scalar values and therefore can be represented in appropriate coordinate systems. These parameters describe the basic material used in sonic art and computer music. The higher up we move within the hierarchy of compositions or sound-generating systems, the less structures and their interrelations can be represented in parameter spaces in an obvious way. Relationships, influences, cross references between elements within systems can still be represented graphically, not in the representation spaces discussed above, but in symbolic notation like flowcharts. Several disciplines have developed terminology to describe such relationships. The terminology always deals with complex entities, systems with properties and with a certain behavior or certain interactions with other systems. Examples for this are cybernetics, various system theories, neural networks, agent-based systems and object-oriented programming. It may astonish the reader that none of these disciplines is much older than half a century. The concept according to which complex structures are composed of many individual, partly identical and often very simple elements is very general and in this form hardly provides much insight. Only the capability of computers to simulate the behavior of such systems has made it necessary to formalize fundamental concepts. Furthermore, only simulations have made it possible to study typical behaviors of complex systems like self-organization and emergence. Unlike the above-mentioned geometrical, topological concepts developed in mathematics and physics, these ideas were developed by disciplines that deal with life and machines: cybernetics comes from biology and technological research, system theories were developed in sociology and engineering sciences and theories of neural networks and agent-based systems had their beginning in artificial intelligence and artificial life studies.

We may not be able to imagine systems in spatial terms, but as social beings we do have a rich experience of interaction with others and thus a good intuition with respect to the behavior of individuals who influence each other and exchange information. The behavior of people in groups such as families, school classes or teams, but also in masses, as in traffic or in an audience, is familiar to us. We know intimately the mechanisms involved. Phenomena like the synchronization of animal behavior or the appearance of geometric structures in chemical reactions are fascinating because there seem to be creative, regulating forces at work. We may not be able to imagine systems, and they cannot be mapped to simple parameter spaces, but we are able to imagine our own actions. The imagination of one's movements in space, the anticipation of motion and situations, the mental testing of behavior and verbal utterance and the anticipation of other people's actions are highly developed, typically human capabilities. Just as humans and animals playfully test and train certain mental and physical capabilities in a delimited and protected setting, so art challenges and develops our perception and those mental processes like gestalt recognition, memory and recognition of commonalities and differences. In more recent art forms like interactive installations, these capabilities help us to understand processes, to influence them in a meaningful way and to communicate with both artificial systems like computer programs and human beings.

Appendix A Fundamentals of Mathematics

The following compilation of basic mathematical operations and concepts is limited to those areas of mathematics necessary for understanding this book. These topics are introduced in the corresponding chapters, not in the Appendix: complex numbers (3.1.2), the Fourier transform (3.1.1), the z-transform (3.2.3), differential equations (3.4.1), the calculus of probability (10.1) and statistics (10.2).

Further Reading: The specialized mathematical literature is not very accessible to non-mathematicians. Books for engineers and scientists are a better choice: Mathematics for Engineers and Scientists *by Alan Jeffrey [89],* Handbook of Mathematics for Engineers and Scientists *by Andrei D. Polyanin and Alexander V. Manzhirov [90] or* Schaum's Outline of Advanced Mathematics for Engineers and Scientists *by Murray R. Spiegel [1].*

A.1 Numbers and Arithmetic Operations

A.1.1 Numbers

In the set of the *natural numbers* $\mathbb{N} = \{1, 2, 3, 4, ...\}$, only addition and multiplication are possible without restriction. The set of *integers* $\mathbb{Z} = \{..., -2, -1, 0, 1, 2, ...\}$ allows unrestricted subtraction.

Numbers like $2/3$, $-1/4$, 0.25 and $1.111...$ belong to the set of *rational numbers* $\mathbb{Q}$. A rational number is any number that can be expressed as the quotient of two integers. The decimal representation of a rational number either terminates after a finite number of digits ($1/4 = 0.25$) or repeats a finite sequence of digits ($3/7 = .42857142857142...$). All four arithmetical operations are possible on rational numbers without restriction, except division by zero.

Numbers that cannot be represented as the quotient of two integers are *irrational numbers*, for example $\sqrt{2}$ and π. The decimal representations of irrational numbers are of infinite length and contain no periodic repeating sequences.

The set of *real numbers* $\mathbb{R}$ fills the entire number range and is usually represented as a straight line. All four arithmetical operations can be carried out on real numbers without restriction, except division by zero.

For complex numbers, of which real numbers are a subset, see Chapter 3.1.2.

A.1.2 Rules of Algebra

If a, b and c are real numbers, the following identities hold:

$a + b = b + a$	commutative property of addition
$ab = ba$	commutative property of multiplication
$a + (b + c) = (a + b) + c$	associative property of addition
$a(bc) = (ab)c$	associative property of multiplication
$a(b + c) = ab + ac$	property of distributivity

The commutative property means that changing the order of the elements in addition and multiplication does not change the result. Below are some examples of the distributive law.

Expansion (distributivity from left to right):

$$2(4 + 7) = 2 \cdot 4 + 2 \cdot 7 = 8 + 14 = 22$$
$$a(a^3 + a^4) = a^4 + a^5$$

Factoring (distributivity from right to left):

$$abc + bcd = bc(a + d)$$
$$a^4 + a^5 = a^4(1 + a)$$

The following rules hold when using fractions:

$$\frac{a}{b} \cdot \frac{c}{d} = \frac{ac}{bd}$$

$$\frac{a/b}{c/d} = \frac{a}{b} \cdot \frac{d}{c} \qquad \text{Example: } \frac{2/3}{3/4} = \frac{2 \cdot 4}{3 \cdot 3} = \frac{8}{9}$$

$$\frac{a}{b} + \frac{c}{d} = \frac{ad + cb}{bd} \qquad \text{Example: } \frac{2}{3} + \frac{3}{4} = \frac{2 \cdot 4 + 3 \cdot 3}{12} = \frac{17}{12}$$

The following rules hold when working with powers:

$$a^m a^n = a^{m+n} \qquad \text{Example: } a^5 a^4 = a^9$$

$$\frac{a^m}{a^n} = a^{m-n} \qquad \text{Example: } \frac{a^5}{a^4} = a$$

A.2 Statements, Sets and Operations on Sets

A.2.1 Statements

A *statement* in logic is a declarative sentence that is either true or false. "Good Day" is not a statement. "Seven is a prime number" or "$1 = 1$" are true statements, "$1 > 3$" is a false statement. The *truth value* (or logical value) of a true statement is written as T, the truth value of a false statement as F. In order to calculate with truth values, F is considered to be equal to zero and T equal to one in Boolean algebra. If a sentence like "x is a prime number" is made a statement by giving x a value, one calls the sentence a *proposition*.

Propositions and statements can be associated with each other using words like "and", "or", etc. The symbols used to express such association are called *logical connectives* or *logical operators*. $\neg A$ means "not A", $A \wedge B$ means "A and B", $A \vee B$ means "A or B" (not exclusive), $A \Rightarrow B$ "if A, then B", and $A \Leftrightarrow B$ "A if and only if B". Below is a truth table for logical connections between two statements A and B, given their logical values.

A	B	$\neg A$	$A \wedge B$	$A \vee B$	$A \Rightarrow B$	$A \Leftrightarrow B$
T	T	F	T	T	T	T
T	F	F	F	T	F	F
F	T	T	F	T	T	F
F	F	T	F	F	T	T

In addition, so-called *quantifiers* are used in the formalization of mathematical theories:

$\bigwedge_{x}$ means "for any x", $\bigvee_{x}$ means "there exists at least one value of x such that".

A.2.2 Sets

A *set* is a collection of distinct *elements* or *members*. If we use the variables $a, b, \ldots$ for the elements and $A, B, \ldots$ for the sets, $a \in A$ means that a is an element of set A, $b \notin X$ that b is not an element of set X. A finite set can be specified by listing all its elements, for example, A = the set of the decimal digits = $\{0, 1, 2, 3, 4, 5, 6, 7, 8, 9\}$. An arbitrary set can be specified by defining the characteristics that elements must exhibit to be included. If $P(x)$ stands for "x is a prime number", then we can specify the set of prime numbers less than 100 as $P_{100} = \{x \mid P(x) \wedge x < 100\}$. A set with no elements is called the *empty set* and written as $\{\}$ or $\emptyset$.

A.2.3 Subsets and Power Set

If all the elements of set A are also elements of set B, then set A is a *subset* of set B. One writes $A \subseteq B$ when A and B can be the same, and $A \subset B$ (A is a *strict subset* of B) when B contains elements not in A. The elements of a set can be sets themselves. A set whose elements are sets is called a *family of sets*. A special case of a family of sets is the power set $P(A)$. It contains all subsets of the set A:

$$P(A) := \{X \mid X \subseteq A\}.$$

A.2.4 Operations on Sets

The *relative complement* of B in A contains all elements of A which do not belong to B:

$$A \setminus B := \{x \mid x \in A \wedge x \notin B\}.$$

If $A \subseteq G$, $A \setminus G$ is the complement of A with respect to G and can be written $\overline{A}$. The *intersection* of two sets consists of those elements that belong to both sets:

$$A \cap B := \{x \mid x \in A \wedge x \in B\}.$$

The *union* of two sets consists of those elements belonging to at least one of the sets:

$$A \cup B := \{x \mid x \in A \vee x \in B\}.$$

A.2.5 The Cartesian Product

The *Cartesian product* of two sets consists of all ordered pairs of elements (a, b) where a is taken from A and b from B:

$$A \times B := \{(a, b) \mid a \in A, b \in B\}.$$

$$\{0, 1\} \times \{0, 1, \ldots\} = \{(0, 0), (1, 0), (0, 1), (1, 1), (0, 2), (1, 2), \ldots\}$$

$\mathbb{Z} \times \mathbb{Z}$ gives the coordinates of all grid points (points having integer coordinates) of a plane. $\mathbb{R}^3 = \mathbb{R} \times \mathbb{R} \times \mathbb{R} = \{(x, y, z) \mid x \in \mathbb{R}, y \in \mathbb{R}, z \in \mathbb{R}\}$ gives the coordinates of all points in 3D space.

A.3 Equations

A.3.1 Definitions and Concepts

Terms are expressions consisting of numbers, constants, variables and operators. For example: $\pi + 1$, x^2, e^x, 6.37, etc. If we declare two terms T_1 and T_2 equal, we have an *equation*. If T_1 and T_2 contain no variables, then the expression $T_1 = T_2$ is a *statement* of equality that is either true or false. If T_1 and T_2 contain variables, the expression $T_1 = T_2$ is a *proposition* which only becomes true or false when the variables are replaced by numbers. For example, the proposition $x + 1 = 4$ becomes a true statement when x is made equal to 3. The set of numbers which transform an equation into a true statement is called its *solution set*. Two equations are called *equivalent* when their solution sets are the same. In an equation in several variables, it is possible to interpret some of the variables as equation variables and some of the variables as *parameters*. One can then solve the equation for the equation variables as a function of the parameters. For example, $x = p + 1$ is both a proposition with the variables x and p and a solution of the equation $x - p = 1$ for x with the parameter p.

A.3.2 Equivalence Transformations

An equation can be solved by transforming it into progressively simpler equivalent equations. The following rules hold:

$T_1 = T_2$ is equivalent to $T_1' = T_2'$ if T_1 is equal to T_1' and T_2 is equal to T_2'. This means in particular that the terms of both sides of an equation can be simplified independently.

$T_1 = T_2$ is equivalent to $T_2 = T_1$, that is, the terms of both sides of an equation can be exchanged.

$T_1 = T_2$ is equivalent to $T_1 + T_3 = T_2 + T_3$ and to $T_1 - T_3 = T_2 - T_3$. This means that the same term can be added to or subtracted from both sides of an equation.

$T_1 = T_2$ is equivalent to $T_1 T_3 = T_2 T_3$ and to $T_1 / T_3 = T_2 / T_3$ (if $T_3 \neq 0$). This means that both sides of an equation can be multiplied or divided by the same term.

A.3.3 Algebraic Equations

Algebraic equations can be written in the form

$$A_n x^n + A_{n-1} x^{n-1} + \ldots + A_1 x + A_0 = 0.$$

The A_i are called *coefficients*, n is the *degree* of the equation. For example, $x^2 + 3x - 1 = 0$ is a second degree algebraic equation with the coefficients 1, 3 and -1, $\sin(p)x^3 - p = 0$ is a third degree algebraic equation with the parameter p, and $4/x = (2x - 1)/(5 - x)$ is an algebraic equation because it is equivalent to $2x^2 + 3x - 20 = 0$ ($x \neq 0 \wedge x \neq 5$). Dividing an algebraic equation by the leading coefficient A_n yields the *normalized form*

$$x^n + a_{n-1} x^{n-1} + \ldots + a_0 = 0.$$

Algebraic equations can be solved by closed formulas only up to the fourth degree. The solution of the quadratic equation $x^2 + px + q = 0$ is

$$x_{1,2} = -p/2 \pm \sqrt{(p/2)^2 - q}.$$

If the left side of a normalized algebraic equation can be factored into so-called *linear factors*

$$x^n + a_{n-1}x^{n-1} + \ldots + a_0 = (x - \alpha_1)(x - \alpha_2)\ldots(x - \alpha_n),$$

then the α_i are the solutions of the equation $x^n + a_{n-1}x^{n-1} + \ldots + a_n = 0$. The fundamental theorem of algebra says that every algebraic equation of nth degree has maximal n real solutions (A4.3.3).

If m simultaneous equations in n variables are given, we speak of a *system of equations*. In general, the values of the variables are only defined when $n = m$. For example, the non-algebraic system of equations $\sin(x)y = -\pi/2 \wedge x + y = 0$ has the solution $x = -\pi/2, y = \pi/2$; the system of equations $x^2 + y^2 = 1 \wedge x + y = 1$ has the solutions $x = 1, y = 0$ and $x = 0, y = 1$.

A.3.4 Systems of Linear Equations and Matrices

If all the equations of a system of equations are linear, we speak of a *system of linear equations*. We obtain the solutions of a system of linear equations by the successive elimination of variables. Given the following system of equations:

$$\begin{array}{rcrcl} x_1 & + & 2x_2 & = & 0 \\ 3x_1 & - & 4x_2 & = & 2 \end{array}$$

If we subtract the first equation three times from the second, we get $-10x_2 = 2$, hence $x_2 = -1/5$. If we put this value back into one of the equations, we obtain $x_1 = 2/5$.

Using matrices and column vectors, linear systems of equations can be written as

$$\begin{pmatrix} 1 & 2 \\ 3 & -4 \end{pmatrix} \begin{pmatrix} x_1 \\ x_2 \end{pmatrix} = \begin{pmatrix} 0 \\ 2 \end{pmatrix} \quad \text{or simply as } \mathbf{Ax} = \mathbf{b}.$$

$$\mathbf{A} = \begin{pmatrix} a_{11} & a_{12} \\ a_{21} & a_{22} \end{pmatrix} = \begin{pmatrix} 1 & 2 \\ 3 & -4 \end{pmatrix} \text{ is the coefficient matrix.}$$

$$\mathbf{x} = \begin{pmatrix} x_1 \\ x_2 \end{pmatrix} \text{ and } \mathbf{b} = \begin{pmatrix} b_1 \\ b_2 \end{pmatrix} = \begin{pmatrix} 0 \\ 2 \end{pmatrix} \text{ are column vectors.}$$

Using indices, we can write the matrix $\mathbf{A}$ as (a_{ij}) and the vector $\mathbf{x}$ as (x_j).

The product of a matrix $\mathbf{A}$ and a vector $\mathbf{x}$ is a vector whose nth element is $b_n = \sum a_{nk} x_k$. Generally, one defines the product of a matrix $\mathbf{A} = (a_{jk})$ with m rows and n columns and a matrix $\mathbf{B} = (b_{jk})$ with n rows and p columns as the matrix $\mathbf{AB} = \mathbf{C} = (c_{jk})$ with m rows and p columns.

$$c_{jk} = \sum_{l=1}^{n} a_{jl} b_{lk}.$$

$$\texttt{MatrixForm}\left[\begin{pmatrix} a_{11} & a_{12} & a_{13} \\ a_{21} & a_{22} & a_{23} \end{pmatrix} \cdot \begin{pmatrix} b_{11} & b_{12} \\ b_{21} & b_{22} \\ b_{31} & b_{32} \end{pmatrix}\right]$$

$$\begin{pmatrix} a_{11}b_{11} + a_{12}b_{21} + a_{13}b_{31} & a_{11}b_{12} + a_{12}b_{22} + a_{13}b_{32} \\ a_{21}b_{11} + a_{22}b_{21} + a_{23}b_{31} & a_{21}b_{12} + a_{22}b_{22} + a_{23}b_{32} \end{pmatrix}$$

`LinearSolve`$\left[\begin{pmatrix} 1 & 2 \\ 3 & -4 \end{pmatrix}, \begin{pmatrix} 0 \\ 2 \end{pmatrix}\right]$ $\{\{\frac{2}{5}\}, \{-\frac{1}{5}\}\}$

One can solve the system of equations above using Mathematica by passing the matrix A and the vector b to the function LinearSolve[ ].

Linear mapping can be done using matrices. For example, we can rotate the vector x by the angle ϕ (in anticlockwise direction) by multiplying the vector and the rotational matrix

$$\begin{pmatrix} \cos(\phi) & -\sin(\phi) \\ \sin(\phi) & \cos(\phi) \end{pmatrix}.$$

$\phi = \pi/4;$ `x =` $\begin{pmatrix} 1 \\ 0 \end{pmatrix}$; `rot =` $\begin{pmatrix} \text{Cos}[\phi] & -\text{Sin}[\phi] \\ \text{Sin}[\phi] & \text{Cos}[\phi] \end{pmatrix}$; `rot.x` $\{\{\frac{1}{\sqrt{2}}\}, \{-\frac{1}{\sqrt{2}}\}\}$

Let $A = (a_{jk})$ be an $n \times n$ matrix. Many problems lead to systems of equations of the form $Ax = \lambda x$.

$$Ax = \lambda x = \begin{pmatrix} a_{11} & a_{12} & .. & a_{1n} \\ a_{21} & a_{22} & .. & a_{2n} \\ .. & .. & .. & .. \\ a_{n1} & a_{n2} & .. & a_{nn} \end{pmatrix} \begin{pmatrix} x_1 \\ x_2 \\ .. \\ x_n \end{pmatrix} = \lambda \begin{pmatrix} x_1 \\ x_2 \\ .. \\ x_n \end{pmatrix} \quad \text{or}$$

$$\begin{array}{lllll} (a_{11} - \lambda)x_1 & + & a_{12}x_2 & ... & a_{1n}x_n & = 0 \\ a_{21}x_1 & + & (a_{22} - \lambda)x_2 & ... & a_{2n}x_n & = 0 \\ ... & & ... & ... & & = 0 \\ a_{n1}x_1 & + & a_{n2}x_2 & ... & (a_{nn} - \lambda)x_n & = 0 \end{array}$$

If for certain values λ_i of λ this system of equations has non trivial solutions, the λ_i are called *eigenvalues* and the corresponding x_i are called *eigenvectors*. With every eigenvector x_i any multiple αx_i is also an eigenvector.

The eigenvalues of the matrix $\begin{pmatrix} 1 & 2 \\ 3 & -4 \end{pmatrix}$ are -5 and 2, with corresponding eigenvectors $\begin{pmatrix} -1 \\ 3 \end{pmatrix}$ and $\begin{pmatrix} 2 \\ 1 \end{pmatrix}$, because $-5 \cdot \begin{pmatrix} -1 \\ 3 \end{pmatrix} = \begin{pmatrix} 1 & 2 \\ 3 & -4 \end{pmatrix} \begin{pmatrix} -1 \\ 3 \end{pmatrix} = \begin{pmatrix} 5 \\ -15 \end{pmatrix}$ and $2 \cdot \begin{pmatrix} 2 \\ 1 \end{pmatrix} = \begin{pmatrix} 1 & 2 \\ 3 & -4 \end{pmatrix} \begin{pmatrix} 2 \\ 1 \end{pmatrix} = \begin{pmatrix} 4 \\ 2 \end{pmatrix}$.

`Eigenvalues`$\left[\begin{pmatrix} 1 & 2 \\ 3 & -4 \end{pmatrix}\right]$ $\{-5, 2\}$

`Eigenvectors`$\left[\begin{pmatrix} 1 & 2 \\ 3 & -4 \end{pmatrix}\right]$ $\{\{-1, 3\}, \{2, 1\}\}$

A.3.5 Transcendental Equations

Equations in which an equation variable appears as the argument of a transcendental function like exp(), log(), sin(), cos(), etc. are called *transcendental equations*. Transcendental equations often can only be solved by approximation procedures. In special cases they can be transformed into algebraic equations. Exponential equations in which the variable only appears in exponents of terms that are not summed can be solved using logarithms. For example,

$$2^x = 3 \rightarrow x \cdot \log(2) = \log(3) \rightarrow x = \log(3)/\log(2).$$

Exponential equations in which the variable only appears in exponents of terms whose bases are powers of the same number can be solved by substitution. For example: In

$$2^x = 4^{x-1} + 8^{x-2}$$

substitute z for 2^x giving

$$z = z^2/4 + z^3/64 \rightarrow z_1 = 0, z_2 = -1 - \sqrt{2}, z_3 = -1 + \sqrt{2}.$$

From this follow the solutions $x_i = \log(z_i)/\log(2)$.

Logarithmic equations in which the variable only appears in the argument of the logarithm and in the same term $T(x)$ can be solved by substitution. For example: In

$$\log(2x) - 1 = \sqrt{\log(2x)},$$

substitute z for $\sqrt{\log(2x)}$, yielding $z^2 - 1 = z$ with the solutions

$$z = (1 \pm \sqrt{5})/2.$$

From this, taking $\log(2x) = z^2$, that is $x = \exp(z^2)/2$, follow the solutions of the equation. Substituting these values back into the equation shows that only $z = (1 + \sqrt{5})/2$ is a valid solution.

Trigonometric equations in which the variable x or nx only appears in the argument of trigonometric functions can be represented using a single trigonometric function. For example:

$$\sin(x) + \cos^2(x) = .5 \Rightarrow \sin(x) + 1 - \sin^2(x) = .5 \Rightarrow \sin^2(x) - \sin(x) - .5 = 0$$
$$\Rightarrow \sin(x) = (1 \pm \sqrt{3})/2.$$

Since $(1 + \sqrt{3})/2$ is greater than 1, only $\sin(x) = (1 - \sqrt{3})/2$ give real solutions, namely

$$x = -.375 \pm n\pi \text{ and } -2.776 \pm n\pi.$$

A.4 Functions

A.4.1 Definition

A *function f* is a rule that associates to every element x of a set A an element y of a set B. This association is also spoken of as mapping the set A into the set B. A is referred to as the *domain* of the function, B as its value domain or *range*. One writes: $f: x \rightarrow y$ or $f: x \rightarrow f(x)$ or $y = f(x)$, where $f(x)$ indicates the value of the function at the point x. x is the *argument* or the *independent variable*, y is the *dependent variable* or the *function value*. A function can be considered a set of ordered pairs (x, y) where each $x \in A$ occurs at most once. For example: $A = \{1, 2, 3, ...\}$, $B = \{0, 1\}$. Let the function f associate all even numbers with 0, and all odd numbers with 1: $f(1) = f(3) = ... = 1$, $f(2) = f(4) = ... = 0$: $f = \{(1, 1), (2, 0), (3, 1), (4, 0), ...\}$. Another example: $A = \mathbb{R}$, $B = \{-1, 0, 1\}$. The function $\text{sign}(x)$ associates all negative numbers with -1 and all positive number with $+1$. 0 is associated with 0. A third example: $A = \mathbb{R}$, $B = \mathbb{R}$. The equation $y = f(x) = 2x + 3$ associates exactly one number y to every value of x. Sequences of numbers (A6.1) are functions with $A = \mathbb{N}$ and $B \subset \mathbb{R}$.

A.4.2 Properties of Functions and Graphical Representation

Functions having a finite domain D can be represented by listing possible arguments and their function values. Functions can be described by a rule specifying how the association of an *independent variable* (input) and a *dependent variable* (output) should take place. Functions can also be specified by an analytical expression or function equation.

Functions that map $\mathbb{R}$ into $\mathbb{R}$ can be represented by a graph associating each pair (x, y) with the point in the number plane with the coordinates (x, y). Since every value in the domain is associated with only one function value, there can be only one value on any vertical line, whereas there can be arbitrarily many values on any horizontal line (figure a below).

Let us illustrate some important concepts of functions using the graphs below. A function whose graph is continuous, that is, without sudden leaps, is called *continuous*. A function that increases in value with increasing argument is called *monotone increasing* (figure b below) (and analogically for decreasing arguments and values). A function that is symmetrical to the y-axis, that is, for which $f(x) = f(-x)$ is true for all $x \in D$, is called an *even* function (figure c). A function whose graph has rotational symmetry with respect to the origin, that is, for which $f(-x) = -f(x)$ is true for all $x \in D$, is called an odd function (figure d). A value of x for which the function value is 0 ($y = f(x) = 0$) is a *zero* or a *root* of the function.

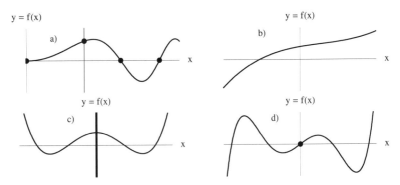

If all the arguments of a function are mapped to different values, that is, $f(x_1) \neq f(x_2)$ when $x_1 \neq x_2$, then the function can be inverted. The definition domain becomes the value domain and vice versa, and we have the so-called *inverse function*, written as f^{-1}, whereby

$$f^{-1}(x) := y \quad \text{if} \quad f(y) = x.$$

The graph of the function f^{-1} is symmetrical to the graph of the function f with respect to the diagonal $y = x$ (see the figures below).

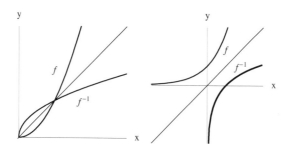

The following figure shows simple manipulation of function graphs. Compared to the graph of the given function $f(x)$ the graph of the function $f(x - d)$ is shifted by d to the right, the graph of the function $f(x) + d$ is shifted up by d (left figure below). The graph of the function $f(cx)$ is compressed along the x-axis, the graph of the function $c \cdot f(x)$ stretched along the y-axis (middle figure below). The graph of the function $f(-x)$ is reflected in the y-axis, the graph of the function $-f(x)$ in the x-axis (right figure below).

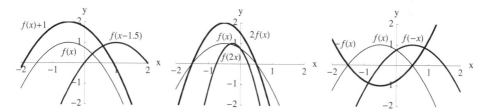

A.4.3 Basic Functions

In what follows, we will discuss some frequently used functions and derive function equations from certain requirements.

A.4.3.1 The graph of a *linear function* is a straight line. The equation for a straight line is $y = a_1 x + a_0$. The term a_0, the y-intercept, is the function value at $x = 0$, and the factor a_1 is the *slope* of the straight line and corresponds to $\tan(\alpha)$. The parameters of the straight line through two points $P_1(x_1, y_1)$ and $P_2(x_2, y_2)$ can be derived from the two equations

$$y_1 = a_1 x_1 + a_0 \quad \text{and} \quad y_2 = a_1 x_2 + a_0 \quad \text{and are}$$

$$a_1 = \frac{y_2 - y_1}{x_2 - x_1} \quad \text{and} \quad a_0 = \frac{x_1 y_2 - x_2 y_1}{x_1 - x_2}.$$

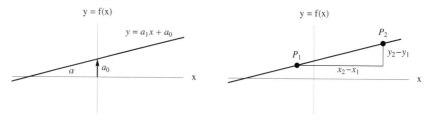

A.4.3.2 The graph of a *quadratic function* $y = a_2x^2 + a_1x + a_0$ is a *parabola* whose axis of symmetry is parallel to the y-axis. The parabola opens up if $a_2 > 0$ and opens down if $a_2 < 0$. It is narrow for large $|a_2|$ and wide for small $|a_2|$. Its position, given by the coordinates of the *vertex A*, is determined by a_2, a_1 and a_0.

$$A\left(-\frac{a_1}{2a_2}, -\frac{a_1^2}{4a_2} + a_0\right).$$

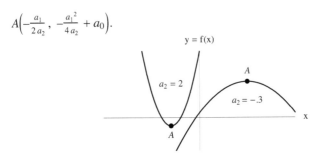

The formulas for determining the coefficients a_2, a_1 and a_0 from three given points $P_1(x_1, y_1)$, $P_2(x_2, y_2)$ and $P_3(x_3, y_3)$ are:

```
Solve[{a0 + a1*x₁ + a2*x₁² == y₁,
       a0 + a1*x₂ + a2*x₂² == y₂, a0 + a1*x₃ + a2*x₃² == y₃}, {a0, a1, a2}]
```

$$\left\{\left\{a0 \to -\frac{-x_2^2 x_3 y_1 + x_2 x_3^2 y_1 + x_1^2 x_3 y_2 - x_1 x_3^2 y_2 - x_1^2 x_2 y_3 + x_1 x_2^2 y_3}{(x_1 - x_2)(x_1 x_2 - x_1 x_3 - x_2 x_3 + x_3^2)},\right.\right.$$

$$a1 \to -\frac{x_2^2 y_1 - x_3^2 y_1 - x_1^2 y_2 + x_3^2 y_2 + x_1^2 y_3 - x_2^2 y_3}{(x_1 - x_2)(x_1 x_2 - x_1 x_3 - x_2 x_3 + x_3^2)},$$

$$\left.\left. a2 \to -\frac{-(-x_1 + x_3)(y_1 - y_2) + (-x_1 + x_2)(y_1 - y_3)}{-(-x_1^2 + x_2^2)(-x_1 + x_3) + (-x_1 + x_2)(-x_1^2 + x_3^2)}\right\}\right\}$$

A.4.3.3 A function that can be written as a sum

$$y = \sum_{i=0}^{n} a_i x^i = a_n x^n + a_{n-1} x^{n-1} + \ldots + a_2 x^2 + a_1 x + a_0$$

with $i \geq 0$ is called a *polynomial function*. The functions discussed above are special cases of polynomial functions with $n = 1$ and $n = 2$ respectively. Polynomial functions are continuous. When n is even, we have $f(x) \to +\infty$ for $x \to \pm\infty$ when a_n is positive, and $f(x) \to -\infty$ for $x \to \pm\infty$ when a_n is negative. When n is odd, we have $f(x) \to \pm\infty$ for $x \to \pm\infty$ when a_n is positive, and $f(x) \to \mp\infty$ for $x \to \pm\infty$ when a_n is negative. Polynomial functions have at most n roots, that is, the equation $f(x) = 0$ has at most n solutions.

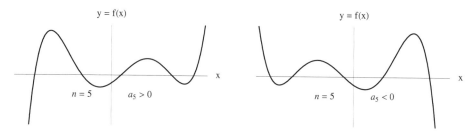

The fundamental theorem of algebra says that the polynomial

$$P(x) = a_n x^n + a_{n-1} x^{n-1} + \ldots + a_2 x^2 + a_1 x + a_0$$

can be written as the product $a_n(x - x_1)(x - x_2) \ldots (x - x_n)$, where the x_i are the roots of the polynomial function $y = P(x)$. The x_i are real when the function has n roots and partly complex when the function has less than n roots. Since the complex solutions are pairwise conjugate, two factors $(x - x_i)$ can be taken together as $x^2 + px + q$ with real p and q. In the lines below, we compute the roots of the function shown at the top left above and compare them with the factorization of the polynomial.

```
a2 = -.73; a1 = .2; a3 = .08; a0 = .3; a4 = .06;
Solve[a0 + a1*x + a2*x² + a3*x³ + a4*x⁴ == 0, x]
```

$\{\{x \to -4.28834\}, \{x \to -0.511612\}, \{x \to 0.881614\}, \{x \to 2.58501\}\}$

```
Factor[a0 + a1*x + a2*x² + a3*x³ + a4*x⁴]
```

$0.06 \, (-2.58501 + x) \, (-0.881614 + x) \, (0.511612 + x) \, (4.28834 + x)$

If we put a larger value for a_0 into the function, the curve is moved up and there are only two roots.

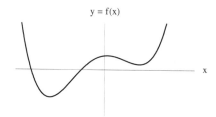

If we now solve the modified equation, we obtain four solutions, two real and two complex. Accordingly, we find a factor of the form $x^2 + px + q$ in the real factorization.

```
a2 = -.73; a1 = .2; a3 = .08; a0 = 1.5; a4 = .06;
Solve[a0 + a1*x + a2*x² + a3*x³ + a4*x⁴ == 0, x]
```

$\{\{x \to -4.12342\}, \{x \to -1.30096\},$
$\{x \to 2.04552 - 0.690046 \, I\}, \{x \to 2.04552 + 0.690046 \, I\}\}$

```
Factor[a0 + a1*x + a2*x² + a3*x³ + a4*x⁴]
```

$0.06 \, (1.30096 + x) \, (4.12342 + x) \, (4.66034 - 4.09105 \, x + x^2)$

A.4.3.4 Functions that can be written as the quotient of two polynomials (of degree n and m respectively) are called *rational functions*:

$$y = \frac{A(x)}{B(x)} = \frac{a_n x^n + a_{n-1} x^{n-1} + \ldots + a_2 x^2 + a_1 x + a_0}{b_m x^m + b_{m-1} x^{m-1} + \ldots + b_2 x^2 + b_1 x + b_0}$$

When the numerator is equal to zero, the value of the function is equal to zero, as long as the denominator is not also zero. Ordinary zeros in the denominator result in so-called *poles* of the function, that is values of x for which the function value is undefined. To generate a function with a zero at $x = 1$ and poles at $x = 2$ and $x = -1.5$, we set

$$f(x) = \frac{x-1}{(x-2)(x+1.5)} = \frac{x-1}{x^2 - .5x - 3}.$$

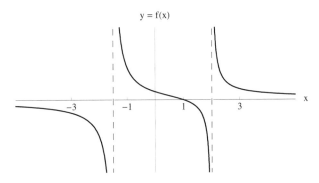

Every rational function can be decomposed into a sum of so-called *partial fractions* of the form

$$\frac{A}{x - p_i^k}, \text{ or for multiple poles } \frac{B + Cx}{(x^2 + px + q)^m}.$$

For the example above we get

$$\text{Apart}\left[\frac{x-1}{x^2 - .5*x - 3}\right] \qquad \frac{0.2857}{-2. + x} + \frac{0.7143}{1.5 + x}$$

We can set a double pole at $x = 2$ using the factor $(x - 2)^2$ and a triple pole at $x = 0$ using the factor x^3 in the denominator.

$$\text{Together}\left[\text{ExpandAll}\left[(x-1)/((x-2)^2 * x^3)\right]\right] \qquad \frac{-1 + x}{4x^3 - 4x^4 + x^5}$$

$$\text{Apart}\left[\frac{-1 + x}{4 x^3 - 4 x^4 + x^5}\right] \qquad \frac{1}{8(-2+x)^2} - \frac{1}{16(-2+x)} - \frac{1}{4x^3} + \frac{1}{16x}$$

A.4.3.5 A function of the form $y = a^x$ ($a > 0$) or more generally $y = ca^x$ is called an *exponential function*. The function $y = (1/a)^x$ is equal to the function $y = a^{-x}$. The base $e = 2{,}7182818\ldots$, Euler's number (B3), is of particular importance (right figure below).

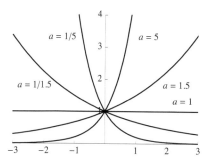
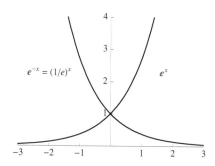

If a computer or a programming language only provides the exponential function to the base e, the function $y = a^x$ can be replaced by $e^{x \cdot \ln(a)}$, since $a = e^{\ln(a)}$. Multiplying an exponential function by a constant c results in stretching the function's graph vertically, or, since $ca^x = a^{x+\log_a(c)}$, in a horizontal displacement of $-\log_a(c)$ (left figure below).

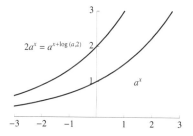
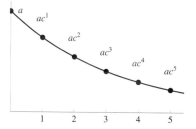

An exponential function increases over equal intervals by the same factor: $a^{x+d} = a^d a^x = ca^x$. Therefore, a discrete time function in which every new value is derived from the last value by multiplication by a constant will be an exponential function (right figure above).

A single point (x_0, y_0) determines an exponential function either of the form

$$a^x = e^{x \cdot \ln(a)} = e^{xk} \ ((x_0, y_0) \neq (0, 1)) \text{ or of the form } ce^x = e^{x+d}.$$

In the first case, the base $a = e^{\ln(y_0)/x_0}$ results from $y_0 = a^{x_0}$, and we have the factor k of the exponent in e^{xk} as $k = \ln(a) = \ln(y_0)/x_0$. In the second case, the factor $c = y_0/e^{x_0}$ results from $y_0 = ce^{x_0}$ (left figure below), as does the displacement $d = \log_a(x_0) - y_0$.

Two points (x_1, y_1) and (x_2, y_2) determine an exponential function of the form $ca^x = e^{xk+d}$, where

$$k = \frac{\ln(y_1/y_2)}{x_1 - x_2} \text{ and } d = \ln(y_1) - x_1 k,$$

as results from the solution of the simultaneous equations $y_1 = e^{x_1 k + d}$ and $y_2 = e^{x_2 k + d}$. In the following example we determine the function describing the exponential increase in frequency of a tone from 100 Hz at time $t = 2$ to 200 Hz at time $t = 5$ (figure at right below).

```
x1 = 2; x2 = 5; y1 = 100; y2 = 200; k = Log[y1/y2] / (x1 - x2); d = Log[y1] - x1*k;
```

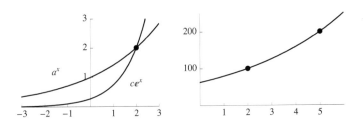

We observe these rules for calculating with powers:

$$a^m a^n = a^{m+n} \qquad \frac{a^m}{a^n} = a^{m-n} \qquad (a^m)^n = (a^n)^m = a^{mn} \qquad a^{\frac{m}{n}} = \sqrt[n]{a^m} = \left(\sqrt[n]{a}\right)^m$$

A.4.3.6 The *logarithmic function* $y = \log_a(x)$ with $a > 0$ is the inverse function of the exponential function $y = a^x$. Hence, the graph of the logarithmic function corresponds to the reflection of the exponential function about the diagonal $y = x$ (A4.2). From the identity $x = a^{\log_a(x)} = b^{\log_b(x)}$ we derive the formula for the conversion of logarithms from one base to another: $\log_a(x) = \log_b(x)/\log_b(a)$. In particular we have $\log_a(x) = \ln(x)/\ln(a)$.

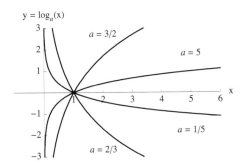

We observe these rules for calculating with logarithms:

$$\log(ab) = \log(a) + \log(b) \qquad \log(\tfrac{a}{b}) = \log(a) - \log(b) \qquad \log(a^b) = b \cdot \log(a)$$

A.4.3.7 The *trigonometric functions* $\sin(x)$ and $\cos(x)$ are discussed at length in Chapter 2.2.1.

In the right triangle, the trigonometric functions are defined as follows:

$$\sin(\alpha) = \frac{\text{opposite leg}}{\text{hypotenuse}} = \frac{a}{c}, \quad \cos(\alpha) = \frac{\text{adjacent leg}}{\text{hypotenuse}} = \frac{b}{c}, \quad \tan(\alpha) = \frac{\text{opposite leg}}{\text{adjacent leg}} = \frac{a}{b} = \frac{\sin(\alpha)}{\cos(\alpha)}.$$

Fundamentals of Mathematics

The trigonometric functions are periodic: $\sin(\alpha)$ and $\cos(\alpha)$ have the period 2π, $\tan(\alpha)$ has the period π.

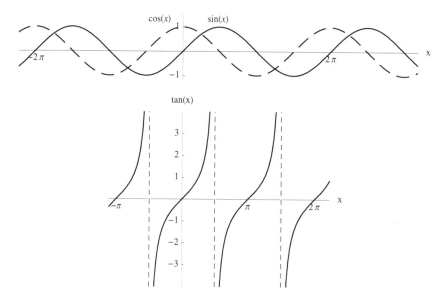

One can shift the graph of the function $\sin(x)$ horizontally by adding a phase constant ϕ to the argument x (figure a below). When ϕ is positive, the graph is shifted to the left, when ϕ is negative, the graph is shifted to the right. The graph can be stretched vertically by multiplying the function by a constant c greater than one. Correspondingly, when the positive constant is smaller than one, the graph is compressed in the vertical direction (figure b). Multiplying the argument by a constant k greater than one compresses the graph along the x-axis, while a constant k smaller than one stretches the graph (figure c).

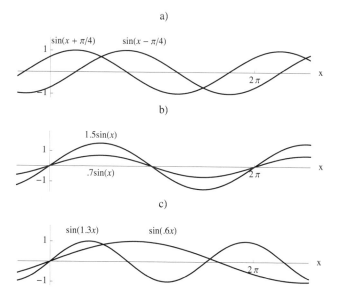

A given point (x_1, y_1) determines a sine function with one parameter (c, k or φ). The function can be derived from the point's coordinates as follows:

```
Solve[y1 == Sin[x1 + φ], φ]              {{φ → -x1 + ArcSin[y1]}}

Solve[y1 == c*Sin[x1], c]                {{c → y1 Csc[x1]}}
```

$Csc(z) \equiv 1 / \sin(z)$.

```
Solve[y1 == Sin[k*x1], k]                {{k → ArcSin[y1]/x1}}
```

Mathematica gives this warning for each calculation:

```
Solve::ifun :
  Inverse functions are being used by Solve, so some
    solutions may not be found.
```

Two given points (x_1, y_1) and (x_2, y_2) determine a sine function with two parameters. The function can be derived from the points' coordinates as follows:

```
Solve[{y1 == Sin[k*x1 + φ], y2 == Sin[k*x2 + φ]}, {φ, k}]

{{φ → (-x2 ArcSin[y1] + x1 ArcSin[y2]) / (x1 - x2),
  k → (ArcSin[y1] - ArcSin[y2]) / (x1 - x2)}}
```

For the points (1, .6) and (2, –.3) this gives (figure a below):

```
x1 = 1; x2 = 2; y1 = .6; y2 = -.3;
NSolve[{y1 == Sin[k*x1 + φ], y2 == Sin[k*x2 + φ]}, {φ, k}]

{{φ → 1.59169, k → -0.948194}}
```

If k and c are to be determined, there is no general solution:

```
Solve[{y1 == c*Sin[k*x1], y2 == c*Sin[k*x2]}, {k, c}]

Solve::tdep :
  The equations appear to involve transcendental functions
    of the variables in an essentially non-algebraic way.
```

Mathematica will compute a numerical solution if one exists (figure b below):

```
x1 = 1; x2 = 2; y1 = 1.3; y2 = 1.6;
NSolve[{y1 == c*Sin[k*x1], y2 == c*Sin[k*x2]}, {k, c}]

{{c → -1.64927, k → -0.907923}, {c → 1.64927, k → 0.907923}}
```

If φ and c are to be determined, Mathematica computes a solution that is three pages long. Here again, one has to rely on a numerical solution (figure c below). (→ Solution computed by Mathematica)

```
Solve[{y1 == c*Sin[x1 + φ], y2 == c*Sin[x2 + φ]}, {φ, c}]

x1 = 1; x2 = 2; y1 = 1.3; y2 = 1.6;
NSolve[{y1 == c*Sin[x1 + φ], y2 == c*Sin[x2 + φ]}, {φ, c}]
```

$\{\{c \to -1.68163, \varphi \to 3.02524\}, \{c \to 1.68163, \varphi \to -0.116349\}\}$

a)

$y = \sin(-.948x + 1.59)$

b)

$y = 1.65 \cdot \sin(.0908x)$

c)

$y = 1.682 \cdot \sin(x - .116)$

Here a numerical example for determining three coefficients from three given points:

```
x1 = 1; x2 = 2; x3 = 3; y1 = 1.3; y2 = 1.3; y3 = -1.6;
NSolve[{y1 == c*Sin[k*x1 + φ],
   y2 == c*Sin[k*x2 + φ], y3 == c*Sin[k*x3 + φ]}, {φ, c, k}]
{{c → -1.9547, k → 1.68644, φ → 2.18273}}
```

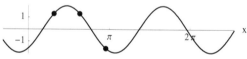

The inverse trigonometric functions are also known as *arc-trigonometric functions*. Because of the trigonometric functions' periodicity, their domains and ranges must be restricted:

Trigonometric function	Domain	Range	
$y = \sin(x)$	$-\frac{\pi}{2} \le x \le +\frac{\pi}{2}$	$-1 \le y \le +1$	$x = \arcsin(y)$
$y = \cos(x)$	$0 \le x \le \pi$	$+1 \ge y \ge -1$	$x = \arccos(y)$
$y = \tan(x)$	$-\frac{\pi}{2} < x < +\frac{\pi}{2}$	$-\infty < y < +\infty$	$x = \arctan(y)$
	Range	Domain	Inverse trigonometric function

arcsin(x)

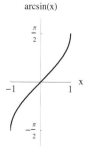

arccos(x)

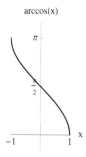

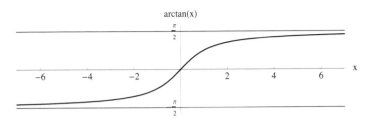

We observe these rules for calculating with trigonometric functions:

$$\sin(\alpha \pm \beta) = \sin(\alpha)\cos(\beta) \pm \cos(\alpha)\sin(\beta)$$

$$\sin(2\alpha) = 2\cdot\sin(\alpha)\cos(\alpha)$$

$$\cos(\alpha \pm \beta) = \cos(\alpha)\cos(\beta) \mp \sin(\alpha)\sin(\beta)$$

$$\cos(2\alpha) = \cos^2(\alpha) - \sin^2(\alpha)$$

$$1 + \cos(\alpha) = 2\cos^2\left(\frac{\alpha}{2}\right)$$

$$\sin(\alpha) \pm \sin(\beta) = 2\cdot\sin\left(\frac{\alpha\pm\beta}{2}\right)\cos\left(\frac{\alpha\mp\beta}{2}\right)$$

$$\cos(\alpha) + \cos(\beta) = 2\cos\left(\frac{\alpha+\beta}{2}\right)\cos\left(\frac{\alpha-\beta}{2}\right)$$

$$\cos(\alpha) - \cos(\beta) = -2\cdot\sin\left(\frac{\alpha+\beta}{2}\right)\sin\left(\frac{\alpha-\beta}{2}\right)$$

$$\tan(\alpha \pm \beta) = \frac{\tan(\alpha) \pm \tan(\beta)}{1 \mp \tan(\alpha)\tan(\beta)}$$

$$\tan(2\alpha) = \frac{2\cdot\tan(\alpha)}{1 - \tan^2(\alpha)}$$

$$\sin(2\alpha) = \frac{2\cdot\tan(\alpha)}{1 + \tan^2(\alpha)}$$

$$\cos(2\alpha) = \frac{1 - \tan^2(\alpha)}{1 + \tan^2(\alpha)}$$

$$\sin^2(\alpha) + \cos^2(\alpha) = 1$$

A.4.4 Composite Functions

Functions can be composed of several simple functions over the entire definition domain. For example, a damped oscillation can be described as the product of an exponential function and of a sine function (2.2.1.2). A piecewise function is defined along a sequence of intervals (5.1.3.1, 5.1.3.2). In order to simplify the description of a complicated time function, different functions are often used for different time intervals.

A.4.5 Parametric Representation

Functions of the form $y = f(x)$ can only produce graphs associating every x-value with a single y-value. Hence drawing a circle is impossible. Circles and more complex curves can be described parametrically by representing the x- and y-coordinates of the points on the curve as functions of an external variable, called a *parameter*. When time is the parameter, the curve represents movement. All relationships implicit in the form $f(x, y) = 0$ can be represented in this way. The figure below shows two spirals. The first moves from the center outward. The equations for the x- and y-coordinates as functions of time are $x = t\cdot\sin(t)$ and $y = t\cdot\cos(t)$. The second spiral moves inward to the center. The corresponding equations for x and y are: $x = e^{-.1t}\sin(t)$ and $y = e^{-.1t}\cos(t)$ (see the example in Chapter 3.4.1.1).

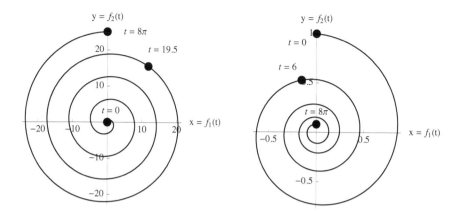

Using only one parameter in three dimensions draws a curve in space. The three equations for the spiral shown on the left below are: $x = \sin(t)$, $y = \cos(t)$ and $z = t$. One can represent surfaces using two parameters. The three equations of the cylinder below are: $x = \sin(t)$, $y = \cos(t)$ and $z = u$. Varying the parameter t for a fixed value of u gives a circle parallel to the xy-plane, varying the parameter u gives the displacement along the z-axis.

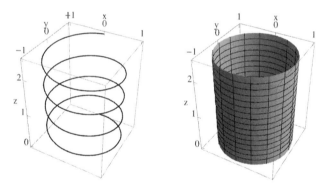

A.4.6 Functions of Several Variables

A function of two variables associates a value with every point (x, y). A continuous function of two variables can be represented as a surface above the xy-plane. The illustration below shows on the left the function $z = \sin(x) + \sin(y)$, on the right the function $z = \sin(x)\sin(y)$.

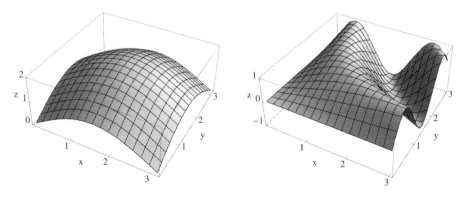

A.4.7 Even and Odd Functions

A function is called even if $f(x) = f(-x)$ and odd if $f(-x) = -f(x)$ over the entire domain of the function. The graph of an even function is symmetrical to the y-axis, the graph of an odd function has rotational symmetry to the origin. The function $\cos(\omega t)$ is even, the function $\sin(\omega t)$ is odd. Every function can be decomposed into an even and an odd component:

$$f(x) = f_e(x) + f_o(x).$$

The even component of f is

$$f_e = \frac{1}{2}(f(x) + f(-x)),$$

the odd component is

$$f_o = \frac{1}{2}(f(x) - f(-x)).$$

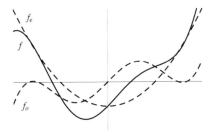

A.4.8 Sequences and Series

For *number sequences* $a_1, a_2, a_3, \ldots$, we can write $\langle a_n \rangle$. Sequences can be defined in many ways: by writing the first numbers, for example 2, 3, 5, 7 ...; by a description, for example a_n = nth prime number; by explicitly indicating the nth element, for example $a_n = n^2$; by recursive instructions showing how to compute the next element from previous elements, for example $a_n = a_{n-1} + a_{n-2}$ with initial values $a_1 = 1$, $a_2 = 1$. A sequence is called *bounded above* when a number exists greater than all the numbers of the sequence *bounded below* when a number exists smaller than all the numbers in the sequence. A sequence is said to be convergent to the limit a if for every arbitrarily small positive number ϵ there exists a natural number n_ϵ such that the relation $|a_n - a| < \epsilon$ holds for all $n > n_\epsilon$. One writes $\lim_{n \to \infty} a_n = a$.

In an *arithmetic sequence* the difference between adjacent elements is constant: $a_n = a_{n-1} + d$.

In a *geometrical sequence* the quotient of successive elements is constant: $a_n = a_{n-1} q$.

The sequence $\langle s_n \rangle$ of the partial sums

$$s_n = a_1 + a_2 + a_3 + \ldots = \sum_{i=1}^{n} a_i$$

of a given number sequence $\langle a_n \rangle$ is called a *series*. If there is a finite number s with

$$\lim_{n \to \infty} \sum_{i=1}^{n} a_i = s,$$

the series is called convergent to the limit s. The geometrical series $a_1 + a_1 q + a_1 q^2 + \ldots$ is convergent for $-1 < q < 1$ and its limit is $s = a_1/(1 - q)$.

A.5 Calculus

A.5.1 The Derivative of a Function

The *derivative* of a function $y = f(x)$ at the point x_0 is equivalent to the slope of the tangent to the curve at the point $P(x_0, y_0)$. To calculate the slope, consider the secant through the point $P(x_0, y_0)$ and another point on the curve $P(x_1, y_1)$. The slope of the secant is

$$\frac{y_1 - y_0}{x_1 - x_0} = \frac{\Delta y}{\Delta x} = \frac{f(x + \Delta x) - f(x)}{\Delta x}.$$

As Δx goes to zero, the secant approaches the tangent and the derivative of the function at x_0 is

$$f'(x_0) = \lim_{\Delta x \to 0} \frac{f(x + \Delta x) - f(x)}{\Delta x}.$$

If the derivative is defined at every point on the curve, one obtains a new function $y' = f'(x)$ which at every point assigns the derivative of $f(x)$ to the independent variable x.

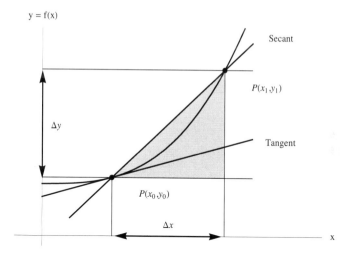

To determine the derivative of a function, one rewrites the so-called differential coefficient $\Delta y/\Delta x$ so that Δx no longer appears in the denominator and the limiting process $\Delta x \to 0$ can take place. The differential coefficient of the function $y = x^2$ is:

$$\frac{\Delta y}{\Delta x} = \frac{(x + \Delta x)^2 - x^2}{\Delta x} = \frac{x^2 + 2x\,\Delta x + \Delta x^2 - x^2}{\Delta x} = 2x + \Delta x.$$

As Δx goes to zero, we have $\dfrac{\Delta y}{\Delta x} = 2x$.

The table below shows a few functions and their derivatives.

Function	Derivative
c (const)	0
x^n	$n\,x^{n-1}$
e^x	e^x
a^x	$a^x \ln(a)$
$\ln(x)$	$1/x$
$\log_a(x)$	$\dfrac{1}{x \ln(x)}$
$\sin(x)$	$\cos(x)$
$\cos(x)$	$-\sin(x)$
$\tan(x)$	$\dfrac{1}{\cos^2(x)}$

Function	nth Derivative
c (const)	0
x^m	$m(m-1)..(m-n+1)\,x^{m-n}$
$\ln(x)$	$(-1)^n (n-1)!\,\dfrac{1}{x^n}$
e^{kx}	$k^n e^{kx}$
$\sin(x)$	$\sin\!\left(x + \dfrac{n\pi}{2}\right)$
$\cos(x)$	$\cos\!\left(x + \dfrac{n\pi}{2}\right)$
$\sin(kx)$	$k^n \sin\!\left(kx + \dfrac{n\pi}{2}\right)$
$\cos(kx)$	$k^n \cos\!\left(kx + \dfrac{n\pi}{2}\right)$

A.5.2 Rules for Differentiation

The following rules are used to differentiate composite functions. a, c, ... are constants, u and v are functions of x.

$(u \pm v)' = u' \pm v'$

$(cu)' = c \cdot u'$

$(uv)' = v \cdot u' + u \cdot v'$ product rule

$(u/v)' = \dfrac{v \cdot u' + u \cdot v'}{v^2}$ quotient rule

For $u = u(v)$ and $v = v(x)$, $\dfrac{du}{dx} = \dfrac{du}{dv} \cdot \dfrac{dv}{dx}$ chain rule

Mathematica can compute derivatives symbolically.

$\mathtt{D\!\left[x^5 + 4\,x^3 + 2\,x^2 + 1,\ x\right]}$

$4\,x + 12\,x^2 + 5\,x^4$

$\mathtt{D\!\left[t * Sin\!\left[t^2 + \sqrt{Cos[t]}\,\right],\ t\right]}$

$t\,\mathrm{Cos}\!\left[t^2 + \sqrt{\mathrm{Cos}[t]}\,\right]\left(2t - \dfrac{\mathrm{Sin}[t]}{2\sqrt{\mathrm{Cos}[t]}}\right) + \mathrm{Sin}\!\left[t^2 + \sqrt{\mathrm{Cos}[t]}\,\right]$

If $y = f(x)$ reaches a *maximum* or a *minimum*, the first derivative y' becomes equal to zero. The second derivative y'', which corresponds to the curvature of the graph of y, is negative when the function reaches a maximum and positive when it reaches a minimum. At the *inflection points* of the function's graph, the curvature is zero and the second derivative y'' disappears.

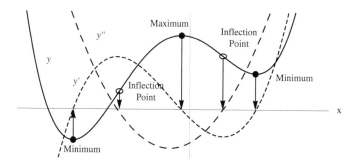

A.5.3 The Indefinite Integral of a Function

Differentiation assigns to a given function exactly one new function, its derivative. The opposite procedure, integration, is not unambiguous, since functions that differ only by a constant term have the same derivative. A function $F(x)$ whose derivative equals $f(x)$ is called an *antiderivative* of $f(x)$. The set of all antiderivatives of $f(x)$ is called the *indefinite integral*

$$\int f(x)dx.$$

Together with $F(x)$, every function $F(x) + C$ is also an antiderivative of the function $f(x)$. The antiderivatives of elementary functions can be found in tables. The table below lists a few functions and their indefinite integrals.

f (x)	F (x)
c	c x
x^p	$\frac{x^{p+1}}{p+1}$, $p \neq -1$
x^{-1}	$\ln(|x|)$
a^x	$a^x / \ln(a)$

f (x)	F (x)
e^x	e^x
sin (x)	−cos (x)
cos (x)	sin (x)
tan (x)	−ln (cos (x))

A.5.4 Integration Formulas

The following formulas can be used to determine the indefinite integral of compound functions. $a, c, \ldots$ are constants, u and v are functions of x.

$\int (u \pm v)dx = \int u\, dx \pm \int v\, dx$

$\int cu\, dx = c \int u\, dx$

$\int u \left(\frac{dv}{dx}\right)dx = uv - \int v \left(\frac{du}{dx}\right) dx$ or $\int u\, dv = uv - \int v\, du$ integration by parts

$\int F(u(x))dx = \int \frac{F(w)}{w'} dw$ where $w = u(x)$ and $w' = \frac{dw}{dx}$ integration by substitution

Mathematica can compute integrals symbolically.

$\int \left(2*x^2 + \text{Sin}[5*x]\right) dx \qquad\qquad \frac{1}{15}\left(10\, x^3 - 3\, \text{Cos}[5\, x]\right)$

A.5.5 The Definite Integral

The *definite integral*

$$\int_a^b f(x)\,dx$$

of a function corresponds to the area under the function between the values $x = a$ and $x = b$. The area can be approximately calculated as a sum of n rectangles under the curve with sides $h = (b - a)/n$ and $f(a + ih)$. As h goes to zero, n goes to infinity, and one obtains the required area.

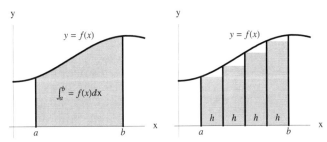

The definite integral of the function $f(x)$ between a and b is defined as

$$\int_a^b f(x)\,dx = \lim_{h \to 0} h[f(a) + f(a + h) + f(a + 2h) + \ldots + f(a + (n-1)h)] .$$

The definite and the indefinite integral are associated to each other through the fundamental theorem of calculus.

If $f(x) = \frac{d}{dx} g(x)$, then $\int_a^b f(x)\,dx = \int_a^b \frac{d}{dx} g(x)\,dx = g(x)\Big[_a^b := g(b) - g(a)$.

In other words, if $g(x)$ is an antiderivative of $f(x)$ (that is, $f = g'$), the integral of $f(x)$ between $x = a$ and $x = b$ is equal to the difference $g(b) - g(a)$. The constant of integration cancels out because of the subtraction: $g(b) + C - (g(a) + C) = g(b) - g(a)$.

Example 1: An antiderivative of x^2 is the function $x^3/3$. The area under the function x^2 between 0 and 1 is equal to $1^3/3 - 0^3/3 = 1/3$.

Example 2: An antiderivative of $\sin(x)$ is the function $-\cos(x)$. The area under the function $\sin(x)$ between 0 and π is equal to $-\cos(\pi) + \cos(0) = 1 + 1 = 2$.

A.5.6 Partial Derivatives

A *partial derivative* is the derivative of a function in several variables $x_1, x_2, \ldots$ with respect to one of these variables. We write:

$$\frac{\partial}{\partial x_i} f(x_1, x_2, \ldots, x_i, \ldots) \text{ or } \frac{\partial f}{\partial x_i} \text{ or } f_{x_i}.$$

One obtains the partial derivative with respect to x_i by treating all other variables as constants:

`D[x₁*x₂² + Sin[x₃], x₁]`	x_2^2
`D[x₁*x₂² + Sin[x₃], x₂]`	$2 x_1 x_2$
`D[x₁*x₂² + Sin[x₃], x₃]`	$\cos[x_3]$

In functions of two variables x and y, the partial derivative with respect to x at (x_0, y_0) corresponds to the slope of the surface $f(x, y)$ at the point $(x_0, y_0, f(x_0, y_0))$ along the x-axis. The illustration below shows the function $f(x, y) = \sin(x)y^3$. The derivative with respect to x is $f_x = \cos(x)y^3$, the derivative with respect to y is $f_y = 3 \cdot \sin(x)y^2$. The point P_1 has the coordinates $(0, -1, 0)$. The slope of the surface at P_1 is -1 in the x-direction and 0 in the y-direction. The point P_2 has the coordinates $(\pi/2, -1, 1)$ The slope of the surface at P_2 is 0 in the x-direction and $3 \cdot \sin(\pi/2)(-1)^2 = 3$ in the y-direction.

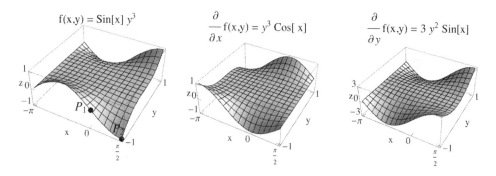

Appendix B   Tables

B.1   Pitches

Frequencies of the equal-tempered tone system when $a^1 = 440$ Hz.

	C^2	C^1	C	c^1	c^2	c^3	c^4	c^5
c	32.703	65.406	130.813	261.626	523.251	1046.50	2093.00	4186.01
cis	34.648	69.296	138.591	277.183	554.365	1108.73	2217.46	4434.92
d	36.708	73.416	146.832	293.665	587.330	1174.66	2349.32	4698.64
es	38.891	77.782	155.563	311.127	622.254	1244.51	2489.02	4978.03
e	41.203	82.407	164.814	329.628	659.255	1318.51	2637.02	5274.04
f	43.653	87.307	174.614	349.228	698.456	1396.91	2793.83	5587.65
fis	46.249	92.499	184.997	369.994	739.989	1479.98	2959.96	5919.91
g	48.999	97.999	195.998	391.995	783.991	1567.98	3135.96	6271.93
gis	51.913	103.826	207.652	415.305	830.609	1661.22	3322.44	6644.88
a	55.000	110.000	220.000	440.000	880.000	1760.00	3520.00	7040.00
b	58.270	116.541	233.082	466.164	932.328	1864.66	3729.31	7458.62
h	61.735	123.471	246.942	493.883	987.767	1975.53	3951.07	7902.13

B.2   Formants of English Vowels

See [60].

	/a/	/e/	/i/	/o/	/u/
f_1 in Hz	622 / 784	392 / 392	262 / 349	392 / 440	349 / 329
BW_1 in Hz	60 / 80	40 / 60	60 / 50	40 / 70	40 / 50
Amp_1 in dB	0	0	0	0	0
f_2 in Hz	1047 / 1175	1661 / 1568	1760 / 1661	784 / 784	587 / 740
BW_2 in Hz	70 / 90	80 / 80	90 / 100	80 / 80	80 / 60
Amp_2 in dB	−7 / −4	−12 / −24	−30 / −20	−11 / −9	−20 / −12
f_3 in Hz	2489 / 2794	2489 / 2794	2489 / 2794	2489 / 2794	2489 / 2794
BW_3 in Hz	110 / 120	100 / 120	100 / 120	100 / 100	100 / 170
Amp_3 in dB	−9 / −20	−9 / −30	−16 / −30	−21 / −16	−32 / −30

B.3 Constants

π is the mathematical constant expressing the ratio of a circle's circumference to its diameter.

3.1415926535897932384626433832795028841971693993751058209749445923078164 0628

e is the mathematical constant, sometimes called Euler's number, which is the base of the natural logarithm.

2.7182818284590452353602874713526624977572470936999595749669676277240 7663035

The Golden Ratio $\left(\sqrt{5} - 1\right)/2$

0.6180339887498948482045868343656381177203091798057628621354486227052 6046281

B.4 Fibonacci Numbers

$a_1 = 1; a_2 = 2; a_n = a_{n-1} + a_{n-2};$

1	2	3	5	8
13	21	34	55	89
144	233	377	610	987
1597	2584	4181	6765	10 946
17 711	28 657	46 368	75 025	121 393
196 418	317 811	514 229	832 040	1 346 269
2 178 309	3 524 578	5 702 887	9 227 465	14 930 352
24 157 817	39 088 169	63 245 986	102 334 155	165 580 141
267 914 296	433 494 437	701 408 733	1 134 903 170	1 836 311 903
2 971 215 073	4 807 526 976	7 778 742 049	12 586 269 025	20 365 011 074
32 951 280 099	53 316 291 173	86 267 571 272	139 583 862 445	225 851 433 717
365 435 296 162	591 286 729 879	956 722 026 041	1 548 008 755 920	2 504 730 781 961
4 052 739 537 881	6 557 470 319 842	10 610 209 857 723	17 167 680 177 565	27 777 890 035 288
44 945 570 212 853	72 723 460 248 141	117 669 030 460 994	190 392 490 709 135	308 061 521 170 129

B.5 Prime Numbers

The first 299 prime numbers.

	0	1	2	3	4	5	6	7	8	9
0		2	3	5	7	11	13	17	19	23
10	29	31	37	41	43	47	53	59	61	67
20	71	73	79	83	89	97	101	103	107	109
30	113	127	131	137	139	149	151	157	163	167
40	173	179	181	191	193	197	199	211	223	227
50	229	233	239	241	251	257	263	269	271	277
60	281	283	293	307	311	313	317	331	337	347
70	349	353	359	367	373	379	383	389	397	401
80	409	419	421	431	433	439	443	449	457	461
90	463	467	479	487	491	499	503	509	521	523
100	541	547	557	563	569	571	577	587	593	599
110	601	607	613	617	619	631	641	643	647	653
120	659	661	673	677	683	691	701	709	719	727
130	733	739	743	751	757	761	769	773	787	797
140	809	811	821	823	827	829	839	853	857	859
150	863	877	881	883	887	907	911	919	929	937
160	941	947	953	967	971	977	983	991	997	1009
170	1013	1019	1021	1031	1033	1039	1049	1051	1061	1063
180	1069	1087	1091	1093	1097	1103	1109	1117	1123	1129
190	1151	1153	1163	1171	1181	1187	1193	1201	1213	1217
200	1223	1229	1231	1237	1249	1259	1277	1279	1283	1289
210	1291	1297	1301	1303	1307	1319	1321	1327	1361	1367
220	1373	1381	1399	1409	1423	1427	1429	1433	1439	1447
230	1451	1453	1459	1471	1481	1483	1487	1489	1493	1499
240	1511	1523	1531	1543	1549	1553	1559	1567	1571	1579
250	1583	1597	1601	1607	1609	1613	1619	1621	1627	1637
260	1657	1663	1667	1669	1693	1697	1699	1709	1721	1723
270	1733	1741	1747	1753	1759	1777	1783	1787	1789	1801
280	1811	1823	1831	1847	1861	1867	1871	1873	1877	1879
290	1889	1901	1907	1913	1931	1933	1949	1951	1973	1979

B.6 Bessel Functions

BesselJ[n, index] for $n = 0$ to $n = 11$ and index = 0 to index = 30, rounded to the nearest thousandth. Example: BesselJ[1, 4] = -.066 (→ plot)

index \ n	0	1	2	3	4	5	6	7	8	9	10	11
0	1.	0.	0.	0.	0.	0.	0.	0.	0.	0.	0.	0.
0.5	0.938	0.242	0.031	0.003	0.	0.	0.	0.	0.	0.	0.	0.
1	0.765	0.44	0.115	0.02	0.002	0.	0.	0.	0.	0.	0.	0.
1.5	0.512	0.558	0.232	0.061	0.012	0.002	0.	0.	0.	0.	0.	0.
2	0.224	0.577	0.353	0.129	0.034	0.007	0.001	0.	0.	0.	0.	0.
2.5	−0.048	0.497	0.446	0.217	0.074	0.02	0.004	0.001	0.	0.	0.	0.
3	−0.26	0.339	0.486	0.309	0.132	0.043	0.011	0.003	0.	0.	0.	0.
3.5	−0.38	0.137	0.459	0.387	0.204	0.08	0.025	0.007	0.002	0.	0.	0.
4	−0.397	−0.066	0.364	0.43	0.281	0.132	0.049	0.015	0.004	0.001	0.	0.
4.5	−0.321	−0.231	0.218	0.425	0.348	0.195	0.084	0.03	0.009	0.002	0.001	0.
5	−0.178	−0.328	0.047	0.365	0.391	0.261	0.131	0.053	0.018	0.006	0.001	0.
5.5	−0.007	−0.341	−0.117	0.256	0.397	0.321	0.187	0.087	0.034	0.011	0.003	0.001
6	0.151	−0.277	−0.243	0.115	0.358	0.362	0.246	0.13	0.057	0.021	0.007	0.002
6.5	0.26	−0.154	−0.307	−0.035	0.275	0.374	0.3	0.18	0.088	0.037	0.013	0.004
7	0.3	−0.005	−0.301	−0.168	0.158	0.348	0.339	0.234	0.128	0.059	0.024	0.008
7.5	0.266	0.135	−0.23	−0.258	0.024	0.283	0.354	0.283	0.174	0.089	0.039	0.015
8	0.172	0.235	−0.113	−0.291	−0.105	0.186	0.338	0.321	0.223	0.126	0.061	0.026
8.5	0.042	0.273	0.022	−0.263	−0.208	0.067	0.287	0.338	0.269	0.169	0.089	0.041
9	−0.09	0.245	0.145	−0.181	−0.265	−0.055	0.204	0.327	0.305	0.215	0.125	0.062
9.5	−0.194	0.161	0.228	−0.065	−0.269	−0.161	0.099	0.287	0.323	0.258	0.165	0.09
10	−0.246	0.043	0.255	0.058	−0.22	−0.234	−0.014	0.217	0.318	0.292	0.207	0.123
10.5	−0.237	−0.079	0.222	0.163	−0.128	−0.261	−0.12	0.124	0.285	0.311	0.248	0.161
11	−0.171	−0.177	0.139	0.227	−0.015	−0.238	−0.202	0.018	0.225	0.309	0.28	0.201
11.5	−0.068	−0.228	0.028	0.238	0.096	−0.171	−0.245	−0.085	0.142	0.282	0.3	0.239
12	0.048	−0.223	−0.085	0.195	0.182	−0.073	−0.244	−0.17	0.045	0.23	0.3	0.27
12.5	0.147	−0.165	−0.173	0.11	0.226	0.035	−0.198	−0.225	−0.054	0.156	0.279	0.29
13	0.207	−0.07	−0.218	0.003	0.219	0.132	−0.118	−0.241	−0.141	0.067	0.234	0.293
13.5	0.215	0.038	−0.209	−0.1	0.165	0.198	−0.018	−0.214	−0.204	−0.027	0.167	0.275
14	0.171	0.133	−0.152	−0.177	0.076	0.22	0.081	−0.151	−0.232	−0.114	0.085	0.236
14.5	0.088	0.193	−0.061	−0.21	−0.026	0.196	0.161	−0.062	−0.221	−0.182	−0.004	0.176
15	−0.014	0.205	0.042	−0.194	−0.119	0.13	0.206	0.034	−0.174	−0.22	−0.09	0.1
15.5	−0.109	0.167	0.131	−0.133	−0.182	0.039	0.208	0.122	−0.098	−0.223	−0.161	0.015
16	−0.175	0.09	0.186	−0.044	−0.203	−0.057	0.167	0.183	−0.007	−0.19	−0.206	−0.068
16.5	−0.196	−0.006	0.196	0.053	−0.176	−0.139	0.092	0.206	0.082	−0.126	−0.22	−0.14
17	−0.17	−0.098	0.158	0.135	−0.111	−0.187	0.001	0.188	0.154	−0.043	−0.199	−0.191
17.5	−0.103	−0.163	0.084	0.183	−0.022	−0.193	−0.088	0.132	0.194	0.045	−0.147	−0.214
18	−0.013	−0.188	−0.008	0.186	0.07	−0.155	−0.156	0.051	0.196	0.123	−0.073	−0.204
18.5	0.077	−0.167	−0.095	0.146	0.143	−0.084	−0.188	−0.038	0.16	0.176	0.011	−0.164
19	0.147	−0.106	−0.158	0.072	0.181	0.004	−0.179	−0.116	0.093	0.195	0.092	−0.098
19.5	0.179	−0.021	−0.181	−0.016	0.176	0.088	−0.131	−0.169	0.009	0.177	0.154	−0.019

20	0.167	0.067	−0.16	−0.099	0.131	0.151	−0.055	−0.184	−0.074	0.125	0.186	0.061
20.5	0.115	0.136	−0.102	−0.156	0.056	0.178	0.031	−0.16	−0.14	0.051	0.185	0.129
21	0.037	0.171	−0.02	−0.175	−0.03	0.164	0.108	−0.102	−0.176	−0.032	0.149	0.173
21.5	−0.049	0.164	0.064	−0.152	−0.107	0.112	0.159	−0.024	−0.174	−0.106	0.085	0.185
22	−0.121	0.117	0.131	−0.093	−0.157	0.036	0.173	0.058	−0.136	−0.157	0.008	0.164
22.5	−0.162	0.043	0.165	−0.014	−0.169	−0.046	0.149	0.125	−0.07	−0.176	−0.07	0.113
23	−0.162	−0.04	0.159	0.067	−0.141	−0.116	0.091	0.164	0.009	−0.158	−0.132	0.043
23.5	−0.124	−0.111	0.114	0.13	−0.081	−0.158	0.014	0.165	0.084	−0.108	−0.167	−0.034
24	−0.056	−0.154	0.043	0.161	−0.003	−0.162	−0.065	0.13	0.14	−0.036	−0.168	−0.103
24.5	0.024	−0.159	−0.037	0.153	0.074	−0.129	−0.127	0.067	0.165	0.041	−0.135	−0.151
25	0.096	−0.125	−0.106	0.108	0.132	−0.066	−0.159	−0.01	0.153	0.108	−0.075	−0.168
25.5	0.144	−0.062	−0.149	0.039	0.158	0.011	−0.154	−0.083	0.108	0.151	−0.001	−0.152
26	0.156	0.015	−0.155	−0.039	0.146	0.084	−0.114	−0.136	0.04	0.161	0.071	−0.106
26.5	0.13	0.087	−0.123	−0.106	0.099	0.136	−0.048	−0.157	−0.035	0.136	0.128	−0.04
27	0.073	0.137	−0.063	−0.146	0.03	0.155	0.027	−0.143	−0.101	0.083	0.156	0.033
27.5	−0.001	0.152	0.012	−0.15	−0.045	0.137	0.095	−0.096	−0.144	0.012	0.152	0.098
28	−0.073	0.131	0.082	−0.119	−0.108	0.088	0.139	−0.028	−0.153	−0.059	0.115	0.142
28.5	−0.126	0.078	0.132	−0.059	−0.144	0.019	0.151	0.045	−0.129	−0.117	0.055	0.156
29	−0.148	0.007	0.148	0.014	−0.146	−0.054	0.127	0.106	−0.076	−0.148	−0.016	0.137
29.5	−0.133	−0.064	0.129	0.082	−0.112	−0.112	0.074	0.142	−0.007	−0.146	−0.082	0.09
30	−0.086	−0.119	0.078	0.129	−0.053	−0.143	0.005	0.145	0.063	−0.112	−0.13	0.025

B.7 Chebyshev Polynomials

ChebyshevT$[n, x]$ for $n = 0$ to $n = 16$.

n	ChebyshevT$[n, x]$
0	1
1	x
2	$-1 + 2x^2$
3	$-3x + 4x^3$
4	$1 - 8x^2 + 8x^4$
5	$5x - 20x^3 + 16x^5$
6	$-1 + 18x^2 - 48x^4 + 32x^6$
7	$-7x + 56x^3 - 112x^5 + 64x^7$
8	$1 - 32x^2 + 160x^4 - 256x^6 + 128x^8$
9	$9x - 120x^3 + 432x^5 - 576x^7 + 256x^9$
10	$-1 + 50x^2 - 400x^4 + 1120x^6 - 1280x^8 + 512x^{10}$
11	$-11x + 220x^3 - 1232x^5 + 2816x^7 - 2816x^9 + 1024x^{11}$
12	$1 - 72x^2 + 840x^4 - 3584x^6 + 6912x^8 - 6144x^{10} + 2048x^{12}$
13	$13x - 364x^3 + 2912x^5 - 9984x^7 + 16640x^9 - 13312x^{11} + 4096x^{13}$
14	$-1 + 98x^2 - 1568x^4 + 9408x^6 - 26880x^8 + 39424x^{10} - 28672x^{12} + 8192x^{14}$
15	$-15x + 560x^3 - 6048x^5 + 28800x^7 - 70400x^9 + 92160x^{11} - 61440x^{13} + 16384x^{15}$
16	$1 - 128x^2 + 2688x^4 - 21504x^6 + 84480x^8 - 180224x^{10} + 212992x^{12} - 131072x^{14} + 32768x^{16}$

Bibliography

The books and articles listed here are all quoted or otherwise referenced in this book. The reader can find extensive bibliographies in *Computer Music* by Charles Dodge and Thomas A. Jerse [2] at the end of each chapter and in the *Computer Music Tutorial* by Curtis Roads [11].

Books (by Author)

[46] Acheson, David. *From Calculus to Chaos: An Introduction to Dynamics.*
Oxford: Oxford University Press, 1998.
[73] Balanov, Alexander, Natalia Janson, Dmitri Postnov and Olga Sosnovtseva.
Synchronization: From Simple to Complex. Berlin: Springer, 2009.
[42] Ballou, Glen M., ed. *Handbook for Sound Engineers.* Boston: Focal Press, 2008.
[97] Baumann, Dorothea. *Music and Space.* Bern: Peter Lang, 2011.
[80] Bilbao, Stefan D. *Wave and Scattering Methods for Numerical Simulation.*
Chichester: John Wiley & Sons, 2004.
[85] Bilbao, Stefan D. *Numerical Sound Synthesis: Finite Difference Schemes and Simulation in Musical Acoustics.* Chichester: John Wiley & Sons, 2009.
[33] Bischof, Norbert. *Struktur und Bedeutung.* Bern: Verlag Hans Huber, 1998.
[45] Blauert, Jens. *Spatial Hearing: The Psychophysics of Human Sound Localization.*
Cambridge, Massachusetts: MIT Press, 1996.
[31] Borucki, Hans. *Einführung in die Akustik.*
Mannheim, Vienna, Zurich: Bibliographisches Institut, 1989.
[40] Boulanger, Richard, ed. *The Csound Book.*
Cambridge, Massachusetts: MIT Press, 2000.
[86] Boulanger, Richard and Victor Lazzarini, eds. *The Audio Programming Book.*
Cambridge, Massachusetts: MIT Press, 2010.
[30] Boulez, Pierre. *On Music Today.*
Translated by Susan Bradshaw and Richard Rodney Bennett.
London: Faber and Faber, 1975.
[12] Breuer, Hans. *dtv-Atlas zur Physik.* Munich: Deutscher Taschenbuch Verlag, 1990.
[62] Breymann, Ulrich. *C++: Einführung und professionelle Programmierung.*
Munich: Hanser Verlag, 2007.
[9] Bronstein, Ilja N. and Konstantin A. Semendjajew.
Taschenbuch der Mathematik. Thun and Frankfurt am Main: Harri Deutsch, 1989.
[84] Chowning, John and David Bristow. *FM Theory & Applications: By Musicians for Musicians.* Tokyo: Yamaha Music Foundation, 1986.
[41] Cook, Perry R. *Music Cognition and Computerized Sound.*
Cambridge, Massachusetts: MIT Press, 1999.
[70] Cook, Perry R. *Real Sound Synthesis for Interactive Applications.*
Natick, Massachusetts: A K Peters/CRC Press, 2002.
[6] Crawford, Frank S. *Waves (Berkeley Physics Course, Vol. 3).*
New York: McGraw-Hill, 1968.
[92] Daniel, Jérôme. *Représentation de champs acoustiques, application à la reproduction de scènes sonores complexes dans un contexte multimédia.*
Thèse de doctorat de l'Université Paris 6, 2000.
[17] DeFatta, David J., Joseph G. Lucas and William S. Hodgkiss.
Digital Signal Processing. New York: John Wiley & Sons, 1988.

[93]   DiStefano, Joseph, Allan Stubberud and Ivan Williams.
       Schaum's Outline of Feedback and Control Systems. New York: McGraw-Hill, 2011.
[2]   Dodge, Charles and Thomas A. Jerse. *Computer Music: Synthesis, Composition and Performance*. New York: Schirmer Books, 1997.
[94]   Doyle, John C., Bruce A. Francis and Allen R. Tannenbaum.
       Feedback Control Theory. New York: Dover Publications, 2009.
[53]   Erne, Markus. *Signal Adaptive Audio Coding Using Wavelets and Rate Optimization*.
       Constance: Hartung-Gorre Verlag, 2000.
[83]   d'Escrivan, Julio and Nick Collins. *The Cambridge Companion to Electronic Music*.
       Cambridge: Cambridge University Press, 2008.
[39]   Feindt, Ernst G. *Computersimulation von Regelungen*.
       Munich and Vienna: Oldenbourg Verlag, 1999.
[38]   Föllinger, Otto. *Nichtlineare Regelungen*.
       Munich and Vienna: Oldenbourg Verlag,1998.
[95]   Frei, Beat. *Digital Sound Generation: Part 1. Oscillators*.
       Online book, http://www.icst.net.
[96]   Frei, Beat. *Digital Sound Generation: Part 2. Filters*.
       Online book, http://www.icst.net.
[49]   Geller, Doris. *Praktische Intonationslehre für Instrumentalisten und Sänger*.
       Kassel: Bärenreiter, 1999
[28]   Gnedenko, Boris W. *The Theory of Probability*.
       The Mathematical Association of America, 2005.
[16]   Golten, Jack. *Understanding Signals and Systems*. London: McGraw-Hill, 1997.
[24]   Haken, Hermann. *Erfolgsgeheimnisse der Natur: Synergetik: die Lehre vom Zusammenwirken*. Stuttgart: Deutsche Verlags-Anstalt, 1986.
[20]   Hall, Donald E. *Musical Acoustics*. Pacific Grove, California: Brooks Cole, 2001.
[91]   Hamming, Richard W. *Numerical Methods for Scientists and Engineers*.
       New York: Dover Publications, 1987.
[36]   Hayes, Monson H. *Schaum's Outline of Digital Signal Processing*.
       New York: McGraw Hill, 2011.
[89]   Jeffrey, Alan. *Mathematics for Engineers and Scientists*.
       Boca Raton: Chapman and Hall/CRC, 2004.
[63]   Jeske, Till. *Nitty-Gritty C++*. Boston: Addison-Wesley, 2002.
[4]   Jetschke, Gottfried. *Mathematik der Selbstorganisation*.
       Frankfurt am Main: Verlag Harri Deutsch, 2009.
[18]   Kammeyer, Karl-Dirk and Kristian Kroschel. *Digitale Signalverarbeitung*.
       Stuttgart: Teubner Studienbücher, 1998 (8th Rev. Edition, Vieweg & Teubner, 2012).
[48]   Karg, Eduard. *Regelungstechnik*. Würzburg: Vogel Buchverlag, 1992.
[35]   Kiencke, Uwe and Holger Jäkel. *Signale und Systeme*. Munich: Oldenbourg, 2010.
[13]   Kinzel, Wolfgang and Georg Reents. *Physics by Computer: Programming of Physical Problems Using Mathematica and C*. Berlin: Springer, 1997.
[81]   Knott, Gary D. *Interpolating Cubic Splines*. Boston: Birkhäuser, 2000.
[59]   Krieger, David J. *Einführung in die allgemeine Systemtheorie*.
       Munich: Wilhelm Fink Verlag, 1996.
[51]   Küppers, Bernd Olaf, ed. *Ordnung aus dem Chaos*. Munich and Zurich: Piper, 1987.
[69]   Lang, Christian B. and Norbert Pucker. *Mathematische Methoden in der Physik*.
       Heidelberg: Spektrum Akademischer Verlag, 2005.
[67]   Leigh, James R. *Applied Digital Control: Theory, Design and Implementation*.
       New York: Dover Publications, 2006.
[76]   Livio, Mario. *The Golden Ratio: The Story of Phi, the World's Most Astonishing Number*. New York: Broadway Books, 2002.

[52] Louis, Alfred K., Peter Maass and Andreas Rieder. *Wavelets: Theory and Applications*. New York: John Wiley & Sons, 1997.
[64] Loy, Gareth. *Musimathics: The Mathematical Foundations of Music. Vol. 1 and 2*. Cambridge, Massachusetts: MIT Press, 2006.
[37] Lutz, Holger and Wolfgang Wendt. *Taschenbuch der Regelungstechnik*. Thun and Frankfurt am Main: Verlag Harri Deutsch, 2002.
[29] Madden, Charles. *Fractals in Music: Introductory Mathematics for Musical Analysis*. Salt Lake City: High Art Press, 2007.
[75] Maeder, Marcus, ed. *Milieux Sonores/Klangliche Milieus: Klang, Raum und Virtualität*. Bielefeld: Transcript Verlag, 2010.
[55] Mallat, Stéphane. *A Wavelet Tour of Signal Processing*. San Diego: Academic Press, 2008.
[25] Mandelbrot, Benoît B. *The Fractal Geometry of Nature*. New York: W. H. Freeman, 1982.
[44] Mathews, Max V. and John R. Pierce, eds. *Current Directions in Computer Music Research*. Cambridge, Massachusetts: MIT Press, 1989.
[78] Mazzola, Guerino. *Geometrie der Töne: Elemente der mathematischen Musiktheorie*. Basel: Birkhäuser, 1990.
[79] Mazzola, Guerino, Stefan Göller and Stefan Müller. *The Topos of Music: Geometric Logic of Concepts, Theory, and Performance*. Basel: Birkhäuser, 2003.
[68] McMahon, David. *Signals and Systems Demystified*. New York: McGraw-Hill, 2006.
[21] Michels, Ulrich. *dtv-Atlas zur Musik*. Munich: Deutscher Taschenbuch Verlag, 1977.
[66] Miller, John H. and Scott E. Page. *Complex Adaptive Systems: An Introduction to Computational Models of Social Life*. Princeton: Princeton University Press, 2007.
[60] Miranda, Eduardo Reck. *Composing Music with Computers*. Oxford: Focal Press, 2001.
[15] Miranda, Eduardo Reck. *Computer Sound Synthesis for the Electronic Musician*. Oxford: Focal Press, 1998.
[8] Moore, F. Richard. *Elements of Computer Music*. Englewood Cliffs, N.J.: Prentice Hall, 1990.
[77] Muzzulini, Daniel. *Genealogie der Klangfarbe*. Bern: Peter Lang, 2006.
[74] Osipov, Grigory V., Jürgen Kurths and Changsong Zhou. *Synchronization in Oscillatory Networks*. Berlin: Springer, 2007.
[3] Papula, Lothar. *Mathematik für Ingenieure und Naturwissenschaftler Band 2*. Braunschweig/Wiesbaden: Vieweg + Teubner, 1997, 2011
[27] Papula, Lothar. *Mathematik für Ingenieure und Naturwissenschaftler Band 3*. Braunschweig/Wiesbaden: Vieweg + Teubner, 1997, 2011
[22] Peitgen, Heinz-Otto, Hartmut Jürgens and Dietmar Saupe. *Fractals for the Classroom: Part 1: Introduction to Fractals and Chaos*. New York: Springer, 1992.
[23] Peitgen, Heinz-Otto, Hartmut Jürgens and Dietmar Saupe. *Fractals for the Classroom: Part 2: Complex Systems and Mandelbrot Set*. New York: Springer, 1992.
[72] Pikovsky, Arkady, Michael Rosenblum and Jürgen Kurths. *Synchronization: A Universal Concept in Nonlinear Sciences*. Cambridge: Cambridge University Press, 2003.
[90] Polyanin, Andrei D. and Alexander V. Manzhirov. *Handbook of Mathematics for Engineers and Scientists*. Boca Raton: Chapman and Hall/CRC, 2006.
[65] Proakis, John G. and Dimitris G. Mamolakis. *Digital Signal Processing*. Upper Saddle River, New Jersey: Prentice Hall International, 2006.
[87] Puckette, Miller S. *The Theory and Techniques of Electronic Music*. Singapore: World Scientific Publishing Company, 2007.
[5] Purcell, Edward M. *Electricity and Magnetism (Berkeley Physics Course, Vol. 2)*.

New York: McGraw-Hill, 1984.
[19] Reinhardt, Fritz and Heinrich Soeder. *dtv-Atlas zur Mathematik*.
Munich: Deutscher Taschenbuch Verlag, 1974.
[26] Rinne, Horst, *Taschenbuch der Statistik*.
Thun and Frankfut am Main: Verlag Harri Deutsch, 2008.
[7] Roads, Curtis, ed. *The Music Machine: Selected Readings from The Computer Music Journal*. Cambridge, Massachusetts: MIT Press, 1989.
[11] Roads, Curtis. *The Computer Music Tutorial*.
Cambridge, Massachusetts: MIT Press, 1996.
[88] Roads, Curtis, Stephen Pope, Aldo Piccialli and Giovanni De Poli, eds.
Musical Signal Processing. Lisse: Swets & Zeitlinger, 1997.
[43] Roads, Curtis and John Strawn, eds. *Foundations of Computer Music*.
Cambridge, Massachusetts: MIT Press, 1985.
[71] Roads, Curtis. *Microsound*. Cambridge, Massachusetts: MIT Press, 2004.
[34] Saunders, Peter Timothy. *An Introduction to Catastrophe Theory*.
Cambridge: Cambridge University Press, 1980.
[1] Spiegel, Murray R. *Schaum's Outline of Advanced Mathematics for Engineers and Scientists (Schaum's Outline Series)*. New York: McGraw-Hill, 2009.
[14] Steeb, Willi-Hans and Albrecht Kunick. *Chaos in dynamischen Systemen*.
Mannheim etc. BI-Wissenschaftsverlag, 1989.
[32] Steiglitz, Ken. *A Digital Signal Processing Primer*.
Menlo Park, California: Addison-Wesley, 1996.
[57] Supper, Martin, ed. *Elektroakustische Musik und Computermusik*.
Darmstadt: Wissenschaftliche Buchgesellschaft, 1997.
[82] Trautmann, Lutz and Rudolf Rabenstein. *Digital Sound Synthesis by Physical Modeling Using the Functional Transformation Method*.
New York: Kluwer Academic/Plenum Publishers, 2003.
[54] Truax, Barry. *Handbook for Acoustic Ecology*.
Vancouver: Cambridge Street Publishing, 1999.
[56] Ungeheuer, Elena, ed. *Elektroakustische Musik*.
Handbuch der Musik im 20. Jahrhundert, Bd V. Laaber: Laaber-Verlag, 2002.
[47] Vongxaya, Bounthong. *Ordnung und Chaos bei nichtlinearen Schwingungen: Ein Computergestütztes Lehr- und Lernprogramm*.
Thun and Frankfurt am Main: Verlag Harri Deutsch, 1995.
[50] Walser, Hans. *The Golden Section*. The Mathematical Association of America, 2001.
[10] Wolfram, Stephen. *The Mathematica Book, Version 3*.
Cambridge: Cambridge University Press, 1996.
[58] Wolfram, Stephen. *A New Kind of Science*. Champaign Ill. Wolfram Media, 2002.
[61] Wolfram, Stephen. *The Mathematica Book, Version 4*.
Cambridge: Cambridge University Press, 1999.

Books (by Number of Reference)

[1] Spiegel, Murray R. *Schaum's Outline of Advanced Mathematics for Engineers and Scientists (Schaum's Outline Series)*. New York: McGraw-Hill, 2009.
[2] Dodge, Charles and Thomas A. Jerse. *Computer Music: Synthesis, Composition and Performance*. New York: Schirmer Books, 1997.
[3] Papula, Lothar. *Mathematik für Ingenieure und Naturwissenschaftler Band 2*.
Braunschweig/Wiesbaden: Vieweg + Teubner, 1997, 2011
[4] Jetschke, Gottfried. *Mathematik der Selbstorganisation*.
Frankfurt am Main: Verlag Harri Deutsch, 2009.

[5]   Purcell, Edward M. *Electricity and Magnetism (Berkeley Physics Course, Vol. 2)*.
      New York: McGraw-Hill, 1984.
[6]   Crawford, Frank S. *Waves (Berkeley Physics Course, Vol. 3)*.
      New York: McGraw-Hill, 1968.
[7]   Roads, Curtis, ed. *The Music Machine: Selected Readings from The Computer Music Journal*. Cambridge, Massachusetts: MIT Press, 1989.
[8]   Moore, F. Richard. *Elements of Computer Music*.
      Englewood Cliffs, N.J.: Prentice Hall, 1990.
[9]   Bronstein, Ilja N. and Konstantin A. Semendjajew.
      Taschenbuch der Mathematik. Thun and Frankfurt am Main: Harri Deutsch, 1989.
[10] Wolfram, Stephen. *The Mathematica Book, Version 3*.
      Cambridge: Cambridge University Press, 1996.
[11] Roads, Curtis. *The Computer Music Tutorial*.
      Cambridge, Massachusetts: MIT Press, 1996.
[12] Breuer, Hans. *dtv-Atlas zur Physik*. Munich: Deutscher Taschenbuch Verlag, 1990.
[13] Kinzel, Wolfgang and Georg Reents. *Physics by Computer: Programming of Physical Problems Using Mathematica and C*. Berlin: Springer, 1997.
[14] Steeb, Willi-Hans and Albrecht Kunick. *Chaos in dynamischen Systemen*.
      Mannheim etc. BI-Wissenschaftsverlag, 1989.
[15] Miranda, Eduardo Reck. *Computer Sound Synthesis for the Electronic Musician*.
      Oxford: Focal Press, 1998.
[16] Golten, Jack. *Understanding Signals and Systems*. London: McGraw-Hill, 1997.
[17] DeFatta, David J., Joseph G. Lucas and William S. Hodgkiss. *Digital Signal Processing*. New York: John Wiley & Sons, 1988.
[18] Kammeyer, Karl-Dirk and Kristian Kroschel. *Digitale Signalverarbeitung*.
      Stuttgart: Teubner Studienbücher, 1998 (8th Rev. Edition, Vieweg & Teubner, 2012).
[19] Reinhardt, Fritz and Heinrich Soeder. *dtv-Atlas zur Mathematik*. Munich:
      Deutscher Taschenbuch Verlag, 1974.
[20] Hall, Donald E. *Musical Acoustics*. Pacific Grove, California: Brooks Cole, 2001.
[21] Michels, Ulrich. *dtv-Atlas zur Musik*. Munich: Deutscher Taschenbuch Verlag, 1977.
[22] Peitgen, Heinz-Otto, Hartmut Jürgens and Dietmar Saupe. *Fractals for the Classroom: Part 1: Introduction to Fractals and Chaos*. New York: Springer, 1992.
[23] Peitgen, Heinz-Otto, Hartmut Jürgens and Dietmar Saupe. *Fractals for the Classroom: Part 2: Complex Systems and Mandelbrot Set*.
      New York: Springer, 1992.
[24] Haken, Hermann. *Erfolgsgeheimnisse der Natur: Synergetik: die Lehre vom Zusammenwirken*. Stuttgart: Deutsche Verlags-Anstalt, 1986.
[25] Mandelbrot, Benoît B. *The Fractal Geometry of Nature*.
      New York: W. H. Freeman, 1982.
[26] Rinne, Horst, *Taschenbuch der Statistik*.
      Thun and Frankfut am Main: Harri Deutsch, 2008.
[27] Papula, Lothar. *Mathematik für Ingenieure und Naturwissenschaftler Band 3*.
      Braunschweig/Wiesbaden: Vieweg + Teubner, 1997, 2011
[28] Gnedenko, Boris W. *The Theory of Probability*.
      The Mathematical Association of America, 2005.
[29] Madden, Charles. *Fractals in Music: Introductory Mathematics for Musical Analysis*.
      Salt Lake City: High Art Press, 2007.
[30] Boulez, Pierre. *On Music Today*. Translated by Susan Bradshaw and Richard Rodney Bennett. London: Faber and Faber, 1975.
[31] Borucki, Hans. *Einführung in die Akustik*.
      Mannheim, Vienna, Zurich: Bibliographisches Institut, 1989.

[32] Steiglitz, Ken. *A Digital Signal Processing Primer*.
      Menlo Park, California: Addison-Wesley, 1996.
[33] Bischof, Norbert. *Struktur und Bedeutung*. Bern: Verlag Hans Huber, 1998.
[34] Saunders, Peter Timothy. *An Introduction to Catastrophe Theory*.
      Cambridge: Cambridge University Press, 1980.
[35] Kiencke, Uwe and Holger Jäkel. *Signale und Systeme*. Munich: Oldenbourg, 2010.
[36] Hayes, Monson H. *Schaum's Outline of Digital Signal Processing*.
      New York: McGraw Hill, 2011.
[37] Lutz, Holger and Wolfgang Wendt. *Taschenbuch der Regelungstechnik*.
      Thun and Frankfurt am Main: Verlag Harri Deutsch, 2002.
[38] Föllinger, Otto. *Nichtlineare Regelungen*.
      Munich and Vienna: Oldenbourg Verlag,1998.
[39] Feindt, Ernst G. *Computersimulation von Regelungen*.
      Munich and Vienna: Oldenbourg Verlag, 1999.
[40] Boulanger, Richard, ed. *The Csound Book*.
      Cambridge, Massachusetts: MIT Press, 2000.
[41] Cook, Perry R. *Music Cognition and Computerized Sound*.
      Cambridge, Massachusetts: MIT Press, 1999.
[42] Ballou, Glen M., ed. *Handbook for Sound Engineers*. Boston: Focal Press, 2008.
[43] Roads, Curtis and John Strawn, eds. *Foundations of Computer Music*.
      Cambridge, Massachusetts: MIT Press, 1985.
[44] Mathews, Max V. and John R. Pierce, eds. *Current Directions in Computer Music Research*. Cambridge, Massachusetts: MIT Press, 1989.
[45] Blauert, Jens. *Spatial Hearing: The Psychophysics of Human Sound Localization*.
      Cambridge, Massachusetts: MIT Press, 1996.
[46] Acheson, David. *From Calculus to Chaos: An Introduction to Dynamics*.
      Oxford: Oxford University Press, 1998.
[47] Vongxaya, Bounthong. *Ordnung und Chaos bei nichtlinearen Schwingungen: Ein Computergestütztes Lehr- und Lernprogramm*.
      Thun and Frankfurt am Main: Verlag Harri Deutsch, 1995.
[48] Karg, Eduard. *Regelungstechnik*. Würzburg: Vogel Buchverlag, 1992.
[49] Geller, Doris. *Praktische Intonationslehre für Instrumentalisten und Sänger*.
      Kassel: Bärenreiter, 1999
[50] Walser, Hans. *The Golden Section*. The Mathematical Association of America, 2001.
[51] Küppers, Bernd Olaf, ed. *Ordnung aus dem Chaos*. Munich and Zurich: Piper, 1987.
[52] Louis, Alfred K., Peter Maass and Andreas Rieder.
      Wavelets: Theory and Applications. New York: John Wiley & Sons, 1997.
[53] Erne, Markus. *Signal Adaptive Audio Coding Using Wavelets and Rate Optimization*.
      Constance: Hartung-Gorre Verlag, 2000.
[54] Truax, Barry. *Handbook for Acoustic Ecology*.
      Vancouver: Cambridge Street Publishing, 1999.
[55] Mallat, Stéphane. *A Wavelet Tour of Signal Processing*.
      San Diego: Academic Press, 2008.
[56] Ungeheuer, Elena, ed. *Elektroakustische Musik*.
      Handbuch der Musik im 20. Jahrhundert, Bd V. Laaber: Laaber-Verlag, 2002.
[57] Supper, Martin, ed. *Elektroakustische Musik und Computermusik*.
      Darmstadt: Wissenschaftliche Buchgesellschaft, 1997.
[58] Wolfram, Stephen. *A New Kind of Science*. Champaign Ill. Wolfram Media, 2002.
[59] Krieger, David J. *Einführung in die allgemeine Systemtheorie*.
      Munich: Wilhelm Fink Verlag, 1996.
[60] Miranda, Eduardo Reck. *Composing Music with Computers*.
      Oxford: Focal Press, 2001.

[61] Wolfram, Stephen. *The Mathematica Book, Version 4.*
Cambridge: Cambridge University Press, 1999.
[62] Breymann, Ulrich. *C++: Einführung und professionelle Programmierung.*
Munich: Hanser Verlag, 2007.
[63] Jeske, Till. *Nitty-Gritty C++.* Boston: Addison-Wesley, 2002.
[64] Loy, Gareth. *Musimathics: The Mathematical Foundations of Music. Vol. 1 and 2.*
Cambridge, Massachusetts: MIT Press, 2006.
[65] Proakis, John G. and Dimitris G. Mamolakis. *Digital Signal Processing.*
Upper Saddle River, New Jersey: Prentice Hall International, 2006.
[66] Miller, John H. and Scott E. Page. *Complex Adaptive Systems: An Introduction to Computational Models of Social Life.* Princeton: Princeton University Press, 2007.
[67] Leigh, James R. *Applied Digital Control: Theory, Design and Implementation.*
New York: Dover Publications, 2006.
[68] McMahon, David. *Signals and Systems Demystified.* New York: McGraw-Hill, 2006.
[69] Lang, Christian B. and Norbert Pucker. *Mathematische Methoden in der Physik.*
Heidelberg: Spektrum Akademischer Verlag, 2005.
[70] Cook, Perry R. *Real Sound Synthesis for Interactive Applications.*
Natick, Massachusetts: A K Peters/CRC Press, 2002.
[71] Roads, Curtis. *Microsound.* Cambridge, Massachusetts: MIT Press, 2004.
[72] Pikovsky, Arkady, Michael Rosenblum and Jürgen Kurths. *Synchronization: A Universal Concept in Nonlinear Sciences.*
Cambridge: Cambridge University Press, 2003.
[73] Balanov, Alexander, Natalia Janson, Dmitri Postnov and Olga Sosnovtseva.
Synchronization: From Simple to Complex. Berlin: Springer, 2009.
[74] Osipov, Grigory V., Jürgen Kurths and Changsong Zhou.
Synchronization in Oscillatory Networks. Berlin: Springer, 2007.
[75] Maeder, Marcus, ed. *Milieux Sonores/Klangliche Milieus: Klang, Raum und Virtualität.* Bielefeld: Transcript Verlag, 2010.
[76] Livio, Mario. *The Golden Ratio: The Story of Phi, the World's Most Astonishing Number.* New York: Broadway Books, 2002.
[77] Muzzulini, Daniel. *Genealogie der Klangfarbe.* Bern: Peter Lang, 2006.
[78] Mazzola, Guerino. *Geometrie der Töne: Elemente der mathematischen Musiktheorie.*
Basel: Birkhäuser, 1990.
[79] Mazzola, Guerino, Stefan Göller and Stefan Müller. *The Topos of Music: Geometric Logic of Concepts, Theory, and Performance.* Basel: Birkhäuser, 2003.
[80] Bilbao, Stefan D. *Wave and Scattering Methods for Numerical Simulation.*
Chichester: John Wiley & Sons, 2004.
[81] Knott, Gary D. *Interpolating Cubic Splines.* Boston: Birkhäuser, 2000.
[82] Trautmann, Lutz and Rudolf Rabenstein. *Digital Sound Synthesis by Physical Modeling Using the Functional Transformation Method.* New York:
Kluwer Academic/Plenum Publishers, 2003.
[83] d'Escrivan, Julio and Nick Collins. *The Cambridge Companion to Electronic Music.*
Cambridge: Cambridge University Press, 2008.
[84] Chowning, John and David Bristow. *FM Theory & Applications: By Musicians for Musicians.* Tokyo: Yamaha Music Foundation, 1986.
[85] Bilbao, Stefan D. *Numerical Sound Synthesis: Finite Difference Schemes and Simulation in Musical Acoustics.* Chichester: John Wiley & Sons, 2009.
[86] Boulanger, Richard and Victor Lazzarini, eds. *The Audio Programming Book.*
Cambridge, Massachusetts: MIT Press, 2010.
[87] Puckette, Miller S. *The Theory and Techniques of Electronic Music.*
Singapore: World Scientific Publishing Company, 2007.

[88] Roads, Curtis, Stephen Pope, Aldo Piccialli and Giovanni De Poli, eds. *Musical Signal Processing*. Lisse: Swets & Zeitlinger, 1997.
[89] Jeffrey, Alan. *Mathematics for Engineers and Scientists*. Boca Raton: Chapman and Hall/CRC, 2004.
[90] Polyanin, Andrei D. and Alexander V. Manzhirov. *Handbook of Mathematics for Engineers and Scientists*. Boca Raton: Chapman and Hall/CRC, 2006.
[91] Hamming, Richard W. *Numerical Methods for Scientists and Engineers*. New York: Dover Publications, 1987.
[92] Daniel, Jérôme. *Représentation de champs acoustiques, application à la reproduction de scènes sonores complexes dans un contexte multimédia*. Thèse de doctorat de l'Université Paris 6, 2000.
[93] DiStefano, Joseph, Allan Stubberud and Ivan Williams. *Schaum's Outline of Feedback and Control Systems*. New York: McGraw-Hill, 2011.
[94] Doyle, John C., Bruce A. Francis and Allen R. Tannenbaum. *Feedback Control Theory*. New York: Dover Publications, 2009.
[95] Frei, Beat. *Digital Sound Generation: Part 1. Oscillators*. Online book, http://www.icst.net.
[96] Frei, Beat. *Digital Sound Generation: Part 2. Filters*. Online book, http://www.icst.net.
[97] Baumann, Dorothea. *Music and Space*. Bern: Peter Lang, 2011.

Articles from the Computer Music Journal (CMJ)

[1-1] Cadoz, Claude, Annie Luciani and Jean-Loup Florens. "Responsive Input Devices and Sound Synthesis by Simulation of Instrumental Mechanisms: The Cordis System." CMJ Vol. 8, No. 3, 1984.
[1-2] De Poli, Giovanni. "A Tutorial on Digital Sound Synthesis Techniques." CMJ Vol. 7, No. 4, 1983.
[1-3] Karplus, Kevin and Alex Strong. "Digital Synthesis of Plucked-String and Drum Timbres." CMJ Vol. 7, No. 2, 1983.
[1-4] Jaffe, David A. and Julius O. Smith. "Extensions of the Karplus-Strong Plucked-String Algorithm." CMJ Vol. 7, No. 2, 1983.
[1-5] Smith, Julius O. "Physical Modeling Synthesis Update." CMJ Vol. 20, No. 2, 1996, also available at: http://ccrma.stanford.edu/~jos/pmupd/.
[1-6] Borin, Gianpaolo, Giovanni De Poli and Augusto Sarti. "Algorithms and Structures for Synthesis Using Physical Models." CMJ Vol. 16, No. 4, 1992.
[1-7] Smith, Julius O. "Physical Modeling Using Digital Waveguides". CMJ Vol. 16, No. 4, 1992, also available at: http://ccrma.stanford.edu//~jos/pmudw/.
[1-8] Woodhouse, James. "Physical Modeling of Bowed Strings." *CMJ* Vol. 16, No. 4, 1992.
[1-9] Djoharian, Pirouz. "Generating Models for Modal Synthesis." *CMJ* Vol. 17, No. 1, 1993.
[1-10] Keefe, Douglas H. "Physical Modeling of Wind Instruments." *CMJ* Vol. 16, No. 4, 1992.
[1-11] Bate, John A. "The Effect of Modulator Phase on Timbres in FM Synthesis." *CMJ* Vol. 14, No. 3, 1990.
[1-12] Kendall, Gary S. "The Decorrelation of Audio Signals and Its Impact on Spatial Imagery." *CMJ* Vol. 19, No. 4, 1995.
[1-13] Kendall, Gary S. "A 3-D Sound Primer: Directional Hearing and Stereo Reproduction." *CMJ* Vol. 19, No. 4, 1995.

[1-14] Malham, David G. and Anthony Myatt. "3-D Sound Spatialization Using Ambisonic Techniques." *CMJ* Vol. 19, No. 4, 1995.

[1-15] Chareyron, Jacques. "Digital Synthesis of Self-Modifying Waveforms by Means of Linear Automata." *CMJ* Vol. 14, No. 4, 1990.

Articles from Other Sources

[2-1] Kramer, Jonathan D. "The Fibonacci Series in Twentieth-Century Music." JMT Vol. 17, No. 1, 1973: pp. 110-148.

[3-1] Bennett, Gerald. "Chaos, Self-Similarity, Musical Phrase and Form." http://www.gdbennett.net/texts/publications.html

[3-2] Gerzon, Michael A. "Periphony: With-Height Sound Reproduction." Journal of the Audio Engineering Society, Vol. 21, No. 1, 1973.

[3-3] Hammer, Øyvind and Henrik Sundt. "Musical Applications of Decomposition with Global Support." Proceedings of the 2nd COST G-6 Workshop on Digital Audio Effects (DAFx99) held at the Norwegian University and Technology NTNU in Trondheim, December 9-11, 1999.

[3-4] Tomisawa, Norio. "Tone Production Method for an Electronic Musical Instrument." US Patent 4,249,447, 1981. http://www.patents.com/us-4249447.html.

[3-5] Neukom, Martin. "Applications of Synchronization in Sound Synthesis." Paper presented at the Sound and Music Computing Conference SMC 6-9 July 2011 in Padova. http://smcnetwork.org/resources/smc2011

[3-6] Neukom, Martin. "Ambisonic Panning." Paper presented at the Audio Engineering Society Convention 123, New York, NY, USA, 2007.

[3-7] Neukom, Martin and Jan C. Schacher. "Ambisonics Equivalent Panning." Proceedings of the 2007 International Computer Music Conference, Copenhagen. The International Computer Music Association, 2007.

[3-8] Neukom, Martin and Jan C. Schacher. "Where's the Beat? Tools For Dynamic Tempo Calculations." Proceedings of the 2007 International Computer Music Conference, Copenhagen. The International Computer Music Association, 2007.

[3-9] Neukom, Martin. "Decoding Second Order Ambisonics to 5.1 Surround." Paper presented at the Audio Engineering Society Convention 123, San Francisco, CA, USA, 2006.

[3-10] Bisig, Daniel and Martin Neukom. "Swarm Based Computer Music: Towards a Repertory of Strategies." Proceedings of the 11th Generative Art Conference. Milano, 2008. Abstract and paper available at: http://www.generativeart.com/

[3-11] Bisig, Daniel, Martin Neukom and Jan C. Schacher. "Composing With Swarm Algorithms: Creating Interactive Audio-Visual Pieces Using Flocking Behaviour." Proceedings of the 2011 International Computer Music Conference, Huddersfield. The International Computer Music Association, 2011.

[3-12] Kuhn-Rahloff, Clemens, Martin Neukom and Mathias S. Oechslin. "The Doppler Effect in Warning Signals: Perceptual Investigations into Aversive Reactions Related to Pitch Changes." Paper presented at Forum Acusticum 2011, Aalborg, 2011.

[3-13] Oechslin, Mathias S., Martin Neukom and Gerald Bennett. "The Doppler Effect: An Evolutionary Critical Cue for the Perception of the Direction of Moving Sound Sources." Paper presented at the International Conference on Audio, Language and Image Processing ICALIP 2008, Shanghai.

[3-14] Bisig, Daniel, Martin Neukom and John Flury. "Interactive Swarm Orchestra: A Generic Programming Environment for Swarm Based Computer Music". Proceedings of the 2008 International Computer Music Conference, Belfast. The International Computer Music Association, 2008.

[3-15] Grey, John M. and John W. Gordon. "Perceptual Effects of Spectral Modifications on Musical Timbres." Journal of the Acoustical Society of America, Vol. 63, Issue 5, 1978, pp. 1493-1500.

Articles Available on the Internet

[I-2] First and Second Order Ambisonic Decoding Equations by Richard W. E. Furse.
http://www.muse.demon.co.uk/ref/speakers.html
[I-3] Schweizerisches Zentrum für Computermusik SZCM.
http://www.computermusic.ch/
[I-5] A large archive of software from NOTAM in Oslo:
http://www.notam02.no/web/category/research-and-development/software/?lang=en
[I-6] Institute for Computer Music and Sound Technology ICST. http://www.icst.net/
[I-7] Swarm based music. http://swarms.cc/

Index

absolute value 3.1.2.1, 3.3.17
absorption 2.2.4.4, 9.1.2.1, 9.2.1.2
   - coefficient 9.1.2.1, 9.2.1.2, 9.2.2.1
acceleration 2.1.1, 2.1.2.3, 8.1.1.1, 8.1.1.1
   10.2.3.2
   - nonlinear 3.4.4.4, 8.1.1.5, 8.1.2.3, 8.1.3.4
ADC (analog-to-digital converter) → converter
address 4.4.1.3
agent 10.3.4
aire column 2.2.1.5
algorithm 10.3.2
aliasing 3.2.1.5
alignment 10.3.4.1
Ambisonics 9.3.1.2, 9.3.2.2
   - decoding 9.3.2.2
   - encoding 9.3.2.2
   - equivalent panning AEP 9.3.2.2
   - order 9.3.2.2
amplitude 2.2.1.1, 2.3.3.1, 8.1.1.1
   - complex 3.1.2.5
amplitude response 3.3.1.4, 3.3.1.5, 5.3.1.1
   - ripple 3.3.2.4, 3.3.3.6
amplitude modulation 2.2.2.3, 5.1.1.2, 6.1.1, 6.2.2.1
analysis 5.2.2
animation 4.3.3, 4.5.2
antiderivative A.5.5
antinode 2.2.1.5
approximation 3.4.1.4, 5.1.3.2, 6.2.3.2
   8.1.2.5, 10.3.2.3, A.3.5
argument A4.1
   - of a complex number 3.1.2.2
array 4.4.1.3
assignment 4.1.1.4
associative property A.1.2
attractor 3.4.2.3, 6.2.3.1, 10.3.3.3
   - chaotic 3.4.3.3
   - periodic 3.4.5.1
   - Rösseler 3.4.3.4
attribute 4.4.1.8
audio signal → signal
autocovariance 10.2.1.3
automata
   - cellular 10.3.1
average
   - moving 3.3.3.5, 5.1.3.4
axioms
   - of Probability Calculus 10.1.2.4

B-format 9.3.2.2,
band-pass → filter
band-stop → filter
bandwidth 2.3.4.1, 3.3.1.9, 5.3.1.1, 5.3.2.2, 8.1.1.4
bar 2.2.1.5, 3.4.2.1, 8.1.4
beat 5.1.3.5
beats 2.2.1.4, 2.3.2.6, 3.1.2.5, 6.1.1.1, 9.1.1.1
bell 2.3.2.4
Bessel function 6.1.2.3, B.6

bifurcation 3.4.2.1, 3.4.3.3, 3.4.5.4
   - diagram 3.4.2.1, 3.4.2.2
   - set 3.4.2.2, 3.4.2.2
bin 5.2.2.2
binomial 10.1.1.6
   - coefficients 10.1.1.6
bit 4.4.1.2
Boids 10.3.4.1
boundary value problem 3.4.1.2, 3.4.1.2
buffer
   - circular 3.3.2.2, 5.1.4.2, 9.2.2.2
   - ring 5.1.4.2
butterfly effect 10.3.3.2
buzz 5.3.2.2
byte 4.4.1.2

C/C++ 1.2.3, 4.1.4.2, 4.4
calculation
   - numerical 3.4.1.4, 4.3.1, 7.2.3.3
calculus A.5
canon 5.1.3.5
capacitance 3.1.4.1
capacitor 3.1.4.1
carrier frequency 6.1.1.2
carrier wave 6.1.1.2, 6.1.2.2
Cartesian product A.2.5
cast operator 4.4.1.4
catastrophe theorie 3.4.2.2
causal 3.3.1.2
cent 2.3.1.3, 9.1.3.1
center frequency 2.3.4.1, 3.3.1.9, 3.4.4.6, 5.2.2.3, 5.3.1.1, 5.3.2.2, 8.1.1.4
chain of pearls 4.5.2, 8.1.3.2
channel 7.2.2
   - band-pass 5.2.2.1
chaos 3.4.3, 3.4.5.5, 6.2.3.1, 10.3.3
   - deterministic 10.3.3.1
   - in differential equation 3.4.3.4
characteristic curve 2.3.3.3, 3.4.1.3, 3.4.2.1, 3.4.4.4
   - logarithmic 3.4.1.3
   - cubic 3.4.2.3
   - distortion 6.1.3.2
Chebyshev polynomials 6.1.3.5, B.7
Chladni figures 8.1.4.3
chorus effect 5.1.4.5
Chowning, John 6.1.2.6
circle of fifths 10.3.5.2
class 4.4.1.8
CLM (Common Lisp Music) 5.1.1.3
coefficient 7.2.1, A.3.3
cohesion 10.3.4.1
combination 10.1.1, 10.1.1.3
combination tone 2.3.4.3
combinatorics 10.1.1
comma 2.3.1.3
   - Pythagorean 2.3.1.3
   - syntonic 2.3.1.3
comment 4.4.1.7, 4.4.3.1
commutative property A1.2

compiler 4.4.1.1
complex A.4.3.3
complexity 10.3.3.1
compressor 6.1.3.6
connection diagram 3.4.4.2, 5.1.1.1, 5.1.1.3
constant 4.1.1.2, 4.4.1.2, B.3
Constant Q Filter Bank 7.2.2
constructor 4.4.1.8, 4.4.3.9
continued fraction 10.3.2.3
control circuits 3.4.4.1, 3.4.4, 3.4.4.3, 3.4.4.6, 5.1.4.4, 6.2.4.1, 6.2.4.4
  - nonlinear 3.4.4.4
control function 3.4.4, 5.1.1.5, 5.1.4.4, 5.2.1.2
control period 4.1.1.1
control rate 4.1.1.1, 4.1.1.10
control signal → signal
control flow 4.1.1.5, 4.4.1.5, 5.1.1.7
controlled variable 3.4.4.1
controller 3.4.4.1, 3.4.4.7
  - adaptive 6.2.4.4
  - differential element 3.4.4.2
  - integral 3.4.4.2, 3.4.4.4, 3.4.4.6
  - nonlinear 3.4.4.7
  - PI-element 3.4.4.2
  - proportional element 3.4.4.2
      - with delay 3.4.4.2, 3.4.4.3
  - reaction time systems 3.4.2.2, 3.4.4.3
convergence 3.4.3.3, A.4.8
converter 3.2.1, 3.2.1.5
convolution 3.3.1.3, 5.1.3.3, 7.1.1.1, 9.2.2.5, 9.3.2.3
  - discrete 3.3.1.3
  - Fourier transform 3.2.2.4
  - fast 3.3.1.3
  - Z-transform 3.2.3.2
Conway, J. H. 10.3.1.3
coordinate system 2.1.1.1, 6.2.1.2
coordinates 2.1.1.1, 9.3.2.2, A.2.5
  - Cartesian 2.1.1.1, 5.2.2.2, 9.3.2.2
  - polar 3.1.2.2, 5.2.2.2, 9.3.2.2
  - spherical 9.3.2.2
correcting variable 3.4.4.1, 3.4.4.6
correlation 9.3.2.3, 10.2.1.3
cosine A.4.3.7
coupling 3.4.5.4, 8.1.6.1, 8.1.2.1
covariance 10.2.1.3
crosstalk 9.3.2.3
Csound 1.2.2, 4.1, 5.1.1.1, 5.1.1.3
curvature 5.1.3.2, A5.2
cusp 3.4.2.2
cutoff frequency 3.3.1.9, 5.3.1.1

DAC (digital-to-analog converter)
    → converter
damping 2.2.1.2, 3.1.4.1, 3.3.3.5, 3.4.1.5, 8.1.1.3, 8.1.1.4, 8.1.2.2, 8.1.3.4, 8.1.5.7, 8.2.1.2, 10.3.2.6
data reduction 5.2.2.1
data type 4.4.1.2
    - combined 4.4.1.3
    - derived 4.4.1.3

decay 2.3.2.3
Decibels 2.3.3.2, 4.1.1.1, 4.1.1.3
declaration 4.4.1.2
decorrelation 9.3.2.3
decrement
  - logarithmic 2.2.1.2
degree of freedom 8.1.5.1
delay 3.2.1.3, 3.2.1.6, 3.3.1.1, 3.3.2.6, 4.1.1.8, 6.2.4.4, 8.2.1.2, 9.2.2.2
delay line 3.3.2.3, 4.1.1.8, 5.1.4, 8.2.1, 8.2.1.1, 8.2.2.1, 9.1.3.1
  - with feedback 5.1.4.4
  - variable 5.1.4.3, 6.2.4.3, 6.2.4.5
density 2.1.2.1, 2.2.1.5, 8.1.4.10
  - diagram 7.2.3.2
density function 10.1.3.2
derivative 3.4.2.1, 3.4.2.2, 3.4.4.2, 5.1.3.2, A.5.1
  - partial A.5.6
destructor 4.4.1.8
determinant 8.1.4.6
DFT → Fourier transform
difference equations 3.4.1.4, 3.4.4.2, 6.2.3.2, 8.1.2.5
difference tone 2.3.4.3, 2.3.2.6
differential equation 3.3.3.5, 3.3.3.5, 3.3.3.6, 3.4.1, 3.4.4.2, 6.2.3.2, 8.1.1.1
  - autonomous 3.4.1.2
  - homogeneous 3.4.1.2, 3.4.1.4
  - inhomogeneous 3.4.1.2, 3.4.1.4, 3.4.3.2
  - linear 3.1.4.2, 3.4.1.2
  - nonlinear 3.4.1.5
  - order 3.4.1.2
  - ordinary 3.4.1.2
  - system of 3.4.1.2, 3.4.3.4, 3.4.5.2, 8.1.2.4, 8.1.3.1
diffraction 2.2.4.4, 2.3.5.2
dimension 10.3.3.4, 10.3.5, 10.3.5.1, 10.3.5.2, 10.3.5.4
  - array 4.4.1.3
  - fractal 10.3.3.3
diphthong 7.1.2.2
direct current (DC) 3.1.1.2, 6.1.3.5, 6.2.4.5, 7.1.2.2, 7.2.3.3, 10.2.3.3
direction field 3.4.1.2, 3.4.4.4, 6.2.3.2
directional characteristics 3.4.1.3, 9.3.2.1
discontinuity 10.3.3.2, 10.3.3.4
dispersion 2.2.4.4
dispersion relation 8.1.3.2, 8.1.3.3
displacement 2.2.1.1, 8.1.1.1
  - function A.4.3.5
  - maximal 3.4.4.7
distance 2.1.1.1, 2.3.5.3, 9.1.1.1, 9.2.2.1, 10.3.4.1, 10.3.5.2
  - of a sound source 2.3.5.3, 9.1.1.2, 9.1.2
distortion 5.1.1.2, 6.1, 6.1.3, 6.2.1.1
distribution
  - binomial 10.1.5.3
  - exponential 10.1.5.6
  - Gamma 10.1.5.7
  - uniform 10.1.5.1
  - normal 10.1.5.5
  - Poisson 10.1.5.4
  - standard normal 10.1.5.5
  - trapezoid 10.1.5.2

Index 589

- Weibull 10.1.5.8
distribution function 10.1.3.3, 10.1.4.2, 10.1.5
 - inversion 10.1.4.3
distributivity property A.1.2,
disturbance variable 3.4.4.1, 3.4.4.6, 6.2.4.1
domain of definition A.3.1, A.4.1
Doppler effect 2.2.4.5, 5.1.4.3, 9.1.1.1, 9.1.3.1, 9.3.1.2
downsampling 3.2.1.3, 7.2.3.3
drum 8.2.1.6
DTF (directionally dependent transfer function) 9.1.1.3
duration 5.1.3.5, 10.1.6.2
dynamic range 2.3.3.2, 3.2.1.4

ear 9.1.1.3, 9.1.3.2
edge 8.1.4.5
eigenfrequency 3.3.3.5, 3.4.3.5, 3.4.5.1, 8.1.4.3, 8.2.3.4
eigenvector A.3.4, 10.2.2.4
eigenvalue A.3.4, 8.1.3.2, 10.2.2.4
elasticity 8.1.1.4, 8.1.4.10
element
 - of a set A.2.2
elementary event 10.1.2.1
elementary wave 2.2.4.4
elongation → displacement
emergence 10.3.4
end
 - closed 2.2.1.5, 8.2.2.2
 - open 2.2.1.5, 8.2.2.2
energy 2.1.3, 2.3.3.1, 3.4.1.5, 3.4.2.1, 3.4.2.3, 7.1.1.3, 9.1.2.1
 - damped oscillation 2.2.1.2
 - kinetic 2.2.1.1, 3.4.1.5
 - potential 2.2.1.1, 3.4.1.5, 3.4.2.2
envelope 2.2.1.2, 2.2.1.4, 4.1.1.7, 7.1.1.1, 8.1.1.3
equalizer 5.2.2.4, 5.3.2
equation 4.3.1, A.3
 - algebraic A.3.3
 - degree A.3.3
 - equivalent A.3.2
 - exponential A.3.5
 - logarithmic A.3.5
 - trigonometric A.3.5
 - quadratic 10.3.2.1
 - system 8.1.2.4, A.3.3
 - transcendental A.3.5
 - linear system A.3.4
Euler's formula 3.1.2.3, 8.1.2.4
Euler's method 3.4.1.4, 3.4.1.5, 3.4.3.2
 - improved 3.4.1.4
Euler's number A.4.3.5, B.3
event
 concatenation 10.1.2.2
event space 10.1.2.2
evasion 10.3.4.1
excitation 3.1.4.1, 3.3.1.1, 3.4.2.3, 8.1.1.2, 8.1.1.4, 8.1.2.2, 8.1.3.6, 8.2.1.5, 8.2.2.3, 8.2.3, 8.2.3.3
 - periodic 3.4.3.1, 3.4.3.5, 3.4.5.3, 8.1.1.4
 - random 3.4.4.7, 8.1.1.2

exit condition 4.4.1.5
expander 6.1.3.6
expected value 10.1.3.6
exponential 2.3.1.1, 4.1.4.5
exponential sequence 3.2.1.2, 3.2.3.1
expression 4.1.1.3, 4.1.1.4, 4.3.1, 4.4.1.4, 4.4.1.5
external 4, 4.2.3

faculty 10.1.1.2
feedback 3.3.2, 3.4.4.4, 4.1.1.10, 4.2.2, 5.1.4.4, 6.1.2.7, 6.2.3, 6.2.4.5, 8.2.3.4, 9.2.2.2, 10.3.1, 10.3.3.2,
Feigenbaum constant 3.4.3.3
Fibonacci numbers 10.3.2, 10.3.2.2, B.4
file 4.3.1
 - binary 4.4.1.9, 4.4.2.1
 - unformatted 4.4.1.9
filter 3.3, 3.3.1.6, 3.4.4.6, 5.3.1, 6.2.4.2, 7.1.1.1, 8.1.1.4, 10.3.1.2
 - adaptive 3.2.2.1, 5.3.1.2
 - all-pass 3.3.1.9, 3.3.3.5, 5.3.1.3, 8.2.1.3, 8.2.1.4, 9.2.2.2, 9.3.2.3
 - anti-notch → peak
 - band-pass 3.3.1.9, 3.4.4.6, 5.2.2.1, 5.2.2.2, 5.3.1.1, 6.2.4.4, 8.1.1.4
 - band-stop 3.3.1.9, 5.3.1.1
 - basic types 3.3.1.9
 - biquad 3.3.3.2, 3.3.3.5
 - Butterworth 3.3.3.6
 - Cauer 3.3.3.6
 - causal 3.3.1.6
 - Chebyshev 3.3.3.6
 - coefficients 3.3.2.1, 3.3.3.2, 3.3.3.3, 6.2.4.2
 - comb 3.3.3.5, 5.3.1.3, 6.2.4.5, 9.1.1.3, 9.2.2.2
 - inverse 5.3.1.3
 - Csound 4.1.1.7
 - DC-blocker 3.3.3.5
 - design 3.3.1.9, 3.3.2.3, 3.3.3.3, 3.3.3.6
 - elliptic 3.3.3.6
 - high-pass 3.3.1.9, 5.3.1.1, 7.2.3.3
 - implementation 3.3.2.2
  - canonical structure 3.3.1.6, 3.3.3.2
 - interpolation 5.1.3.3
 - linear 3.3.1.2, 5.3.1.2
 - linear phase 3.3.1.8, 3.3.2.1
 - low-pass 3.3.1.9, 3.3.2.4, 5.3.1.1, 7.2.3.3
 - moving average 3.3.3.5
 - notch 3.3.1.9, 3.3.3.5, 5.3.1.3
 - non-causal 3.3.2.4, 5.1.3.3
 - nonlinear 5.3.1.2, 7.1.1.3, 10.3.3.4
 - non-recursive (FIR) 3.3.1.6, 3.3.2, 5.1.3.3, 5.3.1.2
 - order 3.3.2.4, 3.3.3.6
 - peak 3.3.3.5
 - recursive (IIR) 3.3.1.6, 3.3.1.7, 3.3.3, 3.3.3.1, 3.4.1.4, 5.3.1.2, 6.2.3
 - reson → resonator
 - stability 3.3.2.1, 3.3.3.1, 3.3.3.4, 3.2.2.1, 3.3.3.5
 - time invariant 5.3.1.2
 - time variant 5.3.1.2, 5.3.1.6
filter bank 5.2.2.1, 7.2.2

fit 5.1.3.2
fixed point 3.4.1.2, 3.4.2.1, 3.4.2.3, 3.4.3.3, 3.4.3.3
flanger 5.1.4.5
flutter echo 9.2.2.2
FOF (Function d'Onde Formantique) 7.1.2.1
fold catastrophe 3.4.2.2
for 4.4.1.5,
force 2.1.2, 3.4.2.1, 3.4.2.2
 - restoring 8.1.1.1
forcing function 3.4.1.2, 3.4.1.4
formant 2.3.2.2, 5.2.1.5, 5.3.2.2, 6.1.2.6, 7.1.1.1, 7.1.1.3, 7.1.2.1, B.2
form parameter 10.1.3.7
Fourier J. B. 2.2.2.5
Fourier integral 3.1.3
Fourier series 3.1.1, 3.1.2.6, 3.3.2.4, 5.2.1.1, 9.3.2.2
 - coefficients 3.1.1.2, 3.1.1.3
Fourier transform 2.2.2.5, 3.1, 3.1.3.1, 7.2.2, 9.3.2.3
 - discrete (DFT) 3.2, 3.2.2, 5.2.2.2
   - coefficients 3.2.2.1
   - complex 3.2.2.4
 - fast (FFT) 2.2.2.5, 3.2, 3.2.2.5, 5.2.2.2
 - inverse 3.3.2.4, 5.2.2.2, 9.3.2.3
 - long-term 5.2.2.3
 - short-term (STFT) 5.2.2.2, 5.2.2.4
fractal 4.5.1, 10.3.2.4, 10.3.3.3, 10.3.3.4, 10.3.5.3
frequency 2.2.1.1, 3.3.1.4, 8.1.1.1, 8.1.1.3, 8.1.1.6, 8.1.2.4, 8.1.2.5, 8.1.3.2, 8.2.1.3, 9.1.3.1, 10.3.5.2
 - angular 2.2.1.1, 2.2.1.1, 3.1.2.5, 3.2.1.2, 3.3.1.4
   - normalized 3.2.1.2, 3.3.1.4
 - damped oscillation 2.2.1.2, 8.1.1.3
 - instantaneous 2.2.3.5, 3.1.2.5, 6.1.2.2, 6.1.2.7
 - modulation 6.1.1.2
frequency domain 3.1, 3.1.4.3, 3.3.1.4, 10.3.5.2
frequency group 2.3.4.1
frequency modulation 2.2.2.3, 5.1.1.2, 6.1.2, 6.2.4.1, 6.2.4.3
frequency ratio 2.3.1.2, 2.3.1.3, 8.1.2.1, 8.1.2.3, 8.1.4.1
 - irrational 10.3.2.6, 10.3.3.4
frequency response 3.3.1.4, 3.3.1.5, 3.3.1.7, 3.4.4.5, 5.3.1.1
frequency sampling method 3.3.2.4, 3.3.3.5
friction 3.4.1.5, 3.4.2.1, 6.2.3.2
fundamental theorem
 - of algebra 3.1.2.4, A3.3, A.4.3.3
 - of calculus A.5.5
function 4.4.1.1, 4.3.3, 4.4.1.4, 4.4.1.6, 4.4.1.8, 7.2.3.1, A.4
 - arc A.4.3.7
 - cosine A.4.3.7
 - Csound 4.1.1.3, 4.1.3
 - cubic 5.1.3.1
 - exponential 5.1.3.1, 5.1.3.2, A .4.3.5, 6.2.1.2
 - even 3.1.3.1, 3.2.1.3, A.4.7

 - inverse 5.1.3.5, 10.1.4.3, A.4.2, A.4.3.7
 - linear A.4.3.1
 - logarithm A.4.3.6
 - odd 3.1.3.1, 3.2.1.3, A.4.7
 - of random values 10.1.3.5
 - of several variables A4.6, 6.2.1.2, 6.2.2.1, 6.2.2.2, 6.2.3.2, 7.2.3.2
 - parametric 4.3.3, 5.1.3.2, A4.5
 - periodic 7.2.1, A.4.3.7
 - piece-wise 5.1.3.1, 5.1.3.2
 - quadratic A4.3.2
 - rational A.4.3.4
 - sine 5.1.3.2, A.4.3.7
 - tangent A.4.3.7
 - trigonometric A.4.3.7
fusion 2.3.2.5, 5.2.1.3

game of chance 10.2.1.1
Game of Life 10.3.1.3
Gauss plane 3.1.2.1
Gauss distribution → normal distribution
generator 5.1.1.1, 5.1.2, 7.1.2.1
 - noise 5.1.2.3
 - recurrence 5.1.2.3
glissando 5.1.3.1, 5.1.4.3, 8.2.3.3, 10.3.3.4
golden ratio 2.2.3.1, 10.3.2, 10.3.3.4
goto 4.1.1.5
gravitation 2.1.1, 8.1.1.4, 8.1.5.6
grid 8.1.4.2, 8.1.4.3, 8.1.6.3
 - 3D 8.1.5.2
 - irregular 8.1.4.8, 8.1.4.9
grain 7.1, 7.1.1.1
granular synthesis 5.1.1.2, 7.1
 - asynchronous 7.1.1.4
 - synchronous 7.1.1.3
graphics 4.3.3, 4.5.1

half-value period 5.1.3.1
harmonic series 2.2.2.5, 2.3.1.2, 2.3.1.2
harmonics 2.3.1.2, 8.1.3.7, 8.2.3.6, 10.3.2.6
header 4.1.1.1, 4.4.1.1, 4.4.2.2, 5.1.1.3
header file 4.4.1.1, 4.4.1.7, 4.4.1.8
HRTF (head-related transfer function) 9.1.1.3, 9.3.1.3
hysteresis 3.4.1.3, 3.4.2.2, 3.4.3.5, 3.4.4.4

if 4.1.1.5, 4.4.1.5
illusion 5.2.1.5, 9
impulse 2.1.3, 3.1.3.2, 3.2.1.2, 3.2.2.2, 3.2.3.1, 7.1.1.1, 8.1.4.6
 - spectrum 2.2.3.3
impulse response 2.3.5.1, 3.3.1.3, 3.3.1.8, 3.3.2.1, 3.3.2.6, 3.4.1.4, 6.2.4.5, 7.1.1.1, 7.1.2.1, 9.2.2.5
 - finite 3.3.2.1, 3.3.3.5
 - infinite 3.3.3.1
 - symmetrical 3.3.1.8
 - antisymmetrical 3.3.1.8
include 4.4.1.1
increment 4.3.1, 4.4.1.5, 5.1.2.1

Index 591

independence
  - stochastic 10.1.2.5
index 4.4.1.3, 5.1.1.4
inductance 3.1.4.1
inertia 8.1.1.1
inflection point A.5.2
information 9.1.3.2, 10.3.3.1
inheritance 4.4.1.8
initial condition 3.4.3.5
initial values
  - delicate dependence 3.4.3.1
input signal 3.1.4.1, 3.3.1.1, 3.4.1.4
instability 3.3.3.1, 3.4.2.1, 3.4.4.4, 6.2.3
instrument 4.1.1.1, 5.1.1.1
  - Csound 4.1.1.1
integral
  - definite A.5.5,
  - indefinite A.5.3
integral element 3.4.4.2
integrate 5.1.4.3
  - numerically 5.1.3.5
integration
  - constant 5.1.4.3, A.5.5
intensity difference 9.1.1.2
interaction 3.4.4, 10.3.3.4
interaural intensity difference IID 2.3.5.2, 9.1.1.2
interaural time difference ITD 2.3.5.2, 9.1.1.1
interference 2.2.4.3
intermittence 3.4.3.3, 3.4.5.5
interpolation 5.1.1.6, 5.1.2.1, 5.1.3.2, 5.1.3.3, 8.2.1.3
intervall 2.3.1.2, 2.3.1.2, 2.3.1.3, 5.1.4.3, 10.3.2.6, 10.3.5.2
  - pure 2.3.1.2
  - micro- 2.3.1.3

Karplus-Strong algorithm 8.2.1.6
keyboard 4.5.1
Koch curve 10.3.3.3
Kolmogorov 10.1.2.4

label 4.1.1.1
ladder network 3.3.3.2
Laplace transform 3.2, 3.2.3, 3.1.4.3, 3.3.3.6
LC circuit 3.1.4.1
leakage effect 3.2.2.3
library 4.4.1.1, 4.5.3
Ligeti, György 10.3.3.4
limit 3.4.3.3, 10.3.2, A.4.8
limit cycle 3.4.2.3, 3.4.3.1, 3.4.4.4, 3.4.5.1
limit distribution vector 10.2.2.3
limiter 3.4.1.3, 6.1.3.6
linear prediction 5.3.2.3
linear factors A.3.3,
linearization method 3.4.2.1
list 4.3.1
location parameter 10.1.3.7
logarithmic 2.3.1.2
logistic equation 3.4.3.3, 3.4.3.4, 6.2.3.1, 10.3.3.2, 10.3.3.2

Lorenz equations 3.4.3.4, 3.4.3.4
loudness 2.3.3, 2.3.4, 2.3.5.3, 5.1.3.1
Lyapunov function 3.4.2.1

macro 4.4.1.1
Mandelbrot set 10.3.3.3, 10.3.5.3
mantissa 4.4.1.2
mapping 10.3.5.6, A.4.1
Markov chain 10.2.2
Markov process 10.2.2
masking 2.3.4.2
mass 2.1.2, 8.1, 8.1.4.8
  - gravitational 2.1.2.1
  - inertial 2.1.2.1
Mathematica 1.2.1, 4.3
matrix 8.1.2.4, 8.1.4.3, 10.2.2.3, A.3.4
Max/MSP 4.1.4.4, 4.2
maximum A.5.2
mean
  - square 4.1.1.7, 5.1.3.3
  - root 5.1.3.3
mean value 7.2.3.1, 10.1.3.6, 10.2.1.3, 10.2.3.3
measure 3.4.4.7
membrane 8.1.4
mental representation 10.3.5, 10.3.5.1
method 4.4.1.8
MIDI 4.3.2
minimum 3.4.2.1, A.5.2
missing fundamental 2.3.2.6
modulation frequency 6.1.1.2
modulation index
  - amplitude modulation 6.1.1.2
  - frequency modulation 6.1.2.2
  - distortion 6.1.3.3
modulation signal 6.1.1.2, 6.1.2.2
modulation techniques 6.1
monotonic A.4.2
mouthpiece 8.2.2.2, 8.2.3.3
multi-agent system 10.3.4

natural frequency 2.2.1.5, 2.3.2.3, 8.1.2.1, 8.1.2.2, 8.1.2.4, 8.1.3.2, 8.1.4, 8.1.4.3, 9.2.2.1
Newton 2.1.2.3, 3.1.4.1, 8.1.1.1, 8.1.1.1
nodal line 8.1.4.3
node 2.2.1.5, 3.4.1.2
noise 2.2.3.2, 5.1.2.3
  - colored 2.2.3.2, 5.3.2.1
  - pink 2.2.3.2, 5.3.2.1
  - white 2.2.3.2, 2.3.4.4, 5.3.2.1, 10.1.6.1, 10.2.1.2, 10.3.3.1, 10.3.3.4
nonlinear techniques 6
nonlinearity 3.4.5.3, 8.2.1.4, 10.3.3.2
normal distribution 10.1.3.7, 10.1.5.5
normal form
  - of an equation A.3.3
normalize 7.2.3.2
note 4.1.2
number A.1
  - complex 3.1.2, A.3.3
    - exponential form 3.1.2.3
  - complex conjugate 3.1.2.1

- floating point 4.4.1.2
- imaginary 3.1.2.1
numerical sequence A.4.8, A.4.1
numerical methods 3.4.1.4, 3.4.3.2
Nyquist frequency 3.2.1.1, 5.1.2.2

object 4.4.1.8
octave 2.3.1.1
- identity 10.3.5.2
operator 4.1.1.4, 4.4.1.4
orchestra 4.1.1, 5.1.1.1
order (not chaos) 10.3.3.1
order 10.1.1, 10.1.1.5, 10.3.2.5
orientation 9.1.3.1
OSC (Open Sound Control) 4.1.4.3, 4.1.4.4, 4.5.3
oscillation 2.2.1.1, 5.1.3.2
- aperiodic 2.2.3, 8.1.2.1
- coupled masses 4.5.2, 8.1, 8.1.2.1
- driven 2.2.1.6, 3.4.3.5
- damped → damping
- exciting 2.2.1.6
- harmonic 2.2.1.1, 2.2.1.1, 5.1.2.1, 8.1.1.1, 8.1.3.3
- longitudinal 2.2.1.5, 2.2.1.5, 8.1.5.1
- periodic 2.2.2, 2.2.2.5, 2.3.2.1, 3.1.1.1, 5.2.1
- quasi-periodic 2.2.3.4
- sum 2.2.1.3
- torsional 8.1.4.5, 8.1.5.2, 8.1.5.3
- transversal 2.2.1.5, 8.1.5.1
oscillator 3.4.2.2, 3.4.4.2, 3.4.2.2, 5.1.2.1, 5.1.3.6, 5.2.1.1
- Chamberlin 5.1.2.1
- chaotic 3.4.5.5
- coupled 3.4.5.4
- coupled-form 5.1.2.1, 5.1.3.6
- Csound 4.1.1.6
- digital sinusoidal 5.1.2.1
- direct form 5.1.2.1
- integrate and fire 3.4.5.1
- nonlinear 3.4.3.5, 3.4.4.7, 4.2.3
- relaxation 3.4.5.1
- Rössler 3.4.5.5
- self-sustained 3.4.5.1
- Van der Pol 3.4.5.2, 3.4.5.5
output signal 3.1.4.1, 3.3.1.1, 3.4.1.4
overblowing 8.2.3.3
overflow 4.4.1.2
override 3.4.2.2, 3.4.4.6, 6.1.3.6
oversampling 5.1.3.3

panorama control 9.1.1.2
parabola A.4.3.2
parameter 4.1.2, 10.3.3.4, 10.3.5, 10.3.5.5, A.3.1
parametric representation 4.3.3, A.4.5
partial 2.3.1.2, 2.2.2.5, 2.3.1.2, 2.3.2.1, 5.3.2.1, 10.3.2.6
partial fraction A.4.3.4
pass-band 3.3.1.9

patch 4.2.2
path 10.2.1.1
pendulum 3.4.1.5, 6.2.3.2, 8.1.6.1
perception 9, 10.3.5.2
- just noticeable difference (JND) 2.3.1.3
- single events 2.3.1.1
period 2.2.1.1
permutation 10.1.1.2
phase 2.2.1.1, 3.1.2.5, 5.1.2.1, 5.1.3.6, 5.2.2.3, 6.2.2.1, 7.1.1.2
- amplitude modulation 6.1.1.3
- constant 2.2.1.1, 2.2.1.3, 3.1.1.1, 5.2.1.1
- difference 2.2.1.3, 2.2.1.4, 2.2.4.3, 3.4.5.3
- frequency modulation 6.1.2.8
- initial 2.2.1.1, 2.3.2.1, A.4.3.7
- instantaneous 2.1.1.1, 2.2.3.5, 3.1.2.5, 3.4.5.3
- locking 3.4.5.3
- response 3.3.1.4, 3.3.1.5, 3.3.3.2, 3.3.3.5, 8.2.1.4
- shift 6.2.2.1, 9.1.1.1
- unwrapped phase 2.1.1.1, 5.1.3.6
- velocity 2.2.4.2, 2.2.4.4
- wrapped phase 2.1.1.1, 5.1.3.6
phase space 3.4.1.1, 3.4.1.5, 3.4.2.1, 3.4.2.3, 3.4.4.4, 3.4.5.1, 6.2.3.2
- extended 3.4.1.1, 3.4.2.3, 3.4.3.2
- three-dimensional 3.4.3.2, 3.4.3.4
- two-dimensional 3.4.3.2
Phase Vocoder 5.2.2.5
phasor 2.2.1.1
phon 2.3.3.3
physical modeling 8, 5.1.1.2
pitch 2.3.1, 2.3.4, 5.1.3.1, 10.1.6.2, 10.3.3.4, 10.3.5.2, B.1
pitch shifting 7.1.2.3
plane
- curved 8.1.4.7, 9.2.1.1, 10.3.5.2, A.2.5
plate 2.3.2.4, 8.1.4, 8.1.4.3
plugin 4
pointer 2.2.1.1, 2.2.1.3, 3.1.2.1, 3.1.2.5, 3.2.1.6, 4.4.1.3, 4.4.1.9, 5.1.2.1, 5.2.2.2
pole 3.1.4.3, 3.3.1.7, 3.3.3.2, 3.3.3.3, 3.3.3.5, 3.3.3.6, A.4.3.4
polar coordinates 3.1.2.2, 5.2.2.2
polynomial 5.1.3.2, 6.1.3.4, 10.1.1.6, A.4.3.3
- coefficient 10.1.1.6
potential (energy, function) 3.4.2.2, 3.4.2.2, 3.4.2.2
power 2.1.3, 2.3.3.1, 5.1.3.4
power (math) A.1.2
precedence effect 9.1.1.1
prediction 10.3.3.2
pressure 2.2.1.5, 8.2.2.2, 8.2.3.4
prime numbers B.5
principle of Huyghens 2.2.4.4, 9.3.1.2
principle of superposition 6.2.4.1
priority (expression evaluation) 4.4.1.4
probability 10.1.2.4
- conditional 10.1.2.5
- function 10.1.3.1
- calculus 10.1.2
process 10.3.2
process of growth 10.3.2.5
Processing (program) 4.1.4.3, 4.5

Index 593

proportion 10.3.2, A.1.2
proportional element 3.4.4.2
pulse train 2.2.2.2, 2.2.2.4, 2.2.2.5, 2.3.1.1,
  5.1.2.2, 5.3.2.1
pulse generator 5.1.2.2

quantization 3.2.1.1, 3.2.1.4
quantifier A.2,
quality factor 3.3.1.9, 5.3.1.1, 7.2.2

random
  - experiment 10.1.2.1, 10.1.2.2
  - generator 3.4.4, 4.1.1.6, 4.1.4.2, 10.3.2.6
  - number 10.1.4
    - pseudo-random 5.1.2.3, 10.1.4.1
    - random sieves 10.1.6.4, 10.2.3.5
  - variable 10.1.3.1
  - walk 10.2.3.2
range A.4.1
recursion 3.3.1.2, 4.4.1.6, 10.3.2.2,
  10.3.2.4, A.4.8
reed 8.1.4.5
reflection 2.2.4.4, 2.3.5.1, 8.2.2.2, 8.2.3.5,
  9.1.1.3, 9.1.3.1, 9.2, 9.2.1, 9.2.2.1, 9.3.2.2
refraction 2.2.4.4
region of convergence
  - Z-transform 3.2.3.1
rejection sampling 10.1.4.4, 10.1.6.3
relay 3.4.4.4, 3.4.4.4, 3.4.4.7
resampling 6.2.4.3, 7.1.1.3
residue pitch 2.3.2.6
resonance 2.2.1.6, 2.3.2.2, 3.3.2.6, 3.3.3.5,
  3.4.3.5, 5.3.1, 8.2.3.3, 9.2.2.2, 10.3.2.6
  - curve 3.4.3.5, 3.4.4.5
resonator 3.3.2.6, 3.3.3.5, 3.4.4.5, 3.4.4.7,
  5.1.2.1, 5.3.1.3, 5.3.2.2, 7.1.1.1, 7.1.2.1,
  8.1.1.4
response 3.1.4.1, 3.3.1.1
restoring force 8.1.1.1
  - cubic 3.4.4.7
result set 10.1.2.1, 10.1.2.2
resynthesis 5.2.2
return value 4.4.1.6
reverberation 2.3.5.1, 2.3.5.3, 3.3.1.3, 9.2,
  9.2.2, 9.3.2.3
  - initial 2.3.5.1, 9.2.2.1
  - radius 9.2.2.1
  - time 2.3.5.1
reverberator 9.2.2.2
Reynolds, C. 10.3.4.1
RIFF (Resource Interchange File Format)
  4.4.2.2
rigid body motion 8.1.4.6, 8.1.5.6,
ring modulation 3.3.1.1, 5.1.1.2, 6.1.1, 6.1.1.4,
  6.2.2.1, 6.2.2.2, 6.2.4.2
ripple 3.3.2.4
Risset, Jean-Claude 5.2.1.5, 10.3.3.4
root mean square (rms) 4.1.1.7
Rössler 3.4.3.4
roughness 2.3.4.1, 6.1.1.1
Runge-Kutta-Method 3.4.1.4

rules for calculating
  - differential calculus A.5.2
  - integral calculus A.5.4
  - logarithm A.4.3.6
  - powers A.4.3.5
  - trigonometry A.4.3.7

s-Transform → Laplace transform
saddle point 3.4.1.2
sample 4.4.2.1
sampling period 3.2.1.1
sampling rate 3.2.1.1, 4.1.1.1, 5.1.1.5
sampling control 3.4.4
scale parameter 10.1.3.7
scaling 3.2.1.3, 3.3.1.1
scattered 9.2.1.2
scope operator 4.4.1.8
score 4.1.2, 4.1.4, 4.3.1, 5.1.1.1
self-similarity 10.3.3.3, 10.3.3.4, 10.3.3.4
self-organization 10.3.3.2, 10.3.4
separatrix 3.4.2.3
series 5.1.3.5, A.4.8
set A.2
  - Mandelbrot 10.3.5.3
settling time 2.2.1.6
Shepard 5.2.1.5, 10.3.3.4
side band 6.1.1.3, 6.1.2.3
Sierpinkski triangle 4.5.1
signal 3
  - analog 3.1
  - aperiodic 3.1.3
  - audio 4.1.1.1, 4.2.1, 4.2.2, 5.1.1.5
  - digital 3.2
  - even 3.2.1.3
  - control 4.1.1.1, 4.1.1.6, 4.2.1, 4.2.2,
    5.1.1.5, 5.1.3
  - correlated 9.3.2.3
  - periodic 3.1.1, 3.2.1.3, 3.2.2.1
  - odd 3.2.1.3
signal-flow graph 3.3.1.6, 3.4.4.3, 3.4.4.4
signum A.4.1
simulation 4.5.2, 10.3.1, 10.3.3.4
skew tent map 3.4.5.5
slope 3.4.1.4, 5.1.3.2, A.5.1
slope field → direction field
solution A.3.1, A.3.3
sonification 10.3.5.6
sound
  - intensity 2.3.3.1, 2.3.4.4, 9.1.2.1
  - level 9.2.2.1
  - localization 2.3.5.2, 9, 9.1.1
  - pitched 2.2.2, 2.3.2.5
  - power 2.3.3.1
  - radiation 8.2.3.5
  - speed 2.2.4.2
sound file 4.4.2.1, 4.4.2.2, 5.1.1.3
sound pick up 8.1.3.5, 8.2.2.1, 8.2.3
sound synthesis
  - additive 5.1.1.2, 5.2
  - subtractive 5.1.1.2, 5.1.2.2, 5.1.2.3, 5.3
  - wave table 5.1.1.2
source code 4.4.1.1

space 2.3.5, 8.1.5, 8.1.6.3, 9, 9.1.3.2, 10.3.5.2, 10.3.5.5, A.2.5
- Euclidian 10.3.5.2
- simulation 9, 9.3.2.2, 9.3.2.3
spectrum 2.2.2.5, 2.3.2.1, 2.3.2.2, 3.1.1.2, 3.1.1.4, 5.1.2.3, 5.1.3.3, 5.2.2, 7.1.1.1, 7.1.1.3, 7.1.2.2, 7.1.1.3, 10.3.3.4
- amplitude modulation 6.1.1.3
- frequency modulation 6.1.2.3
- harmonic 3.3.3.5
- envelope 7.1.1.3, 7.1.2.2, 7.1.1.3, 10.3.3.4
- complex 5.2.2.2
- inharmonic 2.3.2.4
- variable 2.3.2.3, 6.1.2.5
- of a random process 10.2.3.3
speech 7.1.2.2
sphere 9.3.2.2, 10.3.5.2
spherical harmonics 9.3.2.2
spline 5.1.3.2, 5.1.3.5
- Bezier 5.1.3.2
- c- 5.1.3.2
- cubic 5.1.3.2
- Hermite 5.1.3.2
   - blending functions 5.1.3.2
- Kochanek-Bartels 5.1.3.2
spring 3.1.4.1, 3.4.1.1, 3.4.4.2, 4.5.2, 6.2.4.4, 8.1.1.1
spring constant 2.1.2.3
stability 3.3.1.2, 3.3.3.4, 3.4.4.2, 6.2.3, 6.2.4.4, 10.3.2.5
stable spiral 3.4.1.2
standard deviation 10.1.3.6
state vector 10.2.2.3
statement A.2,
- conditional 5.1.1.7
statistics 10.1.5.5
steady state vector 10.2.2.3
step response 3.4.4.2
stereo 9.1.1.2, 9.3.2.1
stereoscopic 3.4.3.4
stochastic process 10.2, 10.2.1.3
stop-band 3.3.1.9
stretching A.4.3.5,
string (alphanumeric) 4.4.1.7
string 2.2.1.5, 8.1.3, 8.1.3.2, 8.1.6.1, 8.2.1.1, 8.2.1.6, 8.2.2.2, 10.3.1.2, 10.3.2.6
- excitation 8.1.3.6
- bowed 8.1.1.4, 8.1.3.6, 8.2.3.4, 8.2.3.6, 10.3.3.4
structure 4.4.1.3
sum
- discrete 5.1.3.5
- of harmonic cosine functions 5.1.2.2
sum tone 2.3.4.3
superposition 2.2.1.3, 2.2.4.3, 3.1.2.5, 3.3.1.2
- of waves 2.2.4.3
surface 9.2.1.2
- of equilibrium 3.4.2.2
sustain 2.3.2.3
Swallowtail Catastrophe 3.4.2.2
swarm simulation 10.3.4
sweep 5.2.2.3
synchronization 3.4.5, 5.1.3.6

- complete 3.4.5.5
- higher order 3.4.5.3
- of beats 5.1.3.5
- phases 5.1.3.6
synergetics 10.3.3.2
system 3, 3.3, 3.4.4.6, 6.2.4.1, 10.3.4, 10.3.5.7
- adaptive 3.3.2.1
- analog 3.3, 3.1.4, 3.3.3.6, 3.4.4.2
- digital 3.3, 3.3.1, 3.3.3.6, 3.4.4.2, 6.2.3.2
- dissipative 3.4.2.3
- dynamic 10.3.3.2, 10.3.3.4
- causal 3.3.1.2
- linear 3.3, 3.3.1.2
- linear phase 3.3.1.8, 3.3.2.1
- Linear-Time-Invariant LTI 3.3.1.3
- nonlinear 3.3, 3.4, 3.4.1.3, 3.4.3.1, 3.4.4.4, 6.1.3
- non-recursive 3.3.1.2, 6.2.1, 6.2.2, 6.2.4.2, 6.2.4.3
- physical 6.2.3.2
- recursive 3.3.1.2, 3.4.1.4, 4.2.2, 6.2.3, 6.2.4.4, 6.2.4.5
- stability 3.3.1.2, 3.3.2.1, 6.2.3.2
- time-discrete 3.4.3.3
- time-variant 6.2.4
system analysis
- stationary 3.4.1.3
system deviation 3.4.4.1, 3.4.4.2, 3.4.2.2, 3.4.4.6
system time 4.4.1.4

table 5.1.1.4, 5.1.2.1, 10.1.4.5
tangent 5.1.3.2, 9.2.1.1, A.5.1
Taylor 3.1.2.3
temperament
- equal 2.3.1.3, B.1
tempo 5.1.3.5
tempo function 5.1.3.5
term A.3.1
threshold 3.4.1.3, 5.2.2.4
- of pain 2.3.3.1
- of hearing 2.3.3.2
timbre 2.3.2, 2.3.4.1, 2.3.4.4, 10.3.5.4
time 10.3.5.3
- domain 3.1, 3.1.4.3
- invariant 3.3.1.2, 3.4.4.2
- reversal 3.3.1.1
- series 10.2.1.1
- streching 7.1.2.3
- variant 3.4.4.2
tolerance 3.3.1.9, 3.3.3.6
topology 10.3.5, 10.3.5.2
torus 10.3.5.2
trajectory 3.4.1.1, 3.4.1.5, 3.4.2.1, 3.4.2.3, 3.4.3.1, 3.4.3.4, 3.4.3.5, 3.4.4.4, 3.4.5.1, 5.1.3.2, 6.2.3.2, 10.2.1.1
- chaotic 3.4.3.5
transfer function (→ frequency response) 3.3.1.4, 3.3.1.5, 3.3.1.7, 3.3.3.2, 3.3.3.4, 9.3.2.3
- distortion 6.1.3.2, 6.1.3.3, 6.1.3.4
- head-related 9.1.1.3, 9.3.1.3
- three-dimensional representation 3.3.1.7
transfer element 3.4.4.2

Index 595

- rational 3.4.4.2
transformation 3.1.4.3, 3.3.3.6
    - linear 3.2.2.4, 3.2.3.2
    → Fourier transform
    → Laplace transform
    → Z-transformation
transient effect 2.3.2.3, 3.4.2.3, 7.1.2.2, 8.1.1.4
transition matrix 10.2.2.3
transition probability 10.2.2.2
transposition 5.1.3.5, 5.1.4.3, 6.2.4.3, 6.2.4.5, 7.1.1.3, 9.1.3.1, 10.3.3.4
tube 2.2.1.5, 8.2.2.2, 8.2.3.3
twelve-tone row 10.1.1.1, 10.1.2.6

unit step 3.2.1.2, 3.2.3.1,
upsampling 3.2.1.3, 3.3.1.4
urn model 10.1.1.1

variable
    - dependent A.4.1
    - independent 10.3.5.3, A.4.1
variance 10.1.3.6, 10.2.1.3
variation 10.1.1, 10.1.1.4
vector A.3.4
velocity 2.1.1, 8.1.1.1, 10.2.3.2
vertex A4.3.2
vibrato 6.1.2.1, 7.1.2.2
voice 5.3.2.2, 7.1.2.1
VOSIM (VOice SIMulation) 7.1.2.2
vowel 5.3.2.2

Walsh synthesis 5.1.1.2, 7.2.1
WAVE (format) 4.4.2.2
wave 2.2.4
    - elementary 2.2.4.4
    - equation 8.2.2.3,
    - form 2.2.2.2, 2.2.2.4, 5.1.2.1
    - front 2.2.4.1
    - length 2.2.4.2, 9.2.1.2
    - longitudinal 2.2.4.1, 2.2.4.1
    - number 2.2.4.2
    - one-dimensional 2.2.4.1
    - propagation 2.2.4.4
    - standing 4.5.2, 2.2.4.3
    - transversal 2.2.4.1
    - traveling 8.1.3.3, 8.2.1.1, 8.2.2.1, 8.2.2.3
    - triangle 2.2.2.2
    - two-dimensional 2.2.4.1
    - sawtooth 2.2.2.2, 5.1.2.1
    - square 2.2.2.2, 2.2.2.4, 2.3.2.1, 3.1.1.3, 3.2.1.5, 3.1.2.6, 6.1.3.1
Wave Field Synthesis WFS 9.3.1.2
wave guide 8.2.2
wave table synthesis 5.1.1.2
wavelet 7.2.3
    - Haar 7.2.3.1, 7.2.3.3
    - Mexican Hat 7.2.3.1
wavelet transform
    - continuous 7.2.3.2
    - discrete 7.2.3.3

    - fast 7.2.3.3
waveshaping → distortion
wind instrument 8.2.2.2
window 3.3.2.5, 5.1.3.4, 5.2.2.2, 5.2.2.4, 7.1.1.1, 7.1.1.3, 7.1.2.3
    - amplitude response 3.3.2.5
    - function 3.3.2.5
        - Blackman- 3.3.2.5
        - Hamming- 3.3.2.5
        - Hann- 3.3.2.5
        - rectangular 3.3.2.5
        - triangular 3.3.2.5
work 2.1.3, 2.1.3.4, 3.4.2.1

Z-plane 3.3.1.7
z-transform 3.2, 3.2.3, 3.3.3.6, 3.4.4.5
zero 3.1.4.3, 3.3.1.7, 3.3.1.7, 3.3.2.3, 3.3.3.2, 3.3.3.3, 3.3.3.6, A.4.2, A.4.3.3, A.4.3.4